Wastewater Treatment Plants

WASTEWATER TREATMENT PLANTS

Planning, Design, and Operation

SYED R. QASIM

The University of Texas at Arlington

Holt, Rinehart and Winston

New York   Chicago   San Francisco   Philadelphia
Montreal   Toronto   London   Sydney
Tokyo   Mexico City   Rio de Janeiro   Madrid

Library of Congress Cataloging in Publication Data

Qasim, Syed R.
  Wastewater treatment plants.

  Includes bibliographies and index.
  1. Sewage disposal plants. I. Title.
TD746.Q37 1985   628.3   84-29758
ISBN 0-03-062449-5

5 6 7 8   038   9 8 7 6 5 4 3 2

CBS COLLEGE PUBLISHING
Holt, Rinehart and Winston
The Dryden Press
Saunders College Publishing

To My Parents

Contents

Preface   xvii
Acknowledgments   xix

1 INTRODUCTION   1

   1-1 CURRENT STATUS   1
   1-2 FUTURE TRENDS   1
   1-3 PLANT DESIGN   3
       REFERENCES   3

2 BASIC DESIGN CONSIDERATIONS FOR
  WASTEWATER TREATMENT FACILITIES   5

   2-1   INTRODUCTION   5
   2-2   INITIAL AND DESIGN YEARS   6
   2-3   SERVICE AREA   7
   2-4   SITE SELECTION   7
   2-5   DESIGN POPULATION   9
   2-6   REGULATORY CONTROLS AND EFFLUENT
        LIMITATIONS   10
   2-7   CHARACTERISTICS OF WASTEWATER   17
   2-8   DEGREE OF TREATMENT   18
   2-9   CHOICE OF TREATMENT PROCESSES, FLOW
        SCHEMES, AND COMPARISON OF
        ALTERNATIVES   18
   2-10 EQUIPMENT SELECTION   19
   2-11 PLANT LAYOUT AND HYDRAULIC PROFILE   19
   2-12 ENERGY AND RESOURCE REQUIREMENTS   19
   2-13 PLANT ECONOMICS   20
   2-14 ENVIRONMENTAL IMPACT ASSESSMENT   20
   2-15 PROBLEMS AND DISCUSSION TOPICS   20
       REFERENCES   21

3 WASTEWATER CHARACTERISTICS   23

   3-1 INTRODUCTION   23
   3-2 MUNICIPAL WATER DEMAND   23
   3-3 WASTEWATER FLOW   26
   3-4 QUALITY OF WASTEWATER   33
   3-5 CHARACTERIZATION OF WASTEWATER   43
   3-6 UNIT WASTE LOADINGS AND POPULATION
       EQUIVALENTS   43

3-7 PROBLEMS AND DISCUSSION TOPICS 43
REFERENCES 45

4 WASTEWATER TREATMENT UNIT OPERATIONS AND PROCESSES, AND FLOW SCHEMES 47

4-1 INTRODUCTION 47
4-2 LIQUID TREATMENT SYSTEMS 48
4-3 SLUDGE PROCESSING AND DISPOSAL 57
4-4 PROBLEMS AND DISCUSSION TOPICS 68
REFERENCES 72

5 PREDESIGN STUDIES 73

5-1 INTRODUCTION 73
5-2 FACILITY PLANNING 74
5-3 PREPARATION OF DESIGN PLANS, SPECIFICATIONS, AND COST ESTIMATES 77
5-4 DISCUSSION TOPICS 80
REFERENCES 80

6 MODEL FACILITY PLAN 83

6-1 INTRODUCTION 83
6-2 PROJECT IDENTIFICATION 83
6-3 OBJECTIVES AND SCOPE 84
6-4 EFFLUENT LIMITATIONS AND INITIAL DESIGN PERIOD 84
6-5 EXISTING AND FUTURE CONDITIONS 85
6-6 FORECAST OF WASTEWATER CHARACTERISTICS FROM SERVICE AREAS 94
6-7 WASTEWATER TREATMENT PROCESS ALTERNATIVES 96
6-8 ENGINEER'S DEVELOPMENT OF THE PROPOSED PROJECT 100
6-9 FINANCIAL STATUS 111
6-10 DESCRIPTION OF SELECTED WASTEWATER TREATMENT FACILITY 112
6-11 ENVIRONMENTAL IMPACT ASSESSMENT 114
6-12 PUBLIC PARTICIPATION PROGRAM 115
6-13 COST BREAKDOWN 115
6-14 TIME SCHEDULE FOR PROJECT MILESTONES 115
6-15 ATTACHMENTS 115
6-16 PROBLEMS AND DISCUSSION TOPICS 119
REFERENCES 121

7 *DESIGN OF INTERCEPTING SEWERS* 123

7-1   INTRODUCTION 123
7-2   COMMUNITY SEWERAGE SYSTEM 123
7-3   DESIGN AND CONSTRUCTION OF COMMUNITY
      SEWERAGE SYSTEM 125
7-4   INFORMATION CHECKLIST FOR THE DESIGN OF
      SANITARY SEWERS 139
7-5   DESIGN EXAMPLE 141
7-6   OPERATION AND MAINTENANCE, AND
      TROUBLESHOOTING IN SANITARY SEWERS 148
7-7   SPECIFICATIONS 151
7-8   PROBLEMS AND DISCUSSION TOPICS 153
      REFERENCES 153

8 *SCREENING* 155

8-1   INTRODUCTION 155
8-2   TYPES OF SCREENS 155
8-3   DESIGN FACTORS FOR BAR RACKS (BAR
      SCREENS) 156
8-4   EQUIPMENT MANUFACTURERS OF BAR RACKS 162
8-5   INFORMATION CHECKLIST FOR DESIGN OF BAR
      RACKS 162
8-6   DESIGN EXAMPLE 163
8-7   OPERATION AND MAINTENANCE, AND
      TROUBLESHOOTING AT MECHANICALLY
      CLEANED BAR RACK 172
8-8   SPECIFICATIONS 174
8-9   PROBLEMS AND DISCUSSION TOPICS 175
      REFERENCES 176

9 *PUMPING STATION* 177

9-1   INTRODUCTION 177
9-2   PUMP TYPES AND APPLICATIONS 177
9-3   TYPES OF PUMPING STATIONS 177
9-4   HYDRAULIC TERMS AND DEFINITIONS
      COMMONLY USED IN PUMPING 180
9-5   CENTRIFUGAL PUMPS 184
9-6   DESIGN OF PUMPING STATION 194
9-7   MANUFACTURERS OF PUMPING EQUIPMENT 196
9-8   INFORMATION CHECKLIST FOR DESIGN OF
      PUMPING STATION 196
9-9   DESIGN EXAMPLE 197
9-10 OPERATION AND MAINTENANCE, AND
      TROUBLESHOOTING AT PUMPING STATION 210
9-11 SPECIFICATIONS 213

9-12   PROBLEMS AND DISCUSSION TOPICS   215
        REFERENCES   218

10 FLOW MEASUREMENT   219

10-1   INTRODUCTION   219
10-2   LOCATION OF FLOW MEASUREMENT DEVICES   219
10-3   FLOW MEASUREMENT METHODS AND DEVICES   221
10-4   FLOW SENSORS AND RECORDERS   228
10-5   EQUIPMENT MANUFACTURERS OF
        FLOW-MEASURING AND FLOW-SENSING
        DEVICES AND RECORDERS   228
10-6   INFORMATION CHECKLIST FOR DESIGN OF
        FLOW-MEASURING DEVICES   228
10-7   DESIGN EXAMPLE   230
10-8   OPERATION AND MAINTENANCE   233
10-9   SPECIFICATIONS   233
10-10   PROBLEMS AND DISCUSSION TOPICS   235
        REFERENCES   236

11 GRIT REMOVAL   238

11-1   INTRODUCTION   238
11-2   LOCATION OF GRIT REMOVAL FACILITY   238
11-3   TYPES OF GRIT REMOVAL FACILITIES   238
11-4   GRIT COLLECTION AND REMOVAL   245
11-5   QUANTITY OF GRIT   245
11-6   GRIT DISPOSAL   245
11-7   EQUIPMENT MANUFACTURERS OF GRIT
        REMOVAL FACILITIES   247
11-8   INFORMATION CHECKLIST FOR DESIGN OF GRIT
        REMOVAL FACILITY   247
11-9   DESIGN EXAMPLE   247
11-10   OPERATION AND MAINTENANCE, AND
        TROUBLESHOOTING AT AERATED GRIT
        REMOVAL FACILITY   257
11-11   SPECIFICATIONS   258
11-12   PROBLEMS AND DISCUSSION TOPICS   260
        REFERENCES   261

12 PRIMARY SEDIMENTATION   263

12-1   INTRODUCTION   263
12-2   TYPES OF CLARIFIERS   263
12-3   DESIGN FACTORS   268

12-4   EQUIPMENT MANUFACTURERS OF
       SEDIMENTATION BASINS   283
12-5   INFORMATION CHECKLIST FOR DESIGN OF
       SEDIMENTATION BASIN   283
12-6   DESIGN EXAMPLE   284
12-7   OPERATION AND MAINTENANCE, AND
       TROUBLESHOOTING AT PRIMARY
       SEDIMENTATION FACILITIES   294
12-8   SPECIFICATIONS   297
12-9   PROBLEMS AND DISCUSSION TOPICS   300
       REFERENCES   301

13   *BIOLOGICAL WASTE TREATMENT*   *303*

13-1   INTRODUCTION   303
13-2   FUNDAMENTALS OF BIOLOGICAL WASTE
       TREATMENT   303
13-3   SUSPENDED GROWTH BIOLOGICAL TREATMENT   304
13-4   ATTACHED GROWTH BIOLOGICAL TREATMENT   326
13-5   EQUIPMENT MANUFACTURERS OF BIOLOGICAL
       WASTE TREATMENT PROCESSES   329
13-6   INFORMATION CHECKLIST FOR DESIGN OF
       BIOLOGICAL TREATMENT AND CLARIFICATION
       FACILITIES   329
13-7   DESIGN EXAMPLE   330
13-8   OPERATION AND MAINTENANCE, AND
       TROUBLESHOOTING AT ACTIVATED SLUDGE
       TREATMENT FACILITY   369
13-9   SPECIFICATIONS   372
13-10 PROBLEMS AND DISCUSSION TOPICS   374
       REFERENCES   377

14   *DISINFECTION*   *379*

14-1   INTRODUCTION   379
14-2   METHODS OF DISINFECTION   379
14-3   DISINFECTION WITH CHLORINE   380
14-4   EQUIPMENT MANUFACTURERS OF
       CHLORINATION FACILITY   390
14-5   INFORMATION CHECKLIST FOR DESIGN OF A
       CHLORINATION FACILITY   390
14-6   DESIGN EXAMPLE   391
14-7   OPERATION AND MAINTENANCE, AND
       TROUBLESHOOTING AT CHLORINATION FACILITY   406
14-8   SPECIFICATIONS   407
14-9   PROBLEMS AND DISCUSSION TOPICS   409
       REFERENCES   411

15 EFFLUENT DISPOSAL 412

15-1 INTRODUCTION 412
15-2 METHODS OF EFFLUENT DISPOSAL AND REUSE 412
15-3 INFORMATION CHECKLIST FOR DESIGN OF
OUTFALL STRUCTURE 418
15-4 DESIGN EXAMPLE 418
15-5 OPERATION AND MAINTENANCE, AND
TROUBLESHOOTING AT OUTFALL STRUCTURE 424
15-6 SPECIFICATIONS 424
15-7 PROBLEMS AND DISCUSSION TOPICS 425
REFERENCES 425

16 SOURCES OF SLUDGE AND THICKENER DESIGN 427

16-1 INTRODUCTION 427
16-2 CHARACTERISTICS OF MUNICIPAL SLUDGE 427
16-3 SLUDGE-PROCESSING SYSTEMS AND
ENVIRONMENTAL CONTROL 428
16-4 SLUDGE THICKENING 429
16-5 EQUIPMENT MANUFACTURERS 434
16-6 INFORMATION CHECKLIST FOR THICKENER
DESIGN 434
16-7 DESIGN EXAMPLE 435
16-8 OPERATION AND MAINTENANCE, AND
TROUBLESHOOTING AT SLUDGE-THICKENING
FACILITY 446
16-9 SPECIFICATIONS 448
16-10 PROBLEMS AND DISCUSSION TOPICS 449
REFERENCES 449

17 SLUDGE STABILIZATION 451

17-1 INTRODUCTION 451
17-2 ANAEROBIC DIGESTION 451
17-3 AEROBIC DIGESTION 460
17-4 OTHER SLUDGE STABILIZATION PROCESSES 460
17-5 EQUIPMENT MANUFACTURERS OF SLUDGE
STABILIZATION SYSTEMS 463
17-6 INFORMATION CHECKLIST FOR DESIGN OF
SLUDGE STABILIZATION FACILITY 463
17-7 DESIGN EXAMPLE 464
17-8 OPERATION AND MAINTENANCE, AND
TROUBLESHOOTING AT ANAEROBIC SLUDGE
DIGESTION FACILITY 480
17-9 SPECIFICATIONS 484
17-10 PROBLEMS AND DISCUSSION TOPICS 486
REFERENCES 488

18   SLUDGE CONDITIONING AND DEWATERING   **489**

   18-1   INTRODUCTION   489
   18-2   SLUDGE CONDITIONING   489
   18-3   SLUDGE DEWATERING   491
   18-4   EQUIPMENT MANUFACTURERS OF
         SLUDGE-CONDITIONING AND DEWATERING
         SYSTEMS   505
   18-5   INFORMATION CHECKLIST FOR DESIGN OF
         SLUDGE-CONDITIONING AND DEWATERING
         FACILITIES   505
   18-6   DESIGN EXAMPLE   506
   18-7   OPERATION AND MAINTENANCE, AND
         TROUBLESHOOTING AT FILTER PRESS FACILITY   515
   18-8   SPECIFICATIONS   516
   18-9   PROBLEMS AND DISCUSSION TOPICS   518
         REFERENCES   519

19   SLUDGE DISPOSAL   **520**

   19-1   INTRODUCTION   520
   19-2   OVERVIEW OF SLUDGE DISPOSAL PRACTICES   520
   19-3   PLANNING, DESIGN, AND OPERATION OF
         MUNICIPAL SLUDGE LANDFILLS   523
   19-4   INFORMATION CHECKLIST FOR DESIGN OF
         SLUDGE LANDFILLS   537
   19-5   DESIGN EXAMPLE   540
   19-6   OPERATION AND MAINTENANCE, AND
         SPECIFICATIONS FOR SLUDGE LANDFILLS   546
   19-7   PROBLEMS AND DISCUSSION TOPICS   547
         REFERENCES   547

20   PLANT LAYOUT   **549**

   20-1   INTRODUCTION   549
   20-2   FACTORS AFFECTING PLANT LAYOUT AND SITE
         DEVELOPMENT   549
   20-3   DESIGN EXAMPLE   554
   20-4   PROBLEMS AND DISCUSSION TOPICS   556
         REFERENCES   557

21   YARD PIPING AND HYDRAULIC PROFILE   **561**

   21-1   INTRODUCTION   561
   21-2   YARD PIPING   561
   21-3   PLANT HYDRAULICS   562
   21-4   DESIGN EXAMPLE   564
   21-5   PROBLEMS AND DISCUSSION TOPICS   581
         REFERENCES   582

22 *INSTRUMENTATION AND CONTROLS*   *583*

22-1 INTRODUCTION   583
22-2 INSTRUMENTATION AND CONTROL SYSTEMS   583
22-3 MANUFACTURERS AND EQUIPMENT SUPPLIERS
OF INSTRUMENTATION AND CONTROL SYSTEMS   592
22-4 INFORMATION CHECKLIST FOR DESIGN AND
SELECTION OF INSTRUMENTATION AND
CONTROL SYSTEMS   592
22-5 DESIGN EXAMPLE   593
22-6 PROBLEMS AND DISCUSSION TOPICS   600
REFERENCES   604

23 *DESIGN SUMMARY*   *605*

24 *ADVANCED WASTEWATER TREATMENT AND UPGRADING SECONDARY TREATMENT FACILITY*   *615*

24-1 INTRODUCTION   615
24-2 OVERVIEW OF ADVANCED WASTEWATER
TREATMENT TECHNOLOGY   616
24-3 PROBLEMS AND DISCUSSION TOPICS   632
REFERENCES   634

25 *AVOIDING DESIGN ERRORS*   *636*

25-1 INTRODUCTION   636
25-2 EXAMPLES OF DESIGN ERRORS AND
DEFICIENCIES   636
25-3 PROCEDURE TO AVOID OR REDUCE COMMON
DESIGN ERRORS AND DEFICIENCIES   639
25-4 DISCUSSION TOPICS   641
REFERENCES   642

APPENDIX A PHYSICAL AND CHEMICAL PROPERTIES OF WATER   *643*

APPENDIX B HEAD LOSS CONSTANTS IN OPEN CHANNELS AND PRESSURE PIPES   *647*

APPENDIX C TREATMENT PLANT COST CURVES   *655*

REFERENCES   692

APPENDIX D   MANUFACTURERS AND SUPPLIERS
OF WASTEWATER TREATMENT
PLANT EQUIPMENT   694

APPENDIX E   SELECTED CHEMICAL ELEMENTS   706

APPENDIX F   ABBREVIATIONS AND SYMBOLS,
USEFUL CONSTANTS, UNIT
CONVERSIONS, DESIGN
PARAMETERS, AND UNITS OF
EXPRESSION FOR WASTEWATER
TREATMENT   708

Index   719

Preface

With the impetus given to water quality improvement through the municipal construction grants program, this country has undertaken an unprecedented building program for new and improved wastewater treatment works. Practicing engineers are involved in planning, design, and construction of wastewater treatment facilities. At present, many programs in civil and environmental engineering at many universities offer courses in wastewater treatment plant design.

Several excellent books have been written in recent years that present the theory and principles of wastewater treatment processes. The author has observed during his years of experience in the wastewater treatment field that no publication has been devoted entirely to the technical aspects of planning, design, and operation of wastewater treatment facilities. The intent of the author in developing this book is twofold: first, to consolidate the developments in planning and design of wastewater treatment facilities that have evolved as a result of technological advances in the field, and as a result of the concepts and policies promulgated by the environmental laws and the subsequent guidelines; and second, to develop step-by-step procedures for planning, design, and operation for a medium-size wastewater treatment plant.

This book is divided into 25 chapters. Chapters 1 through 5 are devoted to the basic facts of wastewater engineering. Current and future trends in wastewater treatment technology, basic design factors and effluent guidelines, wastewater characteristics, treatment processes and process combinations, and requirements of predesign studies and facility planning are discussed in detail. Chapter 6 is devoted to facility planning. A model facility plan for a medium-size wastewater treatment facility is developed. Chapters 7 through 23 are devoted to design and operation of a medium-size wastewater treatment facility. Step-by-step design calculations, equipment selection, engineering drawings, operation and maintenance, and plans and specifications of various treatment units are presented. These units include intercepting sewers, bar screens, pumping stations, flow measurement, grit removal, primary clarification, biological treatment, disinfection, outfall, and sludge processing and disposal systems. Separate chapters have also been devoted to yard piping and hydraulics; plant layout; instrumentation and automatic controls; upgrading of secondary treatment facility and advanced wastewater treatment processes; and avoiding errors in plant designs.

The design procedures given in this book are for illustration and general information only, and are not intended to be used as standard for wastewater treatment plant designs. References made in this publication to specific methods, processes, and equipment do not constitute or imply an endorsement or recommendation. Equivalent or improved equipment may be obtained from many other manufacturers not mentioned in this publication.

This book will serve the needs of students, teachers, consulting engineers, equipment manufacturers, and technical personnel in city, state, and federal organizations who must review designs and specifications. In order to maximize the usefulness of this book, the material has been presented in a simplified and concise format. Many tables have been developed, using a variety of sources. Those tables provide information used extensively in wastewater treatment plant design. Basic properties of water and wastewater, hydraulic design information, cost curves, equipment manufacturers, and other related design data are arranged in several appendixes.

A great deal of emphasis has been given to the planning, design, and operation of a conventional treatment plant (secondary treatment using an activated sludge process). Many other equally important processes, such as stabilization ponds, attached growth reactors, chemical precipitation, land treatment, filtration, carbon absorption, and other advanced wastewater treatment processes, are briefly discussed in Chapters 4, 13, and 24, but in-depth design coverage similar to that for a secondary treatment plant has not been given in this book. Covering design procedures for these processes would take another book of equal size. The author strongly believes that the planning and design principles developed in this book can be easily extended by the students and designers for designing any other treatment process for a new plant or upgrading an existing facility. Therefore, the in-depth coverage and step-by-step design procedure for a secondary wastewater treatment plant with suspended growth biological reactor is the strongest feature of this publication.

Metric units are exclusively used in this book. Since old plants will be upgraded in the future, the U.S. customary units will continue to be in use for some time to come. Therefore, where possible, both units are used and proper conversion factors provided. (A complete conversion table is given in Appendix F.)

Acknowledgments

I am very grateful to those who have helped me prepare this book. First, I must thank Walter Chiang and Mike Morrison for their interest and stimulating discussion and response during the development of this book. They reviewed many chapters and made constructive suggestions in process and equipment design. Mike Morrison prepared the initial draft of the chapter on pumping station.

I am especially grateful to many professionals for their assistance. Eric Schweizer helped in the preparation of chapters on sludge processing. Pete Patel reviewed the chapter on instrumentation and prepared the simplified control loop diagrams. Edward Motley and Jay Ulray reviewed many chapters and suggested changes in design layout, operation and maintenance, and equipment specifications. Max Spindler checked hydraulic calculations through many treatment units. K. Udomsinrot provided valuable assistance by checking the design calculations and reviewing the entire manuscript.

Many professionals and students also reviewed various portions of this book, conducted literature searches, checked calculations, worked out solutions to the problems, and prepared drawings. In particular, I should like to thank Tanveer Islam, Hung Ngo, Bill Davis, Steve Sanders, M. Issa, Steve McCrary, and Jim Humphrey.

Many equipment manufacturers and their local representatives provided valuable information on equipment details and specifications. Their names and addresses are included in Appendix D.

The Department of Civil Engineering, College of Engineering, and Construction Research Center at the University of Texas at Arlington provided assistance with manuscript typing. In particular, I should like to thank Azalee Tatum and Ava Chapman, who typed portions of the manuscript.

Finally, I must acknowledge with deep appreciation the support, encouragement, and patience of my wife and daughters, who tolerated the agonies that generally accompanied long hours of work over a period of several years.

Although portions of this book have been reviewed by professionals and students, the real test will not come until it has been used in classes and used by professionals as a design guide. I shall appreciate it very much if all who use this book will let me know of any errors and changes they believe would improve its usefulness.

Arlington, Texas   Syed R. Qasim

1

Introduction

1-1 CURRENT STATUS

Work on the first sanitary sewer in the United States was begun in Chicago in 1855, 12 years after the world's first sanitary sewerage system was completed in Hamburg, Germany. Early treatment plants were built in the 1870s, and by 1948 wastewater treatment plants served some 45 million Americans out of a total population of 145 million.[1]

In 1956 Congress enacted the Federal Water Pollution Control Act, which established the construction grants program. Under the 1972 Amendments to the Federal Water Pollution Control Act (Public Law 92-500) and Clean Water Act of 1977 (Public Law 95-217), thousands of municipal wastewater treatment facilities are being constructed or expanded across the nation to control or prevent water pollution.[2,3] The law established the National Pollutant Discharge Elimination System (NPDES), which calls for limitations on the amount or quality of effluent, and requires all municipal and industrial dischargers to obtain permits. The interim goal is to achieve water quality in natural waters that provides for the protection and propagation of fish, shellfish, and wildlife, and provides recreation in and on the water. The law authorized billions of dollars for construction grants. The needs survey indicated that in 1980 there were approximately 15,251 wastewater treatment facilities in the United States serving a total population of 157 million people, and in the year 2000 there will be 21,600 treatment facilities serving 247 million people.[4]

1-2 FUTURE TRENDS

Many technological advances have been achieved in the wastewater treatment field as a result of concepts and policies promulgated by the environmental laws and guidelines. A great deal has been learned regarding process design and construction, operation and maintenance, problems associated with improper site selection, and plant design. In the next decade there will be some shift in strategy for new plants and upgrading of existing facilities. Some of these changes in design trends are presented below.

1-2-1 Regional versus Decentralized Treatment Facilities

During the 1970s great emphasis was given to areawide wastewater planning and management in an effort to (1) use best practicable waste treatment technology and (2) produce revenues through use of effluent and sludge. Experience has shown that centralized facilities may require pumping of wastewater long distances from different portions of the service area and thus may be in general energy and resource intensive. Also serious odors, dust, noise, and other environmental problems developed in the community, as a result of processing large quantities of wastewater and sludge at one location.

In coming years, the concept of *satellite* wastewater treatment will be reevaluated in the overall context of economics, innovative energy-efficient technology, and reuse of effluent and sludge locally, or sludge alone being pumped to a central location for processing. Also, individual on-site treatment and disposal systems will receive greater attention.

1-2-2 Environmental Considerations and Public Participation

Wastewater treatment works must not be the ugly duckling in the community, but rather a good neighbor. In the past, odors, dust, noise, erosion, and unsightly conditions created public doubts and uneasiness about the nearby municipal treatment works. It is unacceptable to create new environmental problems.[5] Therefore future planning and designs of wastewater treatment facilities will emphasize techniques to minimize adverse environmental impacts and objections by the neighborhood residents. Furthermore, active public participation programs will be an integral part of the decision-making process at all stages of the planning, design, and construction of wastewater treatment facilities.[6]

1-2-3 Land Treatment of Wastewater

Although land treatment of wastewater has been used for centuries in many parts of the world, this method has received recent renewed interest in this country. Future planning and design of wastewater treatment facilities will receive greater emphasis on land treatment in order to achieve water and nutrients reuse for crop production.

1-2-4 Other Effluent Uses

Many other uses of wastewater effluent will be explored in future designs. Potential uses of effluent include groundwater recharge, recreation lakes, aquaculture, industrial uses, indirect municipal reuse as makeup water for water supply reservoirs, and direct municipal reuse under emergency conditions.

1-2-5 Energy Conservation

Energy conservation in process selection and plant design will receive particular attention throughout the planning and process selection stages of the wastewater treatment facilities. Wastewater treatment alternatives which substantially conserve energy are considered innovative and will be preferred in future designs.

1-2-6 Design Based on Process Kinetics

The wastewater treatment technology is evolving rapidly, and an enhanced knowledge of fundamentals will permit the engineers to adapt to new processes and improve design techniques. The empirical methods commonly used in the past are inadequate for interpreting data and optimizing the processes. Laboratory and pilot plant studies will be utilized to develop process design parameters and kinetic coefficients for industrial and joint industrial–municipal wastewater treatment facilities.

1-3 PLANT DESIGN

The task of planning and designing wastewater treatment works is not simple. It involves understanding of service area, sources of wastewater and the resulting characteristics, plant site, conveyance system, and treatment processes for liquid and residues. Many nontechnical factors such as legal issues, regulatory constraints, and public participation may influence planning and design. Furthermore, most of the facilities are designed to provide service over the plant's life expectancy (20 years or more). During this extended time span, technology may improve, new laws may be passed, new regulations may be issued, and economic factors may change. The engineers must consider these possibilities and should favor processes that are sufficiently flexible to remain useful in the face of changing technology, regulations, economics, and wastewater characteristics.

During the design phase of a plant, it is important to recognize that the overall performance of a wastewater treatment facility is the result of the combined performances of many components utilized in the overall process train. The designers must understand the design implications and performance of the individual processes and how these processes may affect one another under normal and adverse operational conditions. As an example, failure to remove solids produced within the treatment processes will eventually cause degradation of effluent quality. Likewise, hydraulic overload to sludge-processing systems may increase the solids in the sidestream which is returned to the plant as a recirculating load, thus adversely effecting the influent quality to be treated.

The objective of this book is to present the nontechnical and technical issues that are most commonly addressed in the planning and design reports for wastewater treatment facilities prepared by practicing engineers. Topics discussed include facility planning, process description, process selection logic, mass balance calculations, design calculations, concepts for equipment sizing, operation and maintenance, and equipment specifications. Thus delineation of such information for use by students and practicing engineers is the main purpose of this book.

REFERENCES

1. U.S. Environmental Protection Agency, *Building for Clean Water*, Office for Public Affairs (A-107) USEPA, Washington, D.C., August 1975.
2. *Federal Water Pollution Control Act Amendments of 1972* (PL 92-500), 92nd Congress, October 18, 1972.
3. *Clean Water Act* (PL 92-217), 95th Congress, December 27, 1977.
4. U.S. Environmental Protection Agency, *The 1980 Needs Survey, Conveyance, Treatment, and*

Control of Municipal Wastewater, Combined Sewer Overflows, and Stormwater Runoff, Summaries of Technical Data, FRD 23, Office of Water Program Operations (WH-595), Washington, D.C., 430/9-81-008, February 10, 1981.

5. Leffel, R. Ernest, *Direct Environmental Factors at Municipal Wastewater Treatment Works,* U.S. Environmental Protection Agency, EPA-430/9-76-003, MCD-20, Washington, D.C., January 1976.

6. Metcalf and Eddy, Inc., *Wastewater Engineering: Treatment, Disposal, and Reuse,* Second Edition, McGraw-Hill Book Co., New York, 1979.

2

Basic Design Considerations for Wastewater Treatment Facilities

2-1 INTRODUCTION

The planning, design, construction, and operation of wastewater treatment facilities is a complex problem. It involves political, social, and technical issues. Therefore, besides meeting the effluent quality requirements, a wastewater treatment and disposal system must also satisfy many other environmental conditions. Some of these environmental conditions are (1) to prevent unsightliness, nuisance, and obnoxious odors at treatment and disposal sites; (2) to prevent contamination of water supplies from physical, chemical, and biological agents; (3) to prevent destruction of fish, shellfish, and other aquatic life; (4) to prevent degradation of water quality of receiving waters from overfertilization; (5) to prevent impairment of beneficial uses of natural waters (recreation, agriculture, commerce, or industry, etc.); (6) to protect against the spread of diseases from crops grown on sewage irrigation or sludge disposal; (7) to prevent decline in land values and, therefore, not to restrict the community growth and development; and (8) to encourage other beneficial uses of effluent.

The purpose of this chapter is to discuss many important design factors that must be considered during the initial planning and design stages of a wastewater treatment facility. Basic design factors are

1. Initial and design years
2. Service area

5

3. Site selection
4. Design population
5. Regulatory control and effluent limitations
6. Characteristics of wastewater
7. Degree of treatment
8. Selection of treatment processes
9. Equipment selection
10. Plant layout and hydraulic profile
11. Energy and resource requirements
12. Plant economics
13. Environmental impact assessment

Although most of these design factors are covered in greater detail in subsequent chapters, the information presented below is an introduction to the plant design.

2-2 INITIAL AND DESIGN YEARS

It generally takes several years to plan, design, and construct a wastewater treatment facility. Accordingly, most of the plant's components are made large enough to satisfy the community needs for several years in the future. The initial year is the year when the construction is completed and the initial operation begins. The design or planning year is the year when the facility is expected to reach its full designed capacity. Selecting the design year is not a simple task. It requires sound judgments and skills in developing future growth estimates from the past social and economic trends of a community. Design periods* are generally chosen with the following factors in mind:[1,2]

1. Useful life of treatment units, taking into account wear-and-tear and process obsolescence
2. Ease or difficulty in expansion
3. Performance of the treatment facility during the initial years when it is oversized
4. Future growth in population (including shifts in community), service area, commercial and industrial developments, water demands, and wastewater characteristics
5. Interest rates, cost of present and future construction, and availability of funds

The design periods of different components of a treatment facility may also vary. As an example, main conduits, channels, and appurtenances that cannot be expanded readily are designed for periods up to 50 years in the future. On the other hand, treatment units, process equipment, pumps, and sludge-handling and disposal facilities are constructed for shorter periods to avoid construction of oversized units. In such cases, adequate space is left at the plant site for expansion of the facility at different staging periods.† According to the guidelines of the construction grants program, the design period may be divided into staging periods (10, 15, and 20 years), depending on the ratio of wastewater flow expected

---

*Design or planning period is defined as the period from the initial year to the design year.

†Staging periods are time intervals when plant expansions are made.

at the design and initial years.[3] These staging periods are summarized in Table 2-1. Additional discussion on staging period and procedure to calculate staging period are provided in Chapter 6. Review Sec. 6-6-4 for development of staging period of 15 years for the Design Example.

2-3 SERVICE AREA

Service area (also called sewer district) is defined as the total land area that will be eventually served by the proposed wastewater treatment facility. The area may be based on natural drainage, political boundaries, or both. Areas outside the city limit but draining toward the proposed wastewater treatment plant and the area that may become incorporated into the city at future dates should also be considered into the service area. On the other hand, areas requiring sewage pumping may also be included into the service area after careful evaluation of economics and other environmental constraints.

It is important that the design engineer and the project team become familiar with the service area of the proposed project. Site visits and review of engineering data on topography, geology, hydrology, climate, ecological elements, and social and economic conditions may be necessary. Existing zoning regulations and land use or zoning plans and probable future changes that may affect both developed and undeveloped lands should be studied. Such efforts should be carefully coordinated with the state, regional, and local planning agencies and should be in conformance with the development and implementation of area-wide waste management plans.[3]

2-4 SITE SELECTION

Site selection of a wastewater treatment facility should be based on careful consideration of regions' land use and development patterns as well as social, environmental, and engineering constraints. It is important to remember that the selection of a site for a wastewater treatment plant will have long-lasting social, economic, and political repercussions on the affected community and neighborhood. Therefore, public participation in decision making is crucial.

All possible sites for a wastewater treatment plant should be fully evaluated on the basis of topography, environmental impacts, and economics of wastewater collection and

TABLE 2-1 Staging Periods for Plant Expansion over the Design Period

Flow Growth Factor (ratio of flows at design and initial years)	Staging Period (years)
less than 1.3	20
1.3–1.8	15
greater than 1.8	10

treatment. Often an interdisciplinary* team approach may be necessary to cover all these aspects. The following is a list of some basic principles that must be considered during site evaluation.

1. A wastewater treatment plant should be located at low elevation in order to permit gravity flow.
2. The site should be fairly isolated from presently built-up areas or areas that have potential for future developments. All plants should be designed with aesthetic considerations and odors in mind. For instance, the drying beds for sludge require large areas and are potential sources of bad odors.
3. A site on a large land area is helpful in maintaining isolation (buffer areas) and fulfills the needs for future expansion.
4. A site that may provide opportunity for local disposal of end products such as effluent, grit, screenings, sludge, and ash is highly desirable.
5. A site within a flood zone should not be selected unless proper flood protection measures are taken. Such measures include raising the units above flood level or constructing levees around the site. Storm drainage system with pumping equipment to discharge the storm water and effluent above the flood level should be provided. Auxiliary power equipment may be necessary in the event of power outages.
6. The site should have year-round, all-weather access roads. Railroads may be helpful for delivery of bulk chemicals and transport of sludge from large facilities.
7. The site should be near a large body of water or irrigable land capable of accepting the treated effluent.
8. The ability of the ground to support structures without extensive piling is important consideration in site selection. Common foundation problems are low bearing capacity, excessive settlement, differential settlement, and floatation due to high groundwater table.
9. A site with a moderate slope will assist in locating various treatment units in their normal sequence without excavation or filling. This will provide gravity flow, least disruption to the natural topography, and least erosion control measures.
10. The sites should be evaluated and checked for presence of archaeological, historical, or other properties included in or eligible for inclusion in the National Register of Historic Places. The site should also be investigated for the presence of endangered or threatened species of flora or fauna, or their critical habitats.
11. Site selection and planning of wastewater treatment facilities should be done with prime consideration of preservation of shorelines, particularly in urban areas. Shoreline preservation involves public pathways along shore, public parks, recreation facilities, and protection against erosion of banks, siltation of the waterway, and preservation of valuable ecological niches.

   Additional information on site selection and planning for wastewater treatment plant may be found in Chapter 20. In summary, proper site selection for proposed wastewater

---

*The interdisciplinary team may include professionals from sanitary and environmental engineering, urban planning, architecture, biology, government, and civic groups.

treatment facility is important and must not be overlooked by the designers. A team approach to site selection will be helpful in future expansion cost and energy savings, and avoid future complications that may occur due to public opposition. Matters to be investigated include topography, drainage, surface and groundwater, soil type, prevailing winds, temperature, precipitation, seasonal solar angles, wildlife habitats, ecosystems, regional and local land use and zoning, transportation, archaeological and historical features, and other factors.

There is a large amount of existing information available on the above subjects at various local, state, and federal agencies, and at the universities. There are also many types of methodologies available for alternative evaluation and selection of sites. These methodologies offer analytical tools for identifying, measuring, and interpreting the data. Data sources and evaluation methodologies may be found in Refs. 4–7.

2-5   DESIGN POPULATION

The volume of wastewater generated in a community depends on the population and per capita contribution of wastewater. It is therefore important to estimate the population to be served at the design year. Accurate population prediction is quite difficult because many factors influence the growth of the city. Among important factors are industrial growth; state of development of the surrounding area; location with regard to transportation sources; availability of raw material, land, and water resources; local taxes and government activities; migration trends; and so on.

The population data can be obtained from several sources. The U.S. Bureau of Census (Department of Commerce) publishes 10-year census data. For the interim periods, reliable data can usually be obtained from local census bureaus; the state, county, or local planning commissions; the chambers of commerce; voters registration lists; the post office; newspapers; and public utilities (telephone, electric, gas and water, etc.). It is important that the design engineer become familiar with the population data sources and the type of information that can be obtained from these sources.

2-5-1 Methods of Population Forecasting

There are many mathematical and graphical methods that are used to project past population data to the design year. Widely employed methods are

1. Arithmetic growth
2. Geometric growth
3. Decreasing rate of increase
4. Mathematical or logistic curve fitting
5. Graphical comparison with similar cities
6. Ratio method
7. Employment forecast
8. Birth cohort

All these methods utilize different assumptions and therefore give different results. Selection of any method depends on the amount and type of data available and whether the projections are made for the short or long term. The arithmetic, geometric, decreasing rate of increase, and logistic curve-fitting methods are summarized in Table 2-2. The remaining methods are presented in Table 2-3 and Figs. 2-1 to 2-3. Several excellent references may be consulted on population forecasting.[8-11]

2-5-2 Population Density

The average population density for the entire city rarely exceeds 7500–10,000 per km^2 (30–40 per acre). Often it is important to know the population density in different parts of the city in order to estimate the flows and to design the collection network. Density vary widely within a city depending on the land use. The average population densities based on land use characteristics are summarized in Table 2-4.

2-6 REGULATORY CONTROLS AND EFFLUENT LIMITATIONS

2-6-1 Water Pollution Control Legislation

Virtually all of the water pollution control legislation in the United States has been passed and written in the past 35 years. However, much progress has been made during this

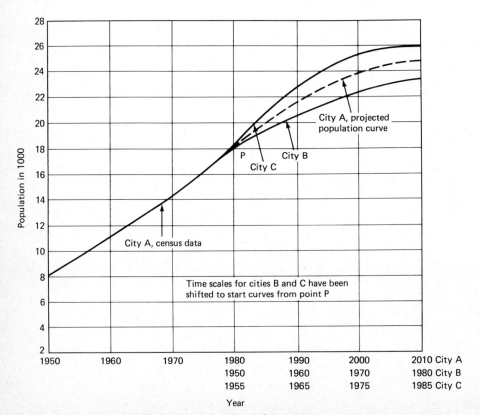

Figure 2-1 Population Estimate by Graphical Comparison.

Fitting

Method	Description	Basic Equations or Procedure[a]	Calculated Values[b]
Arithmetic method	Population is assumed to increase at a constant rate. The method is used for short-term estimates (1–5 yr)	$\dfrac{dY}{dt} = K_a; \; Y_t = Y_2 + K_a(T - T_2)$ $K_a = \dfrac{Y_2 - Y_1}{T_2 - T_1}$	$K_a = 300/\text{yr}$ $Y_t = 21{,}000$
Geometric method	Population is assumed to increase in proportion to the number present. The method is commonly used for short-term estimates (1–5 yr)	$\dfrac{dY}{dt} = K_p Y; \; \ln Y_t = \ln Y_2 + K_p(T - T_2)$ $K_p = \dfrac{\ln Y_2 - \ln Y_1}{T_2 - T_1}$	$K_p = 0.0182$ $Y_t = 21{,}600$
Decreasing rate of increase	Population is assumed to reach some limiting value or saturation point	$\dfrac{dY}{dt} = K_d\,(Z - Y), \; Y_t = Y_2 + (Z - Y_2)(1 - e^{-K_d(T - T_2)})$ $Z = \dfrac{2Y_0 Y_1 Y_2 - Y_1^2(Y_0 + Y_2)}{Y_0 Y_2 - Y_1^2}$ $K_d = \dfrac{-\ln[(Z - Y_2)/(Z - Y_1)]}{T_2 - T_1}$	$Z = 20{,}000$ $K_d = 0.09163$ $Y_t = 19{,}200$
Mathematical or logistic curve fitting	It is assumed that the population growth follows a logistical mathematical relationship. Most common relationship is an S-shaped curve	$Y_t = \dfrac{Z}{1 + ae^{b(T - T_0)}}$ $a = \dfrac{Z - Y_0}{Y_0}$ $b = \dfrac{1}{n}\ln\left[\dfrac{Y_0(Z - Y_1)}{Y_1(Z - Y_0)}\right]$	$Z = 20{,}000$ $a = 1.00$ $b = -0.1099$ $n = 10$ $Y_t = 19{,}287$

[a] dY/dt = rate of change of population with time. Y_0, Y_1, and Y_2 = populations at time T_0, T_1, and T_2. Y_t = estimated population at the year of interest. Z = saturation population. K_a, K_p, and K_d are proportionality constants. a and b = constant. n = constant interval between T_0, T_1, and T_2 (generally 10 yr).
[b] Population for 1990 is estimated using the following census results: $T_0 = 1960$, $Y_0 = 10{,}000$; $T_1 = 1970$, $Y_1 = 15{,}000$; $T_2 = 1980$, $Y_2 = 18{,}000$.

TABLE 2-3   Population Projections by Using Graphical Comparison, Ratio Method, Employment Forecast, and Birth Cohort

Method	Description	Problem Definition	Estimated Population
Graphical comparison	The procedure involves the graphical projection of the past population data for the city being studied. The population data of other similar but larger cities are also plotted in such a manner that all the curves are coincident at the present population value of the city being studied. These curves are used as guides in future projections	Estimate the population of City A by using graphical comparison with cities B and C. The design year is 2010. Year — Population in 1000: City A, City B, City C 1950 — 8.0, 18.0, 16.0 1960 — 11.0, 20.3, 20.0 1970 — 14.0, 22.0, 25.0 1980 — 18.0, 23.2, 25.6 The procedure is illustrated in Figure 2-1.	$Y_{2010} = 24{,}800$
Ratio and correlation	In this method the population of the city in question is assumed to follow the same trends as that of the region, county, or state. From the population records of a series of census years the ratio is plotted and then projected to the year of interest. From the estimated population of the region, county, or state, and the projected ratio the population of the concerned city is obtained	Estimate the population of a city using the ratio method. The design year is 2000. The estimated population of the region in the year 2000 is 988,000. Year — Population in 1000: City, Region, Ratio 1950 — 50, 455, 0.110 1960 — 61, 623, 0.098 1970 — 72, 766, 0.094 1980 — 77, 850, 0.091 The procedure is illustrated in Figure 2-2.	$Y_{2000} = 86{,}944$

Employment forecast or other utility connections forecast.

The population is estimated using the employment forecast. From the past data of population and employment, the ratio is plotted and population is obtained from the projected employment forecast. Procedure is similar to that of the ratio method. Similar procedure can be utilized from the forecast of various utility service connections such as telephone, electric, gas, water and sewers, etc. Utility companies conduct studies and develop reliable forecasts on the future connections. Forecasts of postal and newspaper service points have also been used in population estimates

Estimate the population of a city using employment forecast. The design year is 2000. Use the following data. Employment forecast for the year 2000 is 21,300.

Year	Population in 1000	Employment in 1000	Ratio Population tion employment
1950	20	6.80	2.94
1960	30	10.79	2.78
1970	39	14.77	2.64
1980	46	17.83	2.58

The procedure is illustrated in Figure 2-3.

$Y_{2000} = 54,102$

Birth cohort

A birth cohort is defined by demographers as a group of people born in a given year or period.[a] The existing populations of males and females in different age groups are determined from the past records. From birth and death rates of each group and population migration data, the net increase in each group is calculated. The population data are then shifted from one group to the other until the design period is reached. This procedure is discussed in Ref. 11

[a]Demography is that branch of anthropology which deals with the statistical study of the characteristics of human population with reference to total size, density, number of deaths, births, migration, etc.

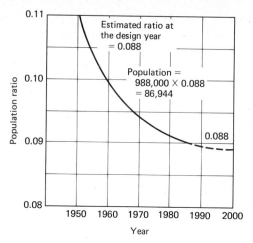

Figure 2-2   Population Estimate by Ratio Method.

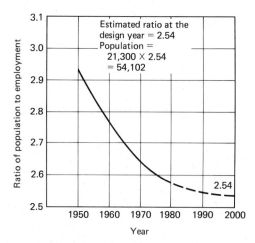

Figure 2-3   Population Estimate by Employment Forecast.

period. Following is the summary of water pollution control legislation in the United States:

- Federal Water Pollution Control Act of 1948
- Federal Water Pollution Control Act of 1956
- Federal Water Pollution Control Act Amendments of 1961
- Water Quality Act of 1965
- Clean Water Restoration Act of 1966
- Water Quality Improvement Act of 1970
- Federal Water Pollution Control Act Amendments of 1972

TABLE 2-4   *Range of Population Densities in Various Sections of a City*

Land Use	Population Range	
	Persons per km²	Persons per acre
Residential areas		
Single-family dwellings, large lots	1250–3700	5–15
Single-family dwellings, small lots	3700–8700	15–35
Multiple-family dwellings, small lots	8700–25,000	35–100
Apartment or tenement houses	25,000–250,000	100–1000 or more
Commercial areas	3700–7500	15–30
Industrial areas	1250–3700	5–15
Total, exclusive of parks, playgrounds, and cemeteries	2500–12,500	10–50

Source: Adapted from *Water and Wastewater Engineering, Vol. 1. Water Supply and Wastewater Removal* by G. M. Fair, J. C. Geyer, and D. A. Okun, 1966. Used with permission of John Wiley & Sons. *Note:* km² × 247.1 = acres.

- Clean Water Act of 1977
- Clean Water Act Amendments of 1980
- Clean Water Act Amendments of 1981

The 1972 and 1977 laws are collectively known as the Clean Water Acts. The ultimate goal of these acts is to eliminate the discharge of pollutants into any surface water by 1985. The 1980 Needs Survey indicated that considerable progress in municipal pollution abatement from publicly owned treatment works, both nationwide and in urban areas, has been made. Existing secondary treatment capacities have been expanded, and less efficient facilities have been upgraded. This progress has been both significant and measurable.[12]

2-6-2 Effluent Standards

Under the Clean Water Acts each state and the area agencies were required to develop a water quality management (WQM) plan that identified sources and severity of pollution, and needed programs to control pollution. Once completed and approved, the plan became the foundation of the WQM process. Each state annually assesses current water quality problems, updates its strategy to solve problems, prepares and carries out a work program to implement solutions, evaluates performance, and revises the plan. As part of the WQM process, the state agency establishes total maximum daily wasteloads for all surface waters throughout the state. The agency classifies bodies of state waters as either "effluent-limited" or "water quality-limited." If state's water quality standards can be met by uniform national discharge limits, the water body is classified as effluent-limited, and all municipal treatment plants need only achieve secondary treatment. (The secondary level of treatment as defined by EPA is summarized in Table 2-5.[13]) Where stricter limits are needed to meet the state's standards, the water body is classified as water quality-limited. The resulting wasteload allocations are generally incorporated into the effluent limitations and compliance schedule in the National Pollutant Discharge Elimination System (NPDES) permit.

2-6-3 Enforcement

The Clean Water Act requirements are backed up by permit program which describe the effluent limitation requirement of the point source discharge and other conditions to be imposed on individual discharges, such as monitoring and schedules for compliance. These permits are part of the National Pollutant Discharge Elimination System (NPDES) and are issued and enforced either by an EPA regional office or by a state water quality agency (most of the states have been approved by EPA to operate the NPDES permit program).

Under the NPDES permitting program, effluent limitations are being established for toxic pollutants for various industrial categories. In a similar manner, industrial pretreatment standards* are being developed for all pollutants that "interfere with, pass through, concentrate in sludge, or otherwise are incompatible" with publicly owned treatment works (POTW). Under the pretreatment regulations, two types of federal pretreatment standards are established: (1) prohibited discharges and (2) categorical standards.

Prohibited discharges are those that cause fire or explosion hazard, corrosion, obstruction, slug discharges, and heat discharge to sewers or POTWs. The categorical

*Pretreatment standards apply to the industrial discharges into the sanitary sewer system.

TABLE 2-5   Secondary Treatment as Defined by EPA

Effluent Parameters	Monthly Average	Weekly Average	Units
BOD₅	30[a]	45[b]	mg/ℓ
Suspended solids	30[a]	45[b]	mg/ℓ
Fecal coliform	200[c]	400[d]	numbers per 100 ml
pH	Between 6.0 and 9.0		

[a]Arithmetic mean for 30 consecutive days. The mean shall not exceed 15 percent of the arithmetic mean of the influent (85 percent removal).
[b]Arithmetic mean for 7 consecutive days.
[c]Geometric mean of 30 consecutive days. Coliform standards were deleted subsequently from the standards.
[d]Geometric mean of 7 consecutive days. Coliform standards were deleted subsequently from the standards.

standards are developed for those pollutants that are incompatible, that is, those that interfere with the operation of, pass through, or contaminate the sludge and other residues from POTWs.

The substances considered for categorical standards are those for which there is substantial evidence of carcinogenicity, mutagenicity, and/or teratogenicity; substances structurally similar to aforementioned compounds; and substances known to have toxic effects on human beings or aquatic organisms at sufficiently high concentration and which are present in the industrial effluents. There are many specific elements or compounds that have been identified as priority pollutants. These include metals, organics, cyanides, and asbestos.[14]

The NPDES permitting and pretreatment programs as they exist today are part of a very complex regulatory scheme. Categorical standards and guidelines for pretreatment and effluent discharges are being established for toxic, conventional, and nonconventional pollutants.

2-6-4 Federal Assistance to Communities

To provide incentives, the Clean Water Act offers federal funds to cover part of the cost of construction of publicly owned wastewater treatment works. In the past, in order to receive construction grants, communities had to meet a series of conditions. Grants were awarded in three steps:

Step 1.   The facility planning phase, when most major decisions leading toward construction of POTWs are made.
Step 2.   The design plans and specifications for the facility are completed.
Step 3.   Actual construction work is performed.

The major changes in the construction grants program that recently occurred are the elimination of grants for Step 1 (planning) and Step 2 (design) of treatment plants; and a reduction in the percentage of funding from 75 to 55 percent beginning October 1, 1984.

2-6-5 Public Participation

The planning procedures for the construction grants program include the broad-based requirements for public participation specified for all Clean Water Act Programs. Public participation is encouraged in the development, revision, effluent limitation, alternatives selection, and other decision-making processes. Public participation is discussed in detail in Chapters 5 and 6.

2-7   CHARACTERISTICS OF WASTEWATER

The characteristics of wastewater are developed in terms of flow conditions and chemical quality. The characteristics depend largely on the water usage in the community, and the industrial and commercial contributions. During wet weather, a significant quantity of infiltration/inflow may also enter the collection system. This will significantly change the characteristics of wastewater. The quantity of infiltration/inflow depends on the condition of sewer system (age and cracks in the pipes, and defective pipe joints and manholes), illegal roof or drain connections, groundwater table relative to sewer position, and the like.

If there is an existing wastewater treatment plant, the characteristics are obtained from the flow records and laboratory data. In the absence of the existing facilities, the data on wastewater characteristics are developed from the population estimates, water usage, and industrial waste discharges.

Wastewater characteristic data are needed for the initial year and for the design year. The data include minimum, average and maximum dry weather flows, peak wet weather flows, sustained maximum flows, and chemical parameters such as BOD_5, total suspended solids, pH, total dissolved solids, total nitrogen, phosphorus, and toxic chemicals.

It is important that reliable estimates on wastewater characteristics be made, as this is what the designed facilities will be treating. Chapter 3 is devoted exclusively to development of wastewater characteristics. Methods of estimating dry weather flows, infiltration and inflow allowances, and various wastewater quality parameters are fully covered in Chapter 3. Detailed procedures for determining infiltration/inflow and wastewater characteristics are presented in Chapter 6.

2-8   DEGREE OF TREATMENT

The degree of treatment required is based on the influent characteristics of the plant and the effluent quality. If the effluent is discharged into the natural water, it should comply with NPDES permit requirements. If used for land irrigation, the plant effluents must also satisfy the health regulations governing the types of crops that are irrigated. Other effluent uses, such as recreational lakes, agriculture, industrial and municipal, may dictate the effluent quality and thus the degree of treatment.

2-9   CHOICE OF TREATMENT PROCESSES, FLOW SCHEMES, AND COMPARISON OF ALTERNATIVES

Wastewater treatment plants utilize a number of treatment processes to achieve the desired degree of treatment. In addition to this, the design engineer must evaluate numerous other important factors in selection of the treatment processes. These factors include constituents treated, effluent limitations, proximity to buildup areas, hydraulic requirements, sludge disposal, energy requirements, and plant economics. The collective arrangement of various treatment processes is called a flow scheme, a flow sheet, or a process train.

Choice of proper treatment processes and development of the flow scheme is not a simple task. It requires understanding of the unit operations and processes, operational capabilities and environmental effects of various treatment components that are arranged to develop the flow scheme for a desired application.

Laboratory and pilot plant studies are often necessary to develop design parameters for physical, chemical, and biological treatment processes used to treat industrial wastewater. The laboratory studies include batch and/or continuous flow reactor studies. Procedures for conducting treatability studies on the industrial wastes are given in Refs. 15–17.

2-10 EQUIPMENT SELECTION

Every wastewater treatment facility will involve manufactured equipment or materials. In fact, many design details are often governed by the dimensions and installation requirements of the selected equipment. It is the responsibility of the design engineer to select the treatment processes and the corresponding types of equipment for achieving the desired results. To do this, a review of the design standards, design procedure, and design assumptions; preliminary design calculations; and careful study of the manufacturers' catalogs may be necessary in advance. Then the manufacturers of the appropriate equipment and their local representatives should be requested to furnish the needed data. It may often be necessary for the engineer to work closely with the equipment supplier, and provide as much data and general information as possible to ensure that the equipment selection is best for the specific application.

The request for information on equipment should not impose any obligation on the engineer to use the equipment or even to include it in the specifications.[18] However, the equipment data should not be requested unless there is some specific application in the design. It may often be necessary to obtain equipment data from more than one supplier. In no case, however, should the request be sent as a routine to all equipment manufacturers. Preparing the needed information on equipment costs money, and supplying such information without obligation can be justified only if there is a prospect for sale.

2-11 PLANT LAYOUT AND HYDRAULIC PROFILE

During early planning and design stages, careful consideration must be given to the existing conditions at the selected site of the proposed wastewater treatment plant. Condition such as topography, available land area, proximity to the developed areas, access roads, flood conditions, need for future expansion, available head, and so on should all be considered in unit selection and layout. Chapters 20 and 21 are devoted exclusively to the topics of plant layout and hydraulic profile. Design engineers should study these chapters in order to include the basic considerations relative to plant layout and hydraulic profile during the preliminary design and process selection.

2-12 ENERGY AND RESOURCE REQUIREMENTS

Because of the recently increased concern about the limited resources available to meet our energy needs, the project planning and design must also include energy conservation. Primary energy is the energy used in the operation of the facility, while secondary energy is needed to manufacture chemicals, other consumable materials, and construction material such as concrete and steel. Under the Clean Water Act of 1977, it is required that the designers encourage wastewater treatment techniques that would reduce total energy requirements. Waste treatment alternatives which substantially conserve energy are considered innovative. Therefore, process energy utilization and conservation should be of particular value throughout the planning, project formulation, and preliminary engineering design. Basic energy needs and energy conservation techniques in municipal wastewater treatment plants may be obtained from Refs. 7 and 19.

2-13 PLANT ECONOMICS

As an integral part of the wastewater treatment plant planning and design, a cost-effective analysis must be performed to ensure that the construction, and the operation and maintenance (O&M) are reasonable and appropriate for the planned level of treatment and process train. A cost-effective solution is one which will minimize total costs of the resources over the life of the treatment facility. Resources costs include capital (land plus construction), operation, maintenance and replacements, and social and environmental costs. Benefits from sludge and effluent sale or reuse will partly offset the resources costs.

Many publications and computer programs are available that are extensively used in making the preliminary cost estimates. When preparing cost estimates, proper assumptions, cost curves, and cost indexes should be used. Discussion on cost curves and cost indexes is given in Appendix C. Procedures for cost estimation and cost-effective evaluation are given in Chapter 6.

2-14 ENVIRONMENTAL IMPACT ASSESSMENT

The National Environmental Policy Act of 1969 (NEPA) was enacted to ensure that federal agencies consider environmental factors in the decision-making process and utilize an interdisciplinary approach in evaluating these issues.[3] The environmental impact assessment must evaluate all impacts, "beneficial" and "adverse," "primary" and "secondary," that may result from the construction of a wastewater treatment facility. The primary impacts are those directly associated with construction and operation of the treatment works. For example, changes in water quality and odors resulting from the plant are the primary impacts. The secondary impacts are indirect, resulting from the growth or change in land use induced or facilitated by the construction of the plant or its associated sewers.

Environmental impact assessment for wastewater treatment facilities is covered in Chapters 5 and 6. In-depth coverage on environmental impact assessment may be found in Refs. 4–7.

It is important that the design engineer work closely with the federal, state, and local regulatory agencies that have responsibilities for planning, design, and operation of the wastewater treatment facilities. The regulation of the state agency (state department of health or state water pollution control agency) usually establishes many basic design considerations and standards. The most widely used standards in the past were the "Ten-States Standards."[20] Now most of the states have developed their own standards. It may often be necessary for the design engineer to deviate from the standards. However, most state officials will take a reasonable approach on these issues if deviations from the state's standards may truly be necessary. Such issues are incorporation of new technology, relative cost savings, or other constraints that may be specific to a particular project.

2-15 PROBLEMS AND DISCUSSION TOPICS

2-1   List the major items of concern for the site selection of a wastewater treatment facility. What agencies would you contact to develop the needed information?

2-2   Visit the wastewater treatment facility in your community. Mark on a map the service area and the plant location. Draw the treatment flow scheme. Briefly summarize the history of wastewater treatment, including major events that helped to bring about improvements. If federal funds were involved under the construction grants program, review the Step 1, Step 2, and Step 3 applications.

2-3   Was an environmental impact report prepared for the most recent expansion of the wastewater treatment facility in your community? List the major environmental issues addressed in this report. Have other major issues been identified since the construction and operation of the facility?

2-4   Discuss various factors that may influence the population growth in a community.

2-5   Estimate the 1990 population for a community by using arithmetic, geometric, decreasing rate of increase, and logistic curve-fitting methods. Use the following census data:

Year	Population in thousands
1960	31.6
1970	36.9
1980	42.3

2-6   The population and employment data for a city are given below. Estimate the 1990 population if the employment projection for 1990 is 9200.

Year	Population	Employment
1960	20,000	5000
1970	21,000	8000
1980	23,000	8800

2-7   Obtain the following information for the wastewater treatment facility in your community:
   (a)   Design year, estimated population, and average flow
   (b)   Initial year, population, and average flow
   (c)   Service area in km^2

Determine (1) design period, (2) population density at the design year (person/km^2), (3) ratio of average design flow to initial flow, (4) does staging period correspond to the value given in Table 2-1? and (5) is the estimated population growth pattern similar to the actual population trends of the community?

REFERENCES

1. The Joint Committee of the Water Pollution Control Federation and the American Society of Civil Engineers, *Wastewater Treatment Plant Design*, WPCF Manual of Practice 8, Water Pollution Control Federation, Washington, D.C., 1977.

2. Fair, G. M., J. C. Geyer, and D. A. Okun, *Water and Wastewater Engineering, Vol. 1, Water Supply and Wastewater Removal*, John Wiley & Sons, New York, 1966.

3. 40 CFR Part 35, Subpart E., "Grants for Construction of Treatment Works," *Federal Register*, Vol. 43, No. 188, September 27, 1978, p. 44022–44097.

4. Planning Research and Design Center, University of Texas at Arlington, *Step-by-Step Procedure for Preparing Environmental Impact Statement for Wastewater Treatment Works*, U.S. Environmental Protection Agency, Washington, D.C., October 1975.

5. Canter, L. W., *Environmental Impact Assessment*, McGraw-Hill Book Co., New York, 1977.

6. Jain, R. K., L. V. Urban, and G. S. Stacey, *Environmental Impact Analysis*, Van Nostrand Reinhold, New York, 1981.

7. U.S. Environmental Protection Agency, *Environmental Assessment of Construction Grants Projects,* FRD-5, EPA-430/9-79-007, January 1979.
8. Clark, J. W., W. Viesseman, and M. J. Hammer, *Water Supply and Pollution Control,* IEP-A Dun-Donnelly, New York, 1977.
9. Metcalf and Eddy, Inc., *Wastewater Engineering: Collection, Treatment and Disposal,* McGraw-Hill Book Co., New York, 1972.
10. Steel, E. W., and T. J. McGhee, *Water Supply and Sewerage,* McGraw-Hill Book Co., New York, 1979.
11. Turk, A., J. Turk, and J. T. Wittes, *Ecology, Pollution, Environment,* W. B. Saunders Co., Philadelphia, 1972.
12. Chamblee, J. A., "Municipal Pollution Abatement Report Card: Significant Progress," *Journal Water Pollution Control Federation,* Vol. 54, No. 5, May 1982, p. 422.
13. 40 CRF, Part 133, "Secondary Treatment Information," *Federal Register,* Vol. 38, No. 159, August 1, 1973, p. 22298–22299.
14. Schwartz, H. G., and J. C. Buzzel, *The Impact of Toxic Pollutants in Municipal Wastewater Systems,* prepared for U.S. Environmental Protection Agency Technology Transfer, Joint Municipal/Industrial Seminar on Pretreatment of Industrial Wastes, Dallas, Texas, July 12, 13, 1978.
15. Metcalf and Eddy, Inc., *Wastewater Engineering: Treatment, Disposal, and Reuse,* McGraw-Hill Book Co., New York, 1979.
16. Qasim, S. R., and M. Stinehelfer, "Effect of a Bacterial Culture Product on Biological Kinetics," *Journal Water Pollution Control Federation,* Vol. 54, No. 3, March 1982, p. 255.
17. Eckenfelder, W. W., *Industrial Water Pollution Control,* McGraw-Hill Book Co., New York, 1966.
18. Hardenburgh, W. A., and E. B. Rodie, *Water Supply and Waste Disposal,* International Textbook Co., Scranton, Pennsylvania, 1960.
19. Wesner, G. M., et al., *Energy Conservation in Municipal Wastewater Treatment Plants,* Report prepared for U.S. Environmental Protection Agency, MCD-32, EPA 430/9-77-011, March 1978.
20. Great Lakes-Upper Mississippi River Board of State Sanitary Engineers, *Recommended Standards for Sewage Works,* Health Education Service, Albany, N.Y., 1971.

3

Wastewater Characteristics

3-1 INTRODUCTION

"Municipal wastewater" is the general term applied to the liquid wastes collected from residential, commercial, and industrial areas and conveyed by means of a sewerage system to a central location for treatment.

It is important that reliable estimates on wastewater characteristics are made, as this is what the treatment plant will be treating in the future. The purpose of this chapter is to develop information that can be utilized in assessing the characteristics of wastewater for the design of treatment facilities. Since under dry weather conditions, municipal wastewater is derived largely from the water supply, the water usage data under present and anticipated future conditions are essential. Therefore, discussion in this chapter is provided in three general categories: (1) municipal water demands, (2) wastewater flow rates, and (3) chemical quality of municipal wastewater.

3-2 MUNICIPAL WATER DEMAND

3-2-1 Components of Municipal Water Demand

Water demand data are very useful in estimating the wastewater characteristics. The average amount of municipal water withdrawn in the United States is approximately 628 liters per capita per day (ℓpcd).*[1,2] This amount includes residential, commercial, light industrial, fire fighting, public uses, and water lost or unaccounted for. There is, however, a wide variation in municipal water withdrawal rate. Factors affecting water withdrawal rates are (1) climate, (2) geographic location, (3) size and economic conditions of the community, (4) degree of industrialization, (5) metered water supply, (6) cost of water, and (7) supply pressure. The average water demand by states in the U.S. is summarized in Table 3-1. Various components of municipal water demand are discussed below.

Residential Water Use. The residential or domestic water demand is the portion of municipal water supply that is used in homes. Typical breakdown of residential water uses

---

*628 ℓpcd = 166 gallons per capita per day (gpcd).

TABLE 3-1   Average Municipal Water Demand by State

State	ℓpcd	gpcd[a]	State	ℓpcd	gpcd	State	ℓpcd	gpcd
Alabama	806	213	Maine	553	146	Oregon	712	188
Alaska	1790	473	Maryland	515	136	Pennsylvania	685	181
Arizona	787	208	Massachusetts	530	140	Rhode Island	462	122
Arkansas	503	133	Michigan	636	168	South Carolina	916	242
California	685	181	Minnesota	473	125	South Dakota	549	145
Colorado	746	197	Mississippi	507	134	Tennessee	488	129
Connecticut	541	143	Missouri	485	128	Texas	587	155
Delaware	700	185	Montana	829	219	Utah	1113	294
Florida	617	163	Nebraska	636	168	Vermont	553	146
Georgia	946	250	Nevada	1154	305	Virginia	420	111
Hawaii	746	197	New Hampshire	485	128	Washington	1200	317
Idaho	897	237	New Jersey	526	139	West Virginia	568	150
Illinois	772	204	New Mexico	772	204	Wisconsin	587	155
Indiana	534	141	New York	609	161	Wyoming	746	197
Iowa	466	123	North Carolina	644	170	District of Columbia	799	211
Kansas	587	155	North Dakota	477	126	Puerto Rico	326	86
Kentucky	314	83	Ohio	594	157	United States	628	166[b]
Louisiana	545	144	Oklahoma	492	130			

[a] 1 gal = 3.785 ℓ.
[b] Average is based on total municipal water demand for the United States divided by the population served.
Source: Adapted from Ref. 1.

TABLE 3-2   Typical Breakdown of Residential
Water Uses

Types of Water Use	Percentage
Toilet flush	41
Washing and bathing	37
Kitchen	11
Drinking, cooking (2–6 percent)	
Dishwashing (3–5 percent)	
Garbage disposal (0–6 percent)	
Laundry	4
Cleaning and general housekeeping	3
Lawn sprinkling	3
Auto washing	1

Source: Adapted in part from Refs. 3–6.

is given in Table 3-2.[3-6] It includes toilet flush, cooking, drinking, washing, bathing, watering lawn, and other uses. The average residential water demand varies from 228 to 456 ℓpcd, while most commonly used numbers are 300–380 ℓpcd.[2,7,8] Typical water uses for various household devices are summarized in Table 3-3.[3-6]

Commercial Water Use.   Commercial establishments include motels, hotels, office buildings, shopping centers, service stations, movie houses, airports, and the like. The commercial water demand depends on the type and the number of commercial establishments. In cities of over 25,000 population, the commercial water demand is about 15–20

TABLE 3-3   Typical Rates of Water Use from Various Devices

Device	Range of Flow
Household faucet	10–20 ℓ/min
Wash basin	4–8 ℓ/use
Shower head	90–110 ℓ/use; 19–40 ℓ/min
Tub bath	60–190 ℓ/use
Toilet flush, tank-type	19–27 ℓ/use
Toilet flush, valve-type	90–110 ℓ/min
Dishwasher	15–30 ℓ/load
Washing machine	110–200 ℓ/load
Lawn sprinkler	6–8 ℓ/min
Continuous flowing drinking fountain	4–5 ℓ/min
Garbage disposal	6000–7500 ℓ/wk, 4–8 ℓ/person per day
Dripping or leaky faucet	10–1000 ℓ/d

Source: Adapted in part from Refs. 3–6.

percent of total water demand.[7,8] Water demands in various types of commercial establishments are given in Table 3-4.[5,9,10]

Industrial Water Use.   Industrial water demands are very large in the United States. Generally, large industries develop their own water supply systems. Only small industries purchase water from the cities and therefore impose demand on local municipal system. Industrial water demand may be estimated on the basis of proposed industrial zoning and type of industries most likely to develop within the city. Typical water demand of small industries is 20–30 percent of total municipal water demand.[7] Some industrial water demand data are summarized in Table 3-5.[11,12]

Public Water Use.   Water used in public buildings (city halls, jails, schools, etc.) as well as water used for public services including fire protection, street washing, and park irrigation is considered public water use. Public water use accounts for 8–15 percent of total municipal water demand.[7,8]

Water Unaccounted For.   In a water supply system there is a certain amount of water that is lost or unaccounted for because of meter and pump slippage, leaks in mains, faulty meters, and unauthorized water connections. In municipal supply systems this may be 8–15 percent of the total water demand.[7]

3-2-2 Variations in Municipal Water Demand

The municipal water demand discussed above is based on annual average daily demand. There are wide variations in seasonal, daily, and hourly water demands. Some of the general observations of municipal water demands are summarized below:

- Working days have higher demand than holidays.
- Hot and dry days have more demand than wet or cold days.
- Maximum months are typically July or August (summer).
- Within a day there are two peak demands. One peak is in the morning as the day's activities start, while the other peak is in the evening.
- The minimum water demand occurs normally around 4 A.M.
- Fluctuations in water use in municipal systems can be estimated using the procedures given in Refs. 7 and 8. Typical hourly variations in a municipal water demand are shown in Figure 3-1. Typical fluctuations are summarized in Table 3-6.

3-3   WASTEWATER FLOW

3-3-1 Relation between Water Supply and Wastewater Flow Rates

Municipal wastewater is derived largely from the water supply. A considerable portion of the water supply, however, does not reach the sewers. This includes water used for street washing, lawn sprinkling, fire fighting, and leakages from water mains and service pipes. A small portion of water may also be consumed in products and manufacturing processes. Also, many homes and other establishments that are not served by a sewerage system may

TABLE 3-4 *Average Water Demand in Residential, Institutional, and Commercial Establishments*

Source	Unit	Flow (ℓ/unit·d)
Residential		
Single-family		
Low-income	Person	270
Medium-income	Person	310
High-income	Person	380
Summer cottage	Person	190
Trailer park	Person	150
Apartment	Person	230
Hotel, motel	Unit	380
Camps	Person	133
Resort (day or night)	Person	190
Institutional		
Hospital	Bed	950
Rest homes	Bed	380
Prison	Inmate	450
Schools		
Boarding	Student	300
Day	Student	76
Commercial		
Country clubs		
Resident	Member	380
Nonresident	Member	95
Restaurant	Customer	30
Cafeteria	Customer	6
	Employee	40
Bar	Customer	8
	Employee	50
Coffee shop	Customer	20
	Employee	40
Dance hall	Person	8
Store	Toilet room	1520
	Employee	40
Department store	m² floor area	8
	Employee	40
Shopping center	m² floor area	6
	Employee	40
Office building	Employee	65

Source	Unit	Flow (ℓ/unit·d)
Barber shop	Chair	210
Beauty salons	Station	1026
Laundries		
Laundromat	Machine	2200
Commercial	Machine	3000
Service station	First bay	3800
	Additional bays	1900
	Employee	190
Theaters[a]		
Drive-in	Car space	19
Movie	Seat	8
Airport[a]	Passenger	10
Car wash[a]	Car washed	209
Industrial building	Employee	55
Factories		
With showers	Employee-shift	133
Without showers	Employee-shift	95

[a]Does not contain per day unit.
Source: Adapted in part from Refs. 5, 9, and 10.

TABLE 3-5   *Typical Industrial Water Demand*

Industrial Use	Quantity
Canning	30–60 m³/metric ton
Milk, dairy	2–3 m³/metric ton
Meat packaging	15–25 m³/metric ton
Cattle	40–50 ℓ/head·d
Dairy	70–80 ℓ/head·d
Chicken	30–40 ℓ/100·d
Pulp and paper	200–800 m³/metric ton
Steel	260–300 m³/metric ton
Tanning	60–70 m³/metric ton raw hides processed

Source: Adapted in part from Refs. 11 and 12.

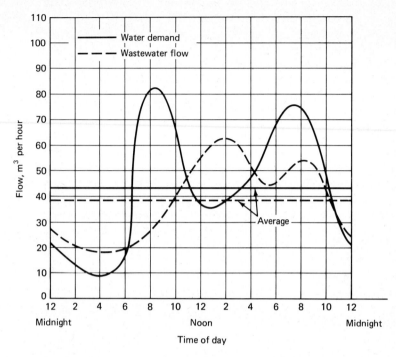

Figure 3-1   Typical Variation in Municipal Water Demand and Wastewater Flow.

use the city water supply but utilize septic tanks and drainage fields for wastewater disposal. On the other hand, infiltration/inflow, and water used by industries and residences that is obtained from privately owned sources may make the quantity of wastewater larger than the public water supply. In general, the average wastewater flow may vary from 60 to 130 percent of the water used in the community. Many designers frequently assume that the average rate of wastewater flow, including a moderate allowance for infiltration/inflow, equals the average rate of water consumption. Average wastewater flows from residential, commercial, industrial, institutional, and other sources may be obtained from careful consideration of local water consumption data or by using wastewater generation rates published in the literature.

TABLE 3-6   Typical Fluctuations in Water Demand in Municipal Water Supply Systems

Condition	Ratio of Annual Average Demand
Maximum day in a year	1.5–2.0
Daily average in maximum week	1.3–1.6
Daily average in maximum month	1.2–1.4
Peak hour within a day	2.7–4.0

3-3-2 Infiltration/Inflow

Infiltration is the groundwater that enters sewers through service connections, cracked pipes, defective joints, and defective pipes and manhole walls. Inflow is the surface runoff that may enter through manhole cover, roof and area drains, and cross-connections from storm sewers and combined sewers.

The amount of infiltration/inflow reaching a sewer system depends on the length and age of the sewers; the construction material, methods, and workmanship; the number of illegal roof or drain connections; the groundwater table relative to the sewer position; and the type of soil, ground cover, and topographic conditions. As part of Federal Construction Grants Program (Chapters 5 and 6), extensive studies of sewer system evaluation and rehabilitation must be undertaken by the municipalities in order to demonstrate that the proposed wastewater treatment facility will not be subject to excessive infiltration/inflow. Procedures for conducting infiltration/inflow evaluation surveys and determining the infiltration/inflow quantities are presented in Chapter 6. The permissible infiltration/inflow allowance is 1394 ℓ per d per cm per km (1500 gpd per in. per mile).[13] If infiltration/inflow exceeds this amount, sewer evaluation and rehabilitation may be necessary.

In older sewer lines infiltration is high because of deterioration of joints and masonry mortar. Newer sewers use joints sealed with rubber gaskets or synthetic material, and precast manhole sections. Therefore, the infiltration rate is significantly smaller. When designing for sewers, allowance must be made for old and new constructions. Average values for infiltration allowance used by designers are 94–9400 ℓ per d per cm per km (100–10,000 gal per d per in. per mile) or 200–28,000 ℓ per ha per d (20–3000 gal per acre per d).[2] Higher rates may be allowed where adverse conditions may exist.

3-3-3 Flow Variations

Like water demand, wastewater flows vary according to the season of the year, weather conditions, day of the week, and time of the day. Under dry weather conditions, the daily wastewater flow shows a diurnal pattern. Figure 3-1 illustrates daily variations in water demand and wastewater flows. The wastewater curve closely parallels that of water demand with a lag of several hours. Also, the fluctuations in wastewater flows are less than that of water supply because of the storage space in the sewers and because of the time required for the wastewater to reach the treatment plant. Also, the commercial and industrial discharges tend to reduce the peak flows. The infiltration/inflow further changes the diurnal flow pattern.

Procedure for developing minimum, average, and peak dry weather flows, infiltration/inflow allowances, and peak wet weather flows are presented in detail in Chapter 6. These procedures are based on long-term flow measurements at the wastewater treatment facility. In the absence of flow records, many designers use the following procedure to develop design flows:

1.   The population estimates, industrial and commercial growths and land use patterns are developed for the initial and design years.* Procedures for making population estimates and development of land use plans are covered in Chapters 2 and 6.

*See Sec. 2-2.

2. Based on the current per capita water demands and future trends, average water usage data for the initial and design years are developed.

3. Average wastewater flow data are developed from water usage. Considerations are given to portions of the water lost due to lawn sprinkling, street washing, and leakage. Also, the portion of wastewater that may be lost due to exfiltration is considered in the analysis. Many designers assume that the average dry weather flow is approximately 80 percent of the water demand. Another method involves determination of wastewater flow rates from different types of residential subdivisions; commercial contribution per employee or per acre; industrial flows based on per employee contribution, unit product, per acre, or industry as a whole; and institutional wastes per person.[14,15]

4. Peak and minimum dry weather flows are estimated from several equations and graphical relationships developed from case studies. The ratios of peak to average and minimum to average flows depend on the population. Larger cities have less deviations from the average than the smaller cities. Several methods for estimating ratios of peak and minimum flows to average flows are given in Refs. 7, 8, and 14. Commonly used equations are given below [Eqs. (3-1), (3-2), and (3-3)]. Ratios of extreme flows to average daily flow (peaking factors) developed from various sources are given in Figure 3-2.[14]

$$M = 1 + \frac{14}{4 + \sqrt{P}} \qquad\qquad (3\text{--}1)^{7,14}$$

$$M = \frac{5}{P^{0.167}} \qquad\qquad (3\text{--}2)^{16}$$

$$P = \text{population in thousands}$$

$$Q_{max} = 3.2 Q_{avg}^{5/6} \qquad\qquad (3\text{--}3)^{17}$$

where

M = ratio of maximum flow (Q_{max}) to average flow (Q_{avg}).

Many designers use the "fixture-unit" load method for estimating peak wastewater flows for facilities such as hospitals, hotels, schools, apartment buildings, and office buildings.[18,19] A procedure for estimating peak flows from the fixture unit method is given in Ref. 14.

5. The peak hourly wastewater flow is the peak dry weather flow plus the infiltration/inflow. In relation to the water supply, this is the peak hourly rate of water demand multiplied by the proportion of the water supply reaching the collection system, plus infiltration/inflow (fire demand does not enter into these calculations).[7,20] In Chapter 6 a procedure for estimating peak-hourly flow from flow-measurement data is provided.

6. Minimum rates of wastewater flow are useful in the design of pumping stations and to investigate the velocity in sewers and channels during the low flows. In the absence of gauging data, minimum flow may be assumed 33 percent and 50 percent of average flows for small- and medium-size communities, respectively.

7. Sustained flows are the flows that persist for various time durations. Peak and minimum flows last only for brief periods (less than 2 h), while sustained flows are those extremes that may persist over longer durations. As an example, extraordinarily dry or hot

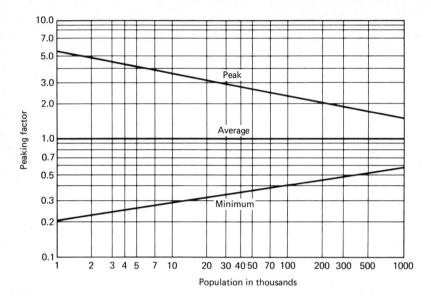

Figure 3-2   *Ratios of Extreme Flows to Average Daily Flow (Peaking Factor) for Municipal Wastewater Under Dry Weather Conditions. [From Ref. 14. Used with permission of Water Pollution Control Federation, and American Society of Civil Engineers.]*

weather may cause sustained low extremes. Likewise, special events in a community such as fairs, exhibitions, conventions, games, and the like may influence large population increases causing high sustained flows. These sustained flows are important in the design of wastewater treatment facilities. A procedure for developing sustained flow-envelope curves from flow records of approximately 46 wastewater treatment facilities throughout the country is given in Ref. 2. The ratios of one-day sustained peak and low flows to average flow are 2.9 and 0.4, respectively.[2]

8.   In the absence of flow measurement records and other pertinent data, many states specify 1500 ℓ (400 gal) per capita per d for design of laterals and submains, and 1300 ℓ (350 gal) per capita per d for mains and trunk sewers. These design flows include normal infiltration/inflow. Additional allowance must be made where industrial wastes are also transported, and conditions favoring excessive infiltration and inflow are present.[7]

3-3-4 Flow Reduction

The construction grants program requires a close examination of various wastewater flow reduction methods as part of facility planning process. Reduction in wastewater volume is achieved by (1) water conservation, (2) reuse of water in homes, and (3) reduction of infiltration/inflow.

Water Conservation.   Water conservation in homes, commercial establishments, and industries is gaining popularity. Water conservation has the beneficial effect of reducing the water and sewer bills. A number of inexpensive devices are now available as part of most faucet combinations. A community awareness program is necessary for the general public to understand the importance of water conservation and the cost and benefits of the water-saving devices. Water savings as high as 20–30 percent can be achieved by using flow reduction devices and by practicing simple water conservation measures in homes, businesses, and industries. Principal flow reduction devices and systems, and the percent reduction that may be achieved over conventional devices are summarized in Table 3-7.[4,21-24] Many water-saving ideas for homes are listed in Table 3-8.[6,23,24] Water conservation efforts require public cooperation, and the efforts of the local governments in the form of plumbing codes.

Reuse of Water in Homes. Another form of in-house water saving might come from reuse of water. The recycled water may be the waste streams from sinks, bathtubs, showers, and laundry which may be treated on site and reused for toilet flushing and lawn sprinkling. Such an effort may provide a 30–40 percent reduction in water consumption and wastewater volume.

Reduction in Infiltration/Inflow. New sewers must utilize tight sewer connections and joints in order to reduce infiltration/inflow. Proper selection of appurtenances is also essential. For existing sewer systems, an effective sewer evaluation and rehabilitation program is necessary to reduce the undesired quantities of ground and surface water. Large volumes of infiltration/inflow under wet weather conditions pose serious hydraulic overload difficulties at the treatment plants.

3-3-5 Flow Measurement

The knowledge of average and diurnal variations in wastewater flows, and sustained flows is essential for design and operation of wastewater treatment facilities. Many types of flow measurement devices are available that can be installed in the intercepting sewers, force mains, or at any location within a plant. Various flow measurement devices, their application and selection criteria, and design procedures are presented in detail in Chapter 10.

3-4 QUALITY OF WASTEWATER

Municipal wastewater contains over 99.9 percent water. The remaining materials include suspended and dissolved organic and inorganic matter as well as microorganisms. These materials give physical, chemical, and biological qualities that are characteristic of residential and industrial wastewaters. In this section, physical, chemical, and biological qualities, and variations in constituents and loadings in wastewater are discussed.

3-4-1 Physical Quality

The physical quality of municipal wastewater is generally reported in terms of temperature, color, odor, and turbidity. These physical parameters are summarized in Table 3-9.

TABLE 3-7 Principal Flow-Reduction Systems and Devices, and Percentage Reduction over Conventional Devices

Flow Reduction System/Device	Description	Percentage Reduction in Water Use over Conventional System/Device
Pressure Reducing Valves	Pressure reducing valves are installed to reduce the city water supply pressure, thus household pressure and flow are reduced. This eliminates unnecessary waste due to spurts from faucet, leaks, and drifts.	20–30
Sink Faucet		
Faucet aerator	Designed to provide smooth and even flow of water. Aerated jet of water increases rising power of water thus reducing amount of wash water used in kitchen and bathroom sink.	1–2
Limiting flow valve	Restrict flow to a constant rate independent of supply pressure.	1–2
Shower		
Flow limiting shower head	The device reduces water consumption by restricting and concentrating water jet.	10–12
Limiting flow valves	Restrict flow to a fixed rate, independent of supply pressure.	8–10
Washing Machine		
Level controller	Level setting use water in accordance to the wash load.	1–4
Toilet		
Shallow trap water closet	Shallow trap toilets are similar to conventional toilets. Due to shallow trap they require less water per flush. These are cost effective for new homes.	30–40

Device	%	Description
Dual-cycle toilet inserts	15–18	Toilet insert devices convert conventional toilets to dual-cycle operation (separate for urine or fecal waste transport).
Dual-cycle toilet	25–30	New types of toilets that have two flush cycles, one for urine and the other for fecal transport. Recommended for new constructions.
Reduced-flush device	12–18	Toilet-tank inserts that reduce the volume of flush by occupying a portion of the tank, or prevent the tank from completely emptying.
Flush valve		The flush valves operate directly from the water supply lines. Controlled amount of water is used for flushing. The mechanism can also be used with shallow trap toilets. Recommended for new homes.
Single-flush valve	10–14	
Dual-flush valve	20–30	
Immersed bottle or brick in toilet tank	2–5	Plastic bottle filled with water and weighted with pebbles occupy tank space and reduce water used per flush (obstruction to float must be avoided). Bricks are also used, but they flake and may clog tubes and valves.
Vacuum-flush toilet system	30–40	The toilets use air and small amount of water or foam to transport waste.
Recirculating toilet system	25–45	The flush water is recirculated for certain period. The system operates on a closed-loop principle in which waste is accumulated in a holding tank. Chemicals are used to suppress the microbiological activity in recirculating water. Other innovations utilize mineral oil as flushing medium. The holding tank is emptied periodically by vacuum trucks.
Urinals	10–15	Use of urinals in homes reduce water consumption. Wall type urinals with flush valve for homes use 5–6 ℓ per flush.

Source: Adapted in part from Refs. 4 and 21–24.

TABLE 3-8   *Water-Saving Ideas for Homes*

Location	Water-Saving Ideas
Bathtub	A full bathtub holds 190 ℓ. One can bathe adequately in one-quarter full bathtub.
Shower	Water consumption in quick shower is less than bathing (40–70 ℓ). Turn off water while soaping up. Use light spray.
Toilet	Do not flush more than necessary. Use water-saving devices discussed in Table 3-7. Water leaks or waste can be detected by adding a few drops of food coloring in the tank. Coloring appears in the toilet if there is a leak.
Washing machine	Use load selector for large or small loads if there is one. Otherwise, wash only full loads. Using cold water saves energy. Buy a machine that uses less water and energy per unit weight of wash. Use suds saver attachment and use less detergent.
Dishwasher	Preclean dishes with used paper napkins. Wash full load. Use less detergent.
Utility sink	Preclean dishes with used paper napkins. Soak overnight with small quantities of low-sudsing detergent. Save rinse water for next soak. Plan ahead to thaw frozen foods in air, not in running water.
Drinking water	Do not run tap waiting for cold water. Use ice cube or keep water pitcher in the refrigerator. Use paper cups at water fountain to avoid waste.
Garbage disposal	Avoid using garbage grinders. Use leftover food for feeding pets. Start a compost pile.
Bathroom sink	Turn off water while shaving or brushing teeth.
Faucet	Check all faucets including outside hose connections for leaks. Replace worn washers O rings, packing, and faulty fixtures. A pinhole leak can waste up to 1000 ℓ water per day. In many areas leaks cause about 95 percent complaints about excessive water bills.
Lawn, garden	Water slowly, thoroughly, and as infrequently as possible. Water at night to minimize evaporation. Keep close watch on wind shifts. Select hardy species and native plants that do not need as much watering. Mulch heavily. Let grass grow higher in dry weather—saves from burning, and less water is needed. Recycle water from bath tub, kitchen sinks, etc. A 1.25-cm garden hose under normal water pressure pours out more than 2300 ℓ/h. A 2-cm hose uses almost 7000 ℓ/h.

Location	Water-Saving Ideas
Backyard pool	Cover when not in use to prevent evaporation, keep clean, and reduce algae growth. Recycle wading pool water for plants, shrubs, lawns.
Carwash	Try to wash car near hedges, shrubs, or lawn. Wash car in short spurts from hose. Use a commercial carwash that recycles water.

Source: Adapted in part from Refs. 6, 23, and 24.

3-4-2 Chemical Quality

The chemical quality of wastewater is expressed in terms of organic and inorganic constitutents. Domestic wastewater generally contains 50 percent organic and 50 percent inorganic matter. Different chemical analyses furnish useful and specific information with respect to the quality and strength of wastewater. The uniformity of procedures prescribed in the Standard Methods made a comparison of different results possible.[25] The typical composition of domestic wastewater and brief descriptions and significance of different tests are summarized in Table 3-10. A general discussion on organic components, total suspended solids, and inorganic salts of wastewater is given below.

TABLE 3-9   Physical Quality of Wastewater

Parameter	Description
Temperature	The temperature of wastewater is slightly higher than that of water supply. Temperature has effect upon microbial activity, solubility of gases, and viscosity. The temperature of wastewater varies slightly with the seasons, but is normally higher than air temperature during most of the year and lower only during the hot summer months.
Color	Fresh wastewater is light gray. Stale or septic wastewater is dark gray or black.
Odor	Fresh wastewater may have a soapy or oily odor, which is somewhat disagreeable. Stale wastewater has putrid odors due to hydrogen sulfide, indol and skatol, and other products of decomposition. Industrial wastes impart other typical odors. Because of odors associated with wastewater treatment facilities, area residents have often vigorously resisted and rejected wastewater treatment plant projects.
Turbidity	Turbidity in wastewater is caused by a wide variety of suspended solids. In general, stronger wastewaters have higher turbidity.

TABLE 3-10 Typical Chemical Quality of Raw Domestic Wastewater

Chemical Quality Parameters	Description	Concentration	
		Range	Typical
Total solids	Organic and inorganic, settleable, suspended and dissolved matter.	375–1800	740
Settleable, mℓ/ℓ	Portion of organic and inorganic solids that settles in 1 h in an Imhoff cone. These solids are approximate measure of sludge that is removed in a sedimentation basin.	5–20	10
Suspended (TSS), mg/ℓ	Portion of organic and inorganic solids that are not dissolved. These solids are removed by coagulation or filtration.	120–360	230
Fixed, mg/ℓ	Noncombustible or mineral components of total suspended solids.	30–80	55
Volatile, mg/ℓ	Combustible or organic components of total suspended solids.	90–280	175
Dissolved (total), mg/ℓ	Portion of organic and inorganic solids that is not filterable. Solids smaller than one millimicron (mμ)[a] fall in this category.	250–800	500
Fixed, mg/ℓ	Noncombustible or mineral components of total dissolved solids.	145–500	300
Volatile, mg/ℓ	Combustible or organic components of total dissolved solids.	105–300	200
BOD₅, mg/ℓ	Biochemical oxygen demand (5-d, 20°C). It represents the biodegradable portion of organic component. It is a measure of dissolved oxygen required by microorganisms to stabilize the organic matter in 5 days.	110–400	210
COD, mg/ℓ	Chemical oxygen demand. It is a measure of organic matter and represents the amount of oxygen required to oxidize the organic matter by strong oxidizing chemicals (potassium dichromate) under acidic condition.	200–780	400
TOC, mg/ℓ	Total organic carbon is a measure of organic matter. TOC is determined by converting organic carbon to carbon dioxide. It is done in a high-temperature furnace in presence of a catalyst. Carbon dioxide is quantitatively measured.	80–290	150

Constituent	Description	Range	Typical
Total nitrogen (TN),[b] mg/ℓ	Total nitrogen includes organic nitrogen, ammonia, nitrite, and nitrate. Nitrogen and phosphorus along with carbon and other trace element serve as nutrients thus accelerate the aquatic plant growth.	20–85	40
Organic (ON) (as N), mg/ℓ	It is bound nitrogen into protein, amino acid, and urea.	8–35	20
Ammonia (NH_3-N) (as N), mg/ℓ	Ammonia nitrogen is produced as first stage of decomposition of organic nitrogen.	12–50	20
Nitrite and nitrate (as N), mg/ℓ	Nitrite and nitrate nitrogens are the higher oxidized forms of nitrogen. Both of these forms of nitrogen are absent in raw domestic wastewater.	0–small	0
Total phosphorus (TP) mg/ℓ	Total phosphorus exists in organic and inorganic form. Phosphorus in natural water is a source of eutrophication.	4–15	8
Organic (as P), mg/ℓ	Organic phosphorus is bound in organic matter.	1–5	3
Inorganic (as P), mg/ℓ	Inorganic form of phosphorus exists as orthophosphate and polyphosphate.	3–10	5
pH	pH is indication of acidic or basic nature of wastewater. A solution is neutral at pH 7.	6.7–7.5	7.0
Alkalinity (as $CaCO_3$), mg/ℓ	Alkalinity in wastewater is due to presence of bicarbonate, carbonate, and hydroxide ion.	50–200	100
Hardness (as $CaCO_3$), mg/ℓ	Hardness in wastewater is primarily due to calcium and magnesium ions. Hardness of wastewater depends on the hardness of water supply.	180–350	240
Chloride, mg/ℓ	Chloride in wastewater comes from water supply, human wastes, and domestic water softeners.	30–100	50
Oils and grease, mg/ℓ	These are soluble portion of organic matter in hexane. Their sources are fats and oils used in foods.	50–150	100

[a] 1 millimicron (mμ) = 10^{-7} cm
[b] TN = ON + NH_3-N + NO_2-N + NO_3-N. Total Kjeldhal nitrogen (TKN) = ON + NH_3 – N.

Organic Components.   Major components of organic matter in domestic wastewater are carbohydrates, proteins and fats, oils, and grease. Carbohydrates and proteins are easily biodegradable. Fats, oils, and grease are more stable, and are decomposed slowly by microorganisms. Wastewater may also contain small fractions of synthetic detergents, phenolic compounds, and pesticides and herbicides. These compounds, depending on their concentrations, may create problems such as nonbiodegradability, foaming, or carcinogenicity.[26,27] The concentrations of these toxic organic compounds in municipal wastewaters are very small. Their sources may be industrial wastes and surface runoffs. These pollutants are subject to "categorical standards" in the industrial pretreatment standards. Discussion on pretreatment standards is provided in Chapter 2.

Measurement of Organic Content

Biochemical Oxygen Demand.   Biochemical oxygen demand (BOD_5) is the most commonly used parameter to express the strength of municipal and industrial wastewaters. The BOD_5 test is important for the design of biological treatment facilities, determining organic loadings to treatment plants, and evaluating the efficiency of treatment systems. The test is also useful in stream pollution control activities. By its use, it is possible to determine the degree of organic pollution in streams at any time. The test is of prime importance in regulatory work and in studies designed to evaluate the purification capacity of receiving waters.

   BOD is defined as the amount of oxygen utilized by a mixed population of microorganisms under aerobic condition to stabilize the organic matter. The BOD test is conducted by placing a measured amount of wastewater in a 300-mℓ standard BOD bottle and filling the bottle with dilution water that contains the essential nutrients and is saturated with dissolved oxygen. Well-acclimated microbial seed may be supplied if sufficient microorganisms are not already present in the wastewater sample. The dissolved oxygen is determined in the diluted sample initially and after incubation at 20°C for 5 d.[25]

   The BOD of wastewater is not a single point, but rather a time-dependent variable. A 5-d incubation has become a standard practice. It takes approximately 20 d to stabilize all biodegradable organic matter and nitrogen contained in a wastewater sample. The ultimate first-stage BOD is the total carbonaceous oxygen demand. The nitrogenous oxygen demand is called second-stage BOD. Discussions on carbonaceous and nitrogenous oxygen demands may be found in many references.[2,7,8,28] A typical BOD curve with time is shown in Figure 3-3. Different BOD relationships are summarized in Eqs. (3-4) through (3-7). The BOD_5 of municipal wastewater is given in Table 3-10. The effluent requirements of BOD_5 for secondary treatment are discussed in Sec. 2-6-2.

$$UOD = L_o + L_n \tag{3-4}$$

$$y = L_o(1 - e^{-Kt}) \tag{3-5}$$

$$L_{o_T} = L_o(0.02T + 0.60) \tag{3-6}$$

$$K_T = K(1.047)^{T-20} \tag{3-7}$$

where

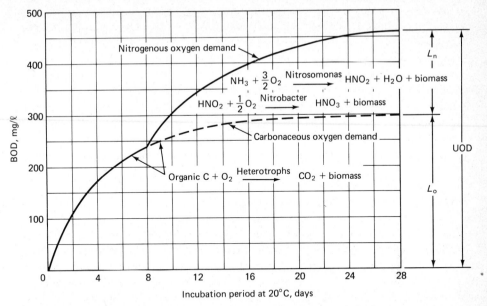

Figure 3-3   *Typical BOD Curve for Domestic Wastewater Showing Carbonaceous and Nitrogenous Oxygen Demands.*

UOD = ultimate oxygen demand, mg/ℓ

L_n = ultimate nitrogenous oxygen demand or second stage BOD, mg/ℓ

L_o = ultimate carbonaceous BOD, or first stage BOD at 20°C, mg/ℓ (for domestic wastewater BOD_5 is approximately equal to $\frac{2}{3}$ L_o)

y = carbonaceous BOD at any time t, mg/ℓ

L_{o_T} = ultimate carbonaceous BOD at any temperature T°C, mg/ℓ

K = Reaction rate constant at 20°C, d^{-1} (for domestic wastewater $K = 0.2$ to 0.3 per d)

K_T = Reaction rate constant at any temperature T°C, d^{-1}

Other Methods.   Although the BOD_5 test is universally used to express the organic strength of wastewaters, this test has many serious limitations. The test requires a 5-d period for a result. Furthermore, preparation of acclimated seed for industrial and toxic wastes is required. In addition, only biodegradable soluble organics are measured, and nitrification may cause serious interference.

There are a number of other tests that are more rapid tools to quantify the organic matter in wastewater samples. Chemical oxygen demand (COD) and total organic carbon (TOC) tests have been presented in Table 3-10. Other tests that are also used to express the organic strength of wastewaters are total oxygen demand (TOD) and theoretical oxygen demand (ThOD). TOD measurement uses the conversion of organic compounds into stable oxides in a platinum-catalyzed combustion chamber. This test is quickly performed if instrumentation is available. ThOD measurement is based on stoichiometric relations. There are certain relations among BOD_5, COD, TOC, TOD and ThOD tests. For each wastewater such relations must be established by laboratory measurement program. The degree of correlation depends on the consistency of the constituents. The typical ratios between different tests for domestic and many industrial wastewaters are available in the literature.[2,29,30]

Total Suspended Solids.   The total suspended solids (TSS) in wastewater may be due to sand, silt, clay, and organic matter. The typical total suspended solids concentration in municipal wastewater is 230 mg/ℓ (Table 3-10). The suspended solids when discharged into the natural waters may increase turbidity of the water and when they settle to the bottom may ruin the spawning and breeding grounds of aquatic animals. Organic solids at the bottom progressively decompose using up dissolved oxygen and produce noxious gases. Therefore, the secondary effluent standards are defined in terms of BOD_5 and TSS. These effluent standards are discussed in Sec. 2-6-2.

Dissolved Inorganic Components.   Total dissolved inorganic components of wastewater fall into two general categories: toxic and nontoxic.

Toxic Inorganic Chemicals.   Toxic inorganic chemicals in municipal wastewater include 13 metals plus cyanides and asbestos.[27] The 13 metals are antimony, arsenic, beryllium, cadmium, chromium, copper, lead, mercury, nickel, selenium, silver, thallium, and zinc. In higher concentrations the metals may cause synergistic or antagonistic effects in terms of toxicity in biological wastewater treatment plants. Generally their concentrations in publicly owned treatment works (POTW) are higher than toxic organics. Metals are not biodegradable and their removal mechanisms depend on physicochemical processes.

Nontoxic Inorganic Chemicals.   Municipal wastewater contains many dissolved salts that may alter chemical qualities such as pH, buffering capacity, salinity, scaling, or corrosiveness of the wastewater. Many compounds include calcium, magnesium, sodium, potassium, iron, and manganese salts of carbonate, bicarbonate, chloride, sulfate, nitrate, and phosphate.[31] To some extent their concentrations depend on the chemical makeup of the water supply, and the use of domestic and industrial water softeners. The mineral makeup of wastewater is important in evaluating the reuse potential of wastewater. Concentrations and significance of nutrients (P and N), pH, alkalinity, total dissolved solids, hardness, and chloride in wastewater are given in Table 3-10. A detailed discussion of the significance and measurement of minerals in wastewater may be found in some excellent references.[2,25,32,33]

3-4-3 Microbiological Quality

The municipal wastewater contains microorganisms that play an important role in biological waste treatment. The principal groups of microorganisms of significance in wastewater treatment include bacteria, fungi, protozoa, and algae. A general discussion on the microbiology of wastewater treatment is provided in Chapter 13. Three excellent sources on the subject may be consulted for additional information.[34,35,36]

3-5   CHARACTERIZATION OF WASTEWATER

Wastewater treatment facilities are normally designed for average loadings of BOD, suspended solids, and other constituents. Designs based on average conditions with no considerations for peak concentrations or average sustained concentrations may result in effluent quality that may often not meet the short-term consecutive-day average limitations. Wastewater quality data also exhibit seasonal, daily, and hourly variations. A sampling program is often necessary to develop peak mass loading and average sustained loading conditions. Some excellent references on sampling, analysis, and data presentation are available in the literature.[2,25,30,33] The data should be used to develop seasonal, daily, and hourly variations, average concentrations, flow-weighted average, mass loadings, and sustained peak mass loadings. Typical ratios of one-day average sustained peak to average mass loading of BOD_5, suspended solids, total Kjeldhal nitrogen, ammonia nitrogen, and total phosphorus are, respectively, 2.5, 2.7, 2.0, 1.5, and 1.6.[2] The typical curves for maximum and minimum mass loading ratios are given in Ref. 2.

3-6   UNIT WASTE LOADINGS AND POPULATION EQUIVALENTS

Loadings of suspended and dissolved solids in municipal wastewater on a per capita basis remain relatively uniform. The variation in constituent loadings per capita per day may be due to industries served, usage of garbage grinders and domestic water softeners, and discharge of septage. In small treatment facilities their effects may be significant. Unit waste loadings in municipal wastewater may be developed from flow rate (liters per capita per day) and concentrations of various constituents in mg/ℓ. Concentrations of many constituents are given in Table 3-10. Important unit waste loadings developed from Table 3-10 are summarized in Table 3-11.

The population equivalent of a waste may be determined by dividing the total mass per day by the per capita mass loadings. Population equivalent has been used as a technique for determining industrial waste treatment costs. Population equivalents may be determined on the basis of flow, BOD_5, COD, TSS, P, N, etc. As an example, an industrial waste of 1000 m^3/d and having a BOD_5 of 500 mg/ℓ would have a population equivalent in terms of $BOD_5 = 0.5$ g/$\ell \times 1/(95$ g/capita $\cdot$ d$) \times 10^3$ m^3/d $\times 1000$ $\ell/m^3 = 5263$ (95 g per capita per d is taken from Table 3-11).

3-7   PROBLEMS AND DISCUSSION TOPICS

3-1   Visit the water treatment plant in your community. Obtain water supply data for one year. Estimate the following:
   (a)   Annual average water demand, ℓpcd.
   (b)   Ratio of average water demand for the month of July or August with respect to the annual average water demand.
   (c)   Ratio of average water demand for the month of January or February with respect to the average annual water demand.

3-2   Obtain the flow data from the wastewater treatment plant in your community. Develop the following information.
   (a)   Determine average wastewater flow under dry weather flow condition, ℓpcd.

TABLE 3-11   Unit Waste Loadings Derived from
Table 3-10

Constituent	Typical Unit Waste Loadings (g/capita·d)
BOD$_5$	95[a]
COD	180
Total suspended solids (TSS)	104
Total nigrogen (as N)	18
Organic nitrogen (as N)	9
Ammonia nitrogen (as N)	9
Total phosphorus (as P)	4

[a]Average flow 450 ℓ/c·d (118 gpcd) × (210 mg/ℓ) × (1 g /1000 mg)
= 94.5 g per capita per day. See Table 3-10 for typical concentrations
of BOD$_5$, COD, TSS, various forms of nitrogen, and total phosphorus.

   (b) Prepare the typical diurnal flow pattern for a typical dry day and for a maximum
        wet day.
   (c) Calculate the ratios of peak dry weather flow to average day.
   (d) Calculate the ratio of peak wet weather flow to average day.
   (e) Compare the peak dry weather flow with those obtained from Eqs. (3-1) and
        (3-2).

3-3 Calculate the average daily water demand and average daily wastewater flow from the
diurnal water demand and wastewater flow data given in Figure 3-1. What percentage of the
water is returned to the wastewater treatment facility?

3-4 A 200-home subdivision is proposed. Calculate the annual water saving that can be
achieved if all homes are installed with suitable water-saving devices given in Table 3-7
instead of conventional plumbing. Make use of information provided in Table 3-2.

3-5 A primary sedimentation tank is designed for an average flow. The design detention time
and overflow rates at average flow are 2 h and 37 m^3/m^2·d. Calculate the detention time and
overflow rates under the peak flow conditions when ratio of peak to average flow is 2.7.

3-6 Calculate the carbonaceous BOD after 7 d at 30°C. Use Eqs. (3-5) through (3-7). BOD$_5$ at
20°C is 120 mg/ℓ, and K at 20°C is 0.25/d.

3-7 Calculate the ratios of total theoretical oxygen demand and total nitrogenous oxygen
demand with respect to total carbonaceous oxygen demand of an industrial waste. The
waste is represented by the chemical formula $C_6N_2H_6O_2$. Assume that nitrogen is converted
to ammonia then to nitrate.

3-8 The ultimate oxygen demand (UOD) of a waste is expressed by the following equation:

UOD (mg/ℓ) = $A \times$ BOD$_5$ + $B \times$ NH$_3$–N

Calculate the factors A and B if the reaction rate K in Eq. (3-5) is 0.21 per d, and the ammonia nitrogen is fully converted to NO$_3$–N.

3-9 An industry is discharging 2500 m^3/d wastewater into a sanitary sewer. The concentrations of BOD$_5$, TSS, TN, and TP are 200, 280, 35, 10 mg/ℓ, respectively. Calculate the population equivalent of the industry based on flow, BOD$_5$, TSS, TN, and TP.

REFERENCES

1. Murrary, C. R., and E. B. Reeves, *Estimated Use of Water in the United States in 1970*, U.S. Geological Survey, Department of the Interior, Geological Survey Circular 676, Washington, D.C., 1972.
2. Metcalf and Eddy, Inc., *Wastewater Engineering: Treatment, Disposal, and Reuse*, McGraw-Hill Book Co., New York, 1979.
3. Dufor, C. N., and E. Becker, *Public Water Supplies of the 100 Largest Cities in the United States*, U.S. Geological Survey, Water Supply Paper 1812, 1964, p. 35.
4. Bailey, J. R., et al., *A Study of Flow Reduction and Treatment of Wastewater from Households*, EPA Water Pollution Control Research Series Report 11050 FKE 12/69, December 1969.
5. Salvato, J. A., "The Design of Small Water Systems," *Public Works*, Vol. 91, No. 5, May 1960.
6. ECOS, Inc. *Water Wheel: Your Guide to Home Water Conservation*, U.S. Environmental Protection Agency, Public Information Center (PM-215), Washington, D.C., 1977.
7. Steel, E. W., and T. J. McGhee, *Water Supply and Sewage*, McGraw-Hill Book Co., New York, 1979.
8. Clark, J. W., W. Viessman, M. G. Hammer, *Water Supply and Pollution Control*, IEP-A Dun-Donnelly, New York, 1977.
9. Culligan International Co., *Engineering Data Handbook*, 8181-70, 1970.
10. Hubbell, J. W., "Commercial and Institutional Wastewater Loadings," *Journal of Water Pollution Control Federation*, Vol. 34, No. 9., September 1962.
11. National Association of Manufacturers, *Water In Industry*, National Association of Manufacturers and Chamber of Commerce of the United States, New York, January 1965.
12. Nordell, E., *Water Treatment for Industrial and Other Uses*, Reinhold Publishing Corp., New York, 1961.
13. U.S. Environmental Protection Agency, *Program Requirement Memorandum 78-10*, Bureau of National Affairs, Washington, D.C., March 17, 1978, p. 45.
14. Joint Committee of the Water Pollution Control Federation and the American Society of Civil Engineers, *Design and Construction of Sanitary and Storm Sewers*, WPCF Manual of Practice No. 9, Water Pollution Control Federation, Washington, D.C., 1970.
15. Chicago Pump, *Hydraulics and Useful Information*, Hydrodynamics Division, Chicago Pump, Chicago, 1966.
16. Gifft, H. M., "Estimating Variations in Domestic Sewage Flows," *Waterworks and Sewerage*, Vol. 92, 175, 1945.
17. Fair, G. M., and J. C. Geyer, *Water Supply and Wastewater Disposal*, John Wiley & Sons, New York, 1961.
18. U.S. Department of Housing and Urban Development, *Minimum Design Standards for Community Sewerage Systems*, FHA 720, Washington, D.C., 1963.

19. International Association of Plumbing and Mechanical Officials, *Uniform Plumbing Code,* Los Angeles, 1973.
20. U.S. Environmental Protection Agency, *Guide for Sewer System Evaluation,* Office of Water Program Operations, Washington, D.C., March 1974.
21. Qasim, S. R., "Wastewater Treatment and Disposal Systems for Special Applications," *Encyclopedia of Environmental Science and Engineering,* Vol. III, Gorden & Beach, New York, 1983.
22. Qasim, S. R., Neil L. Drobny, and Alan Cornish, "Advanced Waste Treatment System for Naval Vessels," *Journal of Environmental Engineering Division, American Society of Civil Engineers,* Vol. 99, No. EES, October 1973, p. 717–727.
23. Metcalf and Eddy, Inc., *Water Pollution Abatement Technology Capabilities and Costs for Publicly Owned Treatment Works.* Report Prepared for National Commission on Water Quality, PB-250 690, National Technical Information Service, March 1976.
24. Sharpe, W. E., "Selection of Water Conservation Devices for Installation in New and Existing Dwellings," *Proceedings, National Conference on Water Conservation and Municipal Wastewater Flow Reduction,* November 28 and 29, 1978, Chicago, EPA/430/9-79-015, August 1979.
25. APHA, AWWA, and WPCF, *Standard Methods for Examination of Water and Wastewater,* American Public Health Association, Washington, D.C., 15th Edition, 1980.
26. *Federal Water Pollution Control Act Amendments of 1972* (PL 92-500) 92nd Congress, October 18, 1972.
27. Schwartz, H. A., and J. C. Buzzeld, *The Impact of Toxic Pollutants on Municipal Wastewater Systems,* EPA Technology Transfer, Joint Municipal/Industrial Seminar on Pretreatment of Industrial Wastes, Dallas, July 12, 13, 1978.
28. Babbitt, H. E., and E. R. Bauman, *Sewerage and Sewage Treatment,* John Wiley & Sons, New York, 1958.
29. Ramaldho, R. S., *Introduction to Wastewater Treatment,* Academic Press, New York, 1977.
30. U.S. Environmental Protection Agency, *Handbook for Monitoring Industrial Wastewater,* Technology Transfer, August 1973.
31. U.S. Environmental Protection Agency, *Quality Criteria for Water,* Prepublication copy EPA-400/9-76-023.
32. Sawyer, C. N., and P. N. McCarty, *Chemistry for Environmental Engineering,* McGraw-Hill Book Co., New York, 1978.
33. U.S. Environmental Protection Agency, *Manual of Methods for Chemical Analyses of Water and Wastes,* Technology Transfer, FPA-625-/6-74-003, 1974.
34. Gaudy, A. F. and E. T. Gaudy, *Microbiology for Environmental Scientists and Engineers,* McGraw-Hill Book Co., New York, 1980.
35. Hawkes, H. A., *The Ecology of Waste Water Treatment,* Pergamon Press, Elmsford, New York, 1963.
36. Cooper, R. C., D. Jenkins, and L. Y. Young, *Aquatic Microbiology Laboratory Manual,* Association of Environmental Engineering Professors, November 1976,

Wastewater Treatment Unit Operations and Processes, and Flow Schemes

4-1 INTRODUCTION

Wastewater treatment units generally fall into two broad divisions: unit operations and unit processes. In the unit operations, the treatment or removal of contaminants is brought about by the physical forces. In the unit processes, however, the treatment occurs predominantly due to chemical and biological reactions. Often the terms "unit operations" and "unit processes" are used interchangeably because many processes are integrated combinations of operations serving a single primary purpose. As an example, activated sludge combines mixing, gas transfer, flocculation, and biological phenomena to remove biodegradable matters.

Wastewater treatment plants utilize a number of unit operations and processes to achieve the desired degree of treatment. The collective treatment schematic is called a flow scheme, flow diagram, flow sheet, process train, or flow schematic. Many different flow schemes can be developed from various unit operations and processes for the desired degree of treatment. However, the most desirable flow scheme is the one that is most cost-effective. Many guidelines have been developed that may aid in evaluation and selection of flow schemes.[1-6] An example of cost effectiveness analysis and selection of a flow scheme is provided in Chapter 6. Basic considerations for developing a flow scheme include (1) characteristics of wastewater and degree of treatment required, (2) requirement of the regulatory agencies, (3) proximity to the buildup areas, (4) topography and site conditions, land area available, and hydraulic requirement, (5) existing facilities, (6) available equipment, (7) preference and experience of the designer, (8) plant economics, (9) quantity and quality of sludge from each process, (10) level of expertise of treatment plant operation personnel, and (11) minimal environmental consequences and maximum environmental improvements.

Wastewater treatment facilities are designed to process the liquid and solid portions of the wastewater. In this chapter the unit operations and processes applicable to the treatment of liquid and solid portions are discussed in separate sections. Many treatment processes, however, are used to treat the liquid and solid portions of the wastewater simultaneously. Examples of such processes are stabilization ponds, aerated lagoons, and land treatment. In these processes many chemical and biological forces act collectively to stabilize the liquid and solid components in the wastewater. These processes are discussed separately in Chapters 13 and 24.

4-2   LIQUID TREATMENT SYSTEMS

4-2-1 Pumping, Flow Measurement, and Flow Equalization

Liquid treatment may utilize pumping, flow measurement, and sometimes flow equalization. Although these systems do not provide any treatment, they are considered as a part of the overall flow scheme. A brief discussion of pumping station flow measurement and flow equalization is provided below.

Pumping Stations.   Treatment plants are normally located at a low point (near a river or a lake, for example) in order to provide gravity flow into the collection systems. At the plant site, the wastewater is pumped to an adequate height when topography dictates, at which time it will flow by gravity through the various treatment units. Pumping stations are also equipped with a wet well which intercepts incoming flows and permits equalization for pump loadings. Pumping stations are usually located after bar screens, and many times are located after grit removal, primary sedimentation, or even complete treatment. The objective is to remove the coarse solids, grit, and organic solids prior to pumping, as these solids often present operational difficulties at the pumping station. However, the cost of construction and operation of these units deep in the ground must be weighed against the cost of pumps that are designed to handle solids, with the treatment units above the ground. Chapter 9 discusses pumping station, pump selection, design criteria, and design procedure.

Flow Measurement.   Flow measurement at wastewater treatment facilities is essential for plant operation, process control, and record keeping. The flow measurement devices may be located in the interceptor sewer, after the pumping station, or at any other location within a plant. Details on various flow measurement devices, their application, design criteria, and design example are given in Chapter 10.

Flow Equalization.   Flow equalization is not a treatment process. It is simply the damping of the flow rate and mass-loading variations. Due to the cyclic nature of wastewater flows and organic strengths, many wastewater treatment units are designed for maximum conditions. By the use of flow equalization, the plants can be designed and operated under a nearly constant ideal flow and mass-loading condition thus minimize shock and achieving maximum utilization of the facilities. The flow equalization process consists of providing storage capacity and adequate aeration and mixing time to prevent odors and solids deposition. The flow equalization basins are in-line or off-line, and generally follow preliminary screening and grit removal to minimize mixing requirements. In in-line equalization, the entire flow is discharged into a completely mixed basin. Controlled flow pumps or gates are used to maintain a constant daily average flow through the plant. In an off-line system an overflow structure is built to bypass the excess flow into an equalization basin. Under low-flow conditions, the stored flow is then routed through the plant. In both systems, nearly constant flow rates are maintained through the plant. However, considerable damping of constituent mass loadings (such as BOD, COD, TSS, etc.) to the downstream processes is achieved with in-line equalization but only slight damping is achieved with off-line equalization.

The volume required for flow equalization is determined by using an inflow mass diagram. Detailed analysis for the volume requirements and damping of flow rate and mass loadings may be found in several excellent references.[7-9]

4-2-2 Unit Operations and Processes

Municipal wastewaters contain approximately 99.1 percent water. The small fraction of solids include organic and inorganic, suspended and dissolved matters. The removal of various contaminants from the liquid depends on the nature of the impurities and their concentrations. The coarse and settleable inorganic and organic solids are generally removed in primary treatment units that include unit operations such as bar screens, grit removal, and sedimentation facilities. The removal of dissolved organics is readily achieved in biological or chemical treatment processes that may be added to the primary treatment. The combined system is called a secondary treatment plant.[*] Many unit operations and processes may be added to the existing primary or secondary treatment systems to achieve the removal of nutrients and other contaminants. This is called tertiary treatment or advanced wastewater treatment.

A summary of different unit operations and processes considered in the design of primary, secondary, and advanced wastewater treatment facilities is provided in Table 4-1. Figure 4-1 illustrates the components of these treatment units. Treatment levels achieved across these units and unit combinations are summarized in Table 4-2. As wastewater is processed at different treatment units, a concentrated waste stream containing various constituents is removed. Table 4-3 provides the characteristics of waste streams indicated in Figure 4-1.

4-2-3 Combination of Unit Operations and Processes into Flow Schemes

Many unit operations and processes described in Table 4-1 and Figure 4-1 can be combined to develop a flow scheme to achieve a desired level of treatment. The level of

---

[*]The secondary treatment as defined by the U.S. Environmental Protection Agency is directed principally toward the removal of BOD_5 and total suspended solids (see Sec. 2-6-2).

TABLE 4-1 *Major Physical, Chemical, and Biological Treatment Unit Operations and Processes Used for Liquid Treatment*

Unit No.	Unit Operations and Processes[a]	Principal Applications	Ref.
A	Screening (UO)	Racks or bar screens are the first step in wastewater treatment. They are used to remove large objects.	Chap. 8
B	Grit removal (UO)	Grit removal facility is used to remove heavy material such as sand, gravel, cinder, eggshell, etc.	Chap. 11
C	Primary sedimentation (UO)	The purpose of primary sedimentation facility is to remove settleable organic solids.	Chap. 12
D	Suspended growth biological reactor (UP)	The process is used to remove dissolved organics. Principal variation is an activated sludge process.	Chap. 13
E	Attached growth biological reactor (UP)	The process is used to remove dissolved organics. Principal variation is a trickling filter.	Chap. 13
F	Secondary clarifier (UO)	Secondary or final clarifier is the sedimentation facility used in conjunction with a biological or chemical treatment process.	Chap. 13
G	Disinfection (UP)	Disinfection facilities are used to reduce the number of water borne pathogens in the effluent. Chlorination is the most common method.	Chap. 14
H	Coagulation (UP)	Coagulation is used for precipitation of suspended solids, BOD and phosphorus. Commonly used chemicals are alum, iron salts, and polymers. The process uses flash mix and flocculation basins.	Chap. 24

I1	Single-stage lime precipitation (UP)	Lime is used for precipitation of suspended solids BOD, and phosphorus. The process is similar to coagulation.	Chap. 24
I2	Two-stage lime precipitation (UP)	Excess lime is used in two stages. Suspended solids, BOD and phosphorus are precipitated.	Chap. 24
J	Nitrification (UP)	Process is used to convert ammonia to nitrate. It is achieved in suspended or attached growth biological reactor.	Chap. 24
K	Denitrification (UP)	Nitrite and nitrate are reduced to nitrogen gas by microorganisms. Denitrification is achieved under anaerobic condition in suspended or attached growth reactors. An organic source such as methanol is needed.	Chap. 24
L	Ammonia stripping (UO)	Ammonia gas is air stripped from the wastewater in a stripping tower.	Chap. 24
M	Breakpoint chlorination (UP)	Ammonia nitrogen is oxidized to nitrogen gas by breakpoint chlorination in a mixing basin.	Chap. 24
N	Ion exchange (UP) (ammonia nitrogen)	Ammonium ion is selectively removed in a bed of clinoptiloite (a zeolite resin). Ammonia removal system from spent reagent is needed.	Chap. 24
O	Filtration (UO)	Filtration is used to polish the effluent. Total suspended solids and turbidity are removed.	Chap. 24
P	Carbon adsorption (UP)	Carbon adsorption is used to remove soluble refractory organics from wastewater effluent.	Chap. 24
Q	Reverse osmosis or ultrafiltration (UO)	It is a demineralization process applicable to production of high-quality water from effluent. The water is permeated through semipermeable membrane at high pressure.	Chap. 24
R	Electrodialysis (UO)	It is a demineralization process. Electrical potential is used to transfer the ions through ion-selective membranes.	Chap. 24

[a]UO = unit operation. UP = unit process.

TABLE 4-2   Degree of Treatment Achieved by Various Unit Operations and Processes

Treatment Units or Combinations	Treatment Units Involved[a]	Stations across the Units[b]	Removal Efficiency (Percent) across the Units[c]					
			BOD$_5$	COD	TSS	TP	ON	NH$_3$-N
Preliminary treatment (PT)	A, PS, B, FM	1–2	small[d]	small[d]	small[d]	small[d]	small[d]	small[d]
Primary sedimentation	C	3–4	30–40	30–40	50–65	10–20	10–20	0
Activated sludge (conventional)	D, F	5–6	80–85	80–85	80–90	10–25	15–50	8–15
Trickling filter (high rate)	E, F	7–8	60–80	60–80	60–85	8–12	15–50	8–15
Disinfection	G	9–10	small[e]	small[e]	small[e]	small[e]	small[e]	small[e]
Coagulation and sedimention after preliminary or secondary treatment	H,F	11–12	40–70	40–70	50–80	70–90	50–90	0
Coagulation in biological treatment process	H, D or E, F	13–14	80–90	80–90	70–90	75–85	60–90	0
Single-stage lime addition after preliminary treatment, or after secondary treatment	I1, F	15–16	50–70	50–70	60–80	70–90	60–90	0
Single-stage lime addition in biological treatment	I1, D or E, F	17–18	80–90	80–90	70–80	75–85	60–90	0
Two-stage lime addition after preliminary treatment, or after biological treatment	I2, F1, F2	19–20	50–85	50–85	50–90	85–95	70–90	0

Nitrification single stage with carbonaceous BOD removal	J1, F	21–22	80–95	80–90	70–90	10–15	75–85	85–95[f]
Nitrification separate stages suspended or attached growth	J2, F	23–24	50–70	50–60	small[e]	—	40–50	90–96
Denitrification separate stages suspended or attached growth[g]	K, F	25–26	small[e]	small[e]	small[e]	small[e]	small[e]	small[e]
Ammonia stripping	L	27–28	0	—	0	0	—	60–95
Breakpoint chlorination	M	29–30	—	—	0	0	—	80–90
Ion exchange (ammonia nitrogen)	N	31–32	0	0	0	0	0	90–95
Filtration	O	33–34	20–50	20–50	60–80	20–50	50–70	0
Carbon adsorption	P	35–36	50–85	50–85	50–80	10–30	30–50	—
Reverse osmosis	Q	37–38	90–100	90–100	—	90–100	90–100	60–90
Electrodialysis	R	39–40	20–60	20–60	—	—	80–95	30–50

[a]For explanation of symbols see Table 4-1 and Figure 4-1.
[b]For station numbering see Figure 4-1.
[c]For definition of terms see Table 3-10.
[d]BOD_5 or COD removal may vary if communitor and/or grit washing is used. With no communitor and no grit washing, BOD_5 removal may be 0–5 percent and TSS removal 5–10 percent.
[e]Removal is normally small and is considered zero.
[f]Nitrate nitrogen (NO_3-N) concentration may reach 15–20 mg/ℓ (as N) in effluent.
[g]NO_3-N removal 85–90 percent.

TABLE 4-3   Characteristics of Waste Streams from Different Unit Operations and Processes Given in Table 4-1 and Figure 4-1

Waste Stream Identification[a]	Treatment Units Involved[b]	Description	Constituents and Processing	Refs.
a	A	Screening	Coarse solids are often comminuted and returned to wastewater stream. Normally screenings are disposed of by landfilling. Quantity 20.0×10^{-3} m³/10^3 m³.	Chap. 8
b	B	Grit	Heavy inorganic solids. Quantity 30.0×10^{-3} m³/10^3 m³. Often washed prior to disposal by landfilling.	Chap. 11
c	C	Primary sludge	Gray and slimy material has offensive odors. Quantity 150–250 g/m³. Solids content 3.0–4.5 percent.	Chap. 12
c'	C	Scum	Consists of floatable materials such as oil, grease, fats, waxes, etc. These are quite odorous. Quantity 8 g/m³.	Chap. 12, 16
d	D, F	Biological sludge (secondary or waste-activated sludge)	Biological solids. Quantity 50–90 g/m³. Solids content 0.3–1.0 percent. Often returned to primary sedimentation basin for concentration.	Chap. 13
e	E, F	Biological sludge (trickling filter sludge)	Biological solids. Quantity 40–60 g/m³. Solids content 3 percent.	Chap. 13
f	H, F	Chemical precipitation sludge (metal hydroxide sludge)	Sludge is slimy or gelatinous. Quantity 80–300 g/m³. Solids content 3–4 percent.	Chap. 24

g	H, D or E, F	Chemical-biological sludge	Solids production 100–150 g/m^3 at 0.8 percent solids from activated sludge and 50–80 g/m^3 at 2 percent solids from trickling filter.	Chap. 24
h	I1, F	Single-stage lime sludge	Sludge is somewhat slimy or gelatinous. Quantity 500–600 g/m^3. Solids content 2–5 percent.	Chap. 24
i	I1, D or E, F	Lime-biological sludge	Solids production 200–300 g/m^3 at 1 percent solids from activated sludge, and 150–200 g/m^3 at 3 percent solids from trickling filter.	Chap. 24
j	I2, F1, F2	Two-stage lime sludge	Sludge is slimy and gelatinous. Quantity 900 g/m^3. Solids content 4–5 percent.	Chap. 13
k	J1, F	Single-stage nitrification	Solids production 70–100 g/m^3 at 0.8 percent solids from activated sludge and 40–70 g/m^3 at 3 percent solids from trickling filter.	Chap. 24
l	J2, F	Separate-stage nitrification	Solids production 10–12 g/m^3. Solids content 0.8–3 percent.	Chap. 24
m	K, F	Separate-stage denitrification	Solids production 20–40 g/m^3. Solids content 0.8–2 percent.	Chap. 24
n	O	Filter backwash	Liquid waste containing suspended solids. This waste stream is normally returned to the head of the plant.	Chap. 24
o	P	Activated carbon backwash	Liquid waste containing suspended solids. This waste stream is normally returned to the head of the plant.	Chap. 24
p	Q	Brine from reverse osmosis system	Dissolved salts require special treatment and disposal.	Chap. 24
q	R	Brine from electrodialysis system	Dissolved salts require special treatment and disposal.	Chap. 24

[a]For explanation of waste stream symbols see Figure 4-1.
[b]For explanation of process symbols see Table 4-1 and Figure 4-1.

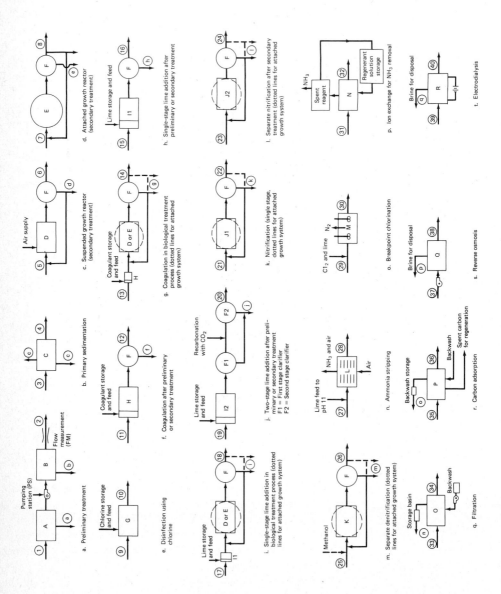

FIGURE 4-1   Unit Operations and Unit Processes Used for Liquid Treatment at Primary, Secondary and Advanced Wastewater Treatment Facilities (see Table 4-1 for definition of unit operations and processes, Table 4-2 for removal efficiencies across the units, and Table 4-3 for characteristics of waste streams).

treatment may range from removal of BOD_5 and TSS, nitrogen, and phosphorus, to complete demineralization. A number of wastewater treatment process combinations and flow schemes and resultant effluent quality are summarized in Table 4-4. Figure 4-2 illustrates several process flow diagrams.

To develop the best possible flow scheme a designer must evaluate many factors that are related to operation and maintenance, process efficiency under variable flow conditions, and environmental constraints.[10,11] In Table 4-5 various factors that are considered important in selection of flow schemes are evaluated.

4-3 SLUDGE PROCESSING AND DISPOSAL

Safe handling and disposal of various residues produced in different treatment units is of equal importance. The solids portion includes screenings, grit, scum, and sludge. The screenings and grit are generally disposed of by landfilling. The sludge (including scum), which may contain solids in concentrations of 0.5–5 percent, offers complex processing and disposal problems. It is odorous and contains large volumes of water. Because the treatment and disposal of sludge is expensive, sludge-handling costs are often the overriding consideration in the design of wastewater treatment plants.[12,13]

4-3-1 Unit Operations and Processes

In general, the sludge-processing and disposal methods include thickening, stabilization, dewatering, and disposal. Many unit operations and processes are utilized at various stages of sludge processing and disposal. To develop a cost-effective system of sludge treatment, the best combination of treatment processes must be chosen. Many of these unit operations and unit processes are illustrated in Figure 4-3.[7,12,13] A summary of different sludge treatment systems commonly used in overall sludge processing scheme is provided in Table 4-6.[7,12,13]

4-3-2 Performance Data

Most of the sludge-processing facilities produce two streams: (1) processed solids and (2) liquids. The liquid streams (also called side streams) must be treated again, and these liquids from various sludge-processing units are returned to the head of the plant. The returned flows often contain high concentrations of suspended solids and BOD. Often equalization facilities are provided to distribute the hydraulic and material loading over 24 hours of operation.[12] To predict the incremental loadings due to returned flow, material mass balance at average design flow must be performed to determine the final loadings to the plant. The procedure for material mass balance analyses is shown in Chapter 13. Table 4-7 provides operational performance data of various sludge-processing facilities.[12,13] This information is used in development of material mass-balance relationships in Chapter 13.

TABLE 4-4   Treatment Process Combinations and Flow Schemes

Treatment Scheme No.	Type of Treatment	Treatment Units and Combinations	Effluent Quality (mg/ℓ)								Comments
			BOD$_5$	COD	UOD^a	TSS	TP	ON	NH$_3$-N	NO$_3$-N	
(1)	Raw wastewater or preliminary treatment^b	A + PS + B + FM	210	400	405	230	11	25	20	0	No treatment or insignificant treatment.
(2)	Primary treatment	A + B + C + G	130	265	285	100	9	21	20	0	Not acceptable effluent quality.
(3)	Activated sludge	A + B + C + D + F + G	20	45	111	20	8	15	18	0–4	Meets secondary effluent quality.
(4)	Trickling filter	A + B + C + E + F + G	30	60	126	30	8	15	18	0–6	Marginal secondary effluent quality.
(5)	Activated sludge with filtration and disinfection	A + B + C + D + F + O + G	10	20	96	10	6	4	18	0–6	Effluent quality better than secondary effluent. Small nutrient removal.
(6)	Activated sludge/ nitrification in single stage	A + B + C + J1 + F + G	10	35	24	20	8	4	2	18	Effluent quality better than secondary effluent. No nutrient removal. Highly nitrified effluent.
(7)	Activated sludge/ nitrification/ denitrification separate stages	A + B + C + D + F + J2 + F + K + F + G	10	25	20	20	8	3	1	1	Effluent quality better than secondary effluent. Nitrogen removal and well-stabilized effluent.

	Process	Flow sheet								Remarks
(8)	Coagulation or lime precipitation, filtration, and carbon adsorption	A + B + H or I1 + F + O + P + G	10	30	105	10	1	3	20	0	Effluent quality better than secondary effluent. No ammonia removal. Excellent phosphorus removal. pH may be high if lime is used. Disinfection may not be required with lime.
(9)	Two-stage lime precipitation, filtration, and carbon adsorption	A + B + I2 + F1 + F2 + O + P	5	20	98	5	1	1	20	0	Effluent quality better than secondary effluent. No ammonia removal. Excellent phosphorus removal. Effluent may require neutralization. Disinfection may not be required with lime.
(10)	Lime or coagulant addition in activated sludge	A + B + C + H or I1 + D + F + G	15	35	99	15	2	4	17	0	Effluent quality better than secondary effluent. Good phosphorus removal.
(11)	Lime or coagulant addition in primary followed by activated sludge/ nitrification (single stage)/filtration/ carbon adsorption	A + B + H or I1 + F + J1 + F + O + P + G	5	10	17	5	1	1	2	18	Effluent quality better than secondary effluent. Excellent phosphorus removal and well-nitrified effluent.

TABLE 4-4   Treatment Process Combinations and Flow Schemes

Treatment Scheme No.	Type of Treatment	Treatment Units and Combinations	Effluent Quality (mg/ℓ)								Comments
			BOD_5	COD	UOD^a	TSS	TP	ON	NH_3-N	NO_3-N	
(12)	Lime or coagulant addition in activated sludge followed by separate stage nitrification/ denitrification/ filtration	A + B + C + H or I1 + D + F + J2 + F + K + F + O + G	5	20	12	5	1	1	1	1	Advanced waste treatment. Effluent low in BOD_5, TSS, and nutrients. Effluent reused.
(13)	Two-stage lime precipitation/ammonia stripping/ filtration/carbon adsorption	A + B + I2 + F1 + L + F2 + O + P	5	25	12	5	1	1	1	0	Advanced waste treatment. Effluent low in BOD_5, TSS, and nutrients. Effluent may require pH adjustment. Effluent reused.
(14)	Activated sludge/ filtration/ reverse osmosis	A + B + C + D + F + O + Q	small	small	small	small	small	small	small	small	Demineralization of wastewater. Effluent reused.

[a] Ultimate oxygen demand, UOD = 1.5 BOD_5 + 4.5 (NH_3-N) [see also Eq. (3-4)].
[b] The effluent quality from the preliminary treatment facility is assumed same as the influent quality at the head of the plant after mixing with the side streams from the sludge-processing areas (see also Table 4-2).

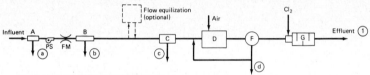

a. Secondary treatment utilizing activated sludge process

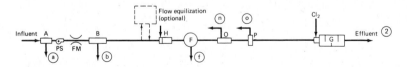

b. Secondary treatment utilizing coagulation, filtration and carbon adsorption

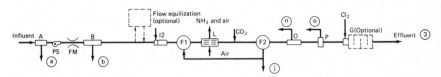

c. Advanced waste treatment utilizing two-stage lime precipitation, ammonia stripping, filtration and carbon adsorption

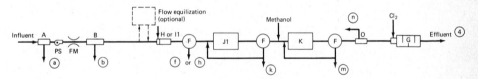

d. Advanced waste treatment utilizing coagulation, single stage activated-sludge-nitrification and denitrification, and filtration.

Parameter mg/ℓ	Influent[a]	Effluent points			
		①	②	③	④
BOD$_5$	210	20	10	5	5
COD	400	45	30	25	20
UOD	405	111	105	12	12
TSS	230	20	10	5	5
TP	11	8	1	1	1
ON	25	15	3	1	1
NH$_3$-N	20	18	20	1	1
NO$_3$	0	0	0	0	1

a. The influent quality is assumed after mixing with the return flow from the sludge processing areas.

FIGURE 4-2.   Process Flow Diagrams of Wastewater Treatment Plants (see Figure 4-1 and Tables 4-1, 4-2, 4-3, and 4-4 for the description of treatment units, unit combination, and efficiencies).

TABLE 4-5 Important Factors Other Than Costs That Are Also Considered in Selection of Wastewater Treatment Unit Operations and Processes

Treatment Units and Combinations	Land Requirements[a]	Adverse Climatic Conditions	Ability to Handle Flow Variations	Ability to Handle Influent Quality Variation	Industrial Pollutants Affecting Process	Reliability of the Process	Ease of Operation & Maintenance	Occupational Hazards[b]	Air Pollution	Waste Products
Preliminary treatment (A+B)	Min.	—	Good	Good	Min.	Very good	Fair		Odors	Screenings and grit
Pumping (PS)	Min.	Freezing	Good	Good	Min.	Very good	Fair		—	—
Primary sedimentation (C)	Mod.	—	Fair	Good	Mod.	Good	Very good		Odors	Sludge
Coagulation and sedimentation (H+F) or (I1+F)	Min.	—	Good	Very good	Max.	Very good	Good	Chemicals	—	Sludge
Trickling filter (E+F)	Max.	Freezing	Good	Fair	Mod.	Very good	Very good		Odors	Sludge
Activated sludge, conventional (D+F)	Mod.	—	Fair	Good	Mod.	Good	Fair			Sludge
Activated sludge with chemicals ((H or I1)+D+F)	Min.	—	Good	Very good	Max.	Good	Good	Chemicals		Sludge

Process									
Dual-media filter (O)	Mod.	—	Good	Poor for TSS	Min.	Very good	Good			Backwash waste
Activated carbon (P)	Mod.	—	Good	Fair	Max.	Good	Good	Fires, explosion	Regenerant gas	Spent carbon
Two-stage lime treatment (I2+F1+F2)	Max.	—	Good	Good	Min.	Very good	Fair	Chemicals		Excess sludge
Biological nitrification (J2+F)	Max.	Cold	Fair	Fair	Mod.	Fair	Fair			Sludge
Biological denitrification (K+F)	Max.	Cold	Fair	Fair	Mod.	Fair	Fair	Chemical		Sludge
Ion exchange (N)	Min.	—	Fair	Good	Max.	Good	Good	Chemicals	Ammonia	Waste regenerant
Breakpoint chlorination (M)	Mod.	—	Good	Poor for TSS	Max.	Very good	Good	Chemicals	Chlorine odor	Adds solids in effluent
Ammonia stripping (L)	Mod.	Cold	Fair	Fair	Min.	Good	Fair		Ammonia	Ammonia
Disinfection (G)	Min.	—	Good	Good	Max.	Very good	Good	Chemicals	Chlorine odor	Adds solids in effluent

[a]Min. = minimum. Max. = maximum. Mod. = moderate.
[b]Occupational hazards due to mechanical equipment and structure are common to all processes
Source: Ref. 5.

TABLE 4-6   Unit Operations and Processes Used for Processing and Disposal of Sludge

Unit No.	Unit Operations and Processes[a]	Principal Applications	Ref.
SA1	Gravity thickening (UO)	Thickening of sludge is done to concentrate solids and reduce the volume. Gravity thickening is used for thickening primary, secondary, or combined sludges.	Chap. 16
SA2	Dissolved air flotation (UO)	Solids are concentrated by flotation. Air is dissolved at high pressure. As the pressure is released, air bubbles rise and float the suspended solid. Dissolved air flotation is generally used for waste-activated sludge or chemically precipitated sludge.	Chap. 16
SA3	Centrifugation (UO)	Solids are thickened or dewatered under the influence of centrifugal force (100–600 times the force of gravity).	Chap. 16
SB1	Chemical Oxidation (UP)	Sludge is stabilized to reduce pathogens, eliminate offensive odors, and control putrefaction. Chemicals such as chlorine, hydrogen peroxide, or ozone are commonly used.	Chap. 17
SB2	Lime stabilization (UP)	Excess lime is used to raise the pH to 12 or higher. A high pH sludge will not putrefy, create odor, or pose health hazard.	Chap. 17
SB3	Aerobic digestion (UP)	Sludge is aerated for extended period (10–15 days). Sludge is stabilized. Used at small plants.	Chap. 17
SB4	Anaerobic digestion (UP)	Sludge is digested under anaerobic condition. Methane gas is recovered as energy source.	Chap. 17
SC1	Chemical conditioning (UP)	Sludge is conditioned to improve its dewatering characteristics. Chemicals such as alum, iron salts, and lime are used.	Chap. 18
SC2	Elutriation (UO)	Elutriation is washing of sludge to remove alkalinity. Elutriated sludge needs less chemicals to condition.	Chap. 18
SC3	Heat treatment or thermal conditioning (UP)	Heating of sludge at 140–200°C conditions and stabilizes the sludge.	Chap. 18

SD1	Drying beds (UO)	Sludge is dewatered in shallow beds of sand. The liquid is removed by underdrain system.	Chap. 18
SD2	Centrifugation (UO)	Sticky chemical sludges are dewatered using centrifuge.	Chap. 18
SD3	Vacuum filter (UO)	Rotary vacuum filters are used for dewatering the sludge.	Chap. 18
SD4	Filter press (UO)	Sludge is pressed between filter cloth held vertically in a frame.	Chap. 18
SD5	Horizontal belt filter (UP)	Sludge is pressed between horizontally mounted continuous belts.	Chap. 18
SE1	Land application (UP)	Digested liquid sludge or sludge cake is applied over farmland.	Chap. 19
SE2	Soil conditioner by composting (UP)	Sludge cake is composted and then used as soil conditioner.	Chap. 19
SE3	Soil conditioner by heat drying (UP)	The sludge cake is dried at a temperature of approximately 370°C. At this temperature part of the volatile matter is lost. The sludge is used as soil conditioner.	Chap. 19
SE4	Sanitary landfilling (UP)	Raw or digested sludge is buried if a suitable site is available within an economical hauling distance. Daily cover of 15–30 cm, and final cover of over 60 cm of compacted soil is required.	Chap. 19
SE5	Incineration (UP)	Incineration involves drying of sludge cake followed by complete combustion of organic matter. Wet-air oxidation is also used.	Chap. 19
SE6	Lime recalcining (UP)	Lime used in single- or two-stage lime precipitation is recovered by recalcining the sludge. Recalcining involves heating of dewatered lime sludge to about 1000°C in multiple-hearth furnace.	Chap. 19
SE7	Pyrolysis (UP)	Pyrolysis or destructive distillation is heating of sludge in oxygen free atmosphere to about 300–700°C. Gas, liquid, solid (charcoal), fractions are produced.	Chap. 19
SE8	Wet oxidation (UP)	Sludge solids are incinerated in liquid phase at high temperature (200–300°C) and high pressure (5–20 MN/m^2).[b]	Chap. 19

[a]UO = unit operation. UP = unit process.
[b]MN = megaN.

TABLE 4-7 Performance Data on Various Sludge-Processing Unit Operations and Processes

Sludge-Processing Systems	Sludge Stream Processed	TSS in Incoming Sludge (%)	TSS Captured in Processed Sludge (%)	TSS in Processed Sludge (%)	Returned Liquid BOD₅ (mg/ℓ)	TSS (mg/ℓ)
Thickener						
Gravity (SA1)	Primary	3–6	85–92	8–10	160–600	300–1000
	Primary + waste-activated sludge	2–5	80–90	4.5–8.5	166–600	300–800
	Waste-activated sludge	0.5–1.2	75–85	2.5–3.5	100–400	200–600
Dissolved air flotation (SA2)	Waste-activated sludge	0.5–1.2	75–85	2–4	160–600	400–1000
Centrifugation (SA3)	Waste-activated sludge	0.5–1.2	75–90	2–6	50–500	100–1000
Stabilization						
Aerobic (SB3)	Thickened combined sludge	3–8.5	—	—	100–1000	1000–10000
Anaerobic (SB4)	Thickened combined sludge	3–8.5	—	—	1000–10000	3000–15,000
Dewatering						
Vacuum filter (SD3)	Digested sludge with chemical conditioning	3–8	90–98	15–25	500–5000	1000–20,000
Pressure filter (SD4, SD5)	Digested sludge with chemical conditioning	3–8	90–98	20–50	500–4000	1000–15,000
Centrifugation (SD2)	Digested sludge with chemical conditioning	3–8	80–95	10–35	1000–10,000	2000–15,000

Source: Refs. 12 and 13.

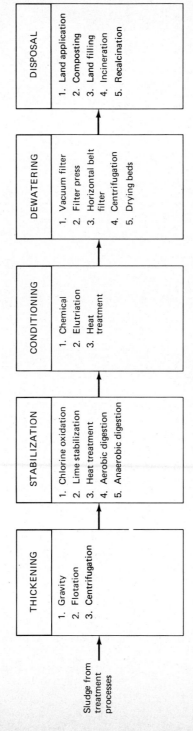

FIGURE 4-3   *Alternative Unit Operations and Processes for Sludge Processing and Disposal.*

4-3-3 Combination of Unit Operations and Processes into Flow Schemes

Proper selection of the sludge-processing equipment is important for trouble-free operation of a wastewater treatment facility. Sludge is quite odorous and may cause serious environmental problems. Therefore, such factors as solids captured, chemical quality of returned flows, ability to handle variable quality of sludge, ease of operation, and odors are often given serious consideration. Some of these factors that must be considered in final selection of appropriate flow schemes are assessed qualitatively in Table 4-8.[5] Several flow schemes for sludge processing are developed in Figure 4-4.

4-4 PROBLEMS AND DISCUSSION TOPICS

4-1   A wastewater plant has the following process train: bar screen, grit chamber, primary sedimentation, trickling filter (high rate), final clarifier, gravity filtration, and chlorine contact basins. Using the average percent removal efficiencies given in Table 4-2 for various units, estimate the effluent quality in terms of BOD_5, COD, UOD, TSS, TP, ON, NH_3-N, and NO_3-N. The influent quality after mixing with the return flows from the sludge-processing areas is as follows: $BOD_5 = 220$, COD $= 450$, UOD $= 425$, TSS $= 255$, TP $= 9$, ON $= 8$, NH_3-N $= 21$, and NO_3-N $= 0$. All units are in mg/ℓ.

4-2   The sludge-processing train of a wastewater treatment facility includes gravity thickener, chemical conditioning, and centrifuge. Calculate the volume of sludge cake (m^3/d) and the concentration of TSS in the overflows from the thickener and centrifuge. Use the following data: Combined sludge is 380 m^3/d at 1.2 percent solids (sp. gr $= 1.01$), solids capture efficiencies of the gravity thickener and the centrifuge are 85 percent each. Gravity-thickened sludge has 4.5 percent solids (sp. gr $= 1.03$). Dry chemicals are added in the thickened sludge at a ratio of 3 percent of the weight of dry solids. The sludge cake has 25 percent solids (sp. gr $= 1.06$). Assume 75 percent chemicals added for sludge conditioning is incorporated into the sludge solids. See Chapter 13 for procedure.

4-3   A gravity thickener receives combined primary and secondary sludge. The combined sludge is 500 m^3/d and contains 1 percent solids. Dilution water at a rate of 350 m^3/d is blended with the sludge for improved thickener operation. Assume that the solids capture efficiency of the thickener is 90 percent and the thickened sludge has 6 percent solids. Calculate the average volume of the supernatant and average concentration of TSS. Specific gravities of combined and thickened sludges are 1.00, and 1.03, respectively.

4-4   A primary wastewater treatment facility uses sludge-drying beds for dewatering of raw primary sludge. The raw wastewater contains 250 mg/ℓ TSS and 210 mg/ℓ BOD_5. The average daily flow to the plant is 2784 m^3/d. In primary sedimentation basin TSS removal is 65 percent, and BOD_5 removal is 35 percent of incoming flow. The primary sludge has 3 percent solids. The solids capture efficiency of the drying beds is 85 percent and the moisture content of the sludge cake is 72 percent. Conduct material mass balance analysis and determine the TSS and BOD_5 in the effluent from the primary treatment facility. Assume specific gravities of the primary sludge and sludge cake as 1.01 and 1.06, respectively.

4-5   Visit your local wastewater treatment plant. Obtain information on various unit operations and processes used to process liquid and residues. Prepare the process train of the entire facility. Using the information provided in Tables 4-2 and 4-7, calculate the average effluent quality.

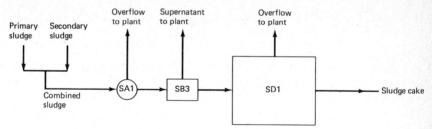

a. Gravity thickening, aerobic digestion, and drying beds used for processing of
primary and biological sludges from small wastewater treatment plants

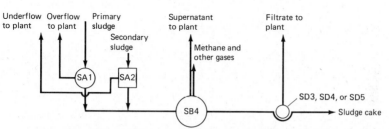

b. Gravity thickening, air flotation, anaerobic digestion, and vacuum filtration used
for processing of primary and biological sludges from large wastewater treatment plants

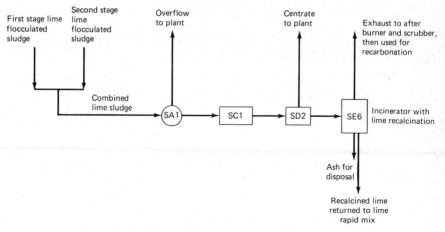

c. Gravity thickening, chemical conditioning, centrifugation and
recalcination of chemical sludge from a two-stage lime treatment
facility used at large wastewater treatment plants

FIGURE 4-4 Typical Sludge Treatment Flow Schemes (see Table 4–6 for process definition).

TABLE 4-8   *Factors Other Than Costs Normally Considered in Selection of Sludge-Processing Systems*

Sludge-Processing Systems	Land Requirements[a]	Adverse Climatic Conditions	Ability to Handle Flow Variations	Ability to Handle Influent Quality Variations	Industrial Pollutants Affecting Process	Reliability of the Process	Ease of Operation and Maintenance	Occupational Hazards	Air Pollution	Waste Products
Thickener										
Gravity (SA1)	Mod.	Freezing	Fair	Good	Min.	Good	Good	—	Odors	Thickened sludge. Liquid returned
Dissolved air flotation (SA2)	Mod.	Freezing	Fair	Good	Min.	Good	Fair	—	—	Thickened sludge. Liquid returned
Centrifuge (SA3)	Min.	—	Good	Good	Min.	Good	Fair	—	—	Thickened sludge. Liquid returned
Stabilization										
Aerobic (SB3)	Max.	Freezing	Good	Good	Max.	Good	Good	—	—	Digested sludge. Liquid returned
Anaerobic (SB4)	Max.	—	Good	Good	Max.	Good	Good	Explosion	—	Digested sludge. CO_2 and methane gas. Liquid returned

Conditioning										
Elutriation (SC2)	Max.	Freezing	Good	Fair	Min.	Fair	Fair	—	Odors	Conditioned sludge. Liquid returned
Heat treatment (SC3)	Mod.	—	Fair	Fair	Min.	Good	Fair	Explosion	Odors	Conditioned sludge
Dewatering										
Drying beds (SD1)	Max.	High rainfall	Good	Good	Min.	Fair	Very good	—	Odors	Sludge cake. Liquid returned
Filtration (SD3, SD4 or SD5)	Mod.	—	Fair	Fair	Min.	Good	Good	Chemicals	Odors	Sludge cake. Liquid returned
Centrifuge (SD2)	Min.	—	Fair	Fair	Min.	Good	Good	Chemicals	—	Sludge cake. Liquid returned
Disposal										
Landfilling (SE4)	Max.	—	Good	Good	Max.	Very good	Good	—	Odors	—
Land application (SE1)	Max.	Freezing	Good	Good	Max.	Fair	Good	—	Odors	—
Incineration (SE5)	Min.	—	Fair	Good	Min.	Very good	Fair	Explosion	—	Ash

[a]Min. = minimum. Max. = maximum. Mod. = moderate.
Source: Ref. 5.

REFERENCES

1. U.S. Environmental Protection Agency, *Alternative Waste Management Techniques for Best Practicable Waste Treatment,* Technical Information Report, U.S. EPA, March 1974.
2. Drobny, N. L., S. R. Qasim, and B. W. Valentine, "Cost-Effectiveness Analysis of Waste Management Systems," *Journal of the Environmental Systems,* Vol. 1, No. 12, June 1971, p. 189–210.
3. 40 CFR Part 35, Subpart E, Appendix A, "Cost-Effectiveness Analysis Guidelines," *Federal Register,* Vol. 38, No. 174, September 10, 1973, p. 24, 639–24 640.
4. Council on Environmental Quality and U.S. Environmental Protection Agency, *Municipal Sewage Treatment: A Comparison of Alternatives,* Contract EQC 316, EPA, Washington, D.C., February 1974.
5. U.S. Environmental Protection Agency, *A Guide to the Selection of Cost-Effective Wastewater Treatment Systems,* Technical Report EPA-430/9-75-002, Washington, D.C., July 1975.
6. U.S. Environmental Protection Agency, *Areawide Assessment Procedures Manual: Performance and Costs,* Appendix H, Municipal Environmental Laboratory, Cincinnati, Ohio EPA 600/9-76-014, July 1976.
7. Metcalf and Eddy, Inc., *Wastewater Engineering: Treatment, Disposal, and Reuse,* Second Edition, McGraw-Hill Book Co., New York, 1979.
8. Clark, J. W., Warren Viessman, and M. J. Hammer, *Water Supply and Pollution Control,* IEP-A Dun-Donnelly, New York, 1977.
9. U.S. Environmental Protection Agency, *Flow Equalization,* Technology Transfer, May 1974.
10. Joint Committee of the Water Pollution Control Federation and the American Society of Civil Engineers, *Wastewater Treatment Plant Design,* MOP/8, Water Pollution Control Federation, Washington, D.C., 1977.
11. U.S. Environmental Protection Agency, *Environmental Pollution Control Alternatives: Municipal Wastewater,* Technology Transfer, EPA-625/5-76-012, 1976.
12. U.S. Environmental Protection Agency, *Process Design Manual, Sludge Treatment and Disposal,* Municipal Environmental Research Laboratory, Office of Research and Development, Center for Environmental Research Information, Technology Transfer, EPA-625/1-79-011, September 1979.
13. U.S. Environmental Protection Agency, *Process Design Manual for Sludge Treatment and Disposal,* Technology Transfer, EPA-625/1-74-006, October 1974.

Predesign Studies

5

5-1 INTRODUCTION

The predesign phase of any project involves a planning process. The planning process is a systematic method of (1) recognizing that a negative situation or problem exists; (2) collecting and analyzing the data about the situation or problem; (3) redefining the situation or problem in the light of analyzed data; (4) establishing objectives which when achieved will remedy the situation or problem; (5) developing alternatives and costs; (6) evaluating the effectiveness of various alternatives, timings, and priorities for achieving the objectives and modifying the plans to meet the changing conditions; and, finally, (7) selecting the alternative that is most cost-effective and least controversial. In requiring this systematic approach, planning for water pollution control projects is similar to planning for any public works project. The end result of the planning process is an engineering report generally called *planning phase report, predesign phase report,* or *phase 1 report.*

Historically, all planning reports for wastewater projects contained discussions on existing problems and the need for expansion, design period, population estimates, wastewater flow and characteristics, degree of treatment and selected process train, preliminary design, and cost estimates.

A facility plan is the first step in a three-step process required to complete treatment works under federal construction grants program. The second and third steps are, respectively, preparation of detailed design plans and specifications, and actual construction of the treatment works. A facility plan is essentially a planning report. The U.S. Environmental Protection Agency (EPA) did much of the groundwork to develop guidelines for preparation of the facility plans. These guidelines are updated periodically. It is believed that the procedures for preparation of the facility plan, and design plans and specifications that have evolved under many concepts and policies promulgated by the environmental laws and subsequent guidelines will continue to be followed in the future. Therefore, in this chapter the general discussions on facility planning (Step 1) and design plans and specifications (Step 2) are provided. A model facility plan for a medium-size wastewater treatment facility is given in Chapter 6. Design procedure and plans and specifications for the same facility are covered in Chapters 7–23.

5-2 FACILITY PLANNING

A facility plan is prepared to identify the water pollution problems in a specific area, evaluate alternatives, and recommend a solution.[1,2] Through the facility plan, the consultants make many decisions that are subsequently used in preparation of the detailed plans and specifications for the wastewater treatment facilities. Specifically, the recommendations are made concerning

- Wastewater characteristics (initial and design) to include minimum average and peak flows; BOD, COD, suspended solids, nutrients, and any other constituents of concern
- Infiltration/inflow problems and sewer rehabilitation
- Existing wastewater treatment facilities and environmental concerns
- Degree of treatment needed in order to meet the effluent and receiving water quality criteria
- Intercepting sewer routing and other steps necessary for sewage collection
- Renovation of existing facilities, flow modifications, new treatment units and processes, and site selection
- Effluent disposal or reuse
- Sludge handling and disposal

The implementing agency (municipality or other) prepares a facility plan to develop a specific project that is cost-effective and environmentally sound.[1,3,4] Public participation in the decision-making process is required through public hearings and citizen advisory committees.

5-2-1 Preliminary Information for Facility Planning

The level of details required in a facility plan will vary according to the nature, size, and location of the undertaking. To make certain that applicants clearly understand the requirements of the construction grants, a preliminary meeting between state agency, the applicant, and the applicant's consultant should be arranged. At this meeting, the major requirements, role of municipality, consultant, the state, EPA, and the extent of planning needed are discussed. Application forms and kits for Step 1 application are available from the state agency and are usually furnished at this meeting.[5,6]

5-2-2 Contents of a Facility Plan

Several guides for preparing facility plans are available that are very helpful.[5-9] Many states and EPA regions have also developed instructional guides for preparing facility plans for specific regions of the country. These guidelines are very valuable and must be consulted to avoid unnecessary efforts and time delays. The contents of a facility plan are discussed below.

Project Need and Service Areas. Describe the project needs in terms of location, service areas, and problems associated with the existing treatment facilities. Discuss in detail the needs for improving or replacing the existing treatment facilities, including such factors as violations of effluent limitations, inability of existing facilities to meet the discharge compliance schedule, or potential public health hazards associated with the existing conditions.

Effluent Limitations. Include the effluent limitations for all discharges and the number of all National Pollutant Discharge Elimination System (NPDES) permits issued to facilities in the planning area. The effluent limitations may be secondary treatment or higher levels of treatment.[10–13]

Existing and Future Conditions. Describe the existing and future environmental conditions of the planning area. The description should be sufficient to provide a basis for analysis of alternatives and determination of impacts of the proposed action. The description should include the following items:[9,14–17]

- Natural environment to include surface and groundwater hydrology, quantity, quality and uses of water, temperature, precipitation, and evaporation.
- Geology and aquatic plant and animal communities
- Air quality and noise, energy production and control
- Economic, demographic and land use, and population projections
- Statutory controls for sewer system
- Industrial wastewater characteristics and composition
- Combined municipal and industrial wastewater characteristics, infiltration/inflow analysis, and sewer system evaluation survey and rehabilitation.

Forecast of Wastewater Characteristics from the Service Area. The future size and capacity of the facility must be estimated over the design period. Based on population growth, industrial growth and flow and waste reduction measures establish the future flow and wasteloads.

Wastewater Treatment Process Alternatives and Alternative Selection. Once the size and scope of the water pollution problem is defined, develop various alternatives for wastewater treatment, effluent disposal, and sludge processing and disposal.[18,19] Screen systematically each alternative to determine those that can meet the federal, state, and local criteria. Then review the principal alternatives to identify those that have cost-effective potential. Develop cost data for the principal alternatives and then evaluate them critically for environmental consequences along with the cost-benefit analysis.[20] Finally, selecting one alternative fully discuss the reasons for its selection and for the elimination of other alternatives.[21,22]

Financial Status. Discuss the legal, institutional, managerial, and financial capabilities of the city. Indicate that the city has adequate funds for its share of construction, operation, and maintenance of the proposed facility. Where more than one municipality and industry are served by the project, it is necessary for the lead municipality to negotiate a service agreement.[23–25]

Description of Selected Wastewater Treatment Facility.   Describe the selected treatment works in detail. Cover all elements, including service areas, collection sewers, interceptors, treatment processes, and ultimate disposal or use of effluent and sludge. Develop preliminary engineering data, including design criteria, detention times, overflow rates, process loadings, removal efficiencies, initial and design flows and reserve capacity, and number of units and dimensions.

Environmental Impact Assessment Report.   Prepare environmental impact report as part of the facility plan. Include only the summary information in the main text. Attach the environmental impact assessment report as an appendix. EPA has developed guidelines and procedure for preparation of environmental impact assessment report.[21,22,26-29] Topics such as environmental setting; historic and archeological sites; floodplanes and wetlands; flood insurance requirements; agricultural lands; coastal zone management; wild and scenic river, fish and wildlife protection; endangered species protection; air and water quality; alternatives selection and direct and indirect impacts and the like, must be fully addressed.

Public Participation.   The public should participate from the beginning in the facility-planning process so that interests and potential conflicts may be identified early and considered as planning proceeds. The planners should define issues and analyze information so that the public will clearly understand the costs and benefits of alternatives considered during the planning process. Public hearings must be held. A report summarizing public participation should be prepared and submitted as part of the facility plan.[30]

Cost Breakdown.   Prepare cost of performing facility plan (Step 1), design plans and specifications (Step 2), and the construction costs of the project (Step 3), and provide them in the facility plan.

Time Schedule for Project Milestones.   Provide in the facility plan the time schedule and project milestones for performing Step 2 and Step 3.

Appendixes.   In order to keep the main text of the facility plan brief, the following information may be arranged in various appendixes and cross-referenced in the main text of the plan:

1.   Preliminary designs, technical data, and cost estimates for alternatives
2.   Infiltration/inflow analyses and sewer evaluation surveys
3.   Environmental impact assessment report
4.   Agreements, resolutions, and comments
5.   Copy of the permits for the facility
6.   Public participation report

5-2-3 Submission and Approval of Facility Plan

Submit the completed facility plan in prescribed format to the regional or state clearing-house, and to EPA for review and comments. The state and EPA criteria for review are presented in several publications.[5,9] The plan should then be revised or amended as necessary.

5-3 PREPARATION OF DESIGN PLANS, SPECIFICATIONS, AND COST ESTIMATES

Once a community's facility plan application is approved by the state water quality agency and EPA, the design phase of the project is undertaken. This phase deals with preparation of detailed engineering design plans, specifications, and estimates. In this section, the basic requirements of design plans, specifications, and estimates are presented.

5-3-1 Preliminary Information for Design Plans and Specifications, and Estimates (Step 2)

The application for Step 2 grant is prepared on specified forms in which each item is completed in accordance with the instructions. Of particular importance are such items as source of local share of project costs (general taxes, sewer revenue funds, etc.), a copy of the resolution authorizing the official representative (mayor, council member, etc.) to act on behalf of the applicant, and a statement regarding the availability of the proposed site. Many other documents needed with the Step 2 application include:

- Proposed contracts or explanation for selection of consulting engineers
- Users charges or resolution indicating that a user charge system will be developed in accordance with EPA regulations
- Letter of agreement from industries served indicating compliance with the industrial cost recovery system developed for the project
- Copy of existing sewer use ordinance or a letter of intent that such an ordinance will be enacted
- Assurance of compliance with Civil Rights Act of 1964
- Statement or resolution indicating compliance with the Uniform Relocation of Land Acquisition Policies Act of 1970
- Copy of service agreements

Details on each of the above documents are available in Refs. 5 and 31–33. Predesign conferences may be necessary by the state and/or EPA to discuss the responsibilities and to finalize the administrative and technical requirements of the project. Following is the list of technical and administrative items that are considered during the Step 2 program.

5-3-2 Project Design Plans and Specifications

The project, in addition to being designed in accordance with sound engineering practice, must take into account those engineering and environmental measures recommended in the approved facility plan. The end product of the Step 2 project is a set of plans (drawings), specifications, and detailed construction cost estimates which are suitable for bidding and construction purposes. The project specifications must comply with state and federal requirements. Key elements of these requirements to be included in the contract documents are highlighted below.[5]

Technical Provisions of Specifications.   The following items must be addressed in the plans and specifications for the project:

*   *Reliability and flexibility:*   The proposed facilities must be reliable and provide for flexibility in operation. These include multiple units, ample pumping capacity and standby units, standby power or provision for wastewater storage.
*   *Bypassing:*   Preventive measures for bypassing during construction and operation.
*   *Mitigative measures:*   Mitigative measures required by environmental impact assessment or impact statement must be complied with. Examples are soil erosion control, hours of operation, backfilling and seeding, structural design for buildings in a flood plain, etc.
*   *Public water supply:*   Public water supplies must be protected by adequate backflow prevention devices (double check valves, air gap, etc.).
*   *Safety precautions:*   Occupational Safety and Health Act (OSHA) and applicable state and local requirements must be complied with.[34,35]
*   *Equipment:*   Except where based on performance specifications, trade names may be specified for major items of equipment. In selecting equipment and components, the consultant must give careful considerations to those which can be operated and maintained with the least effort. Use of mercury in equipment must require special review and approval.
*   *Emergency alarms:*   Adequate alarms must be provided to warn of failures or dangers.
*   *Sewers:*   Sewers must be tested for infiltration. They must maintain minimum scouring velocity and have adequate capacity during peak flow periods.
*   *Pretreatment:*   Incompatible industrial wastes must be pretreated where applicable.
*   *Chemical storage:*   Chemicals must be properly stored in a curbed area large enough to hold the entire volume in the event of an accidental spill. Also, adequate safety protection gear must be provided for plant personnel.
*   *Ventilation:*   Adequate ventilation must be provided in all areas where necessary (such as wet well, dry well, chlorine room, chemical storage area, etc.).
*   *Laboratory facilities:*   Laboratory facilities must be sufficient to give the plant operator control over the operational efficiency of the treatment plant. Additionally, facilities must be adequate to conduct sampling and testing as required by the NPDES permit or the state agency.
*   *Component identification:*   Equipment, piping, switches, instruments, etc., must be clearly marked for ease of identification.
*   *Project sign:*   Proper project sign must be displayed during construction and operation phases.

Supplemental General Provisions of Specifications.   The requirement for supplemental general provisions in the specifications is satisfied by including Appendix C-2 of 40 CFR Part 35 in the specifications.[33] The appendix includes conditions relating to the following subjects:

- Audit and access to records
- Price reduction for defective cost or pricing data
- Contract work hours and safety standards
- Equal employment opportunity
- The specified minimum wage rates
- A covenant against contingency fees; antikickback regulations, gratuities, etc.
- Copyrights
- A clean air and water clause

Bonding/Insurance.   For the construction contracts the bonding and insurance requirements must be a part of the specifications.[6] These include

- Bid bond
- Performance bond and payment bond
- Fire and extended coverage, workmen's compensation, public liability and property damage, and "all risk" insurance as required by local or state law
- Flood insurance as required during and after construction

5-3-3 Project Cost Estimates

The consultant is to prepare detailed construction cost estimates based on the scope of work as reflected in the project plans and specifications. This estimate is used to judge the reasonableness of the bids received.[5]

5-3-4 Continuing Work

While the following items need not be completed until the construction (Step 3) phase of the project is under way, work on them during the design phase should be maintained to ensure their timely completion.[6]

Plan of Operation.   A plan of operation is required for all treatment facilities. The plan should be, in large measure, a sequential listing of actions needed to ready the plant and its personnel for operation when construction is complete. Matters such as staffing and training requirements, operation and maintenance procedures, reports, laboratory testing, and the like must be considered in the plan. In short, the plan must detail the "who, when, and where" of facility operation and maintenance.

The operation and maintenance manual (prepared in conjunction with the plan of operation) is especially important because it provides plant personnel with detailed instructions for assuring efficient operation and proper maintenance of all plant components (including offsite pump stations, etc.). This manual should discuss how the facility is to be operated so as to meet effluent standards contained in the NPDES permit and other state and federal requirements. Portions of the federal grant payments of Step 3 costs may be withheld until the manual is approved.

User Charge and Industrial Cost Recovery Systems.   User charges are fees paid by users of the facilities to cover the operation and maintenance costs of the system.

Industrial, commercial, and residential users are charged a proportionate fee based on the wastewater treatment service provided. Grantees should consult with state agency and EPA before completing a user charge system.

At the time of a Step 2 grant application, applicants must submit a statement or resolution acknowledging their awareness of the need for preparing a user charge system and a procedural schedule for completing such a system. The federal shares of the Step 3 grant may be withheld until evidence of the timely development of the system is received. By the time the project is completed and ready to operate, an approved user charge system must be ready to be implemented.

Industrial cost recovery (ICR) is a system which recovers from industrial users of the wastewater treatment facilities that portion of the federal grant which relates to the industry's proportionate share of the capital cost of the project. A portion of the recovered funds are used by the grantee primarily for expansion and reconstruction, and the other portion is to be returned to the federal government.

At the time of a Step 2 grant application, each industry contributing 10 percent or more of the wastewater must submit a letter of intent to comply with the industrial cost recovery system developed for the project. While industries contributing less than 10 percent of the total are not required to submit letters of intent, all industrial users will be charged their proportionate share. A portion of the federal share of the Step 3 grant may be withheld until evidence of the timely development of the ICR system is submitted (similar to the user charge requirements).

5-4   DISCUSSION TOPICS

5-1   Obtain the most recently updated copy of Ref. 9. Review the chapter on preparation of facility plan. Prepare a detailed table of contents of a facility plan using the information contained in this source.

5-2   Review Refs. 26–28. Prepare a detailed table of contents for an environmental impact document for a wastewater treatment facility.

REFERENCES

1. *Federal Water Pollution Control Act Amendments of 1972* (PL. 92-500), 92nd Congress, October 18, 1972.
2. *Clean Water Act of 1977* (PL 95-217), 95th Congress, December 27, 1977.
3. 40 CFR Part 35, Subpart B, "State and Local Assistance," *Federal Register*, Vol. 38, No. 125, June 29, 1973, p. 17219–27225.
4. U.S. Environmental Protection Agency, *Construction Grants—1982, Municipal Wastewater Treatment*, 430/9-81-020 (CG-82), July 1982.
5. U.S. Environmental Protection Agency, *Guidance for Preparing a Facility Plan*, MCD-46, Revised May 1975.
6. U.S. Environmental Protection Agency, *How to Obtain Federal Grants to Build Municipal Wastewater Treatment Works*, MCD-04, May 1976.
7. U.S. Environmental Protection Agency, *Model Plan of Study*, A Supplement to "Guidance for Preparing a Facility Plan," MCD-08, September 1975.
8. U.S. Environmental Protection Agency, *Model Facility Plan for a Small Community*, A Supplement to "Guidance for Preparing a Facility Plan," September 1975.

9. U.S. Environmental Protection Agency, *Facility Planning 1981, Municipal Wastewater Treatment*, Construction Grants Program, 430/9-81-002 FRD-20, Washington, D.C., March 1981.

10. 40 CFR Part 130, "Policies and Procedures for State Continuing Planning Process," *Federal Register*, Vol. 39, No. 107, June 3, 1974, p. 19634–19639.

11. 40 CFR Part 131, "Preparation of Water Quality Management Basin Plans," *Federal Register*, Vol. 39, No. 107, June 3, 1974, p. 19639–19644.

12. U.S. Environmental Protection Agency, *Guidelines for the Preparation of Water Quality Management Plans*, September 1974.

13. 40 CFR Part 133, "Secondary Treatment Information," *Federal Register*, Vol. 38, No. 159, August 17, 1973, p. 22298–22299.

14. U.S. Environmental Protection Agency, *Alternative Waste Management Techniques for Best Practicable Waste Treatment*, Technical Information Report, March 1974.

15. Guitierrez, A. F., and J. Harry Rowell, "Five Years of Sewer System Evaluation," *Journal of the Environmental Engineering Division, American Society of Civil Engineers*, Vol. 105, No. EE6, December 1979, p. 1149–1163.

16. U.S. Environmental Protection Agency, *Guidance for Sewer System Evaluation*, March 1974.

17. U.S. Environmental Protection Agency, *Design, Operation and Maintenance of Wastewater Treatment Facilities*, Technical Bulletin, September 1970.

18. U.S. Environmental Protection Agency, *Evaluation of Land Application Systems*, Technical Bulletin, EPA-430/9-75-001, March 1975.

19. U.S. Environmental Protection Agency, *Survey of Facilities Using Land Application of Wastewater*, EPA-430/9-73-006, July 1973.

20. Drobny, N. L., S. R. Qasim, and B. W. Valentine, "Cost-Effectiveness Analysis of Waste Management Systems," *Journal of Environmental Systems*, Vol. 1, No. 2, June 1971, p. 189–210.

21. Canter, Larry W., *Environmental Impact Assessment*, McGraw-Hill Book Co., New York, 1977.

22. 40 CFR Part 35, Subpart E, Appendix A, "Cost Effectiveness Analysis Guidelines," *Federal Register*, Vol. 38, No. 174, September 10, 1973, p. 24639–34640.

23. 40 CFR Part 35, Subpart E, Appendix A, "Cost Effectiveness Analysis Guidelines," *Federal Register*, Vol. 38, No. 161, August 21, 1973, p. 22524–22527.

24. 40 CFR Part 128, "Pretreatment Standards," *Federal Register*, Vol. 38, No. 215, November 8, 1973, p. 30982–30984.

25. 18 CFR 704.39, "Discount Rate," *Federal Register*, Vol. 29, No. 158, August 14, 1974, p. 29242. (Published annually under this title by U.S. Water Resources Council.)

26. U.S. Environmental Protection Agency, *Manual for Preparation of Environmental Impact Statements for Wastewater Treatment Works, Facilities Plans, and 208 Areawide Waste Treatment Management Plans*, July 1974.

27. U.S. Environmental Protection Agency, *Instruction for Preparing Environmental Information Documents for Construction Grants Projects*, CG-99T, Region 6, Dallas, April 1980.

28. 40 CFR Part 6, "Preparation of Environmental Impact Statements," *Federal Register*, Vol. 40, No. 72, April 14, 1975, p. 16811–16827.

29. National Park Service, *The National Register of Historic Places*, U.S. Government Printing Office, Washington, D.C., 1972.

30. U.S. Environmental Protection Agency, *Municipal Wastewater Management: Citizen's Guide to Facility Planning*, EPA-430/9-79-006 FRD-6, Office of Water Program Operation, Washington, D.C., 1979.

31. 40 CFR Part 35, "State and Local Assistance," Subpart E, "Grants for Construction of Treatment Works," and Appendix B, "User Charges for Operation and Maintenance of Publicly Owned Treatment Works," *Federal Register*, Vol. 43, No. 188, September 27, 1978.

32. 40 CFR Part 4 "Implementation of the Uniform Relocation Assistance and Real Property Acquisition Policies Act of 1970," *Federal Register,* Vol. 39, No. 54, March 19, 1974.
33. 40 CFR Part 35, Subpart E, Appendix C-2 "Required Provisions: Construction Contracts," *Federal Register,* Vol. 43, No. 188, September 27, 1978.
34. 40 CFR Part II "Occupational Safety and Health Standards," Department of Labor, *Federal Register,* Vol. 36, No. 105, May 29, 1971.
35. 40 CFR Part II, "Occupational Safety and Health Standards," *Federal Register,* Vol. 37, No. 202, October 18, 1972.

6

Model Facility Plan

6-1 INTRODUCTION

A facility plan is prepared to identify the water pollution problems in a specific area, develop design data, evaluate alternatives, and recommend a solution. Most of the data developed in a facility plan are used in preparation of design plans, specifications, and cost estimates of the wastewater treatment facilities.

The purpose of this chapter is to develop a model facility plan for a medium-size community. Efforts have been made to simplify the problem and keep the contents of the facility plan realistic and brief. In actual practice it is not uncommon to find facility plans for a medium-size community that are several hundred pages in length supported by many separate appendixes that are also voluminous. Such an effort is not warranted here. However, a generalized format has been developed for preparation of a facility plan, and for accruing the needed design information for preparing the design plans and specifications for a medium-size wastewater treatment facility in Chapters 7–23.

The names of the town and state, and all other information used to prepare this facility plan, are not real. The information is meant only to be illustrative. It is neither complete nor definitive.

6-2 PROJECT IDENTIFICATION

Owner/applicant:	Modeltown, Anystate
Project name:	Construction of Wastewater Treatment Facilities
Project number:	EPA Federal Grant Number EPA-XXX-000 State Environmental Protection Agency Project Identification Number XXX-000-XXX
Discharge permit number:	EPA NPDES NO. XXX State Pollution Control Permit No. XXX
Final submission date:	January 31, 1983

6-3   OBJECTIVES AND SCOPE

The City of Modeltown currently owns and operates three small wastewater treatment plants that utilize trickling filters, stabilization ponds, and an aerated lagoon. These treatment facilities are inadequate and often are inoperable, causing serious water pollution problems, public nuisance, and health hazards. The existing facilities cannot provide adequate wastewater treatment without renovation or replacement. The objectives of this facility plan are to (1) develop and evaluate alternative wastewater treatment schemes, (2) select a cost-effective plan which will provide compliance with state and federal water quality standards, and (3) develop the preliminary design and cost data for the proposed wastewater treatment facility.

6-4   EFFLUENT LIMITATIONS AND INITIAL DESIGN PERIOD

Pursuant to the state's continuing planning process, all of the water segments in the area have been classified as "effluent limitation segments." Thus, any project for the construction of publicly owned wastewater treatment works must be based on consideration of alternative waste treatment techniques, including physical, chemical, and biological treatment, land application, discharge into natural waters, and reuse of effluent.[1,2] However, a minimum of secondary treatment is required.

The EPA has issued a municipal discharge permit (NPDES) to the city which expires on July 1, 1984. The state pollution control permit issued to the city also expires on November 30, 1984. The current and projected effluent limitations of the state are summarized in Table 6-1. Immediately upon issuance of the new permit from the state, the

TABLE 6-1   Present and Projected Effluent Limitations of the State

Parameter	Current Effluent Concentrations		Projected Effluent Concentrations	
	Average 30 Consecutive Day Sample	Average 7 Consecutive Day Sample	Average 30 Consecutive Day Sample	Average 7 Consecutive Day Sample
BOD_5(mg/ℓ)	30	40	20	30
Total suspended solids, (mg/ℓ)	30	40	20	30
Coliform number/ 100 ml	200	400	100	200
pH	Shall remain between 6.0 and 9.0	Shall remain between 6.0 and 9.0	Shall remain between 6.5 and 8.5	Shall remain between 6.5 and 8.5

permitee must comply to the projected effluent limits within a period of one year. This implies that the upgraded facility must be completed and set into operation by November 1985. Therefore, 1985 is established as the initial year for the project.

6-5 EXISTING AND FUTURE CONDITIONS

6-5-1 Economic, Demographic, and Land Use

Modeltown is a thriving community located at the confluence of the east and west forks of a river. The economy of the town is based on agriculture, allied enterprise, and retail business. The major industries include one slaughter house, one meat-processing and packaging plant, one dairy, and two canning industries. The town also has large business and commercial establishments. There are 12 modern public grade and high schools, and one college that also serves the surrounding farming area.

The planning department of the city, based on census data, has estimated the current and future populations of the town. The census and estimated population data for the town are summarized in Table 6-2 and plotted in Figure 6-1. These population forecasts have been coordinated with the designated management agency for the Modeltown and are in conformance with the areawide wastewater management population forecasts established under the Clean Water Act.

The existing land use within the city has not changed substantially from that

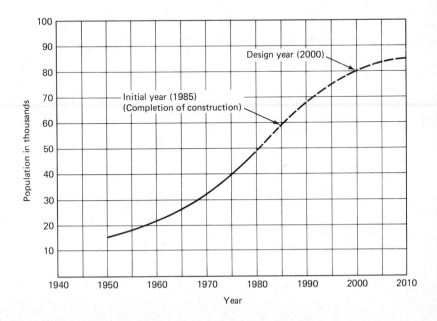

Figure 6-1   Census and Estimated Population Data for the City.

TABLE 6-2   Census and Estimated
Population of Modeltown

Year	Population
1950 (census)	15,000
1960 (census)	22,000
1970 (census)	32,000
1980 (census)	48,000
1985 (estimated, initial year)	58,800
1990 (estimated)	68,000
1995 (estimated)	76,000
2000 (estimated, design year)	80,000

Note: See Chapter 2 for methods of population forecasting.

presented in the 1980 Comprehensive Plan developed by the City Planning Commission. There has been some change in transition of vacant land to single-family dwellings since the plan was developed. The land use plan for 1980 and for the assumed design year (2000) are illustrated in Figure 6-2.*

6-5-2 Natural Environment

The region is subjected to frequent precipitation with the average annual rainfall being approximately 89 cm (35 in.), with high-intensity showers during the late summer and early fall. The climate is mild with minimum and maximum temperatures ranging from –2°C in the winter to 40°C in the summer. Gentle, rolling plains characterize the topography of the area. Soils range from clays to a mixture of clay and silt.

There are no surface water impoundments in the planning area. Groundwater is of high quality and serves as the only source of water supply for the town. The east and west forks of the river are grossly polluted with sewage effluent discharged from the waste-water treatment plants. The 30-d mean low flow and water quality data for the river at the confluence of the east and west forks are summarized in Table 6-3.

---

*The design or staging period is assumed 15 years. This is based on the expected flow growth factor. The flow growth factor for this problem is discussed in Sec. 6-6-4. Additional discussion on design and staging periods may be found in Sec. 2-2.

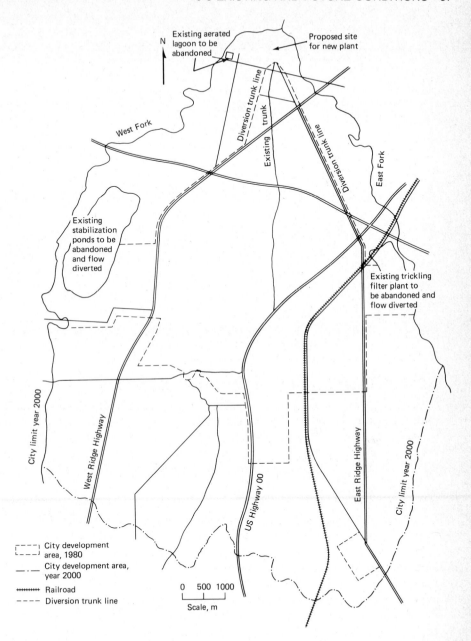

Figure 6-2 Modeltown Existing and Proposed Wastewater Treatment Facilities, Major Roads and City Limits: Years 1980 and 2000, Populations 48,000 and 80,000, Respectively.

TABLE 6-3 30-Day Mean Low Flow and Water
Quality Data for the River at the Confluence of
the East and West Forks

Condition	Value
Mean dry weather flow conditions	
Mean flow (cfs)	54
Mean flow ℓ/s	1535
Average water depth (m)	1.2
Average width (m)	7.5
Average velocity (m/s)	0.15
Water quality at mean flow condition	
BOD_5(mg/ℓ)	6.8
Total dissolved solids (mg/ℓ)	400
DO (mg/ℓ)	4.0
Ammonia nitrogen (mg/ℓ)	2.0
pH	7.3
Total phosphorus (mg/ℓ)	0.4

6-5-3 Wastewater Collection and Treatment Systems

The sanitary sewage collection system for the city was first constructed in 1945 to serve a
small portion of the town. Since then, the sanitary sewer system has been periodically
extended as need occurred to serve the newly developed areas.

The city sewerage system has a total length of 569 km (354 miles). The sewer lines
are 15 cm (6 in.) minimum to 76 cm (30 in.) maximum diameter concrete and vitrified clay
pipes with predominantly cement mortar and compression joints. The equivalent diameter
of the sanitary sewers is 27 cm.* Some parts of the city have storm sewers that were first
constructed in 1955 and then expanded as the need occurred. A large portion of the town
still does not have storm sewers. There are no combined sewers in the entire service area.

Approximately 1250 residents in the southern section of the town are not served by
the sanitary sewer system. These homes have individual septic tanks, absorption trenches,
and open ditches. There is some concern about groundwater contamination and public
health hazard if this practice is allowed to continue. As mentioned earlier, the groundwater is the only source of water for this community.

There are three wastewater treatment facilities operated by the city. These include the

---

*The equivalent diameter of sanitary sewer system is the weighted average obtained from Eq. (6-1):

$$\text{Equivalent diameter} = \frac{d_1 L_1 + d_2 L_2 + d_3 L_3 + \dots}{L_1 + L_2 + L_3 + \dots} \qquad (6\text{-}1)$$

where L_1, L_2, and L_3 are lengths of sewers of diameter d_1, d_2, and d_3. Typical length distribution of average sewer
sizes are 15 cm, 4 percent; 20 cm, 71 percent; 38 cm, 13 percent; 50 cm, 4 percent; 60 cm, 3 percent; 68 cm, 4
percent; and 76 cm, 1 percent. See Table 7-1 for additional details.

stabilization ponds, the trickling filters, and an aerated lagoon. The locations of these treatment facilities are shown in Figure 6-2. A brief description of each of these treatment facilities is given in Table 6-4.

6-5-4 Statutory Controls for Sewer Systems

City ordinance prohibits the discharge of storm water, groundwater, roof runoff, foundation drainage, and other subsurface drainage into the sanitary sewers. Tight house connections are also required under the city ordinance.

TABLE 6-4   Description of Existing Wastewater Treatment Facilities Operated by the City

Treatment Facility	Description
Stabilization ponds	Three stabilization ponds are located on west side of the town which were first constructed in 1945. These ponds serve the residential, commercial, and industrial areas of the southwestern and central parts of the town. Each pond is 7.3 hectares (18 acres) earthen basin. The average water depth is about 1.2 m. A manually cleaned bar screen precedes the ponds, followed by a flow division box that divides the flow equally into three ponds. The effluent is discharged into the west fork by gravity. A lift station pumps the effluent into the west fork under high flow conditions. The pumps have been repaired frequently; major renovation was done in 1970. The ponds, division box, and bar screen are surrounded by a 1.2-m (4-ft) high levee to protect against flooding. The effluent is of poor quality and the area residents have frequently complained against odor problems and mosquito infestation due to these stabilization ponds.
Trickling filter plant	A treatment plant is located on the eastside of the town and utilizes a manually cleaned bar screen and two high-rate trickling filters. This plant was constructed in 1960. The service area for this plant includes residential, commercial, and industrial establishments located on the southeastern and central part of the town. The trickling filters are each 27.5 m diameter (90 ft) with recirculation systems. Two final clarifiers, each 18.3 m diameter (60 ft) provide clarification. A lift station pumps the raw sewage from the wet well into the trickling filters. The raw waste sludge is pumped into the drying beds for dewatering. The sludge cake is landfilled on the plant property. The pumps at the lift station have been repaired frequently. When the pumps are inoperable, raw sewage bypasses the treatment plant and discharges into the east fork. Area residents have frequently complained about the odors from this plant.

TABLE 6-4   *Description of Existing Wastewater Treatment Facilities Operated by the City*

Treatment Facility	Description
Aerated lagoon	An aerated lagoon is located on the north side of the town. The facility serves the residential, commercial, and industrial areas on the northcentral part of the town. The lagoon was constructed in 1970. It has nine floating aerators rated at 9 kW (12 HP). The average water depth and surface area are approximately 4.6 m (15 ft) and 1.2 hectare (3.0 acres), respectively. Effluent is discharged by gravity into the river near the confluence of the east and west forks. The total suspended solids and total BOD_5 in the effluent are considerably higher than the effluent limitations.

All industrial influents to the sewer system are regulated by the city ordinance. The ordinance requires grease traps for all food-handling establishments and pretreatment of all industrial wastes (where necessary) to a level that is compatible with the publicly owned wastewater treatment works.

6-5-5 Industrial Wastewater Characteristics and Composition

There are five industries in town: one slaughter house, one meat-processing and packaging plant, one dairy, and two canning industries. Table 6-5 provides a summary of pretreatment and wasteload information from these industrial operations.

6-5-6 Combined Municipal and Industrial Wastewater Characteristics

Dry Weather Conditions.   Flow data at wastewater treatment plants are essential for planning and design of treatment facilities. Detailed discussion on flow-measuring devices and design procedures are provided in Chapter 10. Three 90° V-notch weirs were installed into the intercepting sewers of the three treatment facilities. Hourly flow measurements were recorded over a period of three months from June 10 through September 9, 1982. Average dry weather flows were plotted at hourly intervals to obtain the diurnal flow patterns.

The typical dry weather diurnal flow curve for the area served by the stabilization pond is shown in Figure 6-3. Similar curves (not shown here) for the other two areas were also developed from the recorded flow data at the trickling filter plant and the aerated lagoon. The average, peak, and minimum daily flows were established from these curves. Based on the estimated connections and resident population, the average per capita flows

TABLE 6-5 Characteristics of Industrial Wastes Discharged into the Municipal Sewers

Industry	Location	Flow (ℓ/s)	Pretreatment Provided	Effluent into Sewers	
				BOD₅(mg/ℓ)	TSS(mg/ℓ)
Slaughterhouse	North industrial park	8.5	Aerated lagoon	250	350
Meat-processing and packaging plants	North industrial park	3.1	Grease trap and aerated lagoon	240	300
Dairy	East industrial park	9.6	Aerated lagoon	200	240
Canning industry 1	West industrial park	6.1	Grease trap and aerated lagoon	230	300
Canning industry 2	West industrial park	7.7	Grease trap and aerated lagoon	240	320
	Total	35.0			

Note: 1 mgd = 1.54 cfs = 43.6 ℓ/s = 3768 m³/d.

were also obtained. These results and calculation procedures are summarized in Table 6-6.

Six 24-h composite wastewater samples of influent were collected at each of the three V notches. These samples were analyzed for pH, BOD₅, total suspended solids, total phosphorus, and total nitrogen. The average values of these analyses are summarized in Table 6-6.

Infiltration/Inflow Analysis   An infiltration/inflow analysis of the entire sewage collection system was completed as one of the requirements for obtaining a wastewater treatment facility grant from the U.S. Environmental Protection Agency and the concerned State Water Pollution Control Agency.[3]* Visual inspections of a large number of key manholes and the lamping and televising of many sewer reaches were made. The manholes were in excellent condition. The sewer reaches that were inspected were free of solids depositions, root intrusions, or disrepair.

During the month of August, high-intensity storms and intermittent showers con-

*An infiltration/inflow report is normally submitted as an appendix to a facility plan. This report contains the results of dry and wet weather flows, and a cost effectiveness analysis to treat the wet weather flows; or to perform a sewer evaluation survey and rehabilitation program.

TABLE 6-6   Dry Weather Influent Characteristics at Three Wastewater Treatment Facilities

Facility	Estimated Number of Connections	Estimated Population Served	Industries Served	Dry Weather Flow				Average Concentrations (mg/ℓ)					
				Average ℓ/s	ℓpcd	Ratio Maximum to Avg.	Ratio Minimum to Avg.	BOD$_5$	Total Suspended Solids	Total Solids	Total P	Total N	pH
Stabilization ponds	5520	18,100	Two canning industries	95	453	2.15[a]	0.52[b]	250	265	910	8.8	49	7.2
Trickling filter	4270	13,100	Dairy industry	67	442	2.00[c]	0.48[c]	255	265	890	9.0	51	7.3
Aerated lagoon	4890	15,550	Slaughterhouse and meat processing and packaging industry	80	445	1.94[c]	0.50[c]	245	251	930	9.2	44	7.1
Septic tanks	320	1250	None	—	—	—	—	—	—	—	—	—	—
	Total = 15,000	Total = 48,000		Total = 242[d]	Avg. = 447	Avg. = 2.03	Avg. = 0.50	Avg. = 250	Avg. = 260	Avg. = 910	Avg. = 9.0	Avg. = 48	Avg. = 7.2

[a] 204.3/95 = 2.15 (max. and average dry weather flows are obtained from Figure 6-3).
[b] 49.4/95 = 0.52 (min. and average dry weather flows are obtained from Figure 6-3).
[c] Obtained from the diurnal dry weather flow data for both service areas. The procedure is similar to that for stabilization ponds given in Figure 6-3. The diurnal dry weather flow and I/I plots for areas served by trickling filter plant, and aerated lagoon are not given in this chapter.
[d] Includes residential, commercial, and industrial flows. The industrial contribution is 35.0 ℓ/s (see Table 6-5); therefore, the residential plus commercial contribution is (242 − 35)ℓ/s = 207 ℓ/s; the per capita residential and commercial contribution = (207 ℓ/s × 60 s/min × 60 min/h × 24 h/d)/(48,000 − 1250) population served = 383 ℓpcd.
Note: ℓpcd = liters per capita per day.

tinued to occur over a 10-d period. The first high-intensity storm occurred on August 4, after a long dry spell. This storm caused a 10.2-cm (4-in.) rainfall in six hours. Hourly flow measurements recorded at the three V notches on this day were used to determine the inflow into the sewer system. Since the groundwater table was low, the infiltration was negligible and the excess flow over the dry weather diurnal flow into the sewer system was the inflow.[4] The inflow data of the August 4 showers for the area served by the stabilization ponds are illustrated in Figure 6-3. Similar plots for the other two areas (not shown here) were also developed. The maximum inflow resulting from the high-intensity showers of August 4 at the three wastewater treatment facilities is given in Table 6-7.

Immediately after the continuous 10-d wet-weather condition, the recorded flow data at the three V-notch weirs provided a measure of the infiltration into the sewer system. The maximum excess flow over the dry weather diurnal flow gave the infiltration values. The hourly variation of recorded infiltration into the interceptor sewer at the stabilization pond is shown in Figure 6-3. Similar plots for the other two areas were also prepared. The maximum infiltration values of the three service areas are summarized in Table 6-7.

The results indicate that the average infiltration/inflow values from three service areas are 152 and 285 ℓpcd, respectively [total I/I = 437 ℓpcd]. This gives a total infiltration/inflow allowance of 1330 liters per d per cm per km (1436 gal per d per in. per

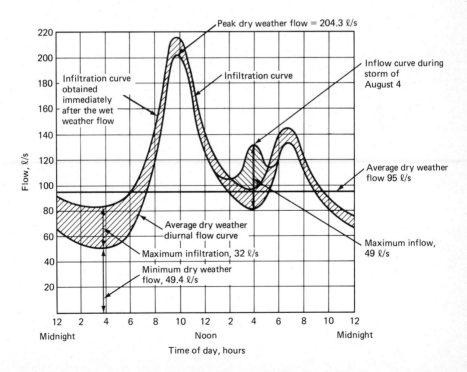

Figure 6-3   Average Diurnal Dry Weather Flow, and Infiltration/Inflow Measured at the 90° V Notch Installed at the Stabilization Ponds. Similar plots were also developed at other plants.

TABLE 6-7   Results of Infiltration and Inflow Study

Treatment Facility and Area Served	Maximum Infiltration[a]		Maximum Inflow[a]	
	ℓ/s	ℓpcd	ℓ/s	ℓpcd
Area served by stabilization ponds	32.0[b]	153[c]	49.0[b]	234
Area served by trickling filter plant	22.8[d]	150	50.2[d]	331
Area served by aerated lagoon	27.7[d]	154	52.2[d]	290
Average	28	152[e]	50	285[e]

[a]These values are in excess over the diurnal dry weather flow.
[b]See Figure 6-3.
[c](32 ℓ/s × 24 h/d × 60 min/h × 60 s/min)/18,100 people = 153 ℓpcd.
[d]These values are obtained from plots similar to Figure 6-3, not shown in this chapter.
[e]Total infiltration/inflow allowance = (152 + 285) ℓpcd = 437 ℓpcd.

mile) of collection system.* This amount is even less than the permissible limits of 1394 ℓ/d per cm per km (1500 gpd per in. per mile) that can normally be expected from typical groundwater infiltration sources.[5] It would be more economical to collect and treat this flow than to rehabilitate the collection system. Therefore, a sewer evaluation survey was not cost-effective and was not considered.

6-6   FORECAST OF WASTEWATER CHARACTERISTICS FROM SERVICE AREAS

6-6-1 Projections under Existing Conditions

A recent residential and industrial survey conducted by the City Planning Board indicated that the per capita municipal wastewater flow will gradually increase approximately 6 percent by the year 2000. This survey also indicated that the area industries anticipate significant industrial growth. Although most of the industries are attempting to reduce the water demand and the resulting wastewater discharges, the portion of the flow contributed by the industries will continue to rise.

6-6-2 Flow Reduction Measures

Conservation of water is gaining popularity among the general public and the regulatory agencies. Water conservation has the beneficial effect of reducing the volume of wastewater that must be collected and treated. There are a wide variety of "water-saving devices"

---

* $\dfrac{437 \ \ell\text{pcd} \times 46{,}750 \text{ population served by sewerage system}}{569 \text{ km (length of sewerage system)} \times 27 \text{ cm (equivalent diam.)}}$ = 1330ℓ/cm · km · d

See Table 6-6 and Sec. 6-5-3 for population and for sewer lengths and equivalent diameters.

that are effective in flow reduction. Many devices are commonly manufactured as part of most faucet combinations and are readily available from most appliance suppliers. Some of these devices are faucet aerators, flow-limiting shower heads, and water-saving toilet devices. Many of these devices and the water savings achieved by their use are discussed in Chapter 3.

Another form of in-house water saving might come from the reuse of water by means of a "closed-loop" system. The recycled water could be the waste streams from wash-basins, bathtubs, and the laundry. These waters would be collectively treated and reused for toilet flushing and lawn sprinkling.

Most local governments have regulatory powers that can be exercised to achieve a reduction in water demands. These regulatory alternatives (also called nonstructural alternatives) may come in the form of plumbing code and zoning ordinances to require water-saving devices in new homes and in commercial and industrial establishments. Another method of achieving flow reduction is to encourage the use of plumbing fixtures and appliances that reduce the water consumption. In all cases, a comprehensive com-munity awareness program is necessary for the general public to understand the impor-tance of water conservation and the costs and benefits of the water-saving devices.

6-6-3 Projections of Wastewater Characteristics for Initial and Design Periods

The social acceptability and the ease with which many of the water-saving devices can be installed in existing and new systems, and the public attitude and cooperation toward use of water saving devices will all affect the flow reduction. However, it is expected that in spite of all available watersaving techniques and all public awareness programs, there will not be any significant reduction in wastewater volumes. It is safe to assume, however, that implementation of all water conservation efforts will prevent a net (ℓpcd) increase in the domestic and commercial flow rates.

The wastewater characteristics for the initial and design years were developed using the present flow records and data obtained from the infiltration and inflow studies. The basic procedure used to develop the initial and design flow characteristics are given below:

1. The average initial dry weather flow (year 1985) is calculated using a wastewater flow of 447 ℓpcd (Table 6-6). This is equal to the current average flow. No increase in residential, commercial, and industrial flow rates is assumed.
2. The average design dry weather flow (year 2000) is calculated using no increase in residential and commercial flow rates and approximately a 244 percent increase in the industrial flow. The projected combined average flow rate for all three service areas is 475 ℓpcd.*

---

*Residential +
commercial flow $= \dfrac{383 \ \ell\text{pcd (see footnote d of Table 6–6)} \times 80{,}000 \text{ population}}{24 \text{ h/d} \times 3600 \text{ s/h}} = 354.6 \ \ell/\text{s}$

Industrial flow $= 35.0 \ \ell/\text{s (Table (6–5))} \times 2.44 = 85.4 \ \ell/\text{s}$

Total flow rate $= \dfrac{(85.4 + 354.6) \ \ell/\text{s} \times (24 \times 60 \times 60)\text{s/d}}{80{,}000 \text{ population}} = 475 \ \ell\text{pcd}$

3. The peak dry weather flows for initial and design years for the individual service areas were obtained using the measured ratio of maximum to average flow.
4. The peak dry weather flow for initial and design years for the combined facility for the three service areas were estimated using Eq. (3-1).[6,7]
5. The wet weather peak flows for initial and design years were obtained by adding the average infiltration/inflow allowance of 437 ℓpcd to the peak dry weather flows (see Table 6-7). This assumption is reasonable as newer sewers will have less infiltration/inflow, while older lines will have an increase in infiltration/inflow.
6. The minimum initial and design flows were obtained from the average measured ratio of minimum to average flows.
7. The quality of raw wastewater produced in the service area is not expected to change significantly over the design period. This assumption is reasonable as the per capita wastewater volume from residential and commercial areas will not increase over the design period, and industrial wastewater discharges into the municipal sewers will be strictly controlled under the city ordinance to meet the industrial pretreatment standards.

   Tables 6-8 and 6-9 have been prepared from the above assumptions to summarize the wastewater characteristics at the initial and design periods.

6-6-4 Check the Design Period

The planning period of a facility may be divided into staging periods (10, 15, or 20 years) depending on the ratio of wastewater flow expected at the design and initial years (see Table 2-1). For the facility under consideration, the growth factor = (440 ℓ/s)/(304 ℓ/s) = 1.5 (see Table 6-9). The design or staging period at a growth factor of 1.5 is 15 years (Table 2-1).

6-7 WASTEWATER TREATMENT PROCESS ALTERNATIVES

A wastewater treatment scheme is a combination of many unit operations and processes to achieve the specific treatment objectives. To achieve secondary level of wastewater treatment, various treatment and disposal alternatives were evaluated as part of the facility-planning process. Many treatment and disposal alternatives that have been evaluated in this facility plan are:

- Wastewater treatment process alternatives
- Effluent disposal alternatives, and
- Sludge-processing and disposal alternatives

   A brief description of each of the treatment and disposal alternatives follows.

TABLE 6-8 Population and Wastewater Flows from Three Service Areas, Initial and Design Periods

Treatment Plant	Year 1985 Initial Conditions					Year 2000 Design Conditions				
	Population Served[a]	Average Flow (ℓ/s)	Peak Dry Weather Flow (ℓ/s)	Minimum Dry Weather Flow (ℓ/s)	Peak Wet Weather Flow (ℓ/s)	Population Served[a]	Average Flow (ℓ/s)	Peak Dry Weather Flow (ℓ/s)	Minimum Dry Weather Flow (ℓ/s)	Peak Wet Weather Flow (ℓ/s)
Stabilization Ponds	24,000	124[b]	267[c]	65[d]	388[e]	35,000	192[f]	413[g]	100[h]	590[i]
Trickling Filter Plant	16,700	86	172	41	257	24,000	132	264	63	385
Aerated Lagoon	18,100	94	182	47	274	21,000	116	225	58	331
Total for the city	58,800	304	—	—	—	80,000	440	—	—	—

[a]Obtained from population projections of each service area.

[b]$\dfrac{24{,}000 \text{ population} \times 447 \text{ ℓpcd}}{(24 \times 3600) \text{ s/d}} = 124$ ℓ/s.

[c]124 ℓ/s × 2.15 (ratio of peak and avg. flows, Table 6–6) = 267 ℓ/s.

[d]124 ℓ/s × 0.52 (ratio of min. and avg. flows, Table 6–6) = 65 ℓ/s.

[e]$267 \text{ ℓ/s} + \dfrac{24{,}000 \text{ population} \times 437 \text{ ℓpcd I/I allowance (Table 6–7)}}{(24 \times 3600) \text{ s/d}} = 388$ ℓ/s.

[f]$\dfrac{35{,}000 \text{ population} \times 475 \text{ ℓpcd}}{(24 \times 3600) \text{ s/d}} = 192$ ℓ/s.

[g]192 ℓ/s × 2.15 (ratio of peak and avg. flow, Table 6–6) = 413 ℓ/s.

[h]192 ℓ/s × 0.52 (ratio of min. and avg. flow, Table 6–6) = 100 ℓ/s.

[i]$413 \text{ ℓ/s} + \dfrac{437 \text{ ℓpcd I/I allowance (Table 6–7)} \times 35{,}000 \text{ population}}{(24 \times 3600) \text{ s/d}} = 590$ ℓ/s.

TABLE 6-9   Characteristics of Combined Wastewater from Three Service Areas at Initial and Design Periods

Design Conditions	Initial Year 1985	Design Year 2000
Total population served	58,800	80,000
Average wastewater flow (ℓpcd)	447	475
Average wastewater flow (ℓ/s)	304	440
Minimum wastewater flow (ℓ/s)	152[a]	220
Maximum dry weather flow (ℓ/s)	669[b]	916
Peak flow	966[c]	1321
Average sewage characteristics		
BOD_5 (mg/ℓ)	250	250
Total suspended solids (mg/ℓ)	260	260
Total solids	910	910
pH	7.2	7.2
Total phosphorus (mg/ℓ)	9.0	9.0
Total nitrogen (mg/ℓ)	48.0	48.0

[a]304 ℓ/s × 0.5 (ratio of min. and avg. flows for three areas, Table 6–6) = 152 ℓ/s.

[b]304 ℓ/s $\left(\dfrac{18 + \sqrt{58.8}}{4 + \sqrt{58.8}} \right)$ = 669 ℓ/s [see Eq. (3-1)].

[c]669 ℓ/s + (58,800 population × 437 ℓpcd I/I allowance, Table 6–7) $\dfrac{1}{(24 \times 3600) \text{ s/d}}$ = 966 ℓ/s.

6-7-1 Wastewater Treatment Process Alternatives

Comparative studies were conducted for *physical–chemical, biological,* and *land treatment* processes to meet the effluent limitations given in Table 6-1. Discussion of various physical–chemical and biological treatment processes, including land treatment systems, have been presented in Chapters 4, 8–19, and 24. Readers are referred to these chapters for process description, effluent quality, and principal advantages and disadvantages of each process. A physical–chemical treatment system utilizing coagulation, clarification, filtration, and carbon adsorption will produce an effluent quality that is superior to the secondary effluent. However, there are many inherent difficulties associated with the physical–chemical treatment systems. These difficulties include (1) large quantities of sludge produced for processing at thickening, digestion, and dewatering facilities, (2) increased plant hydraulic load during filter backwashing, (3) carbon regeneration, (4) higher level of trained personnel needed, (5) concern over scarcity of chemicals, and (6) high energy consumption. Furthermore, the physical–chemical plants are not economical for achieving a secondary level of treatment. Usually, physical–chemical processes have better applications for tertiary or advanced wastewater treatment. A discussion on advanced wastewater treatment facilities may be found in Chapter 24.

An evaluation of principal land treatment systems for primary treated effluent indicates treatment superior to secondary level. However, there are many uncertainties and limitations of the land application system. These include (1) large land areas are needed, (2) accumulation of trace elements in soils, (3) public health effects from pathogens in aerosol droplets and agricultural products, (4) possible groundwater contamination, and (5) disruptions due to equipment failure and severe weather conditions. General discussion on land treatment methods may be found in Chapter 24.

Biological treatment processes of the present day appear to be economical for secondary treatment levels. Due to many uncertainties and inherent problems with physical–chemical and land treatment, the biological treatment system has been selected. Primary treatment facility followed by completely mixed biological reactor or trickling filter and a final clarifier will produce the effluent quality given in Table 6-1.

6-7-2 Effluent Disposal Alternatives

A total of eight methods of effluent disposal were investigated in this facility plan as follows:

1. Natural evaporation
2. Groundwater recharge
3. Irrigational uses (land application)
4. Recreational lakes
5. Aquaculture
6. Municipal uses
7. Industrial uses, and
8. Discharge into natural waters

General discussions on each of the above disposal methods, their suitability and limitations, and comparative evaluation are provided in Chapter 15. In summary, natural evaporation and land application methods require large land areas, are affected by adverse climatic conditions, and may cause groundwater pollution. Exceptionally high effluent quality is needed for recreational lakes and groundwater recharge and municipal uses. No industrial market exists for process and cooling water in the immediate vicinity of the study area. At the same time, secondary treated effluent from the proposed facility will have a quality sufficient to permit its disposal into receiving waters while meeting the proposed state and federal constraints. Therefore, effluent disposal into natural waters has been selected.

6-7-3 Sludge Processing and Disposal Alternatives

The objectives of sludge treatment and disposal are (1) to reduce the organic matter in sludge to relatively stable and inoffensive material, (2) to reduce the volume for ultimate disposal, (3) to destroy or control pathogens, (4) to prevent air, water, and land pollution, and (5) to utilize byproducts. Sludge processing and disposal involves thickening, stabilization, dewatering, and disposal. Various alternatives of sludge processing and disposal have been evaluated in Chapter 4. Design details and equipment selection for sludge-processing systems are also given in Chapters 16–19. Based on an evaluation of various competing sludge-processing alternatives, the following systems have been selected:

1. Gravity thickening of combined primary and secondary sludges
2. Sludge stabilization by anaerobic digestion
3. Filter press for dewatering
4. Use of sludge cake on city properties and park areas, and excess or unused sludge disposal by sanitary landfilling at the plant property.

Disposal of all residues shall be in accordance with the guidelines set under the Resource Conservation and Recovery Act.[8] Systems evaluation procedure and merits and demerits of different systems are provided in detail in Chapters 4 and 16–19.

6-8   ENGINEER'S DEVELOPMENT OF THE PROPOSED PROJECT

A total of five alternative wastewater collection and treatment systems (including no action alternative) were evaluated to select the most cost-effective wastewater treatment scheme. These alternatives are presented in Table 6-10. Preliminary evaluation of these alternatives is given below.

6-8-1 Preliminary Evaluation of Treatment Plant Alternatives

A preliminary evaluation of various alternatives indicate that the alternatives A, B, and C do not meet the basic requirements of water quality compliance, facility centralization and energy conservation. For these reasons these alternatives were not given further considerations. A brief discussion is provided below:

1. The implementation of the no-action alternative (alternative A) would mean discharge of inadequately treated effluent into the receiving waters and violation of state permit conditions. Continued use of septic tanks and absorption trenchs might cause groundwater pollution. Both these conditions would result in a serious health hazard to the area residents. Odors emanating from the stabilization ponds, drying beds, and receiving waters will continue to be environmental issues. Thus the inadequate wastewater treatment facilities would cause a loss of opportunity for orderly development and economic growth of the area.

2. The implementation of alternative B would provide a decentralized approach to wastewater collection and treatment. The decentralized approach reduces the processing of large volumes of wastewater at one location. However, the centralized wastewater treatment works in general permit improved planning and coordination of collection and treatment works, facilitate application of new technology, allow efficient monitoring of effluent by regulatory agencies, reduce the inventory system by providing multiple and compatible equipment, and economize the construction and operating costs. For these reasons, the city and the area regulatory agency have encouraged centralization of the facilities where possible.

3. The implementation of alternative C would require sewage pumping at an average of 192 ℓ/s and 116 ℓ/s across town, and from north to south, respectively. Pumping of large volumes of sewage long distances is not economical. Also, standby pumping units and dual power sources are necessary to avoid pumping disruptions.

TABLE 6-10 *Wastewater Treatment Plant Alternatives*

Alternative	Description
Alternative A—no action	Continue using stabilization ponds, trickling filters, aerated lagoon, and septic tanks.
Alternative B—three plants	Upgrade the trickling filter facility, and construct two new treatment plants to replace the stabilization ponds and aerated lagoon.
Alternative C—one treatment plant and two force mains	Abandon stabilization ponds and aerated lagoon. Construct two pumphouses and force mains to divert flows from stabilization ponds and aerated lagoon to the trickling filter plant. Upgrade and expand the existing trickling filter plant at this location to treat the combined diverted flows.
Alternative D—two treatment plants and one diversion sewer	Abandon stabilization ponds and aerated lagoon. Construct a gravity intercepting sewer to divert wastewater from the stabilization pond to the aerated lagoon site. Construct a new plant at this site to treat the combined flows. Renovate the trickling filter facility.
Alternative E—one treatment plant and two diversion sewers	Abandon all three existing treatment facilities. Construct two gravity-intercepting sewers to divert flows from trickling filter plant and stabilization ponds to the aerated lagoon site. Construct a new treatment plant at this location to treat the combined flow.

Present trends in planning of wastewater treatment facilities are to select alternatives that minimize energy requirements and avoid uncertainties.

6-8-2 Cost Effectiveness Evaluation of Alternatives D and E*

Alternatives D and E minimize pumping of sewage over long distances, utilize a partial or complete centralization approach, and provide acceptable effluent quality. These alternatives are described in great detail to compare and select the most cost-effective system.[†]

Description of Alternative D. This alternative includes diversion of flow from the stabilization pond to the aerated lagoon facility, construction of a new plant at this site, and renovation of the trickling filter facility.

*See Ref. 9.
†Alternatives which meet effluent limitations, such as B and C, could not be dismissed in a facility plan without a more detailed consideration of the economies and environmental effects. In this example, however, only alternatives D and E have been evaluated to illustrate the procedure for selection of the most cost-effective and environmentally acceptable alternative.

Diversion of Flow From Stabilization Pond to Aerated Lagoon. A 91-cm (36-in.) gravity diversion sewer will be constructed along the West Ridge Highway to divert an average design flow of 192 ℓ/s. This sewer will intercept several existing collectors in its proposed length of 3.5 km (2.2 miles).

New Treatment Plant at Aerated Lagoon Site. A new activated sludge facility will be constructed at the aerated lagoon site to treat the combined flow from the stabilization pond and aerated lagoon. The average design flow is 308 ℓ/s (7.0 million gallons per day (mgd)). Table 6-8 provides a summary of flows. The flow scheme includes preliminary treatment,* primary sedimentation, completely mixed aeration, final clarification, and chlorination. The sludge is thickened, anaerobically digested, and dewatered by filter presses. The sludge cake is applied over city land and the excess is disposed of by sanitary landfilling.

Trickling Filter Plant. The existing trickling filter facility will be upgraded to treat an average design flow of 132 ℓ/s (3.0 mgd). Table 6-8 gives a summary of the flow data. The grit removal and primary sedimentation facilities will be added. The existing lift station, trickling filter units, final clarifiers, and chlorination facility will be renovated. The sludge will be digested aerobically and dewatered over the existing drying beds.

Description of Alternative E. This alternative includes diversion of flows from the trickling filter plant and stabilization ponds to the aerated lagoon site. A new plant will be constructed to treat the combined flows. Various components of this alternative are discussed below.

Diversion of Flow From Stabilization Pond to Aerated Lagoon. The diversion trunk sewer is the same as described in alternative D.

Diversion of Flow From Trickling Filter Plant to Aerated Lagoon. A 76-cm (30-in.) gravity diversion sewer will be constructed to convey the flow from the trickling filter plant to the aerated lagoon site. This sewer will be 2.9 km (1.8 miles) long and will intercept many trunk sewers along its length on East Ridge Highway.

Treatment Plant at Aerated Lagoon Site. A treatment plant will be designed for an average flow of 440 ℓ/s (10 mgd) to treat the combined flow from three service areas. Initial and design flow data and wastewater characteristics are given in Table 6-9. The flow scheme is the same as proposed for the treatment facility in alternative D.

Cost Comparison of Alternatives. The construction, operation, and maintenance costs of wastewater treatment alternatives can be developed using many published cost curves.[10-13] Several computer programs are also available for preliminary design and cost estimates of wastewater treatment facilities.[13-16] However, these cost estimates are useful only for evaluation of wastewater treatment alternatives.

The construction, operation, and maintenance cost estimates for alternatives D and E

---

*Preliminary treatment facility includes bar screen, lift station, grit removal, and flow measurement facilities.

were developed using EPA *Areawide Assessment Procedures Manual: Performance and Cost,* Appendix H.[10] Summary of performance details and cost curves for primary and secondary treatment units and processes are provided in Appendix C. In this publication all cost curves are indexed to *ENR* (September 1976). *Engineering News Record* cost indices are used to update the estimates to January 1983 dollars.[17] Tables 6-11 and 6-12 summarize the construction, operation, and maintenance costs of the alternatives D and E, respectively.

The costs of alternatives D and E for cost effectiveness analysis include capital construction costs and annual costs for operation and maintenance. The wastewater treatment systems are evaluated for cost effectiveness over the planning period as defined in the Municipal Wastewater Treatment Works Construction Grants Program (see Sec. 2-2).[3]

The procedure for cost comparison of alternatives includes determination of the present worth and equivalent annual costs for each project alternative.[18] The present worth may be thought of as the sum which, if invested now at a given rate, would provide exactly the funds required to make all necessary expenditures during the planning period. Equivalent annual cost is the expression of a nonuniform series of expenditures used as a uniform annual amount to simplify calculations of present worth. Detailed procedures for making these calculations are well known and explained in many textbooks.[19,20] General procedure is provided in Table 6-13. The interest rate for formulation and evaluation of construction grants projects is published by the Water Resources Council. These are *The Discount Rate* and are periodically revised.[21] The current interest rate (1983) is $7\frac{5}{8}$ percent.

The present worth and equivalent annual costs of alternatives D and E are summarized in Table 6-13. These costs must be updated prior to starting the construction program.

Effectiveness Evaluation of Alternatives   Economic analyses of wastewater treatment facilities were traditionally based on cost considerations. These evaluations utilized total costs incurred and any savings or benefits which could be expressed in monetary units. It has been recognized that combining costs and benefits into single measures will not necessarily indicate the most economically efficient alternatives. Consequently, interest has been developed in evaluating the alternatives on effectiveness and then comparing with system economics.[9] Alternatives D and E have been evaluated in terms of six measures of effectiveness. Many parameters have been grouped together to simplify the analysis. The measures of effectiveness are presented in Table 6-14.

Evaluation Procedure.   To rate the effectiveness, each alternative was scored for every measure of effectiveness by assigning a number 0–4. The scores 0 and 4 indicated low and high effectiveness ratings. Since the measures of effectiveness are not of equal importance, a relative weight was also assigned to each measure of effectiveness. The relative weights varied from 1 to 3, with 3 indicating extremely important. The effectiveness scores (0 to 4) were then multiplied by the relative weights. The final number was then used in the matrix evaluation. Finally, the weighted scores for each alternative were added to give the overall scores for comparison purposes. Such matrix evaluation for two alternatives is given in Table 6-15.

TABLE 6-11   Estimated Construction, and Operation and Maintenance Costs of Alternative D

System Components and Description	Service Life (Years)	Construction Cost (Millions of Dollars)		Annual O&M Cost (Thousands of Dollars)	
		1976[a]	1983[b]	1976[c]	1983[b]
A. 91-cm (36-in.) gravity diversion sewer to divert flow from stabilization ponds to aerated lagoon. Total trunk length 3.5 km (2.2 miles), average design flow 192 ℓ/s (4.4 mgd)	50	1.056	1.690	1.8	2.9
Engineering and construction—supervision 15 percent, and contingencies 15 percent (total = 30 percent)	—	0.317	0.507	—	—
Total costs of gravity diversion sewer	—	1.373	2.197	1.8	2.9
B. Renovation of Trickling filter plant: Design average and peak flows are 132 ℓ/s (3 mgd) and 385 ℓ/s (8.75 mgd), respectively					
Lift station (renovation)	15	0.295[d]	0.472	22.0[d]	35.20
Preliminary treatment (new)	30	0.073	0.117	16.0	25.60
Primary sedimentation (new)	50	0.208	0.333	14.0	22.41
Trickling filter (renovation)	50	0.368	0.589	14.0	22.41
Final clarifier (renovation)	40	0.310	0.496	29.0	46.41
Chlorination (renovation)	15	0.070	0.112	23.0	36.81

Gravity thickener, combined sludge (primary + trickling filter)	50	0.067[e]	0.107	3.4[e]	5.44
Aerobic digester (new)	50	0.220	0.352	18.5	29.61
Drying beds (renovation)	20	0.080	0.128	17.5	28.01
Landfilling of sludge (renovation)	20	0.035	0.056	17.0	27.21
Miscellaneous structures, administrative offices, laboratories, shops, and garage (renovation)	50	0.130	0.208	37.0	59.21
Support personnel	—	—	—	3.0	4.80
Subtotal 1, Cost of trickling filter plant	—	1.856	2.970	214.4	343.12
Piping 10 percent, electrical 8 percent, instrumentation 5 percent, and site preparation 5 percent (total = 28 percent)	—	0.520	0.832	—	—
Subtotal 2	—	2.376	3.802	214.4	343.12
Engineering and construction-supervision 15 percent, and contingencies 15 percent (total = 30 percent)	—	0.713	1.141	—	—
Total cost of trickling filter plant	—	3.089	4.943	214.4	343.12
C. Construction of new activated sludge treatment facility at aerated lagoon site: Design average and peak flows are 308 ℓ/s (21 mgd) and 921 ℓ/s (7 mgd), respectively					
Lift station	15	1.750[f]	2.801	57.0[f]	91.2
Preliminary treatment	30	0.130	0.208	22.0	35.2

TABLE 6-11   Estimated Construction, and Operation and Maintenance Costs of Alternative D

System Components and Description	Service Life (Years)	Construction Cost (Millions of Dollars)		Annual O&M Cost (Thousands of Dollars)	
		1976[a]	1983[b]	1976[c]	1983[b]
Primary treatment	50	0.330	0.528	27.0	43.2
Aeration basin	30	1.080	1.728	75.3	120.5
Final clarifier	40	0.640	1.024	41.0	65.6
Chlorination	15	0.150	0.240	43.0	68.8
Gravity thickener, combined sludge (primary + waste activated sludge)	50	0.105[g]	0.168	5.0[g]	8.0
Anaerobic digester	50	0.485	0.776	21.0	33.6
Filter press	15	0.580	0.928	47.0	75.2
Landfilling of sludge	20	0.067	0.107	26.0	41.6
Miscellaneous structures, administrative offices, laboratories, shops, and garage	50	0.200	0.320	49.0	78.4
Support personnel	—	—		5.4	8.6
Subtotal 1, Cost of activated sludge plant		5.517	8.828	418.7	669.9

Piping 10 percent, electrical 8 percent, instrumentation 5 percent, and site preparation 5 percent (total = 28 percent)		1.545	2.472	—	—
Subtotal 2	—	7.062	11.300	418.7	669.9
Engineering and construction supervision 15 percent, contingencies 15 percent (total = 30 percent)	—	2.119	3.390	—	—
Total cost of Activated sludge plant	—	9.181	14.690	418.7	669.9
Total cost of alternative D	—	13.643	21.830	634.9	1015.92

[a] Capital costs for new units are obtained from the cost curves in Appendix C. Cost of unit renovation are developed from engineering judgment.

[b] Costs, 1983 = costs, 1976 × ENR Index 1983/ENR Index 1976; ENR Index January 1983 = 3961, ENR Index 1976 = 2475.

[c] O & M costs are obtained from cost curves in Appendix C.

[d] $Q_E = (\text{TDH } 11.5 \text{ m}/3.05 \text{ m}) \times Q_{DESIGN} = 3.77\ Q_{DESIGN}$.

[e] Q_E for combined sludges is obtained from thickener loading of 43 kg/m²·d (8.8 lb/sq ft·d) and design sludge mass of 0.22 kg/m³ (1844 lb/10⁶ gal.), $Q_E = 1.53$ Q_{DESIGN} (see Figure C-11).

[f] $Q_E = (\text{TDH } 15.25 \text{ m}/3.05 \text{ m}) \times Q_{DESIGN} = 5.0\ Q_{DESIGN}$.

[g] Q_E for primary and waste-activated sludge is obtained from thickeners loading of 43 kg/m²·d (8.8 lb/sq ft·d) and design sludge mass of 0.22 kg/m³ (1844 lb/10⁶ gal.); $Q_E = 1.53 Q_{DESIGN}$ (see Figure C-11).

TABLE 6-12   Estimated Construction, and Operation and Maintenance Costs of Alternative E

System Components and Description	Service Life (Years)	Construction Cost (Millions of Dollars)		Annual O&M Cost (Millions of Dollars)	
		1976[a]	1983[b]	1976[a]	1983[b]
A. 91 cm (36 in.) gravity diversion sewer to divert flow from stabilization ponds to aerated lagoons. Total trunk length 3.5 km (2.2 miles), average design flow 192 ℓ/s (4.4 mgd)	50	1.056	1.690	1.8	2.9
B. 76 cm (30 in.) gravity diversion sewer to divert flow from trickling filter plant to aerated lagoon. Total trunk length 2.9 km (1.8 miles), average design flow of 132 ℓ/s (3 mgd)	50	0.684	1.095	1.3	2.1
Subtotal of gravity diversion sewer	—	1.740	2.785	3.1	5.0
Engineering and construction supervision 15 percent and contingencies 15 percent (total = 30 percent)	—	0.522	0.836	—	—
Total cost of gravity diversion sewer	—	2.262	3.621	3.1	5.0
C. New Activated Sludge plant: design average and peak flows of 440 ℓ/s (10 mgd) and 1320 ℓ/s (30 mgd) respectively.					
Lift station	15	2.400[c]	3.841	75.0[c]	120.0
Preliminary treatment	30	0.160	0.255	28.0	44.8
Primary sedimentation	50	0.430	0.688	34.0	54.4
Aeration basin	30	1.400	2.241	98.0	156.8
Final clarifier	40	0.800	1.280	57.0	91.2
Chlorination	15	0.180	0.288	57.0	91.2
Gravity thickener, combined sludge (primary + waste-activated sludge)	50	0.139[d]	0.222	6.2[d]	9.9
Anaerobic digester	50	0.600	0.960	24.0	38.4
Filter press	15	0.700	1.120	54.0	86.4
Landfilling of sludge	20	0.080	0.128	30.0	48.0
Miscellaneous structures, administrative offices, laboratories, shops, and garage	50	0.240	0.384	58.0	92.8
Support personnel	—	—	—	7.0	11.2

TABLE 6-12   Estimated Construction, and Operation and Maintenance Costs of Alternative E

System Components and Description	Service Life (Years)	Construction Cost (Millions of Dollars)		Annual O&M Cost (Millions of Dollars)	
		1976[a]	1983[b]	1976[a]	1983[b]
Subtotal 1, cost of activated sludge plant	—	7.129	11.407	528.2	845.1
Piping, 10 percent, electrical 8 percent, instrumentation 5 percent, and site preparation 5 percent, (total = 28 percent)	—	1.996	3.194	—	—
Subtotal 2	—	9.125	14.601	528.2	845.1
Engineering and construction supervision 15 percent, and contingencies 15 percent, (total = 30 percent)	—	2.738	4.380	—	—
Total cost of activated sludge plant	—	11.863	18.981	528.2	845.1
Total cost of alternative E	—	14.125	22.602	531.3	850.1

[a]Capital and O&M costs are obtained from the cost curves in Appendix C.

[b]Cost of 1983 = cost of 1976 $\times \dfrac{\text{ENR Index 1983}}{\text{ENR Index 1976}}$; ENR Index January 1983 = 3961, ENR Index 1976 = 2475.

[c]$Q_E = \dfrac{\text{TDH } 11.5\text{m}}{3.05 \text{ m}} \times Q_{\text{DESIGN}} = 5.0.Q_{\text{DESIGN}}$.

[d]Q_E for combined sludge is obtained from thickeners loading of 43 kg/m^2.d (8.8 lb/sq ft.d) and design sludge mass of 0.22 kg/m^3 (1844 lb/10^6 gal.), Q_E = 1.53.Q_{DESIGN} (see Figure C-11).

The assignment of relative weights to each measure of effectiveness and the rating of each alternative for different measures of effectiveness should be performed independently by a team of at least three professionals. Average values should be used to minimize the preferences and biases of the individuals.

Alternative Selection.   The economic comparison of alternati·?es D and E indicate that alternative E is slightly cheaper than alternative D. However, the effectiveness evaluation of these alternatives in terms of six measures of effectiveness (Table 6-14) indicate that the alternative E is considerably superior to alternative D. These measures of effectiveness include factors such as site conditions, effluent quality, environmental considerations, regionalization, future expansion, energy conservation, operational flexibility and reliability, system compatibility, and sludge handling and disposal. Based on the results of

TABLE 6-13   Present Worth and Equivalent Annual Costs of Alternatives D and E

Alternatives	Total Capital Cost 1983 (Millions of Dollars)	Total Annual O&M 1983 (Thousands of Dollars)	Present Worth of Annual O&M Cost (Millions of Dollars)	Present Worth (Millions of Dollars)	Equivalent Annual Cost (Millions of Dollars)
D	21.83	1015.92	8.898[a]	30.728[b]	3.509[c]
E	22.602	850.1	7.446	30.048	3.431

[a]Interest rate $7\frac{5}{8}$ percent, planning period 15 years, present value of annuity = $[1 - (1 + i)^{-n}]/i = [1 - (1 + 0.07625)^{-15}]/0.07625 = 8.759$. Present worth of annual O&M cost = $(1015.92 \times 8.7590/1000) = 8.898$ millions of dollars.
[b]Present worth = $8.898 + 21.83 = 30.728$ million dollars.
[c]Capital recovery factor = $i(1 + i)^n/[(1 + i)^n - 1] = 0.07625 (1.07625)^{15}/[(1.07625)^{15} - 1] = 0.1142$. Equivalent annual cost = $30.728 \times 0.1142 = 3.509$ million dollars.

TABLE 6-14   Measures of Effectiveness Used in Effectiveness Evaluation of Alternatives D and E

Measures of effectiveness	Description
Plant location	Wastewater treatment plant location should be sensitive to constraints imposed by system design, land cost, and availability for present construction and future expansion, and the nature of the surrounding developments both existing and planned. In this respect, alternative D utilizes two plant locations, while alternative E provides a single site at a remote location. Therefore alternative E is rated higher.
System compatibility	Facilities that have similar and compatible multiple-treatment processes reduce system complexity and operational skills to operate, maintain, and minimize routine daily monitoring activities and reduce the available inventory of spare parts and equipment. Alternative D utilizes two independent treatment processes, namely, activated sludge and trickling filters. Therefore, it is rated lower than alternative E in this category.
Operational flexibility and reliability	Single-treatment process under shock loadings or upset conditions may produce inferior effluent quality. Multiple treatment processes, such as in alternative D, provide a backup system if one process is subjected to operational

TABLE 6-14   Measures of Effectiveness Used in Effectiveness Evaluation of Alternatives D and E

Measures of effectiveness	Description
	difficulties. For this reason in this category, alternative D is rated higher than alternative E.
Environmental quality	Environmental quality control is important for health and aesthetic reasons, which in turn directly affect public opinion and concern. Items of concern here are odors, effluent quality, and surface and subsurface pollution. Trickling filters, in general, provide inferior effluent quality and are also associated with odor problems. Therefore, alternative D is rated lower in this category.
Power requirement	Energy needed to operate a given system are included in power requirements. Low power requirements will reflect favorably in the cost analysis, but since power is a limited resource, low power requirements also contain dimensions which relate purely to effectiveness. Power is needed for pumping and treatment equipment. Trickling filters require lesser power than the activated sludge process to operate; however, the pumping head is greater at a trickling filter plant. Alternative D utilizes two pumping stations as compared with one pumphouse in alternative E. Aerobic sludge digestion for a portion of sludge in alternative D would require higher energy demand and will not yield methane. For these reasons, it is expected that alternative E will have less energy demand.
Sludge handling and disposal	Sludge from wastewater treatment facilities can cause serious odor problems if not handled properly. Alternative D provides sludge drying beds at the trickling filter plant and therefore may cause odor problems. Therefore, alternative E is rated higher in this category.

cost effectiveness evaluation, the alternative E is more cost-effective and has been selected for implementation.

6-9   FINANCIAL STATUS

The city has the legal, institutional, managerial, and financial capability to ensure adequate funds for construction, operation, and maintenance of the project proposed in this facility plan. The city is currently charging a fee on the users of water and sewer service. The sewer service rates are based on sewage flow and strength. The financial and

TABLE 6-15   Matrix Evaluation of Wastewater Treatment Alternatives

Measure of Effectiveness (1)	Assigned Weights to Each Measure of Effectiveness (2)	Alternative D		Alternative E	
		Score (3)	Weighted Score (4)	Score (5)	Weighted Score (6)
Plant location	3.0[a]	2.5[b]	7.5[c]	4.0[b]	12.0[d]
System compatability	2.0	2.5	5.0	4.0	8.0
Operational flexibility and reliability	1.0	3.0	3.0	2.0	2.0
Environmental quality	3.0[a]	3.0	9.0	4.0	12.0
Power requirements	2.5	3.0	7.5	3.0	7.5
Sludge handling	2.5	3.0	7.5	4.0	10.0
Total			39.5		51.5

[a]Plant location and environmental quality have been assigned a weight of 3.0, indicating that these measures of effectiveness are extremely important.
[b]Alternatives D and E, when compared for plant location, received scores of 2.5 and 4.0, respectively, indicating that alternative E ranks much higher in this measure of effectiveness than alternative D.
[c]Product of values in columns (2) and (3).
[d]Product of values in columns (2) and (5).

statistical information about the city is normally provided in an attachment to the facility plan (see Sec. 6-15).

6-10   DESCRIPTION OF SELECTED WASTEWATER TREATMENT FACILITY

6-10-1 Design Considerations

The schematic flow diagram of the selected wastewater treatment facility is shown in Figure 6-4. A summary of the design criteria and unit sizes of various collection and treatment systems is an essential part of a facility plan. Such information is given in Chapter 23. Design analysis and summary of design computations are normally submitted as an appendix to a facility plan (see Sec. 6-15).

6-10-2 Operation and Maintenance

The treatment facility will be operated by trained licensed operators on a 7-d-per-week basis. Supporting staff, including mechanics and electricians on a 24-h basis, will be provided. A laboratory facility for quality control, an administration building, and equipment maintenance shops will also be provided.

Each treatment unit will have an outlet for chlorine solution to wash the walls and

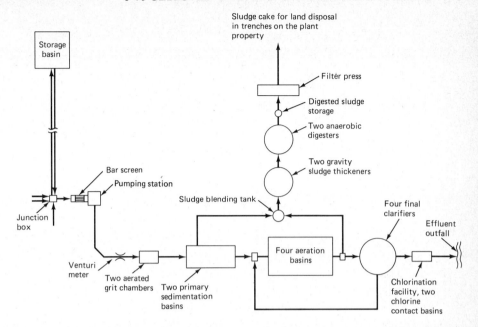

Figure 6-4   Schematic Flow Diagram of the Proposed Wastewater Treatment Plant.

weirs and remove the scum. General discussion on operation and maintenance requirements for the treatment facility is provided in subsequent chapters along with the detailed
design and equipment specifications for each unit.

Control of Bypassing.   The power supply in the area is very reliable. Interruption
records for 8 years at the treatment plants revealed four power outages ranging in durations
from 10 min to 1 h. Prolonged power failure is unlikely. Auxiliary power generators and
multiple electrical primary power lines from different sources are considered unnecessary. In order to protect against overflow and subsequent stream pollution, the existing
aerated lagoons will be used as storage ponds. An overflow structure shall be constructed
to direct raw wastewater into the lagoons for storage. Existing lagoon volume will provide
storage for approximately 35 h. The raw wastewater stored in the lagoon will be routed
through the plant after restoration of power.

All treatment processes have been provided with multiple units, and pumps and
blowers have sufficient standby units in case equipment breakdown occurs.

The proposed treatment facility will not tie into any existing system until the
completion of the interceptors, pumping station, and treatment plant. After construction
of the entire facility, the necessary connections will be made to the existing sewerage
system to divert the wastewater to the new facility.

Flood Hazard Analysis.   All treatment units except bar screen and pumping station will
be constructed above ground level. A levee will be constructed around the entire plant area

to at least 0.5 m above the 100-yr flood for protection against flooding. Provisions will be made for draining the area surrounded by the levees.

Users Charge.   Users charge schedule and industrial cost recovery system shall be implemented in accordance with the federal and state guidelines.

6-11   ENVIRONMENTAL IMPACT ASSESSMENT

An environmental impact assessment report of the project is normally attached as an appendix to a facility plan. Only basic information pertinent to the environmental impact assessment is presented in the main text of the facility plan.

The proposed treatment facilities are expected to reduce health hazard, enhance surface water quality in the receiving stream, eliminate groundwater pollution potential, and aid in orderly development for the area served. It is expected that the proposed project will not have any significant primary adverse impacts on the following areas of the environmental concern:

- Plant and animal communities
- Ecosystem
- Endangered or locally threatened species
- Unique or vulnerable environmental features
- Unique archeological, historic, scientific, or cultural areas, parks, wetlands, or stream corridors
- Community growth pattern and land use pattern
- Air quality
- Aesthetics

The city owns a large tract of land bounded by two streams (Fig. 6-2). Approximately 125 hectare will be allocated to the plant property. No relocation of people or existing structures will be required for the project. Also, no change in the value of the adjacent property is anticipated as a result of this project.

The minor adverse effects which cannot be avoided are those normally associated with the existence and operation of wastewater treatment facilities. The increased noise levels and possible occasional odors emanating from the facility will be minimized by modern design techniques and efficient plant operation. Some degree of disruption of the environment and inconveniences to citizens during construction of interceptors is unavoidable but will be reduced in severity by proper construction scheduling and techniques.

Fuels and other forms of energy will be utilized during construction and operation of facilities. Construction materials will be used for the project and remain permanently. Chemicals such as chlorine, lime, ferric chloride, and organic polymers will be needed for the normal plant operation. The land for treatment facilities will be permanently lost to agricultural production. However, the productivity of the area in current use as an aerated lagoon is minimal.

6-12 PUBLIC PARTICIPATION PROGRAM

During the entire facility-planning process, an active public participation program was maintained. The program utilized public meetings, workshops, mailouts, and press releases to keep the public aware of the progress and to receive their comments before any decisions were made. Various appointed environmental advisory committees held the public meetings to review the work. Notices of each meeting were mailed to the interested citizens and were published in the newspapers at least 45 days prior to each meeting date. Documents considered at that meeting were placed for public review at the library, city hall, and the chamber of commerce at least 30 days before the meeting.[22]

The entire facility-planning process lasted for 57 weeks as was originally proposed during the initial planning. The public participation work program that was conducted in relation to the project decision points is summarized in Table 6-16.

The final meeting will be a public hearing, and all public comments will be included in the future application to conduct the engineering and design of the improvements committed in the facility plan.

6-13 COST BREAKDOWN

The cost of performing facility planning and estimated cost for preparation of design plans and specifications are given in Tables 6-17 and 6-18. The cost estimates for preparation of plans and specifications are developed using EPA Region VI guide.[23] It is estimated that 100 drawings will be needed for detailing the wastewater treatment facility and 30 drawings will be needed for the design and details of diversion sewers. The estimated construction costs of the proposed project are summarized in Table 6-19.

6-14 TIME SCHEDULE FOR PROJECT MILESTONES

The project milestones for preparing design plans and specifications and construction programs are summarized in Tables 6-20 and 6-21. The scheduling of starting and completion periods given in these tables are only tentative.

6-15 ATTACHMENTS

The following information is normally submitted as separate attachments or appendixes to a facility plan.

1. "Letter of clearance" on project review in conformance with OMB circular from the clearinghouse
2. Infiltration/inflow report
3. Design analysis and summary of design computations
4. Environmental impact assessment
5. Effluent discharge permits (NPDES permit)
6. Correspondence from agencies

TABLE 6-16 Public Participation Work Program Conducted During Facility Planning Process

Decision Point	Tasks	Time Schedule in Weeks since Start of Facility Planning	Target Audience
Preparation of public participation program since the start of facility planning	Select engineer, designate public liaison; develop mailing list, public participation work program, and fact sheet; and designate depository for key documents (public library, chamber of commerce, city hall).	1–6	
	Distribute public participation work program and fact sheet to media, depository, and mailing list.	7–8	General public
Assessment of present and future conditions	Send notice of proposed public meeting (press releases and mailout), 45 days prior to meeting.	11–12	General public
	Place relevant documents in designated depository for public review, 30 days before meeting.	13–14	General public
	Contact planning staff and chairpersons of various committees for their advice on key issues, water quality problems, and public participation.	16–17	Planning staff and chairpersons of various committees
	Conduct the workshop with the city council and the environmental advisory committees, and interested citizens (representing neighborhood organizations) to explain major water quality problems, planning process, problems and issues, and community goals and objectives.	18–22	General public
	Publish news releases in local news papers to describe local water quality problems, objectives, goals, issues, and public participation program.	24–25	General public
	Briefings at various local group meetings.	26–28	Local citizen groups

	Prepare agency responsiveness on public comments and place at designated depository.	32–37	EPA and state
Development of alternatives	Prepare the fact sheets describing alternatives, environmental impacts, and cost of each alternative. Distribute fact sheets and notice of upcoming public meeting 45 days prior to the meeting (use the news media and mailing list).	37–39	General public
	Place relevant documents in designated depository for public review 30 days before public meeting.	40–41	General public
	Informal public meeting of various environmental advisory committees to discuss alternatives, answer questions, and identify areas of further investigation.	44–45	General public
Development and submission of final plan to the city	Develop fact sheet to highlight major elements of proposed plan, distribute fact sheet to all on mailing list, and send notice of public hearing in local newspaper and send to all on mailing list (45 days prior to meeting).	48–50	General public
	Place relevant documents in designated depository for public review 30 days before public hearing.	51–52	General public
	Conduct joint meeting of environmental review committees to present final plans and draft EIS.	54–55	General public, key citizens and officials
	Conduct public hearing.	55	General public
	Prepare final responsiveness summary outlining grantees response and evaluation of public hearing to citizen input. Place on file at designated depository.	56	EPA and state
	Publish notice of final decision with press releases and mailout.	57	General public

TABLE 6-17 Cost of Preparing Facility Plan

Item	Description	Estimated Cost (Dollars)
1	Infiltration/inflow analysis	18,000
2	Environmental impact assessment	6,000
3	Facility plans	60,000
4	Subtotal	84,000
5	Legal and administrative costs[a]	1,000
6	Contingencies[b]	4,000
7	Total estimated cost of facility planning	89,000

Total allowance for preparation of facility plan and cost to the city should be determined from the federal and state guidelines.

[a]Legal and administrative costs 1 percent of subtotal.
[b]Contingencies 5 percent of subtotal.

TABLE 6-18 Estimated Cost for Preparation of Design Plans and Specifications

Item	Description	Estimated Cost (Dollars)
1	Proposed diversion sewers	66,000
2	Proposed wastewater treatment plant	158,000
3	Subtotal	224,000
4	Legal and administration[a]	2,240
5	Contingencies[b]	11,200
6	Total estimated cost of preparation of design plans and specifications	237,440

Total allowance for preparation of design plans and specifications and cost to the city should be determined from the federal and state guidelines.

[a]Legal and administrative costs 1 percent of subtotal.
[b]Contingencies 5 percent of subtotal.

TABLE 6-19 Estimated Construction Cost of the Project

Item	Description	Estimated Cost (Dollars)[a]
1	Proposed diversion sewers	3,621,000
2	Treatment plant including pumping station	18,981,000
3	Total estimated construction cost	22,602,000

Total estimated eligible grant amount and total estimated cost to the city should be determined from the federal and state guidelines.

[a]Construction cost of the project must be updated prior to initiating the construction phase of the project.

7. The review comments or approval of relevant state and local agencies indicating compatibility with state and local plans
8. Notice of public hearings, including a brief summary of the public hearings held to consider the proposed project
10. A statement demonstrating that the applicant has necessary legal, technical, financial, institutional, and managerial resources to ensure construction, operation, and maintenance of the proposed treatment works
11. Statement specifying that the requirements on Title VI of the Civil Rights Act of 1964 have been satisfied within the facility plan

6-16 PROBLEMS AND DISCUSSION TOPICS

6-1 From the 1970 and 1980 population data given in Table 6-2, estimate the population for the initial year (1985). Use geometric rate of increase.

6-2 A service area has a total population of 24,000 residents. The total length of the sewerage system is 15 m per resident. The typical length distribution of average sewer sizes are 20 cm, 70 percent; 38 cm, 15 percent; 50 cm, 9 percent; and 60 cm, 6 percent. Calculate the equivalent pipe diameter. If infiltration/inflow is 1400 liters per d per cm per km, express the infiltration/inflow in ℓpcd.

6-3 If the service area in problem 2 is 1600 ha, calculate infiltration/inflow contribution in m^3 per ha per d.

6-4 The existing wastewater treatment facilities for the Modeltown are described in Table 6-4. These include stabilization ponds, trickling filter plant, and an aerated lagoon. Using the dimensions for these facilities and flow and wastewater characteristics given in Tables 6-4 and 6-6, calculate the following:
  (a) Detention time and organic loadings to the stabilization ponds. Assume inner side slope 2 horizontal to 1 vertical.
  (b) Calculate hydraulic and organic loadings to the trickling filters.
  (c) Calculate aeration period and organic loading to the aerated lagoon. Assume side slope 4 horizontal to 1 vertical. Also calculate the oxygen transferred by the floating aerators in kg oxygen per h per kW.

TABLE 6-20 Schedule of Starting and Completion of Design Plans and Specifications

Project Schedule	Estimated Action Time (d)
Initiated after authorization to proceed with design plans and specifications	30
Completed	200
Submitted to the state agency	15
Total	245

TABLE 6-21 Schedule of Starting and Completion of Construction

Project Schedule	Estimated Action Time (d)
Advertise for construction bids within the days indicated after authorization to advertise for bids.	45
Submit bid opening documents within the days indicated after the bid opening date.	30
Award construction contract within the days indicated after authorization to award the contract.	15
Initiation of construction (within the days indicated) after the award of the contract.	30
Complete construction within the days indicated after initiation of construction.	520
Total	640

6-5 Alternatives B and C in Sec. 6-8-1 have not been evaluated in the cost effectiveness analysis. Develop costs of alternatives B and C similar to that of alternatives D and E. Perform cost effectiveness evaluation of all four alternatives (B, C, D, and E). Use six measures of effectiveness summarized in Table 6-15.

6-6 Visit your local wastewater treatment facility. Determine which year the facility was expanded. Obtain a copy of the facility plan. Review the facility plan and the accompanying attachments (see Sec. 6-15). Identify the following:

    (a) How much infiltration/inflow was reaching the plant? Was it excessive? Was rehabilitation survey needed?

    (b) How many alternatives were evaluated? Draw flow schemes of each alternative.

(c) In your opinion, is the selected alternative most cost-effective?

(d) Was an environmental impact report or statement (EIR or EIS) prepared? Were the major environmental factors addressed? Have some controversial issues arisen since the completion of the construction?

REFERENCES

1. 40 CFR Part 131, "Preparation of Water Quality Management Plans," *Federal Register,* Vol. 39, No. 107, June 3, 1974, p. 19639–19644.
2. U.S. Environmental Protection Agency, *Areawide Assessment Procedures Manual,* Volume III, Municipal Environmental Research Laboratory, Cincinnati, Ohio, EPA-600/9-76-014, July 1976.
3. 40 CFR Part 35, Subpart E, "Construction Grants Program," *Federal Register,* Vol. 43, No. 188, September 27, 1978, p. 44022–44097.
4. U.S. Environmental Protection Agency, *Guide for Sewer System Evaluation,* Office of Water Program Operations, Washington, D.C., March 1974.
5. U.S. Environmental Protection Agency, *Program Requirement Memorandum 78-10,* March 17, 1978, published by the Bureau of National Affairs, Washington, D.C., p. 45.
6. Joint Committee of the Water Pollution Control Federation and the American Society of Civil Engineers, *Design and Construction of Sanitary and Storm Sewers,* Water Pollution Control Federation, Washington, D.C., 1970.
7. Fair, G. M., and J. C. Geyer, *Elements of Water Supply and Waste-Water Disposal,* John Wiley & Sons, New York, 1965.
8. 40 CFR Part 257, "Criteria for Classification of Solid Waste Disposal Facilities," *Federal Register,* Vol. 44, No. 179, Sept. 13, 1979.
9. Drobny, N. L., S. R. Qasim, and B. W. Valentine, "Cost-Effectiveness Analysis of Waste Management Systems," *Journal of Environmental Systems,* Vol. 1, No. 2, June 1971, p. 189–210.
10. U.S. Environmental Protection Agency, *Areawide Assessment Procedures Manual: Performance and Cost,* Appendix II, Municipal Environmental Research Laboratory, Cincinnati, Ohio, EPA 600/9-76-014, July 1976.
11. U.S. Environmental Protection Agency, *A Guide to Cost-Effectiveness Wastewater Treatment Systems,* EPA-430-/9-75-002, Office of Water Program Operations, Washingtion, D.C. July 1979.
12. U.S. Environmental Protection Agency, *Construction Costs for Municipal Wastewater Treatment Plants: 1973–1977,* EPA MCD-37, Office of Water Program Operations, Washington, D.C., January 1979.
13. U.S. Environmental Protection Agency, *Analysis of Operations and Maintenance Costs for Municipal Wastewater Treatment Systems,* MCD-39, Office of Water Program Operations, Washington, D.C., May 1978.
14. Eilers, R. G., and Robert Smith, *Wastewater Treatment Plant Cost: Estimating Program Documentation,* Environmental Protection Agency, Distributed by NTIS PB-222 762, U.S. Department of Commerce, Springfield, Va., March 1973.
15. Male, J. W., and S. P. Graef, *Applications of Computer Programs in the Preliminary Design of Wastewater Treatment Facilities,* Short Course Proceedings Section I and II, EPA-600/2-78-185, EPA Cincinnati, Ohio, September 1978.
16. U.S. Army Corps of Engineers and Environmental Protection Agency, *Computer Assisted Procedure for the Design and Evaluation of Wastewater Treatment Facilities (CAPDET),* Waterway Experiment Station, Vicksburg, Mi., 1979.

17. *Engineering News Record,* January 13, 1983.
18. U.S. Environmental Protection Agency, *Guidance for Preparing a Facility Plan,* EPA-430/ 9-76-015, Construction Grants Program Requirements, Washington, D.C., May 1975.
19. Grant, E. L., and W. G. Jresm, *Principles of Engineering Economy,* Fifth Edition, Ronald Press, New York, 1970.
20. James, L. D., and R. Lee, *Economics of Water Resources,* McGraw-Hill Book Co., New York, 1971.
21. Environmental Protection Agency, *Program Requirements Memorandum 80-1,* Bureau of National Affairs, Washington, D.C., November 26, 1979, p. 11.
22. U.S. Environmental Protection Agency, *Municipal Wastewater Management, Citizen's Guide to Facility Planning,* Office of Water Program Operations (WH-547), Washington, D.C., 66A-430/9-79-006, FRD-6, February 1979.
23. U.S. Environmental Protection Agency, *A Technique for Estimating Costs and Analyzing Fees for the Design of Municipal Wastewater Treatment Works,* Technical Report, EPA Region VI.

7

Design of Intercepting Sewers

7-1 INTRODUCTION

Sewers are underground conduits for conveying wastewater to the treatment facilities or to the point of disposal. There are three types of sewers: sanitary, storm, and combined. Sanitary sewers are designed to carry wastewater from residential, commercial, and industrial areas, and a certain amount of infiltration/inflow that may enter the system due to deteriorated conditions of the sewers and manholes. Storm sewers are exclusively designed to carry the surface runoff. Combined sewers are designed to carry both the sanitary and storm flows. Combined sewers are undesirable as heavy precipitation may cause flows far exceeding the designed capacity of the treatment facilities. These days, combined sewers are seldom designed and constructed in the United States.

The characteristics of municipal wastewater and volumes of infiltration/inflow have been discussed extensively in Chapters 3 and 6. In this chapter an overview of the design and construction procedures of sanitary sewers is given. Furthermore, design examples covering step-by-step calculations, sewer profiles, construction details, and operation and maintenance of five sewer lines are also discussed in this chapter. Flow conditions in these lines were developed in the Model Facility Plan (Chapter 6). These sewer lines are (1) diversion sewer from the stabilization pond area, (2) diversion sewer from the trickling filter plant area, (3) existing intercepting sewer from the central part of the town, (4) final sewer line carrying the total flow from the junction box to the bar screen, and (5) the bypass sewer to direct the flow from the junction box to the storage basin during power outages when the pumping equipment is temporarily inoperable.

7-2 COMMUNITY SEWERAGE SYSTEM

The community sewerage system consists of (1) building sewers (also called building connections), (2) laterals or branch sewers, (3) main and submain sewers, (4) trunk sewers, and (5) intercepting sewers.

Building sewers connect the building plumbing to the laterals or to any other sewer lines mentioned above. Laterals or branch sewers convey the wastewater to the main sewers. Several main sewers connect to the trunk sewers that convey the wastewater to large intercepting sewers or to the treatment plant. A typical layout of a municipal sewer system is shown in Figure 7-1.

A sewers except the building sewers are constructed in streets, special easements, or available rights-of-way. They follow the natural ground elevations. The wastewater treatment facility is generally located near the low outskirts of the community if other conditions are acceptable. A tradeoff between the construction cost of laying deeper lines

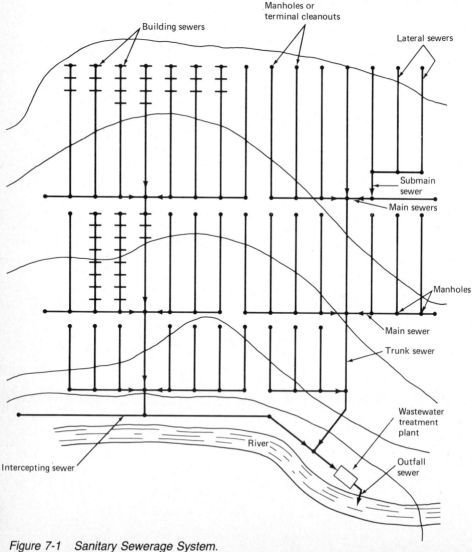

Figure 7-1   Sanitary Sewerage System.

and the cost of lift stations and pumping costs must be established in determining the cost-effective sewerage system.

The diameter of a sewer line is generally determined from the peak flow that the line must carry and the local sewer regulations concerning the minimum sizes of the laterals and building connections. On the other hand, the total length of the sewer lines depends on the layout of the community. Based on the data for 97 cities in 21 states of the United States, the total average length of the sewerage system was 4.3–11.5 m per capita (14–38 ft per capita). The typical distribution of pipe sizes is summarized in Table 7-1.[1,2]

Pressure or vacuum sanitary sewers are also designed these days where gravity lines may not be economically feasible. Examples of such situations are lake communities, rocky terrain, and areas where groundwater table is high. Detailed discussion on pressure and vacuum sewers may be found in Refs. 3–6.

7-3 DESIGN AND CONSTRUCTION OF COMMUNITY SEWERAGE SYSTEM

Designing a community sewerage system is not a simple task. It requires considerable experience and a great deal of information to make proper decisions concerning the layout, sizing, and construction of a sewer network that is efficient and cost-effective. The design engineer needs to generally undertake the following tasks:

1. Define the service area
2. Conduct preliminary investigations
3. Develop preliminary layout plan and profile
4. Review design and construction considerations
5. Conduct field investigations and complete design and final profiles
6. Prepare contract drawings and specifications

Each of these tasks is discussed below.

TABLE 7-1 Typical Distribution of Sewer Sizes

Sewer Diameter (cm (in.))	Distribution (Percent)
10, 12.5, 15 (4, 5, 6)	3.6
20 (8)	73.1
25, 30, 35, 37.5, 40 (10, 12, 14, 15, 16)	13.4
45, 50, 53 (18, 20, 21)	3.8
60 (24)	1.7
67.5 (27) and above	4.4
Total	100.0

Source: Adapted in part from Ref. 2.

7-3-1 Service Area

Service area or sewer district is defined as the total area that will eventually be served by the sewerage system. The service area may be based on natural drainage or political boundaries, or both. It is generally a part of the areawide waste management plan. Additional discussion on the service area can be found in Chapter 2.

7-3-2 Preliminary Investigations

The design engineer must conduct the preliminary investigations to develop a layout plan of the sewerage system. Site visits and contacts with the city and local planning agencies and state officials should be made to determine the land use plans, zoning regulations, and probable future changes that may affect both the developed and undeveloped land. Data must be developed on topography, geology, hydrology, climate, ecological elements, and social and economic conditions. Topographic maps with existing and proposed streets and other utility lines provide the most important information for preliminary flow routing. for preliminary flow routing.

    If reliable topographic maps are not available, field investigations must be conducted to prepare the contours, place bench marks, locate buildings, streets, utility lines, drainage ditches, low and high areas, streams, and the like. All these factors influence the sewer layout.

7-3-3 Layout Plan

Proper sewer layout plan and profiles must be completed before design flows can be established. The following is a list of basic rules that must be followed in developing a sewer layout plan and profile:

1. Select the site for the wastewater treatment plant. Basic considerations for site selection of wastewater treatment plants have been presented in Chapter 2. For gravity system, the best site is generally the lowest elevation of the entire drainage area.
2. The preliminary layout of sewers is made from the topographic maps. In general, sewers are located on streets, or on available right-of-way; and sloped in the same direction as the slope of the natural ground surface.
3. The trunk sewers are commonly located in valleys. Each line is started from the intercepting sewer and extended uphill until the edge of the drainage area is reached, and further extension is not possible without working downhill.
4. Main sewers are started from the trunk line and extended uphill intercepting the laterals.
5. All laterals or branch lines are located in the same manner as the main sewers. Building sewers are directly connected to the laterals.
6. Preliminary layout and routing of sewage flow is done by considering several feasible alternatives. In each alternative, factors such as total length of sewers; and cost of construction of laying deeper lines versus cost of construction, operation, and maintenance of lift station, should be evaluated to arrive at a cost-effective sewerage system.

7. Sewers should not be located near water mains. State and local regulations must be consulted for appropriate separation distance between the sewers and water lines.
8. After the preliminary sewer layout plan is prepared, the street profiles are drawn. These profiles should show the street elevations, existing sewer lines, and manholes. These profiles are used to design the proposed lines.

Finally, these layout plans and profiles are revised after the field investigations and sewer designs are complete. A typical sewer profile is shown in Figure 7-2.[7]

7-3-4 Design and Construction Considerations

Many design and construction factors must be investigated before sewer design can be completed. Factors such as design period; peak, average, and minimum flows; sewer slopes and minimum velocities; design equations; sewer material; joints and connections, appurtenances, and sewer installation; etc., are all important in developing sewer design. Many of the factors are briefly discussed below.

Design Period.   Design period should be based on ultimate tributary population. It is not uncommon to design sewers for a design period of 25–50 years or more.

Design Population.   Population projections must be made for the population at the end of the design year. Discussion on population projection techniques can be found in Chapter 2.

Design Flow.   Sanitary sewers should be designed to carry peak residential, commercial, and industrial flows, and the normal infiltration and inflow where unfavorable conditions exist.[8] Methods of estimating peak, average, and minimum flows and allowance for infiltration/inflow have been presented in Chapters 3 and 6. Laterals and submain sewers are designed for 1500 ℓ per capita per day (400 gpcd); main and trunk sewers for 1330 ℓ per capita per day (350 gpcd), and intercepting sewers for 1000 ℓ per capita per day (260 gpcd). To all these values, industrial contributions and allowance for infiltration/ inflow should also be made.

Minimum Size.   Minimum sewer size recommended by Ten States Standards is 20 cm (8 in.).[9] Many states allow 15-cm (6-in.) lateral sewers.

Velocity.   In sanitary sewers, solids tend to settle under low-velocity conditions. Self-cleaning velocities must be developed regularly to flush out the solids. Most states specify minimum velocity in the sewers under low flow conditions. A good practice is to maintain velocity above 0.3 m/s (1 ft/s) under low-flow conditions. Under peak dry weather condition, the lines must attain velocity greater than 0.6 m/s (2 fps). This way the lines will be flushed out at least once or twice a day. In depressed sewers (inverted siphons) self-cleaning velocities of 1.0 m/s (3.0 fps) must be developed to prevent accumulation of solids.

Velocities higher than 3.0 m/s (10 fps) should be avoided as erosion, and damage may occur to the sewers or manholes.

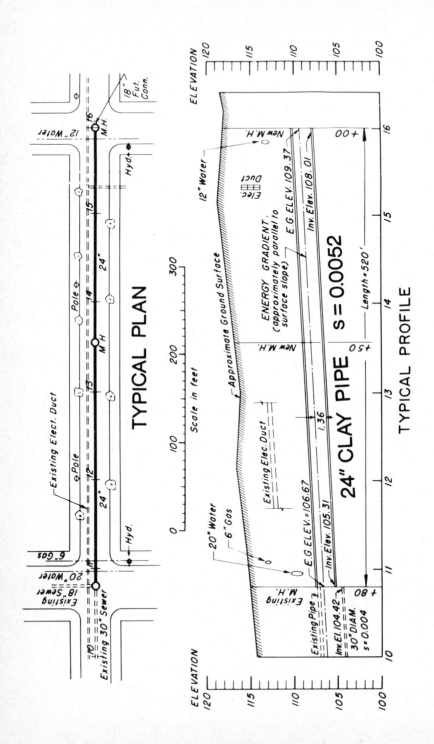

Figure 7-2   Typical Plan and Profile Drawings for Sanitary Sewer. 1 ft = .3048m. (Courtesy National Clay Pipe Institute.)

Slope. Flat sewer slopes encourage solids deposition and production of hydrogen sulfide and methane. Hydrogen sulfide gas is odorous and causes serious pipe corrosion.[10,11] Methane gas has caused explosions. Most states specify minimum slope for the sanitary sewers. These minimum slopes are such that a velocity of 0.6 m/s (2 fps) is reached when flowing full and $n = 0.013$. If sewer slopes of less than the recommended values are provided, the state agencies may require depth and velocity computations at minimum, average, and peak flow conditions. Minimum sewer slopes for different diameter lines are summarized in Table 7-2.[3,9]

Depth. The depth of sewers is generally 1–2 m below the ground surface. Depth depends on the water table, lowest point to be served (ground floor or basement), topography, and the freeze depth.

Appurtenances. Sewer appurtenances include manholes, building connections, junction chambers or boxes, terminal cleanouts, and others. These are discussed briefly below. Discussion of sewer appurtenances can be obtained in Refs. 1, 4, 7, and 11.

Manholes. Manholes for small sewers are generally 1.2 m (4 ft) in diameter. For larger sewers (exceeding 60 cm, or 24 in. in diameter) larger manhole bases are provided, although 1.2 m barrel may still be used. Manholes should be of durable structure, provide

TABLE 7-2   Minimum Recommended Slopes of Sanitary Sewer

Diameter		Slope
in.	mm	(m/m)
6	150	0.0060
8	200	0.0040
10	250	0.0028
12	310	0.0022
14	360	0.0017
15	380	0.0015
16	410	0.0014
18	460	0.0012
21	530	0.0010
24	610	0.0008
27	690	0.00067
30	760	0.00058
36	910	0.00046
42	1050	0.00038
48	1200	0.00032
54	1370	0.00026

Source: Adapted in part from Refs. 3 and 9.

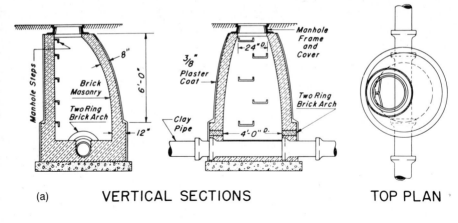

(a) VERTICAL SECTIONS   TOP PLAN

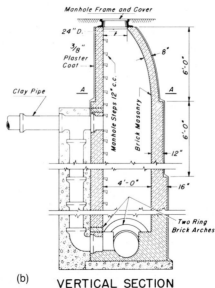

(b) VERTICAL SECTION

Figure 7-3 Typical Designs of Manholes. 1 ft = 0.3048 m. (a) Typical line manhole. (b) Drop manhole details. (Courtesy National Clay Pipe Institute.)

easy access to the sewers for maintenance, and cause minimum interference to the sewage flow. Manholes should be located at the end of the line (called "terminal cleanout"), at the intersection of sewers, and at changes in grade and alignment except in curved sewers.* The maximum spacing of manholes is 90–180 m (300–600 ft) depending on the size of sewer and available size of sewer-cleaning equipment. Manholes, however, should not be located in low places where surface water may enter. If such locations are unavoidable, special water tight manhole covers should be provided. Typical manhole details are shown in Figure 7-3. Details of a terminal cleanout are shown in Figure 7-4.

*Curved sewers are commonly used on curved streets. They eliminate manholes and construction is economical. Minimum radius is 30 m. Modern equipment make cleaning of curved lines possible.

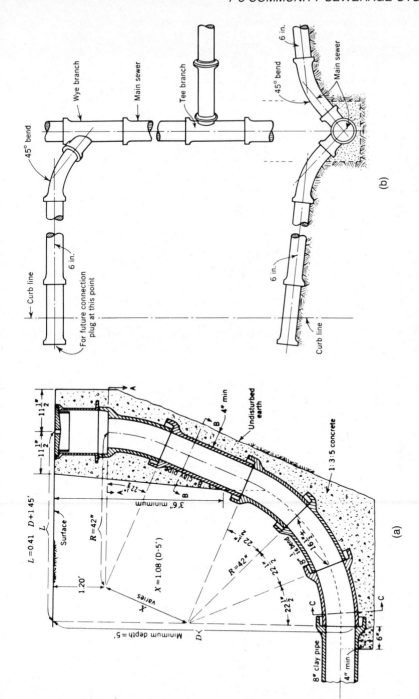

Figure 7-4 Typical Terminal Cleanouts and House Connection. 1 ft = 0.3048 m. (a) Terminal cleanout. (Courtesy National Clay Pipe Institute.) (b) Service connections for shallow sewer. [From Ref. 11. Used with permission of Water Pollution Control Federation and American Society of Civil Engineers.]

Drop Manholes.   The purpose of a drop manhole is to reduce the turbulence in the manhole if the elevation difference between incoming and outflow sewers is greater than 0.5 m (1.5 ft). The turbulence due to sudden drop of wastewater may cause splashing, release of gases that are odorous, and can damage the manhole. Details of drop manholes are shown in Figure 7-3.

Building Connections.   The building sewers are generally 10–15 cm (4–6 in.) in diameter and constructed on a slope of 0.02 m/m. Building connections are also called house connections, service connections, or service laterals. Service connections are generally provided in the municipal sewers during construction. While the sewer line is under construction, the connections are conveniently located in the form of wyes or tees, and plugged tightly until service connections are made. In deep sewers, a vertical pipe encased in concrete (called a chimney) is provided for house connections. Details of house connections are given in Figure 7-4.

Design Equations and Procedure

Design Equations.   Sanitary sewers are mostly designed to flow partially full. Once the peak, average, and minimum flow estimates (including allowances for infiltration/inflow) are made and general layout and topographic features for each line are established, the design engineer begins to size the sewers. Design equations proposed by Manning, Chezy, Gangrullet, Kutter, and Scobey have been used for designing sewers and drains.[1,8,12] The Manning equation, however, has received most widespread application. This equation in various forms is expressed below:

$$V = \frac{1}{n} R^{2/3}S^{1/2}. \qquad \text{(SI unit)} \tag{7-1}$$

$$Q = \frac{0.312}{n} D^{8/3}S^{1/2} \quad \text{(circular pipe flowing full, SI unit)} \tag{7-2}$$

$$V = \frac{1.486}{n} R^{2/3}S^{1/2} \quad \text{(U.S. customary unit)} \tag{7-3}$$

$$Q = \frac{0.464}{n} D^{8/3}S^{1/2} \quad \text{(circular pipe flowing full, U.S. customary unit)} \tag{7-4}$$

where

$V =$ velocity for sewer flowing full, m/s (ft/s)
$R =$ hydraulic mean radius = area/wetted perimeter (circular pipe flowing full, $R = D/4$), m (ft)
$D =$ diameter, m (ft)
$Q =$ flow, m³/s (ft³/s)
$S =$ slope of energy grade line, or invert slope, m/m (ft/ft)
$n =$ coefficient of roughness

Coefficient of roughness depends on the material and age of the conduit. Commonly used values of n for different materials are given in Table 7-3.

TABLE 7-3 Common Values of Roughness
Coefficient Used in the Manning Equation

Material	Commonly Used Values of n
Concrete	0.013 and 0.015
Vitrified clay	0.013 and 0.015
Cast iron	0.013 and 0.015
Brick	0.015 and 0.017
Corrugated metal pipe	0.022 and 0.025
Asbestos cement	0.013 and 0.015
Earthen channels	0.025 and 0.003

Various types of nomographs have been developed for solution of problems involving sewers flowing full. Nomographs based on Manning's equation for circular pipe flowing full and variable n values are provided in Figure 7-5.[7,11-14] Hydraulic elements of circular pipes under part-full flow conditions are provided in Figure 7-6. It may be noted that the value of n decreases with the depth of flow (Figure 7-6). However, in most designs n is assumed constant for all flow depths. Also, it is a common practice to use d, v, and q (lowercase) notations for depth of flow, velocity, and discharge under partial flow condition while D, V, Q (uppercase) notations for diameter, velocity, and discharge for sewer flowing full. Use of Eqs. (7-1) and (7-2) and Figures 7-5 and 7-6 are shown in the Design Example.

Organization of Computation. After the preliminary sewer layout plan and profile are prepared, the design computations are accomplished. Design computations for sewers are repetitious and therefore are best performed in a tabular format. Most textbooks provide computational tables that can be adapted for desired application. Computational procedures may be found in Refs. 1, 4, 11, 13, and 14.

Construction Materials

Sewers. Sewers are made from concrete, reinforced concrete, vitrified clay, asbestos-cement, brick masonry, cast iron, ductile iron, corrugated steel, sheet steel, and plastic. Important factors in selection of sewer material include the following:[11]

1. Chemical characteristics of wastewater and degree of resistance to corrosion against acid, base, gases, solvents, etc.
2. Resistance to scour and flow (friction coefficient)
3. External forces and internal pressures
4. Soil conditions
5. Type of backfill and bedding material to be used
6. Useful life
7. Strength and water tightness of joints required, and effective control of infiltration and inflow

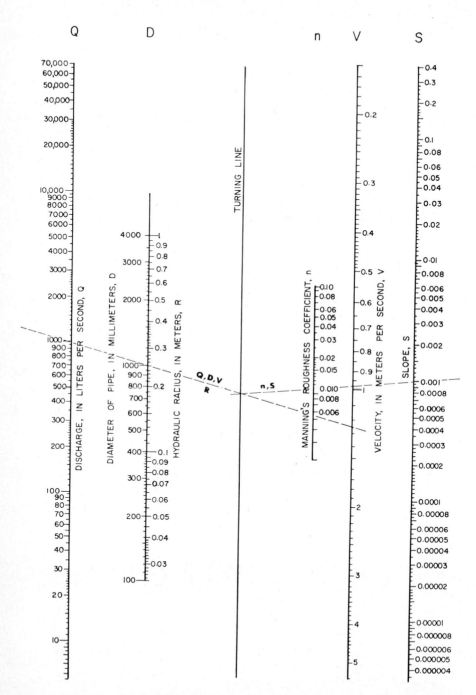

Figure 7-5   Diagram for Solution of the Manning Formula. [From Practical Hydraulics for the Public Works Engineer. Used with Permission of Public Works Journal Corporation Magazine.]

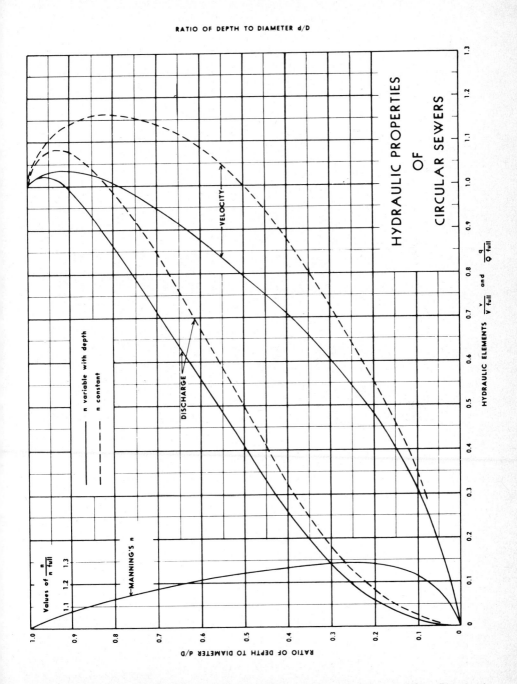

Figure 7-6   Hydraulic Properties of Circular Sewers. (Courtesy National Clay Pipe Institute.)

8. Availability in diameter, length, and ease of installation
9. Cost of construction and maintenance

Basic properties of sewer pipes of different materials are summarized in Table 7-4.

Manholes.   Manholes are generally constructed from (1) brick masonry, (2) precast concrete, (3) cast-in-place concrete, (4) concrete block, and (5) fiberglass. Manhole covers are made of cast iron and should be tightly fitted to eliminate inflow, uncovering, and rattling. They should also have provision to vent gases and admit air. Manhole steps should be placed at intervals of 30–50 cm (12–16 in.) and be designed to prevent corrosion and slipping.

Joints and Infiltration.   The method of making joints should be fully covered in the specifications. Joints should be designed to make sewers water-tight, root-resistant, flexible, and durable. A leakage test should be specified. The leakage shall not exceed 0.5 m^3 per d per cm of pipe diameter per km (500 gpd per in. of pipe diameter per mile).[11] It has been experimentally demonstrated that joints made from rubber gasket and hot-poured bituminous material produced almost no infiltration, whereas cement mortar joints cause excessive infiltration.[15]

Sewer Construction.   Sewer construction involves excavation, sheeting and bracing of trenches, pipe installation, and backfilling. Each of these construction steps is discussed briefly below.

Excavation.   After the sewer alignment is marked on the ground, the trench excavation begins. Machinery such as backhoe, clamshell, dragline, front-end loader, or other specialized equipment is used. Hand excavation may be possible only for short distances. Hard rocks may be broken by drilling; explosives may also be used where situations permit.

Sheeting and Bracing.   Trenches in unstable soil condition require sheeting and bracing to prevent caving. Sheeting is placing planks in actual contact with the trench sides. Bracing is placing crosspieces extending from one side of the trench to the other. Sheeting and bracing may be of various types depending on the depth and width of the trenches and the type of soils supported. Common types are stay bracing, poling boards, box sheeting, vertical sheeting, and skeleton (open) sheeting. Details on sheeting and bracing may be found in Refs. 4 and 11. In many situations pumping may be necessary to dewater the trenches.

Sewer Installation.   After the trench is completed, the bottom of the trench is checked for elevation and slope. In firm, cohesive soils the trench bottom is shaped to fit the pipe barrel and projecting collars. Often granular material such as crushed stones, slag, gravel, and sand are used to provide uniform bedding of the pipe.

The pipes are inspected and lowered with particular attention being given to the joints. The pipe lengths are placed on line and grade with joints pressing together with a

TABLE 7-4 Characteristics of Sewer Pipes Made from Various Construction Materials

Construction Material	Available Sizes (Diam.) cm	in.	Available Section Length m	ft	Reference Standard	Subject to Corrosion or Erosion	Strength	Type of Joint	Application
Concrete						Yes	Good	Bell-and-spigot, or tongue and groove with rubber gasket, hot or cold poured bituminous material, cement mortar	Used for sewers, force mains, submerged outfalls, inverted syphons. Low cost, short lengths, many joints
Plain	10–60	4–24	1.2–7.4	4–24	ASTM[a] C14				
Reinforced	30–360	12–144	1.2–7.4	4–24	ASTM C76				
Vitrified clay	10–122	4–48	1–2	3–6	ASTM C700	No	Brittle	Mortar, rubber gasket, compression	Used for sewers. Short lengths, many joints, susceptible to infiltration
Asbestos-cement	10–110	4–42	—	—	AWWA[b] C400	Yes	Good	Collar with rubber ring	Used for force mains, water distribution lines, sewers. Light weight, easy to transport and handle
Brick masonry	Greater than 91	Greater than 36	Any length	Any length	None	Yes	Good	Mortar	Uncommon these days except for lining of concrete sewers

TABLE 7-4   Characteristics of Sewer Pipes Made from Various Construction Materials

Construction Material	Available Sizes (Diam.)		Available Section Length		Reference Standard	Subject to Corrosion or Erosion	Strength	Type of Joint	Application
	cm	in.	m	ft					
Cast iron	5–120	2–48	Long length	Long length	AWWA C100	Yes	Excellent	Bell and spigot, flanged, mechanical or groove coupled, rubber or neoprene ring-type push on joint, or ball and socket	Used for force mains, water mains, yard pipings, etc. Long laying lengths, tight joints, withstand high pressures. Susceptible to soil corrosion
Ductile iron	5–120	2–48	6.1	20	AWWA C150	Yes	Excellent	Similar to cast iron	Similar to cast iron
Corrugated and plain steel	20–640	8–252	1.2–4.6	4–15	AWWA C200	Yes	Good	Similar to cast iron. Band or bolt, coating or lining, coupling	Used for pressure pipes, and sewers. Light weight
Plastic	10–40	4–15	3.2	10	ASTM D3034	No	Fair	Rubber cement, chemical weld sleeves, compression gasket	Light weight, tight joints, long laying lengths, thin walls, susceptibility to sunlight and temperature which affect shape and strength. Susceptible to long term deflections

[a]American Society for Testing and Materials.
[b]American Water Works Association.

level or winch. The joints are then filled per specifications. A typical pipe installation is shown in Figure 7-7.

Backfilling. The trenches are filled immediately after the pipes are laid. The fill should be carefully compacted in layers of 15 cm (6 in.) deep around, under and over the pipe. After completion of the filling, the surface is then finished.

7-3-5 Field Investigations and Completion of Design

Field work should be conducted to establish bench marks on all streets that will have sewer lines. Soil borings should be conducted to develop subsurface data needed for trenching and excavation. The depth of boring should be at least equal to the estimated depth of the sewer lines. Detailed plans should be drawn showing the following: (1) contours at 0.5-m intervals in map with scale 1 cm equal to 6 m (1 in. = 50 ft), (2) existing and proposed streets, (3) street elevations, (4) railroads, buildings, culverts, drainage ditches, etc., (5) existing conduits and other utility lines, and (6) existing and proposed sewer lines and manholes. The sewer profiles should also be developed showing ground surface and sewer elevations, slope, pipe size and type, and location of special structures and the appurtenances. Profile drawings should be prepared immediately under the sewer plan for ready reference (see Figure 7-2).

7-3-6 Preparation of Contract Drawings and Specifications

It is important that the detailed drawings be prepared and specifications completed before the bids can be requested. The contract drawings should show (1) surface features, (2) depth and character of material to be excavated, (3) the existing structures that are likely to be encountered, and (4) the details of sewer and appurtenances to be constructed.

The specifications should be prepared by writing clearly and completely all work requirements and conditions affecting the contracts. As an example, technical specifications should cover items such as site preparation, excavation and backfill, concrete work, sewer materials and pipe laying, jointing, appurtenances, and acceptance tests (infiltration, exfiltration, smoke or air tests).

7-4 INFORMATION CHECKLIST FOR THE DESIGN OF SANITARY SEWERS

Design of sanitary sewers involves preliminary investigations, a detailed field survey, design calculations, and field drawings. The design engineer should be familiar with the service area, the local and state design criteria, and the design procedures. Adherence to a carefully planned sequence of activities to develop sewer design minimizes project delays and expenditures. A checklist of design activities is presented below. These activities are listed somewhat in their order of performance. However, in many cases separate tasks can be performed concurrently or even out of the order given below.

1. Develop a sewer plan showing existing and proposed streets and sewers, topographic features with contour intervals of 0.5 m, elevations of street intersections, and

(a)

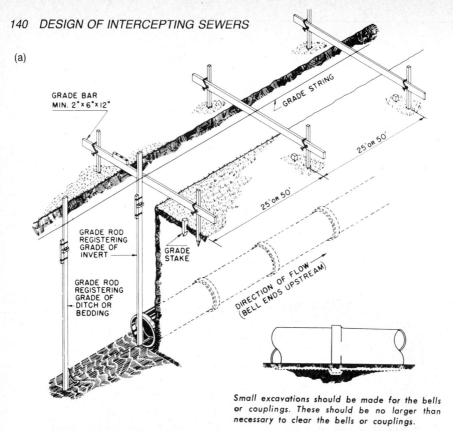

GRADE BAR
MIN. 2"x 6"x 12"

GRADE STRING

25' OR 50'

25' OR 50'

GRADE ROD
REGISTERING
GRADE OF
INVERT

GRADE
STAKE

GRADE ROD
REGISTERING
GRADE OF
DITCH OR
BEDDING

DIRECTION OF FLOW
(BELL ENDS UPSTREAM)

Small excavations should be made for the bells
or couplings. These should be no larger than
necessary to clear the bells or couplings.

(b)

Figure 7-7 Sewer Trench and Installation. (a) Laying clay pipe in line grade. (b) Vitrified clay pipe installation in deep trench. (Courtesy National Clay Pipe Institute.)

location of permanent structures and existing utility lines. Mark the proposed sewer lines and tentative slopes.

2. Locate manholes and number them in accordance with a convenient numbering system.
3. Prepare vertical profile showing ground surface, manhole locations, and elevation at the surface of each manhole.
4. Determine total land surface area that will be eventually served by different sewer lines.
5. Determine expected saturation population densities and average per capita wastewater flow rates.
6. Estimate peak design flow, and peak, average, and minimum initial flows.
7. Review design equations and develop hydraulic properties of the conduits.
8. Obtain state standards, sewer codes, or any design and maintenance criteria established by the concerned regulatory agencies.

7-5 DESIGN EXAMPLE

7-5-1 Design Criteria Used for Sewer Design

Five sewer lines are designed in this example. These lines are

1. Diversion sewer from stabilization pond area.
2. Diversion sewer from the trickling filter plant area.
3. Existing intercepting sewer from the central part of the town.
4. Final sewer line from the junction chamber to the bar screen.
5. By-pass sewer from the junction chamber to the storage basin.

The layout plan of these lines are shown in Figure 6-2 and 6-4, and the initial and design flows were developed in Chapter 6 (Tables 6-8 and 6-9). Following are the basic design criteria used for the design of these lines.

1. All lines shall be designed to flow partially full under peak design flow.
2. The velocity under initial low-flow condition shall not be less than 0.45 m/s (1.5 ft/s).
3. Provide concrete sewers and assume roughness coefficient $n = 0.015$.
4. The peak, average, and minimum flows for initial and design years are summarized in Table 7-5. These values are obtained from Tables 6-8 and 6-9.
5. The sewer layout plan and sewer profiles are shown in Figure 7-8 and 7-9.

7-5-2 Design Calculations

A. Sewer Design.   The design computations for the sewers should be started from the upper reach of the line and carried downward in the direction of the flow. The cumulative flows between manholes are used to determine the sewer diameter and slope, and ground and invert elevations. In this example, the step-by-step calculations for the diversion sewers and intercepting sewers are not shown in their entirety. Instead, the calculations for

TABLE 7-5 *Peak, Average, and Minimum Flows for the Initial and Design Years for Five Sewer Lines[a]*

Sewer Lines		Flow Conditions Initial Year (m³/s)			Flow Conditions Design Year (m³/s)		
		Peak	Avg.	Min.	Peak	Avg.	Min.
i	Diversion sewer from sta-bilization ponds	0.388	0.124	0.065	0.590	0.192	0.100
ii	Diversion sewer from tricking filter plant	0.257	0.086	0.041	0.385	0.132	0.063
iii	Existing intercepting sewer	0.274	0.094	0.047	0.331	0.116	0.058
iv	Final sewer to the plant	0.966	0.304	0.152	1.321	0.440	0.220
v	Bypass sewer	0.966	0.304	0.152	1.321	0.440	0.220

[a]Peak, average and minimum flows were developed in Chapter 6 and summarized in Tables 6-8 and 6-9.

only the last section of these pipes are given with reference to the floor elevation of the rack chamber. This elevation was established from the old sewer line serving the existing aerated lagoon. The design procedure involves the following steps:

1. Select the diameter, slope, and roughness coefficient n for the sewer line.
2. Determine the discharge Q and velocity V from Eqs. (7-1) and (7-2), or from Figure 7-5 when the sewer is flowing full. Q should be larger than the peak design flow.
3. Calculate q/Q ratios at peak design flow and at minimum initial flow.
4. Determine d/D and v/V ratios at these flows from Figure 7-6.
5. Calculate velocity and depth of flow under these flow conditions.
6. The velocity at minimum initial flow should be greater than 0.45 m/s.
7. The design steps are summarized in Table 7-6.

B. Junction Chamber. The junction chamber has internal dimensions of 2.0 m × 2.0 m. Three incoming sewers join at the junction chamber. A single line 1.53 m diameter carries the total flow to the treatment facility. A manually operated sluice gate is provided to close the line and divert the flow into a 1.22-m bypass sewer. The bypass sewer is designed to divert the incoming flow into the storage basin at the time of power outages. Design details of junction chamber are shown in Figures 7-8 and 7-9.

C. Hydraulic Profile. The water surface profiles for all sewers were developed from the water depth, slope of sewer, and the head losses encountered in the appurtenances. The head losses in a manhole or collection chamber are due to (1) exit loss, (2) change in direction, (3) contractions and expansions, (4) transitions, and (5) entrance loss. All these losses may be conveniently expressed by Eq. (7-5).

$$h_L = K \frac{v^2}{2g}$$

(7–5)

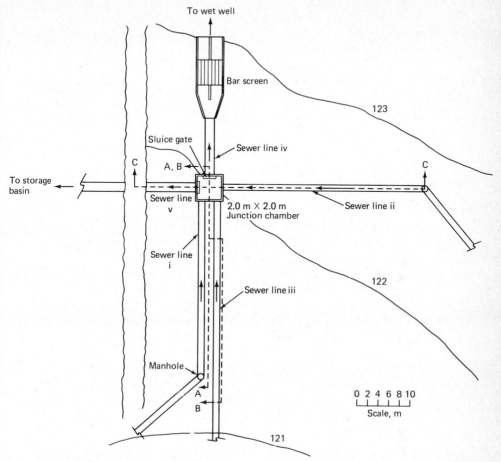

Figure 7-8 *Sewer Layout for Design Example (see Table 7-5 and 7-6 for sewer line identification number and details).*

where

h_L = minor loss due to entrance, exit, or change in direction of the flow, m
v = velocity of flow, m/s
g = acceleration due to gravity, 9.81 m/s²
K = head loss coefficient

The values of K for different conditions of open-channel flow and sewer appurtenances are summarized in Appendix B. The head loss calculations are given below, and hydraulic profiles for different sewer lines are shown in Figure 7-9. These calculations utilize velocities, depths of flow, and sewer slopes developed in Table 7-6. This simplified procedure gives conservative design if higher values of K are selected from Appendix B. An alternative method is to apply the energy equation at two sections and then solve for velocity or depth of flow by a trial-and-error procedure. Beside being tedious and time consuming the accuracy of the result depends on the judgment of the designer in selecting the head loss coefficient which present many uncertainties. The use of energy equation in

TABLE 7-6 Computations for Sewer Design

Design Steps	Diversion Line from Stabilization Pond i	Diversion Line from Trickling Filter Plant ii	Existing Intercepting Sewer from Central Part of Town iii	Final Sewer from the Junction Chamber iv	Bypass Sewer from Junction Chamber v
Select diam (m)	0.91	0.76	0.76	1.53	1.22
Select slope (m/m)	0.00173	0.0020	0.0020	0.00047	0.00147
Select n	0.015	0.015	0.015	0.015	0.015
Q full from Eq. (7-2) (m³/s)	0.673[a]	0.447	0.447	1.402	1.355
V full from Eq. (7-1) (m/s)	1.03[b]	0.98	0.98	0.76	1.16
q peak design flow (Table 7-5) (m³/s)	0.590	0.385	0.331	1.321	1.321
q/Q	0.88[c]	0.86	0.74	0.94	0.98
d/D (Fig. 7-6, n constant)	0.73	0.71	0.64	0.77	0.80
v/V (Fig. 7-6, n constant)	1.15	1.14	1.11	1.156	1.16
d(m)	0.66	0.54	0.49	1.18	0.98
v(m/s)	1.18[d]	1.12	1.09	0.88	1.35
q'minimum initial flow (Table 7-5) (m³/s)	0.065	0.041	0.047	0.152	0.152
q'/Q	0.10	0.09	0.11	0.11	0.11
d'/D (Fig. 7-5, n constant)	0.23	0.21	0.23	0.23	0.23
v'/V (Fig. 7-5, n constant)	0.62	0.58	0.62	0.62	0.62
d'(m)	0.21	0.16	0.18	0.35	0.28
v'(m/s)	0.64	0.57	0.61	0.47	0.72

[a] $Q = (0.312/0.015) \cdot (0.91)^{8/3}(0.00173)^{0.5} = 0.673$ m³/s. Q may also be obtained from Figure 7-5.
[b] $V = (1/0.015)(0.91/4)^{2/3}(0.00173)^{0.5} = 1.03$ m/s. V may also be obtained from Figure 7-5.
[c] $(q/Q) = (0.590/0.673) = 0.88$.
[d] $v = 1.15 \times 1.03 = 1.18$.

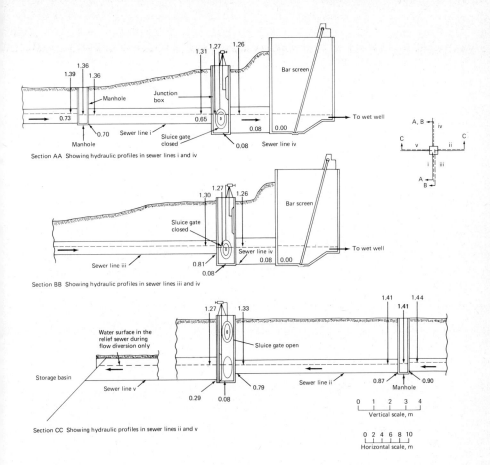

Figure 7-9   *Sewer Profiles for the Design Example (see Tables 7-5 and 7-6 for sewer line identification number and details). All elevations are in meters above the floor of the rack chamber.*

establishing the hydraulic profile is covered in Chapter 8 and in Refs. 1 and 16. Many designers prefer to match the crowns of all sewers meeting at a manhole. Since the outgoing sewer is always the largest, a drop in manhole invert is necessary. Some designers drop the invert of each manhole by about 0.03 m. These drops are generally more than that needed to compensate for the head losses due to the exit, entrance, and others. Therefore the hydraulic profile calculations are not needed.

1. Compute the hydraulic profile from the bar screen to the junction box. Assume that the invert elevation at the bar screen chamber = 0.00 m.
    a. Set differential elevation of incoming sewer line iv at 0.08 m above the invert of the bar screen chamber (see Chapter 8 for more details)

b. Exit loss at the discharge point*

$$= \frac{Kv^2}{2g} = \frac{0.09 \times (0.88 \text{ m/s})^2}{2 \times 9.81 \text{ m/s}^2} = \text{small}$$

c. Depth of flow $= 1.18$ m (Table 7-6)

d. Water surface elevation in the sewer line iv at the exit point $= 0.00$ m $+ 0.08$ m $+ 1.18$ m $= 1.26$

e. Invert elevation of the sewer line iv at the junction

$= 0.08$ m $+$ length $\times$ slope
$= 0.08$ m $+ 8$ m* $\times 0.00047$
$= 0.08$ m $+$ small $= 0.08$ m

f. Water surface elevation in the sewer line iv at the junction box $= 0.08$ m $+ 1.18 = 1.26$ m

g. Entrance loss into the sewer line iv*

$$= \frac{Kv^2}{2g} = \frac{0.3 \, (0.88 \text{ m/s})^2}{2 \times 9.81 \text{ m/s}^2} = 0.01 \text{ m}$$

h. Water surface elevation in the junction box $= 1.26$ m $+ 0.01$ m $= 1.27$ m

2. Compute hydraulic profile from the junction box to the manhole in the sewer line i
    a. Water surface elevation in the sewer line i at the exit point* $= 1.27$ m $+$ exit loss

$$= 1.27 \text{ m} + \frac{Kv^2}{2g}$$

$$= 1.27 \text{ m} + \frac{0.5 \times (1.18 \text{ m/s})^2}{2 \times 9.81 \text{ m/s}^2}$$

$$= 1.27 \text{ m} + 0.04 \text{ m} = 1.31 \text{ m}$$

b. Invert elevation of the sewer line i at the junction box

$= 1.31$ m $-$ water depth
$= 1.31$ m $- 0.66$ m $= 0.65$ m

c. Invert elevation of sewer line i at the manhole

$= 0.65$ m $+$ length $\times$ slope
$= 0.65$ m $+ 30$ m $\times 0.00173$
$= 0.65$ m $+ 0.05$ m $= 0.70$ m

d. Water surface elevation in the sewer line i $= 0.70$ m $+ 0.66$ m $= 1.36$ m

e. Entrance loss in the sewer line i at the manhole

$$= \frac{Kv^2}{2g} = \frac{0.05 \times (1.18 \text{ m/s})^2}{2 \times 9.81 \text{ m/s}^2} = \text{small}$$

f. Water surface elevation in the manhole $= 1.36$ m

*Obtained from Figure 7-8.

g. Water surface elevation in sewer line i upstream of manhole
$= 1.36$ m + losses due to exit and change in direction

$$= 1.36 \text{ m} + \frac{Kv^2}{2g}$$

$$= 1.36 \text{ m} + \frac{0.4 \times (1.18 \text{ m/s})^2}{2 \times 9.81 \text{ m/s}^2}$$

$$= 1.36 \text{ m} + 0.03 \text{ m} = 1.39 \text{ m}$$

h. Invert elevation of sewer line i upstream of manhole
$= 1.39 \text{ m} - 0.66 \text{ m} = 0.73 \text{ m}$

3. Compute the water surface and invert elevations in the sewer line iii
   a. Water surface elevation in sewer line iii
   $= 1.27 \text{ m} + \text{exit loss}$

   $$= 1.27 \text{ m} + \frac{Kv^2}{2g}$$

   $$= 1.27 \text{ m} + \frac{0.5 (1.09 \text{ m/s})^2}{2 \times 9.81 \text{ m/s}^2}$$

   $$= 1.27 \text{ m} + 0.03 \text{ m} = 1.30 \text{ m}$$

   b. Invert elevation
   $= 1.30 - \text{water depth}$
   $= 1.30 \text{ m} - 0.49 \text{ m} = 0.81 \text{ m}$

4. Compute the hydraulic profile in the sewer line ii
   a. Water surface elevation
   $= 1.27 \text{ m} + \text{exit loss}$

   $$= 1.27 \text{ m} + \frac{1.0 (1.12 \text{ m/s})^2}{2 \times 9.81 \text{ m/s}^2}$$

   $$= 1.27 + 0.06 \text{ m} = 1.33 \text{ m}$$

   b. Invert elevation
   $= 1.33 \text{ m} - \text{water depth}$
   $= 1.33 \text{ m} - 0.54\text{m} = 0.79 \text{ m}$

   c. Invert elevation at manhole
   $= 0.79 \text{ m} + \text{length} \times \text{slope}$
   $= 0.79 \text{ m} + 40 \text{ m} \times 0.002$
   $= 0.79 \text{ m} + 0.08 \text{ m} = 0.87 \text{ m}$

   d. Water surface elevation in the sewer line ii at the manhole
   $= 0.87 \text{ m} + 0.54 \text{ m} = 1.41 \text{ m}$

   e. Water surface elevation in the manhole
   $= 1.41 \text{ m} + \text{entrance loss}$

   $$= 1.41 \text{ m} + \frac{Kv^2}{2g}$$

   $$= 1.41 \text{ m} + \frac{0.05 \times (1.12 \text{ m/s})^2}{2 \times 9.81 \text{ m/s}^2}$$

   $$= 1.41 \text{ m} + \text{small} = 1.41 \text{ m}$$

f. Water surface elevation in the sewer line ii upstream of manhole $= 1.41 \text{ m} + \text{losses due to exit and change in direction}$

$$= 1.41 \text{ m} + \frac{0.4\,(1.12)^2}{2 \times 9.81}$$

$$= 1.41 \text{ m} + 0.03 \text{ m} = 1.44 \text{ m}$$

g. Invert elevation of the sewer line ii upstream of manhole $= 1.44 \text{ m} - \text{water depth}$

$$= 1.44 \text{ m} - 0.54 \text{ m}$$

$$= 0.90 \text{ m}$$

5. Compute the hydraulic profile in the sewer line v

The hydraulic profile in the sewer line v during flow bypass is shown in Fig. 7-9.

7-6 OPERATION AND MAINTENANCE, AND TROUBLESHOOTING IN SANITARY SEWERS

7-6-1 Common Problems

A well-designed sanitary sewerage system is expected to provide trouble-free operation. However, serious problems may develop due to lack of a preventive maintenance program or factors beyond the control of the maintenance crew. Common problems that may develop into collection systems include the following:

1. Explosions or severe corrosion due to discharge of uncontrolled industrial wastes.
2. Corrosion of sewer lines and manholes due to generation of hydrogen sulfide gas.
3. Collapse of sewer due to overburden or corrosion.
4. Poor construction or workmanship, or earth shifts may cause pipes to break or joints to open up. Excessive infiltration/exfiltration may occur. Also, tree roots, soil, gravel, etc., may enter the sewer resulting in sewer blockages.
5. Protruding taps in the sewers due to improper workmanship (called plumber taps or hammer taps) are common source of problems. They substantially reduce line capacity and contribute to frequent blockages.
6. Excessive settling of solids in the manhole and sewer line may lead to obstruction, blockage, or generation of undesired gases.
7. The diameter of the sewer line may be reduced by accumulation of slime, grease, and viscous materials on the pipe walls. This may lead to obstruction and blockage of the line.
8. Faulty, loose, or improperly fit manhole covers besides being noisy can be a source of inflow. Ground shifting may cause cracks in the manhole walls or in the pipe joints at the manhole. These cracks become a source of infiltration or exfiltration. Debris (rags, sand, gravel, sticks, etc.) may collect in the manhole and block the lines. Also, tree roots (like in sewer lines) may enter the manholes through the cracks, joints, or faulty cover, and cause serious blockages.

7-6-2 Maintenance Program

Strict enforcement of sewer ordinances and timely maintenance of sanitary sewerage systems is the fastest and most efficient way to prevent interruptions in sewer services. Inspection and maintenance on a regular basis will minimize the possibility of damage to private property by sewer stoppages as well as minimize the legal responsibility of the sewer authority from such damages.

For trouble-free operation of sanitary sewers, a maintenance program is necessary and should be developed. An effective maintenance program involves four steps: (1) sewer ordinance, (2) inspection, (3) preventive maintenance, and (4) repairs. Each of these steps are discussed briefly below.

Sewer Ordinance.   Under the pretreatment regulations of the Clean Water Act of 1977, all industrial users are prohibited from discharging pollutants into the sewerage system that can cause fire or explosion, corrosion, or obstruction to the sewers (see Chapter 2 for pretreatment standards). Also, protruding taps in the sewer system have resulted from lack of inspection during installation of service taps/connections, unauthorized tapping, and ineffective enforcement of plumbing codes. Many cities have passed ordinances designed to protect sewers against injury and blockage. More specifically, these ordinances (1) prohibit discharge of corrosive, inflammable, and explosive liquid, gases, and vapors, and garbage or dead animals, (2) call for installation of grease traps, and (3) prohibit house connections by plumbers, or require city inspection after the connection is made by the plumbers.

Inspection.   Sewer inspection on a regular basis should be performed to determine the general physical condition of the sewers and manholes as well as the need for repairs. Sewers with steep slopes or carrying large flows have adequate velocities and offer little chance for solids to deposit. On the other hand, lines carrying low discharge on flat slopes will accumulate solids and require more frequent cleaning.

In many cities sewer inspections are made and work done only when problems occur. In other cities strict inspection schedules are followed. As an example, sewers on flat grades or lines with history of root intrusions are examined every three months; sewers with no recorded troubles once a year; intercepting sewers once or twice per month; inverted syphons monthly to weekly; and so on. Inspection is made during low flows by observing the condition of the manhole, and looking through the sewer toward an explosion-proof flashlight placed in the downstream manhole. Such observations clearly indicate evidence of corrosion, cracks, and deteriorated conditions of manholes, improperly fitted covers, damaged cover rings, and partial stoppages of the lines. A strong rotten egg odor from manholes indicates hydrogen sulfide gas, which is generated under excessive turbulence, and anaerobic decomposition of settled solids and slimes.

Sewers may also be inspected by close circuit television cameras to determine the actual conditions in the sewer lines. TV inspection is an effective method of detecting exact locations of leaks, corrosion, intrusions, failure, and partial blockages.

Preventive Maintenance.   The sewer maintenance personnel should be properly trained to identify the problem and apply the appropriate remedial measures. Basic information concerning preventive maintenance and repairs is given below. When emergency situations occur (such as collapsed pipes, blockage, sewer backups, etc.), they should receive top priority and prompt action.

Equipment.   The requirements for sewer maintenance equipment depends upon the size and type of sewers serviced. A typical list of equipment for all types of sewers in large cities is provided in Table 7-7.

Sewer Cleaning.   Sewer cleaning is a preventive measure to protect certain lines from stoppages. It can be achieved by the following procedures:

1. Hydraulic flushing is done by attaching hose to fire hydrant or sewer-cleaning machines where high-pressure pumps are used. High-pressure water jet cleans the lines. Care should be taken as high-pressure jet may damage the corroded, cracked, or broken lines. Backing up of sewers should be avoided during flushing operation.

TABLE 7-7   Typical List of Sewer Maintenance Equipment for Municipalities

Major Equipment	Minor Equipment	Safety Equipment
Maintenance truck	Combination wrench set	Safety helmet
Winch truck	Hex key set	Rubber gloves and boots
Back hoe	Three way puller	Gas mask
Sewer ball cleaner	Punch and chisel set	Oxygen deficiency indicator
Hinged sewer cleaner	Pipe wrenches	Hydrogen sulfide detector
Jointed wood cleaning rods	Pipe cutters	Carbon monoxide detector
Rotating steel rodding machine	Tubing cutter set	Chlorine detector
Bucket cleaning machine	Hammers and crow bar	Portable air blower
Spring-cable sewer cleaner	Aluminum level	Protective clothing
High velocity jet cleaner	Hack saw and blades	Safety harness and life lines
(hydraulic)	Shovels, picks, rake and	Fire extinguisher
Manhole-cleaning unit (suction	wheelbarrow	
type)	Hydraulic jack	First aid and life saving kit
Closed circuit television	Water hose and hose nozzles	Explosion proof portable
Propane torch kit	Double branch chain sling	light
Bench vise	Circular drum dollie	Warning signs
High speed drill set	Portable pump with engine	
Ball bearing bench grinder	Portable blower	
Air compressor		

Source: Adapted in part from *Operation and Maintenance Manual for the Wastewater Facilities in Big Spring, Texas,* prepared by Gutierrez, Smouse, Wilmut & Assoc., Inc., Consulting Engineers, Dallas, Texas.

2. Sewer balls are effective means of removing deposits, slime, and grease from the lines. The ball is inserted in the upstream manhole and is pulled downstream by a rope. The ball adjusts itself to the irregularities of the pipe and water held behind the ball rushes around the ball thus flushing grease, slime, and deposits. Roots up to pencil size are broken off and removed.[4]

3. Greater accumulations are removed by special buckets or scoops. The bucket is pulled down the line until the operator feels that it is full. It is then pulled backwards, removed, and emptied. A bucket does not work well on pipes that are out of line due to differential settlement.

Stoppage Clearing. Stoppage occurs when a line is partially or completely blocked. Stoppage may result from large objects being dropped into manholes; accumulation of sand, grit, or grease; tree roots; or collapsed line. Most stoppages (except collapsed lines) can be cleared by a rodding machine. Flexible metal rods rotated by a motor on the ground turn various cleaning and cutting devices inserted into the lines in the direction of the flow. The equipment rods should not be operated upstream as they can damage Wyes and service connections. Rodding-collapsed pipes can damage the cleaning equipment.

Odor Control or Prevention of Hydrogen Sulfide Gas. Several physical and chemical methods have been utilized in controlling odors and generation of H_2S in collection systems. These methods include (1) avoidance of flat slopes of sewer when designing; (2) avoidance of excessive turbulence such as that in drop manholes; (3) ventilation and aeration of sewers and manholes; (4) application of chemicals such as solution of chlorine, hydrogen peroxide, ozone, or other chemicals. Detailed discussion on odor and hydrogen sulfide control may be found in Refs. 10 and 17.

Repairs. Collapsed and severely cracked pipes need excavation and replacement. This constitutes major repairs. Grouting and insertion-renewal (slip-lining) techniques are effective methods of rehabilitating old, corroded, and damaged pipes. Slip lining is done from manhole to manhole with plastic pipes made especially for this purpose.[1,4] Defective joints that admit roots and infiltration should be uncovered and repaired. All protruding taps must be redone as they reduce the capacity of a line and contribute to frequent blockages.

Manhole covers having cracks or improper fit should be repaired or replaced. If the cover ring or gasket is damaged, they should also be replaced. The manhole steps should be repaired or replaced whenever they weaken or show excessive deterioration. Damaged or missing steps can cause serious accidents. Corroded, cracked, or damaged manhole walls may require patching with cement bricks or cement mortar. If damage to manhole walls is extensive, the entire manhole should be replaced.

7-7 SPECIFICATIONS

General specifications for wastewater collection system are briefly summarized below. These specifications are meant only to be illustrative of the design completed. They are neither complete nor definitive.

7-7-1 Excavations

The contractor shall excavate all trenches to the depths shown on the drawings. Excavated material not required for backfill shall be disposed of by the contractor. The contractor shall do everything needed to keep excavations free of water (may need pumping), providing at all times adequate barricades and policing to prevent outsiders and employees from harm. The contractor shall do all bracing, sheeting, and shoring necessary to protect all excavations as required for safety. Where pipes cross roads and walks, excavations shall be filled before nightfall and proper barricades and lighting shall be maintained at all times until the work has been completed.

The contractor shall carefully backfill and compact all excavations, and replace all paving for parking areas, roadways, and walks, including curbs, to the condition that existed before the work began. The collection system shall be laid out so that major trees will not be disturbed, but where it is necessary to temporarily remove shrubs or small trees, the contractor shall remove, ball, and then replace them.

Sewer installation between manholes must proceed with utmost rapidity, and testing shall be done in segments in accordance with the work plan.

7-7-2 Sewer Pipe

Reinforced concrete sewer pipes shall be constructed, tested, and inspected at the point of manufacture in accordance with the specified test requirements and shall bear the initials or name of the manufacturer, the date of manufacture, and the testing laboratory's stamp of approval.

The connecting joints of the reinforced concrete sewer pipe shall be manufactured to ensure accurate and concentric bell and spigot joint. All joints shall be made with approved flexible compression-type rubber gasket. The rubber gasket shall truly conform in size to the spigot end of the pipe. In laying the rubber gasket joints in concrete pipes, care shall be exercised in handling the pipe to prevent damaging the rubber gaskets and ends of the pipes. The inside of the bells and the outside of the spigots shall be clean of dirt or foreign matter. The grooves or bell end shall be coated with approved lubricating material, then pipes pulled with sufficient force to cause the spigot to enter the bell and be forced completely home. On all sizes larger than 61 cm (24 in.) in diameter the inside annular space shall be completely filled with plastic portland cement to effect a smooth flow.

All sewer lines shall be tested as specified for tightness, infiltration, and exfiltration between the consecutive manholes. All joints shall be tight under a 1.2-m head of water.

7-7-3 Manhole

All manholes shall be of precast reinforced concrete. Shop drawings of all manholes shall be approved before manufacture and shall be constructed as shown in the respective drawings.

All steps or ladders in manholes shall be galvanized wrought iron or cast steps 2.5 cm round minimum. Manhole covers shall be new, heavy duty, and constructed in accordance with standard guidelines. The covers shall be gas-tight and with machined bearing surfaces and neoprene gaskets.

7-7-4 Junction Chamber

The junction chamber shall be 2.0 m × 2.0 m, be brought to the depth as shown in the drawing, and be constructed in-place of reinforced concrete. The concrete shall be of approved aggregates and cement in specified proportions, and mixed in an approved type power operated batch mixer to yield the specified ultimate compressive strength at the age of 28 days. Reinforcing steel shall conform to the standard specifications and shall conform in size and position to the requirements of the drawings. The forms shall conform to the lines, dimensions, and shapes of concrete indicated on the drawings. They shall be tight to prevent leakage of concrete, shall be securely braced and maintained in position to prevent movement after concrete is poured, and shall have adequate strength to safely support the concrete. All concrete shall be placed with the aid of mechanical vibration equipment, and curing and finishing shall be in accordance with the ACI standards.* The corners, dead spaces, and the bottom of the junction chamber shall be contoured properly for free flow of incoming wastewater into the exit sewer.

7-8   PROBLEMS AND DISCUSSION TOPICS

7-1   A 60-cm diameter (24-in.) sewer is designed to carry a peak design flow of 0.01 m³/s at a slope of 0.0008 m/m. Calculate the actual depth of flow and velocity. Assume constant value of $n = 0.015$.

7-2   Repeat Problem 1 for variable value of n. Assume $n = 0.015$ when the line is flowing full.

7-3   Calculate the depth of flow and velocity into the sewer line iv in the Design Example, at average and minimum design flows. The diameter and slope of the line are given in Table 7-6. The flow conditions are provided in Table 7-5.

7-4   Calculate the diameter of a trunk sewer to carry 50 ℓ/s when flowing 60 percent full. The velocity in the line at partial full condition must be 0.6 m/s ($n = 0.013$).

7-5   Discuss the basic design factors that must be considered in developing the preliminary design and layout of the sanitary sewerage system for a community.

7-6   Prepare a checklist of an effective maintenance program for the sanitary sewerage system for a community.

7-7   An intercepting sewer is 400 mm in diameter and is serving a population of 5000 residents. The average wastewater flow is 450 ℓpcd. Calculate the depth of flow and velocity at minimum flow. The slope of the line is 0.003. Assume $n = 0.013$, and minimum flow is one-third of the average flow.

7-8   An intercepting sewer is 61 cm in diameter and has a slope of 0.0013. It carries a flow of 113.3 ℓ/s. The sewer enters a standard manhole that is 76 cm in diameter. The outlet sewer is also 61 cm in diameter and has a slope of 0.0014. Using $n = 0.015$, draw the hydraulic profile and determine the invert elevation of the incoming sewer. Assume that the coefficients of exit and entrance are 0.4 and 0.1, respectively, and the manhole invert is in line with the outlet sewer.

REFERENCES

1. Metcalf and Eddy, Inc., *Wastewater Engineering: Collection and Pumping of Wastewater*, McGraw-Hill Book Co., New York, 1981.
2. Metcalf and Eddy, Inc., *Water Pollution Abatement Technology Capabilities and Costs for*

*American Concrete Institute.

Publicly Owned Treatment Works, Report Prepared for National Commission on Water Quality, PB-250 690, National Technical Information Service, March 1976.

3. U.S. Environmental Protection Agency, *Innovative and Alternative Technology Assessment Manual,* Office of Water Program Operations, U.S. Environmental Protection Agency, Washington, D.C., 430/9-78-009, CD-53, February 1980.

4. Steel, E. W., and T. J. McGhee, *Water Supply and Sewerage,* McGraw-Hill Book Co., New York, 1979.

5. Bowne, W. C., *Pressure Sewers,* Environmental Research Information Center, Office of Research and Development, U.S. Environmental Protection Agency, Cincinnati, Ohio, July 1979.

6. Carcich, I. G., L. J. Hetling, and R. P. Farrell, The Pressure Sewer: A New Alternative to Gravity Sewers, *Civil Engineering, American Society of Civil Engineers,* New York, May 1974.

7. National Clay Pipe Institute, *Clay Pipe Engineering Manual,* National Clay Pipe Institute, Washington, D.C., 1978.

8. Parker, H. W., *Wastewater System Engineering,* Prentice-Hall, Englewood Cliffs, N.J., 1975.

9. Great Lakes-Upper Mississippi River, Board of State Sanitation Engineers, *Recommended Standards for Sewage Works,* Health Education Service, Albany, N.Y., 1971.

10. U.S. Environmental Protection Agency, *Process Design Manual for Sulfide Control in Sanitary Sewerage System,* U.S. EPA Technology Transfer, Washington, D.C., October 1974.

11. Joint Committee of the American Society of Civil Engineers and the Water Pollution Control Federation, *Design and Construction of Sanitary and Storm Sewers,* MOP 9, Water Pollution Control Federation, Washington, D.C., 1970.

12. Chow, V. T., *Open-Channel Hydraulics,* McGraw-Hill Book Co., New York, 1959.

13. Clark, J. W., W. Viessman, Jr., and M. K. Hammer, *Water Supply and Pollution Control,* Third Edition, IEP-A Dun-Donnelly, New York, 1977.

14. American Iron and Steel Institute, *Modern Sewer Design,* American Iron and Steel Institute, Washington, D.C., 1980.

15. Santry, I. W., Jr., "Infiltration in Sanitary Sewers," *Journal of Water Pollution Control Federation,* Vol. 36, No. 10, p. 1256, October 1964.

16. Benefield, L. D., J. F. Judkins, and A. D. Parr, *Treatment Plant Hydraulics for Environmental Engineers,* Prentice-Hall, Inc., Englewood Cliffs, N.J., 1984.

17. Cole, C. A., P. A. Paul, and H. P. Brewer, "Odor Control with Hydrogen Peroxide," *Journal Water Pollution Control Federation,* Vol. 48, No. 2, February 1976, p. 297.

Screening

8

8-1 INTRODUCTION

Screening is normally the first unit operation used at wastewater treatment plants. The general purpose of screens is to remove large objects such as rags, paper, plastics, metals, and the like. These objects, if not removed, may damage the pumping and sludge removal equipment, hang over weirs, and block valves, nozzles, channels, pipelines, and appurtenances, thus creating serious plant operation and maintenance problems. A fine screen is sometimes used as a treatment device to remove suspended solids.

In this chapter different types of screening devices and the design criteria for bar racks are covered. The step-by-step design procedure, operation and maintenance, and equipment specifications for bar racks are presented in the Design Example.

8-2 TYPES OF SCREENS

Screening devices can be broadly classified as coarse or fine. Screens may be manually or mechanically cleaned. Various types of coarse and fine screens are discussed below.

8-2-1 Coarse Screens

Coarse screens are used primarily as protective devices and therefore are used as the first treatment unit. Common types of protective devices include bar racks (or bar screens), coarse woven-wire screens, and comminutors. A brief description of racks, woven-wire screens, and comminutors is provided in Table 8-1.[1-3] Installations of bar racks, coarse screens, and comminutors are shown in Figures 8-1 and 8-2.

8-2-2 Fine Screens

Fine screens have openings 2.3–6 mm, with some installations utilizing openings less than 2.3 mm.[2] They are usually mechanically cleaned. The purpose of fine screens is to provide pretreatment or primary treatment. Various types of microscreens have been developed in recent years that are used for upgrading effluent from secondary treatment plants.

Fine screens consist of fixed and movable screens. The fixed or static screens are permanently set in vertical, inclined, or horizontal position, and must be cleaned by rakes, teeth, or brushes. Movable screens are cleaned continuously while in operation. Both

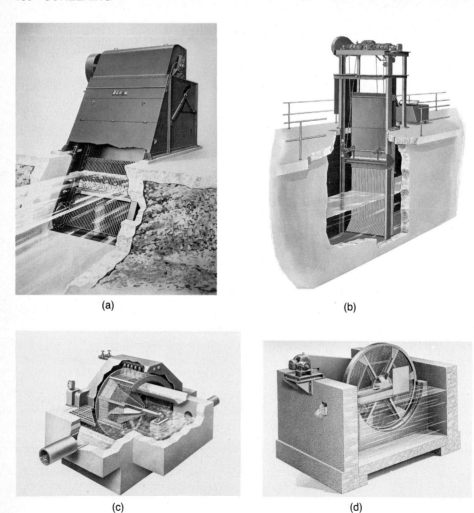

(a)

(b)

(c)

(d)

Figure 8-1   Details of Bar Rack and Coarse Screen. (a) Inclined rack, front cleaned. (b) Cable operated bar rack. (Courtesy Envirex Inc., a Rexnord Company.) (c) Revolving drum screens. (d) Revolving disc screen. (Courtesy FMC Corporation, Material Handling Systems Division.)

types of screens are capable of removing 20–35 percent suspended solids and BOD_5.[1] They also remove grease and tend to increase the DO levels of the wastewater. The moving screens, however, exhibit less head loss, but require more power for operation than the fixed screens. Different types of moving screens are described in Table 8-2.

8-3   DESIGN FACTORS FOR BAR RACKS (BAR SCREENS)

Bar racks are the most commonly used devices at medium- and large-size wastewater treatment facilities. They contain a screen chamber with inlet and outlet structures, and a screening device that has an arrangement for cleaning and removing the screenings. The

TABLE 8-1   Description of Coarse Screens

Type	Location	Description
Bar racks or bar screens	Ahead of pumps and grit removal facilities	Bar racks may be manually cleaned or mechanically cleaned. Manually cleaned racks are provided at small wastewater treatment facilities.
Coarse woven-wire media screens	Behind racks or ahead of trickling filters	These are flat-, basket-, cage-, or disk-type screens used to remove relatively smaller particles. The screens are cleaned by removing them from the channel. The new types of coarse screens use moving screens. They are similar to fine screens in design. They have vertical or drum-type arrangement with wire mesh or screen cloth. The solids are continuously removed into the receiving troughs. The openings may vary from 3 to 20 mm depending on the needs for solids removal.
Comminutor	Used in conjunction with coarse screens	Comminutors are grinders that cut up the materials retained over screens. They utilize cutting teeth or shredding devices on a rotating or oscillating drum that pass through stationary combs, screen, or disks. Large objects are shredded that pass through thin openings or slots 0.6–1 cm. The comminutors are almost submerged. Manufacturers' rating tables are available for different capacity ranges, channel dimensions, submergence, and power requirements. Provision to bypass the comminutor is always made. Comminutor installation is shown in Figure 8-2.

design velocity, bar spacing, bar size, angle of inclination, and allowable head losses through the racks are summarized in Table 8-3. The screen chamber (with inlet and outlet arrangements), methods for calculating head losses through the screen, equipment for cleaning and removing the screenings, and methods of estimating the quantities of screenings are described below.

8-3-1 Screen Chamber

A screen chamber consists of a rectangular channel. The floor of the channel is normally 7–15 cm lower than the invert of the incoming sewer. Also, the channel floor may be flat (horizontal) or at a desired slope. The screen channel is designed to prevent the accumula-

TABLE 8-2   Types of Moving Screens

Type	Description	Ref.
Band screens	Band screen consists of an endless perforated band which passes over upper and lower rollers. A brush may be installed to remove the material retained over the screen. Water jet is also used to flush the debris.	4
Wing or shovel screens	These screens consist of circular perforated radial vans that slowly rotate on a horizontal axis. The vans scoop through the channel.	4
Strainers or drum screen	The strainers consist of a rotating cylinder that has screen covering the circumferential area of the drum. The liquid enters the drum and moves radially out. The solids deposited are removed by a jet of water from the top and discharged into a trough. The screen openings may range from 0.02 to 3 mm. The microstrainers are used for polishing secondary effluent. Discussion on microstrainers may be found in Chapter 24.	5

tion of grit and other heavy materials into the channel.[1,3] The channel is normally provided with a straight approach, perpendicular to the screen, to assure uniform distribution of screenings over the entire screen area.

At least two bar racks, each designed to carry the designed peak flow, must be provided in case one is out of service. Arrangements for stopping the flow and draining the

TABLE 8-3   Design Factors for Manually Cleaned and Mechanically Cleaned Bar Racks

Design Factor	Manually Cleaned	Mechanically Cleaned
Velocity through rack (m/s)	0.3–0.6	0.6–1.0
Bar size		
Width (mm)	4–8	8–10
Depth (mm)	25–50	50–75
Clear spacing between bars (mm)	25–75	10–50
Slope from horizontal (degrees)	45–60	75–85
Allowable head loss, clogged screen (mm)	150	150
Maximum head loss, clogged screen (mm)	800	800

Source: Adapted in part from Refs. 1, 3, and 6

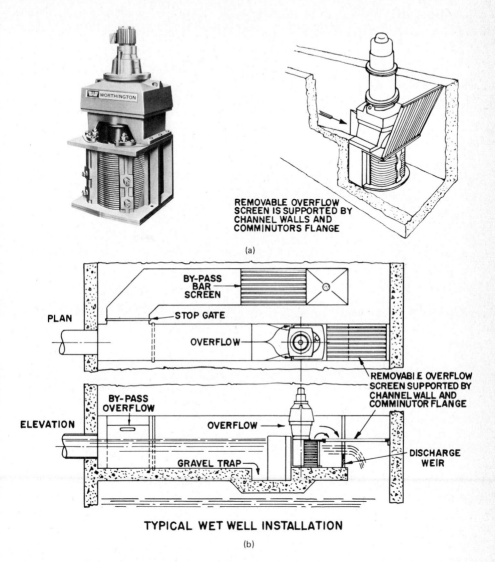

REMOVABLE OVERFLOW
SCREEN IS SUPPORTED BY
CHANNEL WALLS AND
COMMINUTORS FLANGE

(a)

PLAN

BY-PASS
BAR
SCREEN

STOP GATE

OVERFLOW

REMOVABLE OVERFLOW
SCREEN SUPPORTED BY
CHANNEL WALL AND
COMMINUTOR FLANGE

ELEVATION

BY-PASS
OVERFLOW

OVERFLOW

GRAVEL TRAP

DISCHARGE
WEIR

TYPICAL WET WELL INSTALLATION

(b)

Figure 8-2   Details of Comminutor. (a) Comminutor assembly. (b) Typical wet well installation. (Courtesy Worthington Pump Corporation.)

channel should be made for routine maintenance. The entrance structure should have a
smooth transition or divergence in order to minimize the entrance losses as wastewater is
discharged from the interceptor into the channel, and prevent settling and accumulation of
grits. Likewise, the effluent structure should have uniform convergence. The effluent
from individual rack chambers may be combined or kept separate as necessary. The bar
rack chambers as well as influent and effluent structures are shown in Figure 8-3. In all
cases, however, an account must be made for head losses due to exit, bends, expansion,
contraction, and entrance.

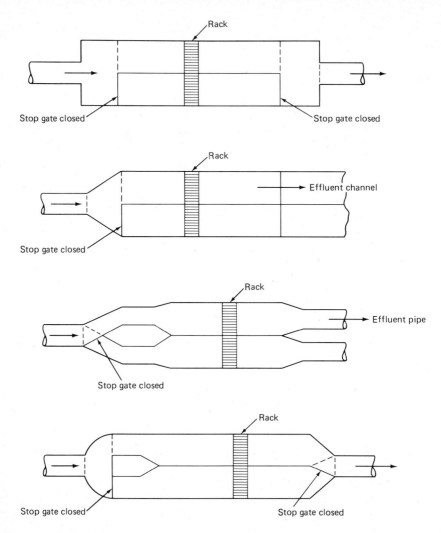

Figure 8-3   Double Chamber Bar Rack, and Influent and Effluent Arrangements.

8-3-2 Head Loss

The head loss through the bar rack is calculated from Eqs. (8-1), (8-2), or (8-3).[1,2] Equation (8-1) is used to calculate head losses through clean or partly clogged bars, while Eq. (8-2) is used to calculate head loss through clean screen only.[1,2] Equation (8-3) is the common orifice formula and is also used to calculate head loss through fine screens.[1]

$$h_L = \frac{V^2 - v^2}{2g} \left(\frac{1}{0.7}\right) \qquad\qquad (8\text{--}1)$$

$$h_L = \beta \left(\frac{W}{b}\right)^{4/3} h_v \sin\theta \qquad\qquad (8\text{--}2)$$

$$h_L = \frac{1}{2g}\left(\frac{Q}{CA}\right)^2 \tag{8–3}$$

where

h_L = head loss through the rack, m
V, v = velocity through the rack and in the channel upstream of the rack, m/s
g = acceleration due to gravity, 9.81 m/s^2
W = maximum cross-sectional width of the bars facing the direction of flow, m
b = minimum clear spacing of bars, m
h_v = velocity head of the flow approaching the bars, m
θ = angle of bars with horizontal
Q = discharge through screen, m^3/s
A = effective submerged open area, m^2
C = coefficient of discharge = 0.60 for clean rack
β = bar shape factor. The values of bar shape factors for clean rack are summarized as follows:

Bar type	β
Sharp-edged rectangular	2.42
Rectangular with semicircular upstream face	1.83
Circular	1.79
Rectangular with semicircular upstream and	
downstream faces	1.67
Tear shape	0.76

8-3-3 Removal of Screenings

Manually cleaned bar racks have sloping bars that facilitate hand raking. The screenings are placed on a perforated plate for drainage and storage.

The mechanically cleaned bar racks are front-cleaned or back-cleaned. In both cases the traveling rake moves the screenings upward and drops them into a collection bin or a conveyor. The back-cleaned raking device has the advantage that it does not jam easily due to obstruction at the base of the screen. In both types the raking is performed continuously by means of endless chains operating over sprockets. The operation can be made intermittent by means of a time clock or actuated by a preset differential head loss across the screen.

8-3-4 Quantities and Composition of Screenings

The quantity of screenings depends on type of wastewater, geographic location, weather conditions, and type and size of screens. The quantity of screenings removed by bar racks vary from 3.5 to 80 m^3/10^{6}m^3 (0.5–11 ft^3/million gallons), with an average of about 20 m^3/10^6 m^3 (2.7 ft^3/million gallons).[2] The average and maximum amounts of screenings collected from mechanically cleaned bar racks with respect to size of the opening are shown in Figure 8-4.

The screenings contain approximately 80 percent moisture and normally weigh 960

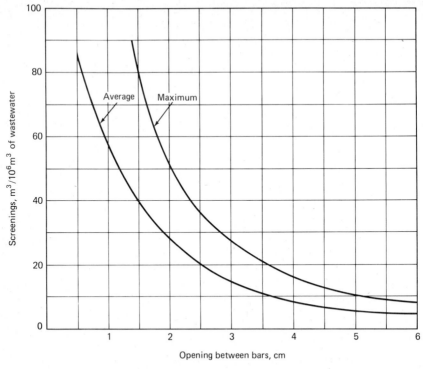

Figure 8-4 Quantities of Screenings Collected from Mechanically Cleaned Bar Racks. (Courtesy Envirex Inc., a Rexnord Company.)

kg/m^3 (60 lb/ft^3). The screenings are odorous and attract flies. Disposal of screenings is achieved by landfilling or incineration. Often screenings are discharged into grinders where they are ground and returned into the wastewater treatment plant.

8-4   EQUIPMENT MANUFACTURERS OF BAR RACKS

A number of equipment manufacturers supply mechanically cleaned bar racks. The names and addresses of many manufacturers of screening devices is given in Appendix D. The design engineer must make many decisions in advance of selecting the equipment. Information such as width and water depth in the channel, clear spacings between the bars, velocity through screens, type of cleaning equipment (that is, front-cleaned or back-cleaned), etc., are necessary for making these decisions. Once these decisions are made, the design engineer must work in cooperation with the equipment manufacturers to select the equipment best suited for the project needs. Important considerations for equipment selection are also covered in Sec. 2-10.

8-5   INFORMATION CHECKLIST FOR DESIGN OF BAR RACK

Bar rack design should not be started until many important design decisions are made and the necessary data developed. The following is a listing of important items:

1. Flow data including peak wet weather, peak dry weather, and average design flows
2. Hydraulic and design data for the influent conduit
3. Treatment plant design criteria prepared by the concerned regulatory agencies
4. Velocities through the bars
5. Equipment manufacturers and equipment selection guide (catalog)
6. Information on existing facility if the plant is being expanded
7. Existing site plan with contours
8. Bar spacings and the head loss constraints through the rack and through the entire plant
9. Velocities through the screen chamber

8-6   DESIGN EXAMPLE

8-6-1 Design Criteria Used

The following design criteria are used for the design of bar racks:

1. Provide two identical bar racks, each capable of handling maximum flow conditions and each equipped with a mechanical cleaning device, $\theta = 75°$.
2. One screen chamber could be taken out of service for routine maintenance without interrupting the normal plant operation.
3. Bar spacing (clear)   = 2.5 cm
4. Peak design wet weather flow   = 1.321 m³/s
   Maximum design dry weather flow = 0.916 m³/s
   Average design dry weather flow   = 0.441 m³/s
5. Provide approximately the following velocities through the rack at different flows:
   Velocity through rack at peak design wet weather flow   = 0.9 m/s
   Velocity through rack at maximum design dry weather flow = 0.6 m/s
   Velocity through rack at average design dry weather flow   = 0.4 m/s

8-6-2 Design Calculations

A. Flow Conditions in the Incoming Conduit
1. Diameter of the conduit   = 1.53 m (Table 7-6)
2. Slope of the conduit   = 0.00047 (Table 7-6)
3. Velocity at peak design flow, v = 0.88 m/s (Table 7-6)
4. Depth of flow in the conduit at
   peak design flow, d = 1.18 m (Table 7-6)

B. Design of Rack (Screen) Chamber
1. Compute bar spacings and the dimensions of the bar rack chamber.

   The rack chamber is designed for peak wet weather flow. The velocities through the rack and channel, and depth of flow in the channel, are also checked for average and minimum design flows.

Assume that the depth of flow in the rack chamber is the same as that in the incoming conduit (1.18 m).

a.   Clear area through the rack openings $= \dfrac{\text{peak design flow}}{\text{velocity through rack}}$

$= \dfrac{1.321 \text{ m}^3/\text{s}}{0.9 \text{ m/s}} = 1.47 \text{ m}^2$

b.   Clear width of the opening at the rack $= \dfrac{\text{area}}{\text{depth of flow}}$

$= \dfrac{1.47 \text{ m}^2}{1.18 \text{ m}} = 1.25 \text{ m}$

c.   Provide 50 clear spacings at 25 mm

d.   Total width of the rack chamber $= 50 \times 25\text{mm} \times \dfrac{1}{1000 \text{ mm/m}} = 1.25 \text{ m}$

e.   Total number of bars $= 49$

f.   Provide bars with 10 mm width

g.   Width of the chamber $= 1.25 \text{ m} + 10 \text{ mm} \times 49 \times \dfrac{1}{1000 \text{ mm/m}}$

$= 1.74 \text{ m}$

h.   The plan longitudinal section, the cross-section of the chamber including datum, water surface at peak design flow, bar arrangement, and spacings are shown in Figure 8-5.

2.   Calculate the efficiency coefficient.

Efficiency coefficient $= \dfrac{\text{clear opening}}{\text{width of the chamber}}$

$= \dfrac{50 \times 25 \text{ mm}}{1740 \text{ mm}} = 0.72$

(Many manufacturers for a given clear spacing and bar thickness give an efficiency coefficient. This coefficient may also be used to calculate the width of the chamber.)

3.   Compute the actual depth of flow and velocity in the rack chamber at peak design flow.

a.   The actual depth of flow in the rack chamber upstream of the bar rack is calculated by using the energy equation. Energy equation with respect to two sections of a channel is expressed by Eq. (8-4) [see sections (1) and (2) in Figure 8-5(b)]

$$Z_1 + d_1 + \frac{v_1^2}{2g} = Z_2 + d_2 + \frac{v_2^2}{2g} + h_L \qquad (8\text{--}4)$$

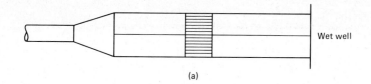

(a)

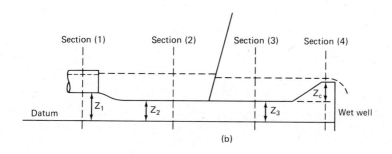

(b)

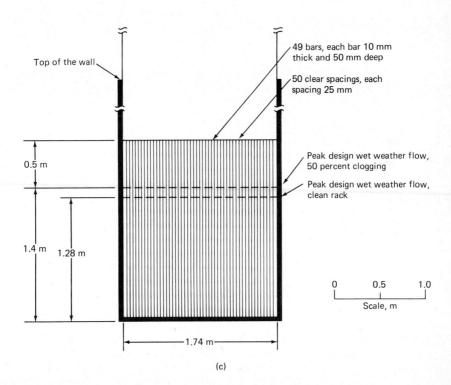

(c)

Figure 8-5   Details of Rack Chamber. (a) Plan, (b) Longitudinal section through the rack chamber, (c) Cross section showing bar arrangement, channel section, and depths of flow.

where

$$Z_1 \text{ and } Z_2 = \text{height above datum, m}$$
$$v_1 \text{ and } v_2 = \text{velocity at sections (1) and (2), m/s}$$
$$h_L = \text{total head loss, m}$$
$$d_1 \text{ and } d_2 = \text{depth of flow at sections (1) and (2), m}$$

In this example

$$h_L = K_e \left(\frac{v_1^2}{2g} - \frac{v_2^2}{2g} \right) \tag{8-5}$$

where

$$K_e = \text{coefficient of expansion (see Appendix B)}$$

b. Using the energy equation between section (1) in the incoming conduit and section (2) in the rack chamber upstream of the rack [Figure 8-5(b)], the actual depth of flow and velocity in the chamber are determined. The energy equation is written using the following conditions:

1. The floor of the chamber is horizontal
2. Reference datum is at the floor of the chamber ($Z_2 = 0$)
3. The invert of the incoming conduit is 8 cm above the reference datum
4. $K_e = 0.3$

$$0.08 \text{ m} + 1.18 \text{ m} + \frac{(0.88 \text{ m/s})^2}{2 \times 9.81 \text{ m/s}^2}$$

$$= d_2 \text{ m} + \frac{\left(\dfrac{1.321 \text{ m}^3/\text{s}}{1.74 \text{ m} \cdot d_2 \text{ m}} \right)^2}{2 \times 9.81 \text{ m/s}^2}$$

$$+ 0.3 \left[\frac{(0.88 \text{ m/s})^2}{2 \times 9.81 \text{ m/s}^2} - \frac{\left(\dfrac{1.321 \text{ m}^3/\text{s}}{1.74 \text{ m} \cdot d_2 \text{m}} \right)^2}{2 \times 9.81 \text{ m/s}^2} \right]$$

c. Simplifying this equation:

$$d_2^3 - 1.288 d_2^2 + 0.021 = 0$$

d. Solving by trial and error:

$$d_2 = 1.28 \text{ m, and } v_2 = \frac{1.321 \text{ m}^3/\text{s}}{1.74 \text{ m} \times 1.28 \text{ m}} = 0.59 \text{ m/s}$$

4. Compute velocity V through the clear openings at the bar rack;

$$V = \frac{\text{flow}}{\text{net area at the rack}} = \frac{1.321 \text{ m}^3/\text{s}}{1.25 \text{ m}^{a*} \times 1.28 \text{ m}}$$

$$= 0.83 \text{ m/s}$$

*50 spaces × 0.025 m clear spacings = 1.25 m.

The actual velocity through the rack at peak design flow is 0.83 m/s. This is slightly less than the design value of 0.9 m/s. Some designer may redesign the rack with 48 or 49 clear openings. The width of the chamber will be reduced and higher velocity through the screen will be encountered (see assignment problem 8-2, Sec. 8-9).

5. Compute head loss through the bar rack.
    The head loss through the bar rack for clean condition is calculated from Eq. (8-1) or (8-2). The head loss calculations from these equations are given below.

$$h_L \text{ [using Eq. (8–1)]} = \frac{(0.83 \text{ m/s})^2 - (0.59 \text{ m/s})^2}{2 \times 9.81 \text{ m/s}^2} \times \frac{1}{0.7} = 0.025 \text{ m}$$

$$h_L \text{ [using Eq. (8–2)]} = 2.42 \times \left(\frac{49 \times 10 \text{ mm}}{50 \times 25 \text{ mm}}\right)^{4/3}$$

$$\times \frac{(0.83 \text{ m/s})^2}{2 \times 9.81 \text{ m/s}^2} \times \sin 75°$$

$$= 0.024 \text{ m}$$

6. Compute the depth of flow and velocity in the rack chamber below the rack.
    a. The depth and velocity in the chamber is calculated by trial and error from the energy equation:

$$d_2 + \frac{v_2^2}{2g} = d_3 + \frac{v_3^2}{2g} + h_L$$

    where

    d_2 and v_2 = depth of flow and velocity in the chamber upstream of the rack

    d_3 and v_3 = depth of flow and velocity in the chamber downstream of the rack

    h_L = head loss through the rack

$$1.28 \text{ m} + \frac{(0.59 \text{ m/s})^2}{2 \times 9.81 \text{ m/s}^2} = d_3 \text{ m} + \frac{\left(\frac{1.321 \text{ m}^3/\text{s}}{1.74 \text{ m} \cdot d_3 \text{ m}}\right)^2}{2 \times 9.81 \text{ m/s}^2} + 0.025 \text{ m}$$

    b. Simplifying this equation:

$$d_3^3 - 1.273 \, d_3^2 + 0.029 = 0$$

    c. Solving this equation by trial and error, $d_3 = 1.25$ m and $v_3 = 0.61$ m/s.

7. Compute the head loss through the rack at 50 percent clogging.
    a. At 50 percent clogging of the rack, the clear area through the rack is reduced to half and the head loss through the rack is obtained from the energy equation.

$$d_2' + \frac{v_2'^2}{2g} = d_3 + \frac{v_3^2}{2g} + h_{50}$$

where

d_2' and v_2' = depth of flow and velocity in the chamber upstream of the rack at 50 percent clogging

h_{50} = head loss through the rack at 50 percent clogging

b. The velocity and depth of flow in the channel below the rack are governed by the condition in the outlet channel (the channel that carries the screened wastewater). Since the conditions in the outlet channel do not change, it can be assumed that d_3 and v_3 are the same as calculated for the clean rack.

c. $h_{50}' = \dfrac{\text{(velocity through rack opening)}^2 - v_2'^2}{2g} \times \dfrac{1}{0.7}$

d. Velocity through rack openings at 50 percent clogging $= \dfrac{1.321 \text{ m}^3/\text{s}}{1.25 \text{ m} \times 0.5 \times d_2' \text{ m}} = \dfrac{2.114}{d_2'} \text{ m/s}$

e. $v_2' = \dfrac{1.321 \text{ m}^3/\text{s}}{1.74\text{m} \times d_2' \text{ m}} = \dfrac{0.759}{d_2'} \text{ m/s}$

f. $d_2' \text{ m} + \dfrac{\left(\dfrac{0.759}{d_2'}\text{m/s}\right)^2}{(2 \times 9.81 \text{ m/s}^2)}$

$= 1.25 \text{ m} + \dfrac{(0.61 \text{ m/s})^2}{2 \times 9.81 \text{ m/s}^2}$

$+ \dfrac{\left[\left(\dfrac{2.114 \text{ m/s}}{d_2'}\right)^2 - \left(\dfrac{0.759 \text{ m/s}}{d_2'}\right)^2\right]}{2 \times 0.7 \times 9.81 \text{ m/s}^2}$

g. Simplifying this equation:

$$d_2'^3 - 1.269 d_2'^2 - 0.254 = 0$$

h. Solving this equation by trial and error:

$d_2' = 1.40 \text{ m}$

$v_2' = \dfrac{1.321 \text{ m}^3/\text{s}}{1.74 \text{ m} \times 1.40 \text{ m}}$

$= 0.54 \text{ m/s}$

i.   The head loss under 50 percent clogging:

$$h_{50} = 1.40 \text{ m} - 1.25 \text{ m}$$

$$= 0.15 \text{ m}^*$$

j.   Velocity V' through rack openings:

$$V' = \frac{2.114}{1.40} \text{ m/s}$$

$$= 1.51 \text{ m/s}$$

k.   Many designers use approximate analysis. When the screen becomes half-plugged, the area of flow is reduced to half, and velocity through the screen is doubled:

$$h_{50} = [(2 \times 0.83 \text{ m/s})^2 - (0.59 \text{ m/s})^2] \frac{1}{2 \times 9.81 \text{ m/s}^2 \times 0.7}$$

$$= 0.18 \text{ m}$$

The head loss calculated by this method is slightly larger than that previously obtained (0.15 m). Higher allowance of head loss gives more conservative design.

l.   The increase of head loss is about 15 cm (5.9 in.) as the screen becomes plugged. The need for accurate control of the cleanup cycle, and protection against surge loads, is thus demonstrated. A freeboard of over 0.5 m is normally provided for protection against flooding and overflow. In this design the bar screen is deep in the ground, and therefore flooding and overflow is not a consideration.

m.   A summary of depth of flow, velocity, and head loss through the bar rack under clean and 50 percent clogging is given in Table 8-4.

C. Bottom Slope of the Channel Below the Rack.   In the above calculations the depth of flow and velocity at sections (1), (2), and (3) [Figure 8-5(b)] were calculated by assuming normal flow conditions in the channel. Since the channel has a free fall into the wet well, the actual depth into the channel will be significantly smaller than the normal depths calculated earlier. Furthermore, the velocities through the screen will also be significantly larger. The bottom of the channel therefore must be raised in order to maintain the previously calculated depths into the screen chamber. The calculation procedure is presented below.

1.   Compute the critical depth of flow.

a.   The critical depth in the rectangular channel is calculated from

$$Q = A_c \sqrt{g \, d_c} = b \sqrt{g} \, d_c^{3/2} \tag{8-5}$$

*h_{50} can also be calculated from Eq. (8–1).

$$h_{50} = \frac{(1.51 \text{ m/s})^2 - (0.54 \text{ m/s})^2}{2 \times 9.81 \text{ m/s}^2 \times 0.7} = 0.15 \text{ m}$$

TABLE 8-4   *Summary of Depth of Flow, Velocity, and Head Loss through Bar Rack at Design Peak Flow*

Conditions	Upstream Channel		Velocity through Rack (m/s)	Downstream Channel		Head Loss (m)
	Depth of Flow (m)	Velocity (m/s)		Depth of Flow (m)	Velocity (m/s)	
Clean rack	1.28	0.59	0.83	1.25	0.61	0.03[a]
50 percent clogging	1.40	0.54	1.51	1.25	0.61	0.15

[a]Actual head loss = 0.025 m.

where

d_c = critical depth, m
b = width of the channel, m
A_c = area of cross-section at critical depth, m^2
Q = flow in the channel, m^3/s

b. At peak design flow of 1.321 m^3/s and channel width of 1.74 m, critical depth d_c is calculated as below:

$$d_c = \left(\frac{1.321 \text{ m}^3/\text{s}}{1.74 \text{ m } \sqrt{9.81 \text{ m/s}^2}} \right)^{2/3} = (0.242)^{2/3}$$

$$= \ 0.39 \text{ m}$$

c. Critical velocity, $v_c = \dfrac{1.321 \text{ m}^3/\text{s}}{1.74 \text{ m} \times 0.39 \text{ m}} = 1.95$ m/s

It has been experimentally found that the critical depth generally occurs at a distance 3–10 times d_c from the point of free fall.[7]

2. Calculate the elevation of the channel bottom near free fall into the wet well.

a. The approximate elevation of the channel floor is calculated by applying the energy equation at sections (3) and (4) (Figure 8-5):

$$Z_3 + d_3 + \frac{v_3^2}{2g} = (Z_3 + Z_c) + d_c + \frac{v_c^2}{2g} + h_L$$

b. Neglecting the head loss due to friction (h_L) and using the values of d_3, v_3, d_c, and v_c previously calculated,

$$0 + 1.25 \text{ m} + \frac{(0.61 \text{ m/s})^2}{2 \times 9.81 \text{ m/s}^2} = Z_c + 0.39 \text{ m} + \frac{(1.95 \text{ m/s})^2}{2 \times 9.81 \text{ m/s}^2}$$

$$Z_c \ = \ 0.69 \text{ m}$$

The floor of the channel is therefore raised by 0.69 m as shown in Figure 8-6.

c. Raising the floor of the channel will produce an impoundment that will encourage deposition of solids during low flows. This can be overcome by manually raking the sloping bottom from the top of the channel and pushing the settled solids into the wet well. Another arrangement may utilize a 0.2-m-diameter orifice pipe at the bottom of the channel and sloping the pipe downward into the wet well. The opening of the pipe would be controlled by a sluice gate operated manually from the top. Many engineers have successfully designed Parshall flume or proportional weir below the bar screen and thus took advantage of the free flow resulting from the drop into the wet well. Such arrangements, however, constitute minor design detail and should be left to the design engineer to make his own selection. In Chapter 14 the design procedure of orifice pipe, proportional weir, and Parshall flume is covered in detail (see Problems 8-6 and 8-7).

D. Hydraulic Profile through the Bar Rack.   The hydraulic profile through the bar rack at peak design flow is illustrated in Figure 8-6. The elevation of the channel floor is assumed 0.00. A total of 0.03 m head loss occurs at peak design flow when the rack is clean. At 50 percent clogging the total head loss through the rack is 0.15 m.

E. Quantity of Screenings.   The quantity of screenings is obtained from Figure 8-4. For clear spacing of 2.5 cm, the average amount of screenings produced at the design average and peak flows are 20 and 36 m^3 per million m^3 of the flow. The design average flow is 0.441 m^3/s.

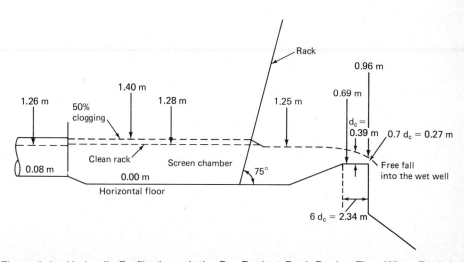

Figure 8-6   *Hydraulic Profile through the Bar Rack at Peak Design Flow When Rack Is Clean and at 50 Percent Clogging. All elevations are with respect to the datum at the floor of the rack chamber.*

a. Average quantity
    of screenings $= 20 \text{ m}^3/10^6 \text{ m}^3 \times 0.441 \text{ m}^3/\text{s} \times 3600 \text{ s/h} \times 24 \text{ h/d}$
                     $= 0.76 \text{ m}^3/\text{d}$
b. Maximum quantity
    of screenings $= 36 \text{ m}^3/10^6 \text{ m}^3 \times 0.441 \text{ m}^3/\text{s} \times 3600 \text{ s/h} \times 24 \text{ h/d}$
                     $= 1.37 \text{ m}^3/\text{d}$

F. Disposal of Screenings.   The screenings will be disposed of by sanitary landfilling. General discussion, design details, land requirements and equipment selection for sanitary landfill are provided in Chapter 19.

G. Design Details.   The design details of the bar rack are provided in Figure 8-7.

8-7   OPERATION AND MAINTENANCE, AND TROUBLESHOOTING AT MECHANICALLY CLEANED BAR RACK

Debris in wastewater if not removed properly will damage equipment and interfere with normal operation of treatment processes. Often high repair records of many processes and equipment may be attributed to poor performance of the bar rack. Common operation and maintenance problems that may develop at a bar rack and the procedure to correct them are summarized below.

8-7-1 Common Operating Problems and Suggested Solutions

The following items may be considered as troubleshooting guides:[8]

1. Obnoxious odors, flies, and other insects around the bar rack indicate prolonged storage of screenings at the facility. Increase frequency of removal and disposal of screenings.
2. Excessive screen clogging is an indication of (a) an unusual amount of debris in the wastewater, (b) low velocity through the rack, or (c) the automatic clock-operated screen rakes do not remove the debris fast enough. Possible solutions include identifying the source of waste causing excessive discharge of debris and stopping it at the source; providing a coarser rack; and resetting the timer cycle or installing level controller override.
3. Excessive grit accumulation in the chamber is due to low velocity in the channel. The possible solutions to this problem are: remove bottom irregularity; reslope the bottom; rake the channel; or flush regularly with a hose.
4. A jammed raking mechanism may render the mechanical rake inoperable, and the circuit breaker will not reset. Remove the obstruction immediately.
5. A broken chain or cable, or a broken limit suited will render the rake inoperable, but the motor will run. Inspect chain and switches, and replace them as necessary.
6. A defective remote control circuit or motor will render the rake inoperable without any visible problem. Check remote control circuit and motor, and replace them as necessary.

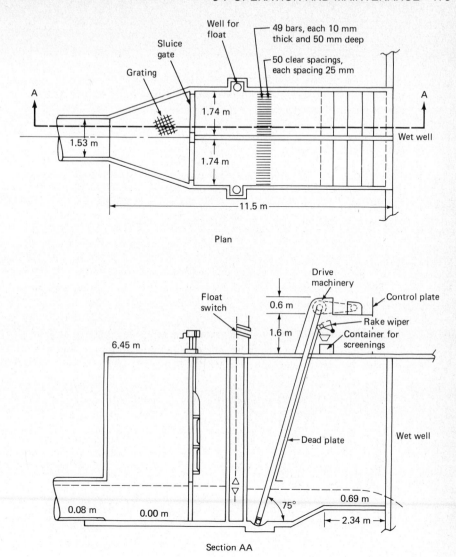

Figure 8-7 *Design Details of Bar Rack. All elevations are with respect to the datum at the floor of the rack chamber.*

8-7-2 Facility Maintenance

Screenings removed from municipal wastewater treatment plants are odorous and attract flies and insects. The screenings should be stored in covered containers and hauled daily for disposal. The area should be thoroughly hosed off daily with chemical solution (chlorine or hydrogen peroxide).

The bar screen raking mechanism (raking chain; sprocket, teeth, and other moving parts) should be inspected daily. All moving parts should be lubricated and adjusted as recommended by the manufacturer.

Each bar screen should be taken out of service for maintenance on a routine basis. The unit should be dewatered and components checked for painting, cable, chain, or teeth replacement, removal of obstructions, and straightening of bent bars etc.

8-8   SPECIFICATIONS*

Brief specifications of several components of bar racks are presented in this section. The purpose of these specifications is to describe many components that could not be fully covered in the Design Example. These specifications should only be used as a guide for understanding the equipment.

8-8-1 General

The manufacturers shall furnish and deliver ready for installation two identical, mechanically cleaned bar racks suitable for operation in a channel 1.74 m wide by 6.45 m deep from the bottom of the channel to the operating floor level. Each screen shall be capable of handling 1.321 m^3/s sanitary flow with a maximum depth in the channel at 1.40 m and shall have a hoisting capacity as specified. The equipment shall remove and elevate all material retained. No screenings once retained shall be allowed to carry over to the downstream side of the bar rack.

Each unit shall consist of bar rack, side frame, cleaning rakes, dead plate, rake-cleaning device, head section, and drive unit. The manufacturer shall have sufficient experience in the manufacture of mechanically cleaned bar racks of a similar size and design.

8-8-2 Bar Rack

The bar rack shall be inclined 75° from the horizontal and shall be held firmly in channel. The steel bars 10 mm thick and 50 mm wide formed straight and true are held firmly and accurately in place with 25-mm clear openings by means of welded spacers at each end. The bar rack shall extend from the bottom of the channel to a height of 0.5 m above the maximum water level.

8-8-3 Cleaning Rake

The cleaning rake shall be front-cleaned jam proof,[†] entering the bar rack from the upstream side, mounted on two strands of chain running over two sets of sprocket wheels at a speed of about 2 rpm. The cleaning rakes shall be of steel plate with teeth of suitable shape to effectively clean the top and side of the bars. The screenings shall be conveyed over dead plate and deposited into a suitable receptacle within the head section.

---

*Adapted in part from Refs. 6 and 8.

†Some designers prefer back-cleaning devices because they have less tendency to jam at the bottom.

8-8-4 Wiper and Dead Plate

An automatic rake wiper shall be provided for the screen, designed and placed so that the screenings do not wrap around the rake or the wiper during screening removal. The dead plate shall be constructed of at least 6-mm-thick steel, shall be bolted to the side frames of the screen, and shall extend from a point 23 cm above maximum liquid level to 75 cm below the center line of the dead shaft in the enclosed housing. A 6-mm-thick steel lip shall be provided at the discharge point of the dead plate.

8-8-5 Screen Chains, Sprockets, Shafting, and Drive Unit

Screen chains shall be of specified ultimate strength with proper attachment links, heat-treated steel pins, and rivets.

The sprockets for drive and screen chains shall be semisteel, cast in a chill of specified hardness and number of teeth, accurately ground to fit the chain. The driving sprockets shall be keyed firmly to shafts, and the foot shaft sprocket shall be suitable for underwater service under gritty conditions. All shafting shall be of steel, straight and true, of sufficient size, properly keyed to transmit power as required.

The drive unit shall consist of a speed reducer and an electric motor assembly with each bar screen having a separate drive unit. The motor shall be totally enclosed, fan-cooled, with a constant speed of ample power for starting and operating the mechanism under normal operating conditions without overloading. The motor shall conform to the specified volt, phase, and cycle current requirements.

8-8-6 Control System

The mechanically cleaned bar screens shall be operated by an independent, motor-driven time clock adjustable to give cycles ranging from 0 to 150 min. The control panel shall be enclosed and shall contain starters, circuit breaker, control relays, and transformers. In addition, the control panel shall have the following pushbuttons: hoist, lower, reset, manual, automatic, stop and start, for proper operation and resetting. A float shall be provided in a stilling well upstream of the rack for an automatic high water level override to start the raking device should the liquid in the channel reach a predetermined depth.

8-9   PROBLEMS AND DISCUSSION TOPICS

8-1.   All calculations in the Design Example were performed at the peak design flow of 1.321 m^3/s. Check the depth of flow and velocity in the chamber upstream and downstream of the bar rack at the average design flow of 0.441 m^3/s. Also calculate the velocity through the clear openings at the bar rack and compare your result with the design criteria given in Sec. 8-6-1.

8-2.   Redesign the bar rack in the Design Example using 47 bars instead of 49. Calculate the velocity through the clear openings at the bar rack and compare your results with those calculated in the Design Example.

8-3.   The average design flow at a wastewater treatment facility is 1.0 m^3/s. The bar rack is mechanically cleaned with an average clear spacing of 1.5 cm. Estimate the average quantity of screenings that will be collected each day.

8-4.   A bar rack precedes a grit channel. The outlet of the grit channel is controlled by a proportional weir such that the depth of flow in the channel is directly proportional to the rate of flow. A 91-cm (36-in.) intercepting sewer brings the flow to the screen chamber. The depth of flow and velocity in the sewer at peak design flow of 0.43 m^3/s are 57.5 cm and 1.0 m/s, respectively. The bar rack has 47 bars. Each bar is 6.4 mm ($\frac{1}{4}$ in.) wide and clear spacing is 2.5 cm (1 in.). The floor of the bar rack chamber is horizontal and is 7.6 cm below the invert of the intercepting sewer. Calculate the depth and velocity in the bar rack chamber upstream and downstream of the rack and through the rack. Also calculate the head loss when the rack is clean and 50 percent clogged.

8-5.   Prepare a list of operation and maintenance problems at various wastewater treatment units that may be directly attributable to the screenings that may pass through a bar rack.

8-6.   Read Sec. 8-6-2, step C2c, and review the designs of the proportional weir and Parshall flume in Chap. 14. What are the advantages of providing a proportional weir or Parshall flume at a screen chamber upstream of a wet well?

8-7.   Study the design of the proportional weir given in Chap. 14. Design a proportional weir at the outfall of the screen chamber at the wet well of the Design Example. Calculate the dimensions of the proportional weir using the following information. Maximum depth of flow into the chamber downstream of the bar rack at peak design flow of 1.321 m^3/s is 1.25 m. The crest of the proportional weir is flushed with the bottom of the chamber.

REFERENCES

1. Joint committee of the Water Pollution Control Federation and the American Society of Civil Engineers, *Wastewater Treatment Plant Design,* MOP/8, Water Pollution Control Federation, Washington, D.C., 1977.
2. Metcalf and Eddy, Inc., *Wastewater Engineering: Treatment, Disposal, and Reuse,"* Second Edition, McGraw-Hill, New York, 1979.
3. "The 1971 Environmental Wastes Control Manual and Catalog File," *Public Works Journal Corporation,* Ridgewood, N.J., 1971.
4. Babbitt, H. E., and E. R. Baumann, *Sewerage and Sewage Treatment,* John Wiley & Sons, New York, 1958.
5. Kim, B. C., S. R. Qasim, and R. Verma, "Modified Filtration Theory for Microstrainer Design," *Journal Korean Institute of Chemical Engineers,* Vol. 9, No. 3, 1971, pp. 149–153.
6. *Rex Water Quality Control Equipment,* Envirex Inc., a Rexnord Company, Binder No. 315, Vol 1, Waukesha, Wisconsin.
7. Daugherty, R. L., and J. B. Franzini, *Fluid Mechanics with Engineering Applications,* McGraw-Hill, New York, 1977.
8. U.S. Environmental Protection Agency, *Performance Evaluation and Troubleshooting at Municipal Wastewater Treatment Facilities,* MO-16, EPA-430/9-78-001, Office of Water Program Operation, Washington, D.C., 1978.

Pumping Station

9

9-1 INTRODUCTION

Wastewater pumping stations are used to lift or elevate the liquid from a lower elevation to an adequate height at which it can flow by gravity or overcome hydrostatic head. There are many pumping applications at a wastewater treatment facility. These applications include pumping of (1) raw or treated wastewater, (2) grit, (3) grease and floating solids, (4) dilute or well-thickened raw sludge, or digested sludge, (5) sludge or supernatant return, and (6) dispensing of chemical solutions. Pumps and lift stations are also used extensively in the collection system. Each of the various pumping applications is unique and requires specific design and pump selection considerations.

The subject matter on pumping, pump selection, and design is very broad. Many books and technical papers have been written on the subject that cover the theory and practice in detail.[1-8] In this chapter various types of pumps and their applications are briefly discussed. Because centrifugal pumps are most commonly used for raw wastewater pumping, discussions on centrifugal pumps and pump selection, design procedure for raw wastewater pumping station, operation and maintenance, and specifications of pumping equipment and pumphouse are presented in detail.

9-2 PUMP TYPES AND APPLICATIONS

According to the Hydraulic Institute, all pumps may be classified as kinetic energy pumps or positive displacement pumps.[1] Brief descriptions and application of many types of pumps in these two classes are given in Table 9-1. A typical pump application chart is illustrated in Figure 9-1. Basic configurations of many types of pumps are also shown in Figure 9-2. Detailed discussions on the subject may be obtained in Refs. 1–8.

9-3 TYPES OF PUMPING STATIONS

The pumping stations are classified as wet well and dry well. The wet-well stations employ either suspended or submersible pumps. Suspended pumps have the motor mounted above the liquid level in the wet well while the pump remains submerged. Submersible pumps have integral motors with special seals suitable for operation below liquid level. The suspended and submersible pumping arrangements in wet well are illustrated in Figures 9-2(a) and (b).

TABLE 9-1   *Pump Types and Major Applications in Wastewater*

Major Classifications	Pump Type	Brief Description	Major Pumping Applications
Kinetic	Centrifugal	Consists of an impeller enclosed in a casing with inlet and discharge connections. The head is developed principally by centrifugal force	Raw wastewater, secondary sludge return and wasting, settled primary and thickened sludge, effluent
	Peripheral (torque-flow or vortex)	Consists of a recessed impeller in the side of the casing entirely out of the flow stream. A pumping vortex is set up by viscous drag	Scum, grit, sludge and raw wastewater
	Rotory	Consists of a fixed casing containing gears, vanes, pistons, cams, screws, etc., operating with minimum clearance. The rotating element pushes the liquid around the closed casing into the discharge pipe	Lubricating oils, gas engines, chemical solutions, small flows of water and wastewater
Positive displacement	Screw	Uses a spiral screw operating in an inclined case	Grit, settled primary and secondary sludges, thickened sludge, raw wastewater

Diaphragm	Uses flexible diaphragm or disk fastened over edges of a cylinder	Chemical solutions
Plunger	Uses a piston or plunger that operates in a cylinder. The pump discharges a definite quantity of liquid during piston or plunger movement through each stroke	Scum, and primary, secondary, and settled sludges. Chemical solutions
Airlift	Air is bubbled into a vertical tube partly submerged in water. The air bubbles reduce the unit weight of the fluid in the tube. The higher unit weight fluid displaces the low unit weight fluid forcing it up into the tube	Secondary sludge circulation and wasting, grit
Pneumatic ejector	Air is forced into the receiving chamber which ejects the wastewater from the receiving chamber	Raw wastewater at small installation (100 to 600 ℓ/min)

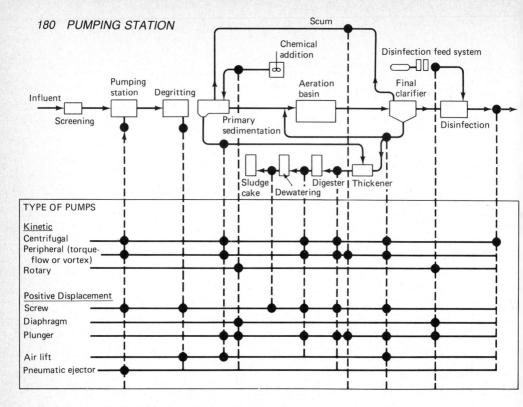

Figure 9-1   Typical Wastewater Pumping Applications.

The dry-well stations employ either dry-well or self-priming centrifugal pumps. The dry-well centrifugal pumps operate within a dry well adjacent to the wet well. The volute of the pump is positioned below the low water level in the wet well to ensure a positive suction or prime as shown in Figures 9-2(c). In the other case a recessed dry section may be located within a wet well as shown in Figure 9-2(d). The pumps are specially designed to provide a suction lift and are self-priming. Other types of pumping stations utilize screw, air lift pumps, or pneumatic ejectors. These types of pumping systems are also illustrated in Figures 9-2(e)–(g).

9-4   HYDRAULIC TERMS AND DEFINITIONS COMMONLY USED IN PUMPING

Common terms used in pumping and pump analysis include (1) head, (2) capacity discharge or flow rate, and (3) work power, and efficiency. These terms are briefly discussed in this section.

9-4-1 Head

Head describes hydraulic energy (kinetic or potential) equivalent to a column of liquid of specified height above a datum. Head and pressure may be expressed in terms of the other

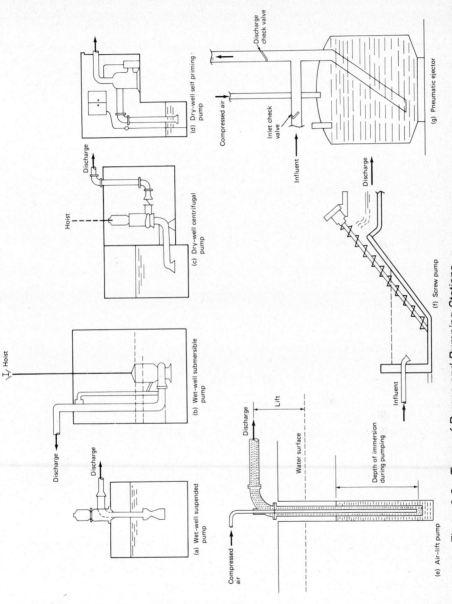

(a) Wet-well suspended pump

(b) Wet-well submersible pump

(c) Dry-well centrifugal pump

(d) Dry-well self priming pump

(e) Air-lift pump

(f) Screw pump

(g) Pneumatic ejector

Figure 9-2   Types of Pumps and Pumping Stations.

(1 m of water = 9.81 kPa; or 1 ft of water = 0.433 psi). Many head terms in pumping include static suction lift (or static suction head), static discharge head, and total static head. These terms are illustrated in Figure 9-3.

The total dynamic head (TDH) of a pump is the sum of total static head, the friction head (including minor losses), and the velocity head. The friction head consists of loss of head in piping (suction and discharge). It is calculated from Darcy-Weisbach or Hazen-Williams equations. The minor losses are produced due to fittings, valves, bends, entrance, exit, etc., and are normally calculated as a function of velocity head. Some of these terms are defined in Eqs. (9-1)–(9-6):

$$\text{TDH} = H_{\text{stat}} + h_f + h_m + h_v \tag{9-1}$$

$$h_f = \frac{fLV^2}{2gD} \text{ (Darcy–Weisbach)} \tag{9-2}$$

$$h_f = 6.82 \left(\frac{V}{C}\right)^{1.85} \times \frac{L}{D^{1.167}} \text{ (Hazen–Williams, SI units)} \tag{9-3}$$

$$h_m = K \frac{V^2}{2g} \tag{9-4}$$

$$h_v = \frac{V^2}{2g} \tag{9-5}$$

$$h_L = h_f + h_m + h_v \tag{9-6}$$

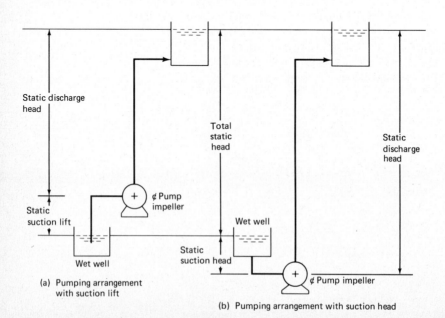

Figure 9-3   Head Terms Used in Pumping.

The Hazen–Williams equation is most commonly used for force mains. This equation in various forms is given below:

$$V = 0.355 \, C \, D^{0.63} \left(\frac{h_f}{L} \right)^{0.54} \quad \text{(SI units)} \tag{9–7}$$

$$V = 0.550 \, C \, D^{0.63} \left(\frac{h_f}{L} \right)^{0.54} \quad \text{(U.S. customary units)} \tag{9–8}$$

$$Q = 0.278 \, C \, D^{2.63} \left(\frac{h_f}{L} \right)^{0.54} \quad \text{(SI units)} \tag{9–9}$$

$$Q = 0.432 \, C \, D^{2.63} \left(\frac{h_f}{L} \right)^{0.54} \quad \text{(U.S. customary units)} \tag{9–10}$$

where

TDH = total dynamic head (or total head, TH), m (ft)
H_{stat} = total static head, m (ft) (see Figure 9–3)
h_f = total friction head loss in suction and discharge pipes, m (ft)
h_L = total head loss, m (ft)
h_m = minor losses, m (ft)
h_v = velocity head, m (ft)
V = velocity in the pipe, m/s (ft/s)
Q = flowrate, m³/s (ft³/s)
f = coefficient of friction (the value of f depends on the Reynolds number and the relative roughness and diameter of the pipe. It may range from 0.01 to 0.10)
C = coefficient of roughness (the value of C depends on the material and age of the pipe. It may range from 60 to 140)
D = equivalent diameter of the pipe, m (ft)
g = acceleration due to gravity, 9.81 m/s² (32.2 ft/s²)
K = head loss coefficient (see Appendix B)
L = length of pipe, m (ft)

9-4-2 Capacity (Discharge or Flow Rate)

The capacity, discharge or flow rate of a pump is the volume of liquid pumped per unit of time. It is expressed as cubic meters per second (m³/s), liters per second (ℓ/s), gallons per minute (gpm), cubic feet per second (ft³/s), etc.

9-4-3 Work Power and Efficiency

The work done by a pump is proportional to the product of the specific weight of the fluid being discharged and the total head against which the flow is moved. The pump efficiency is the ratio of the useful pump output power to the input power. Pump output power and efficiency are expressed by Eqs. (9-11)–(9-13):

$$P_w = K'Q(\text{TDH})\gamma \tag{9–11}$$

$$E_p = \frac{P_w}{P_p} \tag{9-12}$$

$$E_e = \frac{P_p}{P_m} \tag{9-13}$$

where

P_w = power output of the pump (water power), kW (HP)
P_p = power input to the pump (brake power), kW (HP)
P_m = power input to the motor (electrical energy), kW (HP)
Q = capacity, discharge, or flow rate, m³/s (ft³/s)
TDH = total dynamic head, m (ft)
γ = specific weight of the liquid pumped kN/m³ (lb/ft³)
E_p = pump efficiency, usually 60–85 percent
E_e = motor efficiency, usually 90–98 percent
K' = constant depending on the units of expression
(TDH = m, Q = m³/s, γ = 9.81 kN/m³, P_w = kW, K' = 1 kW/kN · m/s)
(TDH = ft, Q = ft³/s, γ = 62.4 lb/ft³, P_w = HP (horse power), K' = 1/550 HP/ft · lb/s)

9-5   CENTRIFUGAL PUMPS

Centrifugal pumps are most commonly used for water and wastewater pumping. Because of their extensive use in water supply and wastewater treatment, the remaining chapter is devoted to the principles of centrifugal pumps, their design, specifications, and pumping station operation.

9-5-1 Action of a Centrifugal Pump

In centrifugal pumps the head is developed principally by centrifugal force. The inlet to the pump is axial and the outlet is tangential. The flow is accelerated by the rotating impeller which imparts to it both radial and tangential velocity; the ratio of the two depends on the design of the impeller. The increase in cross-section of the volute (casing) produces the change from velocity head to pressure head.

9-5-2 Effect of Varying Speed

The pressure and discharge of a pump varies with pump speed. The specific speed of a pump is defined as the speed at which a pump will discharge a unit flow under a unit head at maximum efficiency. It is expressed by

$$N_s = \frac{NQ^{1/2}}{H^{3/4}} \tag{9-14}$$

where

N_s = pump specific speed*
N = pump rotative speed, rev/min (rpm)
Q = flow at optimum efficiency, m³/s (gpm)
H = total head, m (ft)

The specific speed is an index number descriptive of suction characteristics of a given impeller. As a general rule, pumps with high specific speeds are more efficient than those with low. Also, high specific speed is associated with high discharge at low head. Use of specific speed in pump selection is given in Sec. 9-5-10.

9-5-3 Pump Characteristics Curves

The discharge of a centrifugal pump is a function of not only the specific speed but also of the pressure conditions under which the pump operates. The pump manufacturers conduct tests where pump capacity is changed by throttling a valve in the discharge piping and measuring the flow and pressures at the suction and discharge lines. The power inputs are also measured. The results of the tests are represented by a series of curves representing head, power, and efficiency versus flow at constant speed. These curves are shown in Figure 9-4. The shape of head capacity curves is important in selecting pumps for specific applications. These shapes of head capacity curve may be rising head, flat, steep, or drooping head. Such curves are a function of specific speed and pump design.

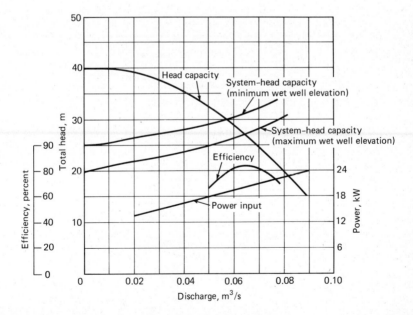

Figure 9-4   Characteristic Curves of a Centrifugal Pump and System-Head Capacity Curves

*N_s in U.S. customary units = 51.6 × N_s in metric units.

Manufacturers frequently present pump data in the form shown in Figure 9-5. A pump casing can accommodate impellers of several sizes; therefore, a particular pump casing may be used for different head and flow applications. This feature is helpful because the existing pump can be expanded by replacement of the impeller and motor. Use of the pump curves in pump selection is discussed in the Design Example. For such modifications, if all other variables are kept equal, Q, TDH, and P can be varied with the choice of pump speed and impeller diameter in accordance with the following relations:

Pump speed N variable

1. Q varies as N
2. TDH varies as N^2
3. P_w varies as N^3

Pump impeller diameter D variable

1. Q varies as D^3
2. TDH varies as D^2
3. P_w varies as D^5

9-5-4 System Head Capacity

A system head capacity curve is prepared by plotting total dynamic head (TDH) at various flow conditions. This curve is prepared by the designer for the suction and discharge

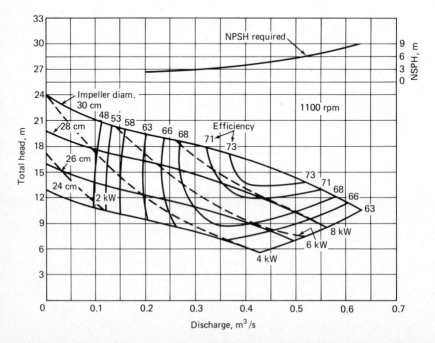

Figure 9-5   Characteristic Curves of a Centrifugal Pump with Several Impeller Sizes.

piping arrangements. The intersection point of pump head capacity and system head capacity curves is important as this point represents the operating head and capacity of the pump for the selected piping system. For good pump selection this point (pump-operating point) should be as close as possible to the peak efficiency. If the wet-well levels fluctuate considerably, it is usual to prepare two system curves using the minimum and maximum static heads based on wet-well water surface elevations. The system head capacity curve is shown in Figure 9-4.

9-5-5 Net Positive Suction Head (NPSH)

For a centrifugal pump to operate, the liquid must enter the center of the impeller under pressure, usually atmospheric pressure. This pressure is referred to as net positive suction head (NPSH). There are two values of NPSH: the available and the required. The available NPSH depends on the location and design of the intake system and is calculated by the engineer. It is the minimum suction head required at the inlet of the impeller to prevent boiling of the liquid under the reduced pressure condition created at the impeller and smooth operation of the impeller without cavitation. Equation (9-15) is used to calculate the available NPSH. The required NPSH is determined by the pump manufacturer based on extensive testing programs. The available NPSH should exceed the required NPSH with a margin of safety (1 meter or more).[5]

$$NPSH_{av} = H_{abso} + H_s - h_L - H_{vp} \qquad (9\text{--}15)$$

where

$\quad NSPH_{av}$ = available net positive suction head, m

$\quad\quad H_{abso}$ = absolute pressure on the surface of the liquid in the suction well, m (ft). This is the barometric pressure at the given altitudes) (See Appendix A.)

$\quad\quad\quad H_s$ = the static head of the liquid above the center line of the pump, m (ft). If the liquid level is below the centerline, H_s is negative.

$\quad\quad\quad h_L$ = total head loss due to friction, entrance, valves, fittings and specials in the suction piping, m (ft).

$\quad\quad H_{vp}$ = absolute vapor pressure of the liquid at the pump temperature, m (H_{vp} at various temperatures are given in Appendix A).

The calculation procedure for available NPSH is illustrated in the Design Example.

9-5-6 Cavitation

Cavitation is the phenomenon of cavity formation or collapse of cavities. Cavities develop when the absolute pressure in a liquid reaches the vapor pressure related to the liquid temperature. This may happen under the following design and operating conditions[5]:

1. When impeller under high speed (revolutions per minute, rpm) travels faster than the liquid can enter or move
2. When suction is restricted

3.   When required NPSH is equal to or greater than available NPSH
4.   When specific speed is too high for optimum design parameters
5.   When the temperature of the liquid is too high for suction conditions
6.   When the pump operates at extreme capacities below or above the *best efficiency point* (BEP)

Cavitation causes reduction in flow and under serious conditions the pump may lose its prime, cause pitting of the impeller surface, a rattling or pinging noise, and eventual breakdown of the pumping equipment.

9-5-7 Pump Combinations

Pumps may be connected in series or in parallel. In parallel operation, for a given head the total discharge is added up for all the pumps. In series operation, the total head for all pumps is added up for a given discharge. Parallel and series combinations are shown in Figures 9-6(a) and (b).

9-5-8 Modified Pump Head-Capacity Curve

When two or more pumps are arranged in parallel, these pumps will discharge into a common header or force main. The procedure for determining the pump-operating point is based on development of modified pump head capacity curve. This curve is prepared by subtracting from the pump head capacity curve the head losses in the suction and discharge piping of each pump. The combined head capacity curve is prepared using the modified curves of each pump. The point of intersection of the combined curve with the system head capacity curve gives the total capacity of the pump combination and modified head for each pump. Note that only the losses in that part of the system common to all pumps are included in the system head capacity curve. The modified head capacity curve is shown in Figure 9-6(c). Procedure for preparation of modified curve for pump combination is presented in the Design Example (see Sec. 9-9-2, step B1). Detailed discussion may be found in Ref. 2.

9-5-9 Pump Classification

Centrifugal pumps are classified in many ways. These classifications or types are generally based on the manner in which the liquid enters and leaves the casing and the manner in which the impeller or vanes impart the energy to the water to transform the velocity into the head. Based on the impeller configuration and specific speed, pumps are traditionally divided into three major classes: radial flow, mixed flow, and axial flow. There are, however, many other variations as presented in Table 9-2. Basic characteristics of the radial flow, mixed flow, and axial flow pumps are given below.

Radial Flow Pumps

1.   The pressure head is developed basically by the action of centrifugal force.
2.   The liquid normally enters the impeller at the hub and flows radially to the periphery.

TABLE 9-2   Classification of Centrifugal Pumps

Types or Variations	Description
Suction	
Single-suction	A pump equipped with one or more single-suction impellers
Double-suction	A pump equipped with one or more double-suction impellers
Number or stages	
Single-stage	A pump in which total head is developed by one impeller
Multistage	A pump having two or more impellers acting in series in one casing
Position of the shaft	
Horizontal	A pump with the shaft normally in horizontal position
Vertical pump	
Dry-pit type	A pump with vertical shaft located in a dry well
Submerged type	A pump with vertical shaft located in a wet well
Casing	
Volute	A pump having a casing made in the form of a spiral or volute. The velocity is transformed into head by gradually increasing area of the water passage. Dry-well pumps generally have volute case.
Circular	A pump having a casing of constant cross-section, concentric with the impeller. Curved vanes are used to transform velocity into head. Wet-well pumps have circular casing.
Diffusion (turbine)	A pump that transforms velocity into head by diffusion vanes. Diffusion casing is used with wet-well pumps.
Impeller	
Enclosed (front and back shroud)	A pump that has enclosed impeller with shroud. Commonly used for pumping sanitary sewage.
Open or semienclosed	A pump that is constructed with the vanes attached to an upper or inbound shroud. Lower shroud is not used. The pump is best suited for pumping large volumes with intermittent service (storm water).

3. The impellers used may be single or double suction.
4. The impeller may have vanes which may be straight or double curvature, such as Francis-Vane.
5. Pump shaft may be horizontal or vertical.
6. The pumps have specific speeds below 80 (4200 in U.S. customary units) with single impeller and 115 (6000 in U.S. customary units) with double-suction impeller.
7. For raw wastewater pumping, the nonclog type is used (Sec. 9-5-11).

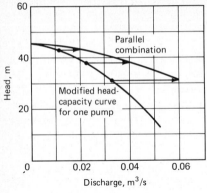

(a) Parallel pump combination
(two identical pumps)

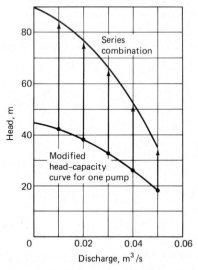

(b) Series pump combination
(two identical pumps)

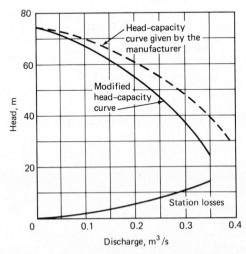

(c) Modified pump head–capacity curve (the station
losses are subtracted from the manufacturer's
head-capacity curve to obtain modified
head-capacity curve.)

Figure 9-6   Pump Combinations in Parallel and Series and Modified Pump Curve.

Mixed Flow Pumps

1. Head is developed partly by centrifugal force and partly by the lift of the vanes on the liquid.
2. The liquid enters axially and discharges in an axial and radial direction.
3. The casing may be either the volute or the diffusion type. This gives two variations of mixed flow pumps, namely, mixed flow volute, and mixed flow propeller pumps.
4. The pump may have single or double suction, and maybe single or multistage.
5. The specific speed of the pump is in the range of 80–120 (4200–6250 U.S. customary units).
6. As specific speed increases, mixed flow pumps become more like axial flow pumps.
7. The mixed flow pumps are applicable for medium-head application 7–16 m and for medium to large capacities.
8. The pumps require positive submergence, although with proper selection of rotative speeds they may be used for limited suction lift applications.
9. Common application is for pumping of raw wastewater and storm water flow. For raw wastewater pumping, a nonclog-type pump must be used (Sec. 9-5-11).

Axial Flow Pump

1. The pumps are also called propeller pump. They develop most of the head by propelling or lifting action of the vanes on the liquid.
2. They have single inlet impeller with the flow entering axially and discharging axially.
3. Specific speed is usually above 200 (10,000 in U.S. customary units).
4. These pumps are customarily used for large flows (over 600 ℓ/s) and low head (below 10 m) installations. Best applications include irrigation work and storm water pumping. They should not be used for pumping raw wastewater, sludge, or unscreened storm water.

9-5-10 Comparison of Specific Speed and Pump Class

Pump comparison based on impeller configuration and specific speed is best illustrated in Figure 9-7 and use of an example. Suppose a designer needs to select a pump type for pumping of raw wastewater for a total head of 30 m and the maximum anticipated capacity of 0.25 m³/s at an operating drive of 1200 rpm. The specific speed is calculated from Eq. (9-14):

$$N_s = \frac{1200 \times (0.25)^{1/2}}{(30)^{3/4}} = 46.81$$

Referring to Figure 9-7, this application falls within the radial flow, Francis-Vane area, with an anticipated pump efficiency of 88 percent.

9-5-11 Nonclog Pump

The term nonclog pump has been used for radial flow and mixed flow pumps. Nonclog pumps differ from conventional units in arrangement, size, smoothness, and contour of channels and impellers to permit passage of clogging material. These pumps have open passages (often slightly less than that of the discharge pipe) and a minimum number of vanes (two in smaller and not more than four in large-size pumps). The peripheral (torque flow or vortex) pumps also offer unique nonclog features as the blades are recessed entirely into the curved shroud; thus the flow does not pass through the impeller. Experience has shown that in pumping of raw wastewater, such materials as rope, string, rags, sticks, cans, rubber goods, and grease present the greatest problem. For pumps

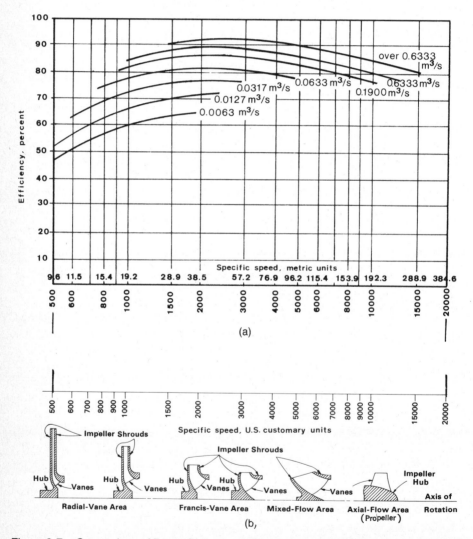

Figure 9-7 Comparison of Pump Capacity, Efficiency, and Impeller Designs with Specific Speed. (a) Pump efficiency versus specific speed. (Courtesy Worthington Pump Corporation.) (b) Profile of several pump impeller designs placed according to where each design fits on the specific-speed scale. (Courtesy Hydraulic Institute.)

smaller than 25 cm in size that handle sanitary sewage, nonclog units are used exclusively. Pumps smaller than 10 cm in size are not recommended for such applications.

9-5-12 Sludge Pumping

Pumps commonly used for sludge transport are nonclog centrifugal, peripheral (torque-flow or vortex), screw, and plunger pumps. The head loss in the pipelines transporting sludge is generally calculated from the Hazen–Williams equation [Eq. (9-3)]. The Hazen–Williams coefficient C is decreased because sludge is more difficult to pump than wastewater. The suggested apparent values of Hazen–Williams coefficient for different types of sludges are summarized below:

1. Waste activated sludge (total solids up to 1 percent) $C = 90$
2. Primary sludge (total solids up to 4.5 percent) $C = 55$
3. Thickened or digested sludge (total solids up to 8 percent) $C = 35$

It is a good design practice to provide sludge piping larger than 15 cm (6 in.) in diameter. Smaller piping may be used for well-homogenized sludge, or if the piping is glass lined. For small sludge flows, the pumps are operated on a time cycle to achieve larger flows and velocities during pumping cycle. A liberal number of cleanouts and hose gates should be provided in the piping for cleaning the stoppages.

9-5-13 Constant and Variable Speed Pumps

Constant-speed pumps are generally used where continual flow is not essential. These pumps generally operate on level controls. These pumps are simplest, most reliable, and low in cost. However, if a continuous flow is desired from the pumping station, a variable-speed drive may be required to adjust the pumping speed to the influent flow rate. The control system varies the head capacity curve of the pump in relation to the station head curve. Therefore, while the flow is continuous from the station, the flow rate will vary with the water surface elevation of the wet well.

Pumping stations having large variations in the flow rate would require more pumping units under the constant-speed system than those under the variable-speed system. On the other hand, the constant drive units generally have better efficiency than the variable-speed units because the variable-speed units reach the best efficiencies when operating at maximum speed.

9-5-14 Pump Drive

The pumping drives include electric motors or internal-combustion engines. Electric motors are most commonly used where reliable electrical power is available. Electric motors are either constant speed or variable speed. Constant-speed motors may be either the squirrel cage, synchronous, or wound rotor induction type. Variable-speed electric drives are also classified in three general groups: wound rotor induction, brush-shift, and slip-coupling combinations. Variable-speed electric drives are used with variable-speed pumps to match the pump output with the influent conditions. They have several other benefits, such as (1) reduction in wet-well size, and (2) water hammer is controlled by slowing the pump gradually before shutdown. For wastewater pumping applications the motor speed is generally in the range of 500–1700 rpm.

Internal combustion engines are installed where (1) wastewater gas is available for fuel, (2) electrical power may not be available, and a (3) backup system for electric power outages may be necessary. The possible use of combustion engines as opposed to electric motors is limited due to noise, vibration, space, maintenance, and operating attendance. Detailed discussion of pump drive units may be obtained in Refs. 2–4 and 7.

9-6   DESIGN OF PUMPING STATION

The design of a pumping station involves selection of site, pumping station, and pumps and controls. Each of these considerations is briefly discussed below.

9-6-1 Selection of Site

Proper site selection for a pumping station is essential for good design and operation of the facility. Important site considerations include (1) all weather roads and parking, (2) protection from 100-year flood stage, (3) location relative to population areas due to odor problems (provision for odor control facilities, such as aeration, chlorination, or hydrogen peroxide treatment, should also be made), and (4) dual-power source for continuous operation in the event of a primary power failure.

9-6-2 Selection of Pumping Station

Wet-well and dry-well pumping stations were discussed in Section 9-3. Wet-well pumping stations utilize suspended and submersible pumps, while dry-well stations utilize self-priming pumps or pumps positioned below the liquid level in the wet well. Basic design considerations for wet- and dry-well pumping stations are given below:

1. The wet and dry wells should be separated by a water- and gas-tight wall with separate entrances provided to each well.
2. Stairway should have nonslip steps and should be provided in dry well, and in wet well where bar screens or other equipment requiring routine inspection and maintenance may be located. Manhole steps with an entrance hatch should be avoided. Circular or spiral stairway with walk-in door should be preferred.
3. As much as possible, all equipment requiring routine inspection should be installed in the dry well.
4. Ventilation should be provided for all dry wells below ground surface and in all wet wells housing equipment.
5. As a general rule, ventilation equipment should provide at least six air changes per hour or the minimum standard of the applicable building codes.
6. The bottom of the wet well should have a minimum slope of 1 : 1 to the pump intake. There should be no projections which may allow solids accumulation. Antivortex baffling should be considered for the pump suction in large pumping stations.

7. Bell mouth inlets are better than straight inlets because they eliminate sharp edges on which solids can accumulate, they minimize head loss, and there is less possibility of vortex forming in the wet well when turned-down bellmouth inlets are used. The bell should not be more than $D/2$ and not less than $D/3$ above the floor (where D is the diameter of the suction bell). In order to obtain scouring inlet velocities and still obtain nearly optimum hydraulic entrance conditions, velocities at the entrance should be kept less than 3 m/s.

8. Wet wells are generally expensive because they normally involve excavations and a great deal of structural concrete. Therefore, storage volume must be established with respect to pump type and operation (constant-speed or variable-speed drive). Consideration should also be given to the submergence of the pump suction inlet to prevent vortexing and smooth transition of flow. Constant-speed pumps require a larger wet well volume to prevent short cycling of the pump (frequent starting and stopping). For variable-speed pumps sufficient volume, i.e., time, should be provided for the change in capacity when a pump is started and brought up to full speed before the start or stop point of the next pump is reached. Many regulatory agencies limit maximum retention time at average design flow. This criteria is to minimize septic condition and odor problems.

9. Wet wells should be designed to have two compartments such that each compartment could be isolated for maintenance purposes without stopping the operation of the pumping station.

10. Mechanical lifting facilities should be provided where pump, motor, valve, meters, piping, and the like weigh more than 90 kg (198 lb). Hoisting equipment such as a crane should be permanently installed if any single item is expected to weight more than 230 kg (510 lb).

11. Pump suction piping should be oversized to facilitate future expansion of the pumping station. Reducers in suction piping should be of the eccentric type so that a level top elevation can be maintained. This will prevent entrapment of air into the suction piping.

9-6-3 Selection of Pump and Control

Pumping requirements can be determined from the knowledge of head and flow. Selection of specific unit requires examination of manufacturers rating curves and comparison of these with the system-head capacity curve. If pump selection is left to the manufacturer, the engineer must check its suitability. The following considerations may be given to pump selection:

1. All raw wastewater pumps should have nonclog impellers and should be able to pass 6.5-cm minimum-size objects. The suction and discharge lines should not be less than 10 cm. Inspection and cleanout plates on the bowl of the pump or a head hole at the first fitting connected to the suction line should be provided.

2. The capacity of the pumping station should be such that the expected peak flow can be pumped with the largest pumping unit out of service.

3. The size and number of pumping units for larger stations should be selected so that the

range of inflow can be met without starting and stopping pumps too frequently, and without requiring excessive wet well storage capacity. Pumping stations discharging directly to the treatment plant should not cause surging effect. This can be avoided by providing automatically controlled variable-capacity pumps matching the inflow rate. Occasionally, the capacity and depth of the wet well can be coordinated with the constant-speed pumping units so that the rise and fall of the water level in the wet well results in a variable-pumping capacity approximating the inflow rate. This is possible because all centrifugal pumps have the inherent characteristic that as the head decreases the capacity increases.

4. No single pumping unit should have a capacity greater than the peak design capacity of the wastewater treatment plant.
5. It is considered good practice to provide space for additional pumps for future needs.
6. For an effective control system a minimum of 0.6 m differential water head between low and high water level in the wet well should be provided. The air bubbler system is the most reliable system for control. Bubbler system should be equipped with two air compressors or with a standby cylinder of inert gas (CO_2, N_2, etc.) in case the air compressor fails. The bubbler system should also be designed with an automatic purge system to prevent clogging of the bubbler line. Float control systems may be fouled due to grease and floating objects. Considerations should be given to those systems that can help in removal of scum and floating matter.
7. Only self-priming pumps or pumps with an acceptable priming system must be used where suction head is negative. A self-priming pump should include means for venting the air back to the wet well when the pump is priming. Self-priming pumps have caused serious problems; therefore, their use in raw wastewater pumping is not recommended.

9-7   MANUFACTURERS OF PUMPING EQUIPMENT

A list of the manufacturers of pumping equipment is provided in Appendix D. The responsibilities of the design engineer for equipment selection are given in Sec. 2-10.

9-8   INFORMATION CHECKLIST FOR DESIGN OF PUMPING STATION

Before designing a pumping station, the design engineer must develop design data and make many decisions that are necessary for good design. Some of the important items are listed below:

1. Characteristics of the liquid which must be pumped (suspended solids, floating solids, maximum size of the objects, density, temperature pressure, etc.).
2. Expected flow range (minimum, average, and peak design flows).
3. Site plan, piping schematic, and hydraulic profile from wet well to the receiving facility.
4. Minimum and maximum water surface elevations in the wet well, and elevation of water surface at receiving facility. These values are established after evaluating the liquid elevation in influent sewer, plant layout, and hydraulic considerations

through the plant, flood protection, site plan considerations, and operating and maintenance requirements. A great deal of engineering judgment and experience is required in establishing these elevations.

5. Type of pumping station (wet-well or dry-well) and suction conditions [including limits of submergence, suction head, suction lift, and available net positive suction head (NPSH)].
6. System head capacity curves.
7. Initial pumping unit selection should include types and number of pumps, constant- or variable-speed drive, specific speed, etc.
8. Drive unit (electrical motor or engine) and expected power requirements.
9. Design criteria for pumping station prepared by the concerned regulatory agencies.
10. Equipment manufacturers and equipment selection guide (catalog).

9-9   DESIGN EXAMPLE

9-9-1 Design Criteria Used

1. Design the pumping equipment for peak, average, and minimum design flows of 1.321, 0.440, and 0.220 m^3/s (see Table 6-9).
2. The site plan, piping schematic, and hydraulic profile developed for the Design Example are shown in Figures 9-8 and 9-9.*
3. Based on the preliminary examination of the plant layout, piping, and hydraulic considerations, the following elevations have been assigned:

   • Floor level elevation of wet and dry well are assumed = El.   0.00 m
   • Minimum water surface elevation in the wet well   = El.   2.44 m
   • Maximum water surface elevation in the wet well   = El.   3.66 m
   • Normal water level elevation in the grit chamber   = El. 13.57 m

4. As an initial selection, provide a total of five pumping units of equal size including one unit as a standby. Select dry-well centrifugal pumps with variable-speed drive.

9-9-2 Design Calculations

A. Preparation of System Curves.   System losses from the common pump header (Figures 9-8 and 9-9) to the point of application at the grit removal facility should be calculated by making an inventory of all system valves, fittings, and specials. Also friction losses through force main must also be calculated. The system losses and friction losses should be calculated for different flow conditions covering the entire anticipated operating flow ranges. These calculations are given below:

1. Compute system losses for valves, fittings, and specials.

*In this design a single header force main is provided from the pumping station to the grit chamber. Many designers prefer two separate lines, or a loop around the flow measurement device. Such arrangement provides added safety in case the flow measurement device needs repairs. A retention basin is provided in this design (see Chap. 6) to bypass and store the flow should such an emergency develop.

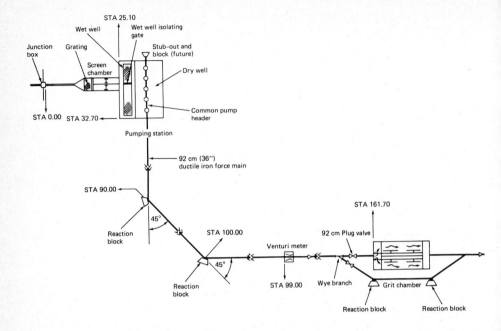

Figure 9-8   Piping Schematic and Layout of Bar Screen, Pumping Station, Flow Meter, and Grit Chamber. Reference station STA 0.00 is at the junction box. See Figure 20-4 for details of yard pipings.

The system losses are calculated using Eq. (9-4). The values of head loss coefficients for different fittings and calculated values of head losses for six different flow conditions are summarized in Table 9-3. The common header (force main) is a ductile iron pipe 0.92 m (36 in.) in diameter, and therefore all fittings are compatible with this pipe.

2. Compute head loss in the force main.

   Friction losses through the force mains are difficult to predict because the observed Hazen–Williams C value may vary substantially with different materials and service lines. It has been customary to use a C value of 140 for new pipes and 100 for old pipes. These considerations are important because they will affect the pump selection and operating conditions.

   The head losses through the proposed force main for two extreme conditions of Hazen–Williams C of 140 and 100 are calculated for several assumed flows. These values are given in Table 9-3.

3. Compute total head losses.

   Total head losses at different flow conditions at C of 140 and 100 are obtained by adding the head losses in the force main, fittings, valves and specials, and velocity head. These values are also given in Table 9-3.

4. Develop system head curve. The total dynamic heads (TDH) at maximum and minimum static heads (due to fluctuations in the wet-well elevations) are obtained by adding total head losses to the respective static heads. A summary of these calculations is given in Table 9-4. The total dynamic head losses given in Table 9-4 are

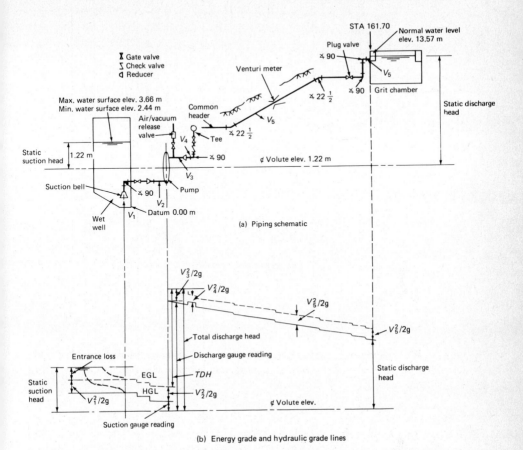

Figure 9-9 *Piping Schematic Profile of Energy Grade and Hydraulic Grade Lines. All elevations are with respect to assumed datum at the floor level of the wet well. See Figure 9-8 for STA location.*

plotted against assumed pumping rates in Figure 9-10. It should be noted that system head curves are plotted for minimum and maximum wet-well water levels and at Hazen-Williams C values of 140 and 100. It should also be noted that only the head losses in the common header (that part of pipe system common to all pumps) are included in the system head curve computations.

B. Preparation of Pump Characteristic Curves. Once the system head curves are plotted, the individual pump characteristic curves are superimposed on the system head curves (Figure 9-10). The pump's characteristic performance varies with the type of pumps and the manufacturers. It is customary to evaluate the characteristic performance of several different pumps supplied by various manufacturers. The pump that exhibits the best efficiencies under the system curves must be selected. The pump that exhibited the best performance and efficiency characteristics under the given system curve for the Design Example is shown in Figure 9-11. This pump has a 54-cm-diameter (21¼-in.)

TABLE 9-3 System Head Loss Calculations[a]

Item No.	Valves, Fittings, and Specials in 92 cm Diameter Force Main	Numbers Provided	K Value	Assumed Flow 0.253 m³/s (4000 gpm)		Assumed Flow 0.507 m³/s (8000 gpm)		Assumed Flow 0.761 m³/s (12,000 gpm)		Assumed Flow 1.013 m³/s (16,000 gpm)		Assumed Flow 1.267 m³/s (20,000 gpm)		Assumed Flow 1.521 m³/s (24,000 gpm)	
				Velocity (m/s)	Head Loss (m)	Velocity (m/s)	Head Loss (m)	Velocity (m/s)	Head Loss (m)	Velocity (m/s)	Head Loss (m)	Velocity (m/s)	Head Loss (m)	Velocity (m/s)	Head Loss (m)
1	90° Elbow	2[b]	0.30	0.38	0.004	0.76	0.018	1.14	0.040	1.52	0.071	1.91	0.112	2.29	0.16
2	45° Elbow	2[c]	0.20	0.38	0.003	0.76	0.012	1.14	0.026	1.52	0.047	1.91	0.074	2.29	0.107
3	22½° Elbow	2[b]	0.15	0.38	0.002	0.76	0.009	1.14	0.020	1.52	0.035	1.91	0.056	2.29	0.080
4	Venturimeter (see Chapter 10)	1[d]	0.14	1.52	0.017	3.05	0.066	4.58	0.150	6.10	0.266	7.62	0.414	9.15	0.597
5	Wye branch	1[c]	1.00	0.38	0.007	0.76	0.030	1.14	0.066	1.52	0.118	1.91	0.186	2.29	0.267
6	Plug valve	1[d]	1.00	0.38	0.007	0.76	0.030	1.14	0.066	1.52	0.118	1.91	0.186	2.29	0.267
7	Outlet (velocity head)	1[d]	1.00	0.38	0.007	0.76	0.030	1.14	0.066	1.52	0.118	1.91	0.186	2.29	0.267
	Subtotal	—	—	—	0.047	—	0.195	—	0.434	—	0.773	—	1.214	—	1.745

Item No.				Assumed Flow 0.253 m³/s		Assumed Flow 0.507 m³/s		Assumed Flow 0.761 m³/s		Assumed Flow 1.013 m³/s		Assumed Flow 1.267 m³/s		Assumed Flow 1.521 m³/s	
				Head Loss C = 100	Head Loss C = 140	Head Loss C = 100	Head Loss C = 140	Head Loss C = 100	Head Loss C = 140	Head Loss C = 100	Head Loss C = 140	Head Loss C = 100	Head Loss C = 140	Head Loss C = 100	Head Loss C = 140
8	Head losses in 92 cm force main, station 32.70 m to station 161.70 m (length = 129.00 m)[e]			0.032	0.017	0.117	0.063	0.247	0.132	0.420	0.225	0.641	0.344	0.896	0.481
	Total			0.079	0.064	0.312	0.258	0.681	0.566	1.193	0.998	1.855	1.558	2.641	2.226

[a]All calculations are for the common header from pump station to grit camber.
[b]Shown in Figure 9-9.
[c]Shown in Figure 9-8.
[d]Shown in Figures 9-8 and 9-9.
[e]Head loss is calculated using Hazen–Williams Eq. (9-3).

TABLE 9-4   *Summary Computations of Total Dynamic Head (System Curve)*

| Assumed Flow (m³/s) | Friction Lossª (m) | | Total Discharge Head (m) | | Total Dynamic Head (TDH) (m) | | | |
| | | | | | Min. Wet Well | | Max. Wet Well | |
	$C = 100$	$C = 140$	Min. Wet Well	Max. Wet Well	$C = 100$	$C = 140$	$C = 100$	$C = 140$
0.253	0.08	0.06	11.13	9.91	11.21	11.19	9.99	9.97
0.507	0.31	0.26	11.13	9.91	11.44	11.39	10.22	10.17
0.761	0.68	0.57	11.13	9.91	11.81	11.70	10.59	10.48
1.013	1.19	1.00	11.13	9.91	12.32	12.13	11.10	10.91
1.267	1.86	1.56	11.13	9.91	12.99	12.69	11.77	11.47
1.521	2.64	2.23	11.13	9.91	13.77	13.36	12.55	12.14

ªTotal head losses from Table 9-3 (rounded to second place of decimal).

impeller and variable-speed drive. The procedure and steps involved in selection of this pump are given below:

1. Prepare modified pump curve.
   The manufacturer's head capacity curve is developed from the gauge reading at suction and discharge lines. It is necessary to account for the losses in the individual pump pipings. These losses are computed from the suction inlet to the station header for varying pumping rates through the individual pump. Calculations are presented in Table 9-5. These individual pump losses are subtracted from the manufacturer's pump curve as shown in Table 9-6. Modified pump curves for the selected pump is shown in Figure 9-10.
2. Prepare pump combination curves.
   Four identical pumps are combined in parallel to provide the peak design flow. Therefore, combined modified pump curves with two, three, and four pumps in parallel are developed using data from Table 9-6. These plots represent combined modified pump curves showing multiple pump operations in parallel.
3. Determine pumps operating head and capacity.
   The pumping head and capacity of the station are determined from the intersection of system head curves and the combined modified pump curve (Figure 9-10). These values vary with the water level in the wet well. Many station-operating conditions are summarized in Table 9-7. Important operating conditions are discussed as follows: (a) The point A is important to describe the operating conditions of the pumps when the wet well is near maximum level (minimum static head) and all four pumps are operating. Under this condition each pump will have operating head and capacity corresponding to point A'. On the other hand, when the wet well has minimum level (maximum static head), only one pump will be pumping and the operating head and capacity are indicated by point D. These are two extreme conditions under normal operation of the pump station and should be used to

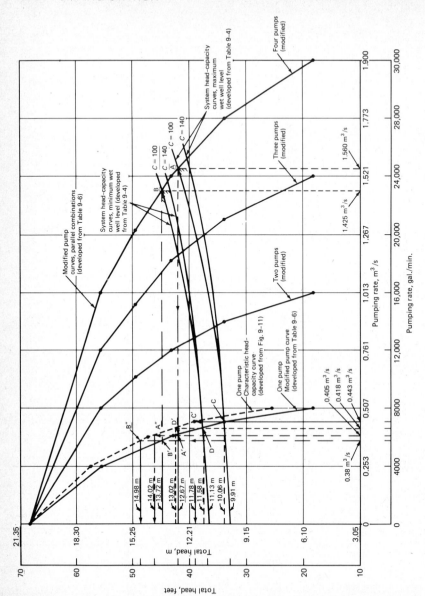

Figure 9-10   System Head Capacity, Pump-Modified Head Capacity, and Pumps Parallel
Combination Curves Showing Operating Heads and Capacities.

TABLE 9-5 Determination of Individual Pump Losses for Preparation of Modified Pump Curve

Item No.	Valves, Fittings, and Specials	No. Provided	K Value	Assumed Flow 0.253 m³/s (4000 gpm)		Assumed Flow 0.317 m³/s (5000 gpm)		Assumed Flow 0.380 m³/s (6000 gpm)		Assumed Flow 0.443 m³/s (7000 gpm)		Assumed Flow 0.507 m³/s (8000 gpm)	
				Velocity (m/s)	Head Loss (m)	Velocity (m/s)	Head Loss (m)	Velocity (m/s)	Head Loss (m)	Velocity (m/s)	Head Loss (m)	Velocity (m/s)	Head Loss (m)
1	Entrance, suction bell (32 in.) 81 cm	1[a]	0.04	0.49	0.001	0.615	0.001	0.737	0.001	0.860	0.002	0.984	0.002
2	90° Elbow (24 in.) 61 cm	1[a]	0.30	0.87	0.012	1.085	0.018	1.300	0.026	1.516	0.035	1.735	0.046
3	Gate valve (24 in.) 61 cm	1[a]	0.19	0.87	0.007	1.085	0.011	1.300	0.016	1.516	0.022	1.735	0.029
4	Reducer (14 in.) 35.5 cm	2[a]	0.25	2.56	0.167	3.203	0.262	3.839	0.376	4.476	0.511	5.122	0.669
5	Check valve (20 in.) 51 cm	1[a]	2.50	1.24	0.196	1.552	0.307	1.860	0.441	2.168	0.599	2.482	0.785
6	90° Elbow (20 in.), 51 cm	1[a]	0.30	1.24	0.024	1.552	0.037	1.860	0.053	2.168	0.072	2.482	0.094
7	Gate valve (20 in.) 51 cm	1[a]	0.19	1.24	0.015	1.552	0.023	1.860	0.034	2.168	0.046	2.482	0.060
8	Tee (20 in. × 20 in.) 51 cm × 51 cm	1[b]	1.80	1.24	0.141	1.552	0.221	1.860	0.317	2.168	0.431	2.482	0.565
	Total	—	—	—	0.563	—	0.880	—	1.264	—	1.718	—	2.250

[a] Shown in Figure 9-9.
[b] The tee connects the pump discharge pipe to the common station header. Location of the tee is shown in Figure 9-9.
Note: The friction loss in the connecting piping is small and therefore is not included in the preparation of modified pump curves.

TABLE 9-6   *Calculations for Preparation of Modified Pump and Parallel Pump Combination Curves*

Pump Characteristic Head-Capacity Curve		Individual Pump Head Loss[b] (m)	Pump-Modified Head (m) (One Pump)	Pump Capacity, Parallel Combination		
Capacity (m³/s)	Total Head,[a] (m)			2 Pumps (m³/s)	3 Pumps (m³/s)	4 Pumps (m³/s)
0.253	17.22	0.56	16.66	0.506[c]	0.759[d]	1.012[e]
0.317	16.00	0.88	15.12	0.634	0.951	1.268
0.380	14.33	1.26	13.07	0.760	1.140	1.520
0.443	11.89	1.72	10.17	0.886	1.329	1.772
0.507	7.62	2.25	5.37	1.014	1.521	2.028

[a]Values taken from Figure 9-11 for 21¼ in. diam. (54cm) impeller corresponding to assumed pump capacity.
[b]Values taken from Table 9-5
[c]2 × 0.253 = 0.506
[d]3 × 0.253 = 0.759
[e]4 × 0.253 = 1.012
Note: It is a customary not to extend the manufacturers' curve. Points beyond the manufacturers' curve may be associated with excessive power demand and cavitation. Point corresponding to pump capacity of 0.507 m³/s (8000 gpm) is outside the manufacturer's curve for 21¼ in. diameter impeller (Figure 9-11).

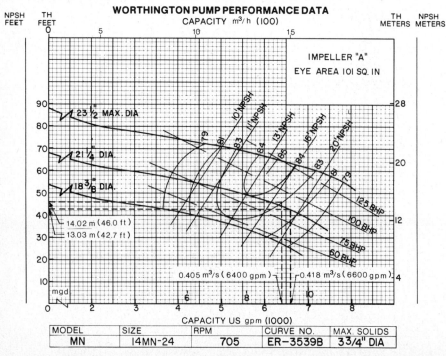

Figure 9-11   *Centrifugal Pump Selected in the Design Example. (Courtesy Worthington Pump Corporation.)*

TABLE 9-7 Summary of Pump Operating Heads and Capacities

Condition and Description	Reference Point in Figure 9-10	Operating Head		Operating Capacity		Remarks
		m	ft	m³/s	gpm	
At minimum static head (maximum wet well level), $C = 100$, four pumps in parallel combination	A	12.67	41.6	1.560	24,700	Maximum station capacity with four pumps in parallel operation
Project horizontally from point A to the individual modified pump curve	A'	12.67	41.6	0.405	6,400	Operating condition of each pump when four pumps are arranged in parallel
Project vertically from point A' to the individual pump characteristic curve	A''	14.02	46.0	0.405	6,400	Line A'A'' represents station losses (1.35 m)[a]
At maximum static head (minimum wet well level), $C = 100$, four pumps in parallel combination	B	13.72	45.0	1.425	22,500	Minimum station capacity with four pumps in parallel operation
Project horizontally from point B to the individual modified pump curve	B'	13.72	45.0	0.380	6,000	Operating condition of each pump with four pumps in parallel operation
Project vertically from point B' to the individual pump characteristic curve	B''	14.98	49.2	0.380	6,000	Line B'B'' represents station losses (1.26 m)[a]

TABLE 9-7 Summary of Pump Operating Heads and Capacities

Condition and Description	Reference Point in Figure 9-10	Operating Head		Operating Capacity		Remarks
		m	ft	m³/s	gpm	
At minimum static head (maximum wet well level), C = 100, one pump in operation (modified pump curve)	C	10.06	33.0	0.443	7,000	Operating condition of one pump
Project vertically from point C to pump characteristic curve	C'	11.78	38.7	0.443	7,000	Line C'C represents station losses (1.72 m)[a]
At maximum static head (minimum wet well level), C = 100, one pump in operation (modified pump curve)	D	11.58	38.0	0.418	6,600	Operating condition of one pump
Project vertically from point D to pump characteristic curve	D'	13.02	42.7	0.418	6,600	Line DD' represents station losses (1.44 m)[a]

[a]Station losses are the head losses in the suction and discharge piping of each pump (see Sec. 9-5-8 and Table 9-3).

determine the range of pump efficiency.* The pump efficiency for these two points are marked in Figure 9-11. (b) The capacity of pumping station is 1.56 m³/s (point A) when the wet well is at maximum level.[†] This capacity is considerably higher than the peak design flow of 1.321 m³/s. Furthermore, the minimum station capacity of 1.425 m³/s (point B) is also kept slightly higher than the peak design flow. Due to uncertainties in forecasting friction losses through the pipings and uncertainties in predicting flow rate, higher station capacity at maximum static head is preferred. In actual practice, the variable-drive units are adjusted to match the actual flow conditions.

C. Determination of Pump Efficiency and Power Input. The pump efficiencies corresponding to minimum and maximum wet-well levels are shown in Figure 9-11. The pump efficiencies, power output of the pump, power input to the pump, and motor power are summarized in Table 9-8. Maximum motor power required is 76.3 kW. Provide a 93-kW (125-HP) motor having specified full-load motor efficiency of 88 percent (see specifications).

TABLE 9-8 Summary Calculations of Pump Efficiency, Power Output of the Pump, Power Input to the Pump, and Motor Power

Operating Conditions for a Single Pump	Maximum Wet-Well Level (Minimum Static Head, Four Pumps Operating)	Minimum Wet-Well Level (Maximum Static Head, One Pump Operating)
Pump operating head	14.02 m (46.0 ft) (point A'')	13.02 m (42.7 ft) (point D')
Pump operating discharge	0.405 m³/s (6400 gpm) (point A'')	0.418 m³/s (6600 gpm) (point D')
Pump operating efficiency	83.0 percent (Figure 9-11)	81.8 percent (Figure 9-11)
Power output of the pump, water power, Eq. (9-11)	55.7 kW (74.4 HP)[a]	53.4 kW (71.2 HP)
Power input to the pump, brake power, Eq. (9-12)	67.1 kW (89.6 HP)[b]	65.3 kW (87.0 HP)
Motor power, Eq. (9-13)	76.3 kW (101.8 HP)[c]	74.2 kW (99.0 HP)

[a] $14.02 \text{ m} \times 0.405 \text{ m}^3/\text{s} \times 9.81 \text{ kN/m}^3 = 55.7 \text{ kW}\left(\dfrac{46.0 \text{ft} \times 6400 \text{ gpm} \times 62.4 \text{ lb/ft}^3}{7.48 \text{ gal/ft}^3 \times 60 \text{ s/min} \times 550 \text{ ft} - 1\text{b/HP}}\right) = 74.4 \text{ HP.}$

[b] 55.7 kW/0.83 = 67.1 kW (74.4 HP/0.83 = 89.6 HP).

[c] 67.1 kW/0.88 = 76.3 kW (89.6 HP/0.88 = 101.8 HP). Provide 93-kW motor.

*It is important that the motor power selected be at least 10–15 percent greater than the maximum power input to the pump. This will prevent motor overloading over the entire range of pump curve and enhance motor life. Additionally, the motor will operate at its most efficient point which is usually around 80 percent of its full load. The 93 kW (125 HP) motor is the next standard available motor. This will give adequate reserve power.

[†] Only one pump operates at low wet well level, and all four pumps operate at maximum wet well elevation.

D. Determination of Available NPSH.   The NPSH is calculated from Eq. (9-15). Calculation steps are given below.

$$\text{NPSH}_{av} = H_{abso} + H_s - h_L - H_{vp}$$

Assumed H_{abso} at an altitude of 500 m above sea level is approximately 9.75 m (Appendix A). The barometric pressure is further reduced by 0.4 m of water due to storm activity. Therefore $H_{abso} = 9.35$ m of water.

H_s = minimum suction head = 1.22 m (Figure 9-9)

H_f = suction pipe losses to suction gauge at 0.418 m³/s are calculated as follows* (see Table 9-5 for additional information):

Entrance (81 cm) $= \dfrac{0.04\ V^2}{2g} = \dfrac{0.04\ (0.81\ \text{m/s})^2}{2 \times 9.81\ \text{m/s}^2} = 0.001$ m

One 90° elbow (61 cm) $= \dfrac{0.3\ V^2}{2g} = \dfrac{0.3\ (1.430\ \text{m/s})^2}{2 \times 9.81\ \text{m/s}^2} = 0.031$ m

One gate valve (61 cm) $= \dfrac{0.19\ V^2}{2g} = \dfrac{0.19\ (1.430\ \text{m/s})^2}{2 \times 9.81\ \text{m/s}^2} = 0.020$ m

One reducer (35.5 cm) $= \dfrac{0.25\ V^2}{2g} = \dfrac{0.25\ (4.223\ \text{m/s})^2}{2 \times 9.81\ \text{m/s}} = 0.227$ m

Velocity head (35.5 cm) $= \dfrac{V^2}{2g} = \dfrac{1\ (4.223\ \text{m/s})^2}{2 \times 9.81\ \text{m/s}^2} = 0.909$ m

Total   = 1.188 m

H_v at operating temperature of 40°C = 55.324 mm of mercury (mm Hg) (Appendix A)

$$= \dfrac{55.324\ \text{mm Hg}}{1000\ \text{mm/m}} \times 13.58^\dagger$$

$$= 0.75\ \text{m of water}$$

NPSH = 9.35 m + 1.22 m − 1.19 m − 0.75 m = 8.63 m

NPSH required in normal pumping range from the manufacture data (Figure 9-11) = 5.5 m (18 ft)

E. Details of Pumping Station.   Important design considerations for wet-well, dry-well, and pump arrangement were presented in Secs. 9-6-2 and 9-6-3. Layout of pumping station and piping arrangements are shown in Figures 9-8 and 9-9. Design details are shown in Figure 9-12. Manufacturer's details and dimensions of selected pumps are given in Figure 9-13. Pump installation details are shown in Figure 9-14. A total of five identical

---

*At minimum suction head the wet-well elevation will be at a minimum and only the lead pump will be operating at an average flow of 0.418 m³/s.
†Specific gravity of mercury is 13.58.

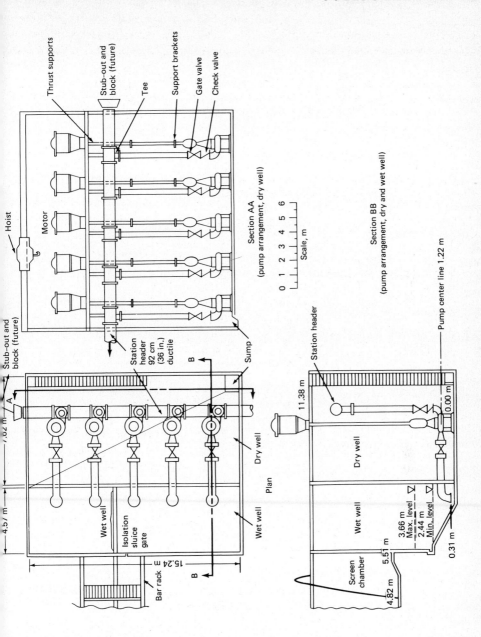

Figure 9-12 Layout of the Pumps and Piping Arrangements. All elevations are with respect to assumed datum at floor of the dry well or wet well.

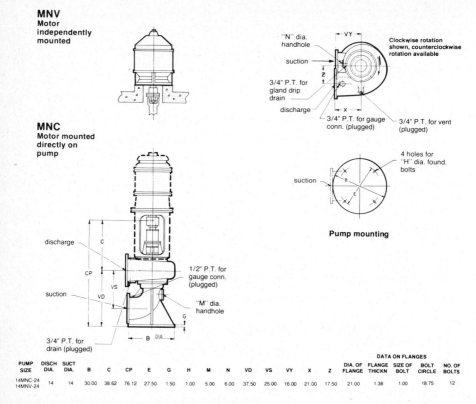

Figure 9-13   *Installation Details and Dimensions of the Selected Pump in the Design Example. (Courtesy Worthington Pump Corporation.)*

pumps are arranged in parallel. One pump serves as a standby unit. Details of pumps, motor, and air bubbler control system are provided in the specifications.

9-10 OPERATION AND MAINTENANCE, AND TROUBLESHOOTING AT PUMPING STATION

The designed pumping station has a standby pump to maintain continuous service when one unit is temporarily out of service. Also, variable-speed pumps have been provided to match the incoming flows and reduce the plant surges. Explosion-proof equipment, ventilation of equipment in dry well, separation of wet well, isolation gate in wet well, bar screen ahead of wet well for protection of pumps, and manual and automatic control systems for the pumps have been provided to achieve trouble-free operation and maintenance. Common operation and maintenance problems that may develop at pumping stations, and procedures to correct them are summarized below.[9,10]

9-10-1 Equipment Maintenance

The following routine maintenance steps should be taken to avoid major breakdowns:

(a)

(b)

Figure 9-14   Typical Pump Installation Details. (a) Vertical elbow suction pumps with motor directly mounted on the top (b) Vertical pumps discharging into a common header. The motor is directly mounted on the top.

1.   Daily observations of pump operations should be conducted and a log should be kept of these inspections.
2.   Inspect stuffing boxes for free movement of glands and examine for leaks on weekly basis.

3. Check temperature of the pump casing by placing a hand on the unit. If it is hot, check stuffing box and bearings. Investigate further for possible problems.
4. Lubricate the motors at least once each week, taking care to avoid overlubrication of the top bearings (overflow onto the commutator or armature windings may be damaging).
5. Lubricate the thrust bearings of each pump every three to six months, depending on frequency of operation.
6. Lubricate yokes and splined slip joints of flexible shafting at least once a month in accordance with the manufacturer's recommendations.
7. Listen for unusual noises.
8. Completely overhaul pumps in accordance with manufacturer's instructions as determined by pump operation conditions.
9. Standby unit shall be exercised on regular basis for uniform wear.

9-10-2 Common Operational Problems and Suggested Solutions

There are a variety of common operating problems which may occur periodically at pumping stations. Some of the common problems and suggested corrective measures are listed below:

1. Surging of the plant influent is indicated by flooded weirs, flow meter records showing intermittent high and low peak flows, and drop in treatment plant efficiency. Surging during dry weather flow may be due to malfunctioning pump controls, insufficient hydraulic capacity of the plant, or illegal connections to the system. The solution to the problem is to check and adjust the pump controls, install a surge tank, and remove illegal connections. Surging during wet weather is an indication of excessive infiltration and inflow. This problem may be overcome by checking and repairing manhole seals and covers, broken sewer lines, and illegal roof connections.
2. Improper liquid levels in the wet well indicate coating on liquid level probes, hangups in the level detectors, leaks in the floats, and fouling of bubbler control. The problem can be overcome by cleaning and repairing probes, level detectors, floats, and bubbler.
3. Accumulation of solids or scum in the wet well may be due to scum blanket in the wet well and improper operation of level-sensing equipment. The extent of the problem can be determined by sounding the wet well with a pole to determine the solids level and measurement of wet-well draw-down during pumping cycle. The problem may be corrected by starting the pump manually and lowering the liquid level in the wet well to the lowest possible level without breaking the suction. The scum should then be broken up with a high-pressure water hose.
4. Odor problems in the wet well may arise from long storage in the wet well or flat grade in the collection system. The severity of the problem is indicated by hydrogen sulfide emission, corrosion of iron work and concrete, and black color observed in liquid or solids. The problem may be corrected by proper operation of lift station, addition of hydrogen peroxide or chlorine solution in the wet well or collection

lines, installation of air diffusers in the wet well, or installation of blower and gas scrubber for oxidation of gases exhausting to the atmosphere.

5. Pump may not start due to blown fuses, defective control, or defective motor. Check and correct the following conditions: (a) fuses and their ratings, (b) corroded or shorted contact switches, (c) loose or broken terminal connections, (d) automatic control mechanism, (e) switches that are not set for operation, (f) contacts of the controls that may be dirty or arching, (g) wiring that may be short-circuited, and (h) motor that may be shorted or burnt out.

6. Pump may not run or circuit breakers may not reset due to clogged pump suction, discharge pipes, or closed valve.

7. Pump may be running but with reduced discharge due to the following reasons: (a) pump not primed or pump that is air-bound, (b) clogged impeller due to grease or other obstructions, (c) speed of motor too low due to improper wiring or defects, (d) discharge head too high, (e) suction lift too high, (f) discharge or suction lines clogged, (g) air leaks in suction line or in packing box, (h) valve partially closed, (i) incorrect impeller adjustment or damaged impeller, (j) worn-out or defective packing, couplings, or wearing rings.

8. High power bill may be due to many causes. These include (a) clogged pump, (b) misaligned belt drive, (c) speed of rotation too high, (d) operating head lower than rating for which the pump was designed, (e) check valves open or force main draining back into the wet well, (f) pump shaft bent, (g) rotating elements binding, (h) packing boxes too tight, (i) wearing rings worn or binding, and (j) impeller rubbing.

9. Excessive wear or damage to the pumps may be due to grit or grease accumulations in the wet well.

10. Noisy pump may be due to the following: (a) cavitation, (b) pump not completely primed, (c) inlet clogged, (d) inlet not submerged, (e) pump not lubricated properly, (f) worn out bearings or impellers, (g) insecure foundation, and (h) defects in the pumps.

9-11 SPECIFICATIONS

Brief specifications of various components of pumping stations are presented in this section. These specifications are written as an integral part of the design, in order to describe many components that could not be fully covered in the Design Example. Many minor details have been omitted; therefore these specifications should be used only as a tool to fully understand the design, and to be used only as a guide in developing the specifications for a real design. The design engineer must work closely with the equipment manufacturers to develop the detailed specifications for the selected equipment.

9-11-1 General

Contractor shall furnish and install five vertical shaft, dry-pit, mixed flow, centrifugal pumping units with variable-speed drive as detailed on the drawings. These pumps shall be identical and arranged in parallel to discharge into a 92-cm (36-in.) common header. All pumps shall be installed complete with all appurtenances necessary for proper operation of the equipment.

9-11-2 Pumps

Casing.   The pump casing shall be of one-piece volute with integral discharge flange. The casing shall be made of close-grained cast iron of sufficient strength, weight, and thickness to provide accurate alignment and prevent deflection. The casing shall be designed to permit removal of the rotating elements without disturbing the suction or discharge connections, and be provided with a cleanout to permit inspection and cleaning of the pump interior. The casing shall be hydrostatically tested to one and one-quarter times the maximum shutoff pressure, and provided with 2-cm vent, drain, and gauge connections.

Suction Head.   The suction head shall be designed to provide equal flow distribution to the impeller eye. It shall be of the same material as the casing and shall have flanged connection, a handhole with removable cover, and a 2-cm gauge tap connection.

Impeller.   The impeller shall be single-stage, end-suction, mixed-flow, enclosed, non-clog type with a minimum number of vanes, designed to pass 8-cm solids. The impellers shall be made of specified material of given strength, machined and polished, statically and dynamically balanced prior to assembly. The impeller shall be secured to the shaft with a shear key such that clogging of the impeller will not damage the pump. Also, the pump shall be provided with reverse rotation protection.

Wearing Rings.   Removable wearing rings of specified material shall be furnished on the impeller and suction head. They shall be securely fastened to prevent any relative rotation and designed to compensate for a minimum of 7-mm wear.

Pump Shaft, Sleeve, and Stuffing Box.   The pump shaft shall be of sufficient size and strength to transmit the full power with liberal safety factor. The shaft shall be of specified material, machined over the entire length, protected from wear in the stuffing box by shaft sleeve sealed to prevent leakage between the sleeve and the shaft.

Bearings.   The bearings shall be of specified material, arranged to eliminate all radial play, and designed for a minimum life of 100,000 h. The bearings shall be sealed to prevent entrance of contaminants, grease lubricated, and provided with tapped openings for addition of lubricant and draining. The bearing frame shall be arranged to provide for the axial adjustment of the wearing rings by the use of jacking screws and removable shims between the bearing frame and the stuffing box head.

Shop Testing.   Each pump shall be fully tested on water in the manufacturer's shop in accordance with the standards of the Hydraulic Institute to determine compliance with the rated conditions.[1] Certified curves shall be submitted for approval prior to shipment.

Variable-Speed Drives.   The variable-speed drives shall be of eddy current type consisting of stationary frame, a constant-speed and an adjustable-speed magnetic member, bearings, bearing brackets, housing, and such other components as are necessary to provide complete operating unit.

Motors.   Motors shall be weatherproof with rodent screens suitable for use outdoors. Motor windings shall be insulated, and motors shall operate continuously at rated voltage and frequency with a temperature rise not to exceed 40°C above ambient when operating at 115 percent of the rated power. Motors shall be rated for a minimum of 93 kW (125 HP) and have a full load efficiency of not less than 88 percent. All motors shall have a full-load power factor of not less than 75 percent. The locked rotor torque shall be not less than 100 percent of full-load torque. The breakdown torque shall be not less than 200 percent of full-load torque. All motor bearings shall be of the antifriction type suited for a 10-year minimum life.

Liquid Level Bubbler Control.   A liquid level bubbler control designed to operate with the adjustable frequency power supply shall be supplied. The bubbler control shall include compressor and air tank with gauge of specified rated capacity, drain valve, bleed valve, pressure maintenance switch, pressure regulator gauge, air filter with drain, air flow regulator of suitable range, level, gauge, electronic pressure transducer, pressure relief valve, needle valve, power on–off switch and power-on indicating light, pneumatic piping and fitting to connect to the bubbler tube.

The pressure signal from the bubbler shall be provided on an analog output directly proportional to the wet-well level above the bottom of the bubbler tube. Contacts to indicate a high and low wet-well level shall be provided. The adjustable speed drive system shall be designed to operate within the wet-well parameters given in Table 9-9. Wet well dimensions shall be length 15.24 m, width 4.57 m, and total depth 11.38 m. Influent flow conditions are minimum flow = 0.152 m^3/s, and maximum flow = 1.321 m^3/s.

Certified Drawings.   Certified prints of the proposed equipment shall be furnished for approval. These shall include a combined electrical drawing showing pumps, motors, driving equipment and coupling, a pump selection drawing with list of materials, performance curves, and liquid level bubbler control system. Written description of sequence of operation of pump drive system under normal automatic mode, normal manual mode, emergency-automatic mode, and manual operation mode shall be provided.

9-11-3 Painting

All ferrous metal surfaces shall receive a protective coating of rust-inhibitive primer and finished coat of approved paint.

9-12 PROBLEMS AND DISCUSSION TOPICS

9-1.   A centrifugal pump is operating at a speed of 1200 rpm and discharges 2.5 m^3/min. The total head of 120 kPa is measured on a pressure gauge located in the discharge pipe. The power required is 7.0 kW. Calculate (a) the pump efficiency, and (b) the discharge, head, and power if the pump speed is changed to 1800 rpm.

9-2.   A pump with a Francis–Vane impeller is selected for operation at best efficiency of 87 percent. The operating capacity is 0.2 m^3/s against a total head of 16 m. Determine the operating speed and the specific speed.

TABLE 9-9 Operating Conditions of the Pumping Station

Condition	Elev. above Floor of the Wet Well (m)	3rd Lag Pump			2nd Lag Pump			1st Lag Pump			Lead Pump		
		Max. Speed[a]	Min. Speed[b]	Stop	Max Speed	Min. Speed	Stop	Max. Speed	Min. Speed	Stop	Max. Speed	Min. Speed	Stop
End of bubbler tube	4.44	X			X			X			X		
High-level alarm	4.16	X			X			X			X		
High level in the wet well	3.66	X			X			X			X		
3rd lag pump minimum speed	3.50		X		X			X			X		
3rd lag pump stopped	3.35			X	X			X			X		
2nd lag pump minimum speed	3.20			X		X		X			X		
2nd lag pump stopped	3.05			X			X	X			X		
1st lag pump minimum speed	2.89			X			X		X		X		
1st lag pump stopped	2.74			X			X			X	X		
Lead pump minimum speed	2.59			X			X			X		X	
Minimum level in the wet	2.44			X			X			X		X	
Low-level alarm and shut down	2.16			X			X			X			X
End of bubbler tube	1.88												X

[a]Maximum speed 705 rpm (Figure 9-11).
[b]Variable-speed drive pumps generally have minimum speed at half the maximum speed (approximately 350 rpm).

9-3. A centrifugal pump is operating at an elevation of 1829 m above sea level. The pump requires 30 kPa net positive suction head (NPSH) when delivering the water at 20°C. What is the allowable suction lift of the pump if entrance and friction losses in the suction lines are 15 kPa? Assume the reduction in barometric pressure due to weather change is 4.1 kPa.

9-4. Water is pumped from reservoir A to B. The water surface elevations in reservoirs A and B are 300 m and 318 m, respectively. The suction line is 9.5 cm, which is short enough that the head losses may be neglected. The force main is 400 m long and 79 cm in diameter. There are three 90° and two 45° bends. Calculate the annual power bill for pumping 0.30 m³/s. Unit power cost is 5 cents per kW·h. Assume $C=100$, and combined pump and motor efficiency is 78 percent.

9-5. The data for characteristic curves of a variable speed pump supplied by a manufacturer are given below. Draw the modified curve and *H–Q* curve for two pumps operating in parallel. Also draw the system head curves. Use the following data:

Piping, valves, fittings, and specials from wetwell to common header
Pipe diameter = 46 cm, and straight length is small.
2 gate valves 46 cm, K = 0.19
1 elbow 90° 46 cm, K = 0.30
2 reducers 31 cm, K = 0.25
1 check valve 46 cm, K = 2.50
1 Tee 46 cm, K = 1.80
1 Entrance suction bell 62 cm, K = 0.04

Piping, valves, fittings and specials from common header to point of discharge
Pipe diameter = 76 cm, and straight length is 65 m. C = 100
2 elbows 90°, K = 0.3
2 elbows 45°, K = 0.2
1 Venturi meter, K = 0.14
1 plug valve K = 1.00
1 Wye branch K = 0.38

The pump volute elevation is 300.00 m. Maximum and minimum water surface elevations in the wet well are 303.00 and 301.00 m. The water surface elevation in the discharge unit is 315.00 m. The maximum and minimum station discharges are 0.65 and 0.20 m³/s. Calculate the operating head and capacity of the pump system.

Pump characteristic curve data supplied by the manufacturer

Head, m	Discharge, m³/s
20	0
15	0.30
10	0.45
5	0.50

9-6. A centrifugal pump delivers 0.15 m³/s at 30 m TDH when operating at a speed of 2000 rpm. Determine the discharge, total dynamic head, and power input if the impeller diameter is reduced from 30 cm to 24 cm.

9-7. Determine the available NPSH of a pump that delivers water at 20°C. The pump is operating at an elevation of 40 m above sea level. The water surface elevation in the wet well is 38 m and the water surface elevation in the discharge unit is 60 m. The capacity of a single pump is 0.2 m³/s. The details of piping, valves, fittings, and specials from wet well to the common pump header are

$$C = 100$$
$$\text{Pipe diameter} = 25 \text{ cm}$$
$$\text{Length of pipe} = 10 \text{ m}$$
$$\text{Two } 45° \text{ elbows, 25 cm, } K = 0.2$$
$$\text{Two } 90° \text{ elbows, 25 cm, } K = 0.25$$
$$\text{One entrance suction bell, 46 cm, } K = 0.04$$
$$\text{One gate valve, 25 cm, } K = 0.1$$

9-8. Different types of pumps are listed below. Write various applications of each type of pump: (a) screw, (b) rotary (c) centrifugal, (d) diaphragm (e) air-lift, (f) plunger, and (g) pneumatic ejector.

9-9. Define force main, wet well, dry well, casing, stuffing box, TDH, and specific speed.

9-10. The modified head-capacity curve, and system-head capacity curve for minimum wet well level are given below. Using these data determine the operating heads and capacities at minimum and maximum wet well levels for one pump, two pumps in parallel, and two pumps in series. The difference between maximum and minimum wet well elevations is 2.0 m.

Head-Capacity Curve		System-Head Capacity Curve	
Head, m	Capacity, m³/s	Head, m	Capacity, m³/s
40	0	30	0
39	0.02	32	0.02
36	0.04	34	0.04
29	0.06	37	0.06
20	0.08	40	0.08
8	0.10	44	0.10

REFERENCES

1. *"Hydraulic Institute Standards for Centrifugal, Rotary and Reciprocation Pumps,* 14th Edition, Cleveland, Ohio, 1983.
2. Metcalf and Eddy, Inc, *Wastewater Engineering: Collection and Pumping of Wastewater,* McGraw-Hill, New York, 1981.
3. Hicks, T. G., and P. E. Edwards, *Pump Application Engineering,* McGraw-Hill , New York, 1971.
4. Karrassik, I. J., and R. Carter, *Centrifugal Pumps,* F. W. Dodge, New York, 1960.
5. Walker, R., *Pump Selection: A Consulting Engineers Manual,* Ann Arbor Science Publishers, Michigan, 1972.
6. Karrassik, I. J., et al., *Pump Handbook,* McGraw-Hill, New York, 1977.
7. Stratton, C. H., "Raw Sewage Pumps," *Sewage and Industrial Wastes,* Vol. 26, No. 12, 1954.
8. Bartlett, R. E., *Pumping Stations for Water and Sewage,* Applied Science Publishers, Ltd., London, 1977.
9. *Operation of Wastewater Treatment Plant,* Water Pollution Control Federation, WPCF Manual of Practice No. 11, 1970.
10. Culp, G. L., and N. F. Heim, *Field Manual for Performance Evaluation and Troubleshooting at Municipal Wastewater Treatment Facilities,* U.S. Environmental Protection Agency, MO-16, January 1978.

10

Flow
Measurement

10-1 INTRODUCTION

Flow measurement at wastewater treatment facilities is essential for plant operation and process control. The average and diurnal variations in flow are needed to determine how much chemicals to add, how much air to supply into the aeration basins, and how much sludge to return into the biological reactors. Some additional reasons for keeping flow measurement records are as follows:

1. Most of the regulatory agencies require that the wastewater treatment plants maintain daily flow records.
2. Flow records are invaluable for future reference, particularly when plant expansion is needed.
3. Substantial increase in daily dry weather flow may be due to population growth, infiltration, or industrial waste discharge into the sewers.
4. Increase in flow during the wet weather is a measure of infiltration/inflow.

In this chapter various types of flow-measuring devices are discussed. In addition, step-by-step design procedure, details, and specifications for a Venturi meter are presented in the Design Example. Design procedure for the Parshall flume is covered in Chapter 14.

10-2 LOCATION OF FLOW MEASUREMENT DEVICES

There are many locations at a wastewater treatment facility where a suitable flow-measuring device can be installed. These locations may include (1) within an interceptor or manhole, (2) at the head of the plant, (3) downstream of bar screen, grit channel, or primary sedimentation, (4) in the force main of pumping station, or (5) before the outfall. Each location has advantages and disadvantages, and will serve some specific need. Often more than one flow measurement devices may be necessary at different locations. In Table 10-1, the operating characteristics of flow-measuring devices are compared when installed at various locations at a wastewater treatment facility.

TABLE 10-1   Comparison of Operational Characteristics of Flow Measurement Devices Installed at Different Locations

Alternative Location of Flow Measurement Device	Sensitive to Fluctuations in Flow	Measurement Represents the Avg. Flow Treated	Affected by Debris	Affected by Silt or Other Settleable Solids	Measurement Useful for Plant Operations	Measurement Useful for Effluent Receiving Source
Within intercepting sewer or manhole	yes	no	yes	yes	yes	no
At the head of the plant	yes	no	yes	yes	yes	no
Below bar screen	yes	no	no	yes	yes	no
Below grit removal or sedimentation facility	no	no	no	no	yes	no
Before outfall	no	yes	no	no	no	yes

10-3 FLOW MEASUREMENT METHODS AND DEVICES

10-3-1 Types

A wide variety of flow measurement methods and devices are available that can be used for the determination of wastewater discharges. The selection of the proper measuring method or device will depend on such factors as cost, type and accessibility of the conduit, hydraulic head available, and type and characteristics of liquid streams. In general, flow measurement systems fall into two categories: (1) pressure pipes and (2) open channel flow. However, some systems are applicable to both. For wastewater flow measurement, a system in which the rate of discharge is related to one easily measurable variable is preferred. In such systems, the direct discharge values can be obtained from the rating curves developed for the system. Table 10-2 provides a list of many different types of methods and devices applicable to fluid flow measurement. Devices that are commonly used for flow measurement of municipal and industrial waste streams are indicated in this

TABLE 10-2 *Types of Flow Measurement Devices Available for Determining Liquid Discharges*

Flow Measurement Devices	Principle of Flow Measurement
1. For pressure pipes	
a. Venturi meter[a]	The differential pressure is measured
b. Flow nozzle meter[a]	The differential pressure is measured
c. Orifice meter[a]	The differential pressure is measured
d. Pitot tube	The differential pressure is measured
e. Electromagnetic meter[a]	Magnetic field is induced and voltage is measured
f. Rotameter	The rise of float in a tapered tube is measured
g. Turbine meter[a]	Uses a velocity driven rotational element (turbine, vane, wheel)
h. Acoustic meter[a]	The sound waves are used to measure the velocity
i. Elbow meter	The differential pressure is measured around a bend
2. For open channels	
a. Flumes (Parshall, Palmer-Bowlus)[a]	Critical depth is measured at the flume
b. Weirs[a]	Head is measured over a barrier (weir)
c. Current meter	Rotational element is used to measure velocity
d. Pitot tube	The differential pressure is measured
e. Depth measurement[a]	Float is used to obtain the depth of flow
f. Acoustic meter[a]	Uses sound waves to measure velocity and depth

TABLE 10-2 Types of Flow Measurement Devices Available for Determining Liquid Discharges

Flow Measurement Devices	Principle of Flow Measurement
3. *Computing flow from freely discharging pipes*	
1. *Pipes flowing full*	
a. Nozzles and orifices	The water jet data is recorded
b. Vertical open-end flow	The vertical height of water jet is recorded
2. *Pipe partly flowing full*	
a. Horizontal sloped open-end pipe	The dimensions of freefalling water jet are obtained
b. Open flow nozzle[a] (Kennison nozzle or California pipe method)	The depth of flow at freefalling end is determined
4. *Miscellaneous method*	
a. Dilution method	A constant flow of a dye tracer is used
b. Bucket and stopwatch	A calibrated bucket is used and time to fill is noted
c. Measuring level change in tank	Change in level in a given time is obtained
d. Calculation from water meter readings	Total water meter readings over a given time period give average wastewater flow
e. Pumping rate	Constant pump rate and pumping duration

[a]Flow-measuring devices commonly used at wastewater treatment facilities.

table. A brief description of devices suitable for wastewater applications is given below. Detailed discussions on these devices may be obtained in several excellent references.[1-4] Several commonly used flow measurement devices are illustrated in Figures 10-1 and 10-2.

Venturi Tube Flow Meter. The Venturi tube flow meter utilizes the principle of differential pressure. A commercially available Venturi meter consists of a converging section (called approach), a throat, and a diverging recovery section. Due to converging section, the velocity at the throat is increased; as a result, the piezometric head is decreased. The difference in piezometric head between the throat and the beginning of the approach is measured. The difference in these two heads is analyzed by electrical or electromechanical instruments. The difference in head is proportional to the flow.

Flow Nozzle Meter. The flow nozzle meters are similar to a Venturi meter and work on the same principle. This device is actually a Venturi meter without the recovery section. The nozzle is constructed inside a pipe, where the approach is made much shorter than in a Venturi meter. The pressure is measured upstream and downstream of the nozzle. The flow nozzle provides a much less expensive installation, but it has a higher head loss than the Venturi meter. Flow nozzle meters are also commercially available.

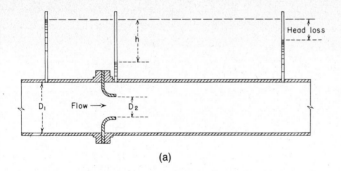

(a)

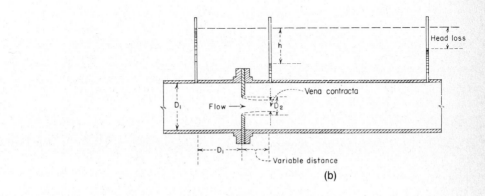

(b)

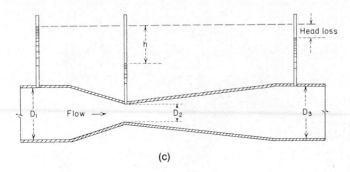

(c)

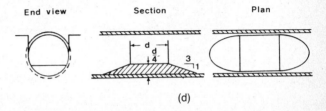

(d)

Figure 10-1   Various Flow Measuring Devices Applicable to Pressure Pipes and Open Channels. (a) Flow nozzle. (b) Orifice meter. (c) Venturi meter. (d) Palmer-Bowlus flume. [From Refs. 2 and 3]

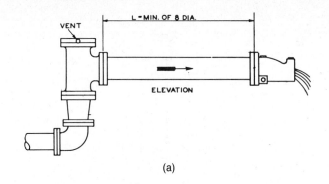

(a)

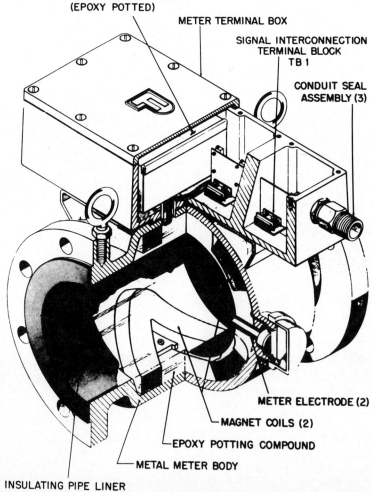

CALIBRATION COMPONENTS
(EPOXY POTTED)

METER TERMINAL BOX

SIGNAL INTERCONNECTION
TERMINAL BLOCK
TB 1

CONDUIT SEAL
ASSEMBLY (3)

METER ELECTRODE (2)

MAGNET COILS (2)

EPOXY POTTING COMPOUND

METAL METER BODY

INSULATING PIPE LINER

(b)

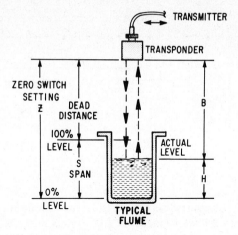

S = MAXIMUM HEAD
H = ACTUAL HEAD

(c)

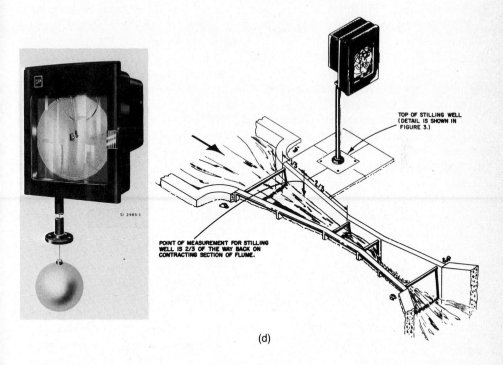

(d)

Figure 10-2   Various Flow Measuring Devices and Typical Installations. (a) Kennison nozzle. (Courtesy BIF.) (b) Magnetic flow meter. (Courtesy Fischer & Porter.) (c) Ultrasonic open channel liquid level indicator. (Courtesy Fischer & Porter.) (d) Float actuated liquid level recorder, and typical installation with Parshall Flume. (Courtesy Fischer & Porter.)

Orifice Meter.   The orifice meter (also called orifice plate) is commercially available and operates in the same manner as the Venturi meter and flow nozzle meter. The orifice acts as an obstacle placed in the path of flow in a pipe with no approach or recovery. Because of its ease of installation and fabrication, the orifice meter is the least expensive of all the differential pressure devices. Unfortunately, this device is also the least efficient and it clogs easily.

Electromagnetic Meter.   An electromagnetic flow meter utilizes Faraday's law to measure the flow. This law states that if a conductor is passed through a magnetic field, a voltage will be produced which is proportional to the velocity of the conductor. In electromagnetic meters, the wastewater acts as a conductor, an electromagnetic coil creates the magnetic field, and two electrodes measure the induced voltage. This type of meter offers exceptional accuracy and can measure a large range of flows. Additionally, because it is essentially a straight piece of pipe, it has no additional head loss. The performance is unaffected by temperature, conductivity, viscosity, turbulance, and the presence of suspended solids. The greatest disadvantage of this type of meter is its initial cost and the need for trained personnel to handle routine operation and maintenance.

Turbine Meter.   The turbine meter utilizes a rotating element (turbine) whose rotational velocity is proportional to the velocity of water. The use of this device is usually limited to pipes running full, under pressure and liquids low in suspended solids. These meters offer good accuracy and a good range of flows.

Acoustic Meters.   The acoustic meters utilize sound waves to measure the flow rates. The meter may be a *sonic meter* or an *ultrasonic meter* depending on whether the sound waves are in or above audible frequency range. The acoustic meters determine the liquid levels, area, and actual velocity (the difference of time it takes for the sound wave to travel same distances upstream and downstream). From velocity and area, the discharge is determined.

The advantages of acoustic meters are low head loss, excellent accuracy, ability to be used in any pipe size, no fouling with solids, and wide flow ranges. The disadvantages are the same as those given for electromagnetic meters.

Parshall Flume.   The Parshall flume is based on fundamental open channel flow principles. It consists of three parts; a converging section, a throat, and a diverging section. The Parshall flume creates a change in flow pattern by a decrease in width and a simultaneous drop in water surface elevation at the throat. Because the throat width is constant, the free discharge is obtained from a single upstream measurement of depth. However, if the flume is operating under submerged conditions, the downstream depth must also be measured. Calibration curves are prepared for converting depth readings to discharge. Parshall flumes are self-cleaning and offer small head losses. The design procedure for the Parshall flume is given in Chapter 14.

Palmer–Bowlus Flume.   The Palmer–Bowlus flume creates a change in the flow pattern by decreasing the width of the channel without changing its slope. It is installed in a sewer at a manhole. It backs up the water in the channel. By measuring the upstream depth, the discharge is read from the calibrated curve prepared for that unit. It has lower head loss than the Parshall flume, but less accuracy.

Weirs (Rectangular, Cipolletti, Triangular, or V-Notch). Weirs are used to measure the flow in open channels. They are easy to install in partly filled pipes, channels, or streams, and they act as a dam or obstruction. The water depth over the weir crest measured at a given distance upstream from the weir is proportional to the flow.

Commercially available weirs are rectangular, trapezoidal (Cipolletti) or triangular (V-notch) cut in a vertical plate with a sharp crest. Proper discharge equations are used to determine the flow. The head over the weir must be measured accurately by a float, hook gauge, or level sensor. Equations of rectangular weirs and V notches are given in Chapters 11–14. The weirs are accurate, simple to install, and relatively inexpensive. Their main disadvantages include large amounts of head loss and settling of solids upstream of the weir.

Depth Measurement. The method requires field measurement of the depth of flow and slope of the sewer. Several types of floats for recording the depth in sewers or manholes are commercially available. Using the Manning equation, the flow can be estimated from depth of flow and slope of the line between two manholes. This method gives approximate flow data; however, it is frequently used for making flow estimations in interceptors. The most desirable feature is the portable float, which can be easily installed in the manhole and then removed for use at other locations.

Open Flow Nozzle. Open flow nozzles are crude devices used to measure flow at the end of freely discharging pipes. The major limitation of this device is that it must have a section of pipe that has a length of at least six times the diameter with a flat slope preceding the discharge. The most common examples of flow nozzles are the Kennison nozzle and the California pipe. Open flow nozzles create large head losses due to free discharge.

10-3-2 Selection of Proper Flow Measurement Systems

In the design of wastewater treatment systems, the designer must be careful to select a proper flow measurement device to suit the particular need. Unfortunately, there is no device that is perfect for all situations. In fact, there is no device that is perfect for any situation. The design engineer has to evaluate and weigh the advantages and disadvantages of the available devices and choose the one which offers the fewest number of disadvantages.

In large, pressure conduits (force mains), the installation of solids bearing Venturi, nozzle, or orifice meters should be considered. The Venturi meter is accurate, offers little head loss, is free from solids accumulation, but is relatively expensive. An orifice meter is inexpensive, exhibits greater flexibility in covering the different flow ranges, but has the disadvantages of high pressure loss and possible accumulation of settled solids. The characteristics of the flow nozzle meter fall between those of the Venturi and orifice meters. Electromagnetic flow measurement device is a promising development for monitoring wastewater with high suspended solids.

For sewers and open channels, flumes and weirs are commonly used. Flumes can handle wastewater with high suspended solids, and offer small head losses. Weirs are relatively inexpensive but require more maintenance and have large head losses. Flow calculations using the friction formulas may give large errors due to inaccurate determination of slope and coefficient of roughness.

An evaluation of operating characteristics of various devices commonly used for wastewater flow measurement is provided in Table 10-3. These evaluations are made primarily for storm and combined sewer applications and will also apply for selection of a flow measurement device at a wastewater treatment facility.[1]

Venturi meters, Parshall flumes, and Palmer–Bowlus flumes have received wide application for flow measurement at wastewater treatment facilities. Details and design procedure for the Venturi meter are presented in Sec. 10-7. Design of the Parshall flume is covered in Chapter 14. The procedure for developing rating curves for a Palmer–Bowlus flume may be obtained in several other references.[4-7]

10-4 FLOW SENSORS AND RECORDERS

Automatic flow recording is normally required at medium- and large-size plants. Normally, the flow-sensing signals from the point of measurement are transmitted to a central panel where they are recorded on a chart. The subject of automatic flow recording is complex, requiring engineer's background in design, analysis, and application. There are a wide variety of systems commercially available that utilize principles of hydraulics, pneumatics, and electronics. In all cases a sensor signals the differential pressure, liquid level, or change in voltage, etc., to a detector that may convert the reading to the flow units for display or recording. Most modern flow measurement devices have an indicator, a recorder, and an on-line integrator (totalizer) mechanism which enables the daily volume to be read off directly, thus eliminating the tedious job of manually integrating the data on the recorder chart. Discussion on instrumentation and controls may be found in Chapter 22.

10-5 EQUIPMENT MANUFACTURERS OF FLOW-MEASURING AND FLOW-SENSING DEVICES AND RECORDERS

A list of the manufacturers of flow-measuring devices, including flow sensors and recorders, is provided in Appendix D. The design engineer should work closely with the local representatives of the manufacturers in selection of equipment for specific needs. Shelley and Kirkpatrick conducted a state-of-the-art assessment of flow-measuring devices.[1] Their publication also contains a detailed manufacturers' survey of flow-metering devices, flow sensor probes, recording devices, specifications, and costs. Responsibilities of the design engineer and important considerations for equipment selection are also covered in Sec. 2-10.

10-6 INFORMATION CHECKLIST FOR DESIGN OF FLOW-MEASURING DEVICE

The following information is necessary prior to design and selection of a suitable flow measurement system:

TABLE 10-3   Evaluation of Various Types of Devices Commonly Used for Wastewater Flow Measurement

	Application		Flow Range and Accuracy			Effect of Solids in Wastewater	Head Loss	Power Requirement	Simplicity and Reliability	Unattended Operation	Maintenance Requirement	Ease of Calibration	Cost	Portability	Application
	Pressure Flow	Open Channel	Range	Accuracy, Max. Flow %											
Venturi meter	Y	N	10:1	±0.5		H^a	L	L	G	G	M	G	H	N	Force main, wastewater
Flow nozzle meter	Y	N	5:1	±0.5		H	M	L	G	G	L	G	M	N	Force main, wastewater
Orifice meter	Y	N	5:1	±0.5		H	H	L	G	G	H	G	L	Y	Force main, wastewater
Electromagnetic meter	Y	N	20:1	±1.0		S	L	M	F	G	M	G	H	N	Force main, wastewater, sludge
Turbine meter	Y	N	15:1	±0.5		H	M	L	F	G	H	G	H	N	Force main, wastewater
Acoustic meter	Y	Y	20:1	±0.5		M	L	M	F	G	M	G	H	N	Force main, channel, wastewater, sludge
Parshall flume	N	Y	20:1	±5		S	L	L	G	G	L	G	M	Y	Channel, wastewater, sludge
Palmer–Bowlus flume	N	Y	20:1	±5		S	L	L	G	G	L	G	L	Y	Interceptor, manhole, channel, wastewater, sludge
Weirs	N	Y	20:1	±5		H	H	L	G	G	M	G	L	Y	Manhole, treatment unit, wastewater
Depth measurement	N	Y	10:1	±50		M	L	L	G	M	L	P	L	Y	Interceptor, wastewater, sludge
Open flow nozzle	N	Y	20:1	±1		S	H	L	G	G	M	F	L	Y	Outfall, discharge point, wastewater, sludge

[a]Effect of solids is substantially smaller if solids bearing, or continuous flushing type, Venturi meter is used.

F = fair, G = good, H = high, L = low, M = medium, N = no, P = poor, S = slight, Y = yes.   Source: Adapted in part from Ref. 1.

1. The characteristics of the fluid for which flow measurement device is needed (suspended solids, density, temperature, pressure, etc.)
2. Expected flow range (maximum and minimum)
3. Accuracy desired (how much error in flow measurement can be tolerated)
4. Any constraints imposed by the regulatory agency
5. Location of flow measurement device and piping system (force main, sewer, manhole, channel, or treatment units)
6. Atmosphere of installation (indoors, outdoors, corrosive, hot, cold, wet, dry, etc.)
7. Head loss constraints
8. Type of secondary elements (level sensors, pressure sensors, transmitters, and recorders)
9. Space limitations and size of device
10. Compatibility with other flow measurement devices if already in operation at the existing portion of the treatment facility
11. Equipment manufacturers and equipment selection guide (catalog)

10-7 DESIGN EXAMPLE

10-7-1 Design Criteria Used

The following design criteria shall be used for the Design Example:

1. A Venturi meter will be provided in the force main.* The force main is 92 cm (36 in.) in diameter.
2. The tube beta ratio (diameter of throat/diameter of the force main) shall be equal to 0.5.
3. Maximum and minimum flow ranges are 1.321 and 0.152 m^3/s respectively.
4. The flow measurement error shall be less than ±0.75 percent at all flows.
5. The head loss shall not exceed 15 percent of the meter readings at all flows.
6. The selected Venturi meter shall be capable of measuring flows of solids bearing liquids.

10-7-2 Design Calculations

A. Design Equations. All differential pressure meters utilize Bernoulli's principle to measure flows. The Bernoulli energy equation for two sections of a pipe is given below:

$$\frac{P_1}{\gamma} + Z_1 + \frac{V_1^2}{2g} = \frac{P_2}{\gamma} + Z_2 + \frac{V_2^2}{2g} + h_L \tag{10-1}$$

where

P_1, P_2 = internal pressures in the pipe, kN/m^2
γ = density of the fluid, kN/m^3
Z_1, Z_2 = elevations of the centerline of the pipe, m
V_1, V_2 = velocity of the fluid, m/s
h_L = head loss, m

*Some designers may prefer a bypass line around the Venturi meter for maintenance purposes. In this design a bypass line is not provided, since the flow can be diverted to a storage lagoon.

Since the differential pressure in the Venturi meter is measured between small lengths of pipes (above approach section and middle of the throat), the head loss (h_L) is negligible. Also, for a horizontal pipe, $Z_1 = Z_2$. Therefore, the equation can be simplified and rearranged as follows:

$$\frac{P_1}{\gamma} - \frac{P_2}{\gamma} = \frac{V_2^2}{2g} - \frac{V_1^2}{2g} \tag{10--2}$$

Replacing P_1/γ and P_2/γ by H_1 and H_2, substituting these values in Eq. (10-2), and using continuity equation ($V_1A_1 = V_2A_2 = Q$),

$$Q = \frac{A_1A_2\sqrt{2g\ (H_1 - H_2)}}{\sqrt{A_1^2 - A_2^2}} = \frac{A_1A_2\sqrt{2gh}}{\sqrt{A_1^2 - A_2^2}} \tag{10--3}$$

where

Q = pipe flow, m³/s
H_1 = upstream piezometric head, m
H_2 = throat piezometric head, m
A_1 = force main area, m²
A_2 = throat area, m²
h = ($H_1 - H_2$), m

Under actual operating conditions and for standard meter tubes, including allowance for friction, Eq. (10-3) is reduced to

$$Q = C_1KA_2 \times \sqrt{2g}\ \sqrt{h} \tag{10--4}$$

where

C_1 = velocity, friction, or discharge coefficient (dimensionless)
K = coefficient (dimensionless)

$$= \frac{1}{\sqrt{1 - (A_2/A_1)^2}} = \frac{1}{\sqrt{1 - (D_2/D_1)^4}} \tag{10--5}$$

D_1, D_2 = diameter of the pipe and the throat, m

For standard Venturi meter the diameter of the throat is one-third to one-half of the pipe diameter and the value of K lies 1.0062 and 1.0328. The value of C_1 generally range from 0.97 to 0.99.[4] The typical variation in C_1 with respect to the Reynolds number is given in Refs. 7–9. The value of C_1 is normally provided by the manufacturer. Figure 10-1(c) illustrates the Venturi meter arrangement and various notations used in the above equations.

B. Unit Sizing and Calibration Curve

1. Determine constant K
   The venturi tube has $D_2/D_1 = 0.5$
   Throat diameter $D_2 = 46$ cm (18 in.)

$$K = \frac{1}{\sqrt{1 - (0.5)^4}} = 1.0328$$

2. Develop calibration equation from Eq. (10-4).
Assume

$$C_1 = 0.985$$

$$Q = 0.985 \times 1.0328 \times \frac{\pi}{4} \ (0.46 \text{ m})^2 \times \sqrt{2 \times 9.81} \text{ m/s}^2 \sqrt{h}$$

$$= 0.7489 \sqrt{h} \text{ m}^3/\text{s} \tag{10-6}$$

3. Develop calibration curve
Assigning different values of differential head recorded by the meter, the pipe discharge can be obtained from Eq. (10-6). At maximum peak design and minimum initial flows of 1.321 and 0.152 m³/s the differential meter readings will be 3.111 and 0.041 m, respectively (122.48 and 1.61 in.). The calibration curve is shown in Figure 10-3. If mercury is used in the glass tube, then the differential pressure readings must be adjusted for the specific weight of mercury (13.58).

C. *Head Loss Calculations.* In a Venturi tube, due to gradual contraction of the approach section, the head loss is considered negligible. Likewise, due to short length of the throat, the head loss in this section can also be neglected. The head loss in the recovery section is estimated from Eq. (7-5).

$$h_\text{L} = K V_2^2/2g$$

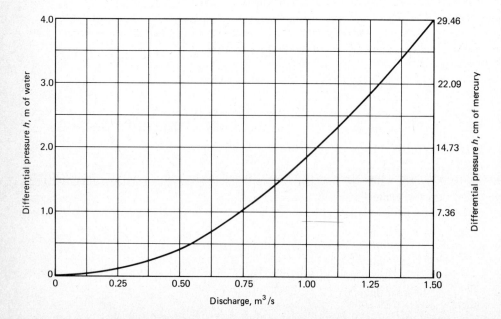

Figure 10-3 Calibration Curve of the Venturi Tube Meter Used in the Design Example.

where

h_L = head loss through the Venturi meter, m

K = 0.14 for angles of divergence of 5° (Appendix B)

At maximum and minimum flows of 1.321 and 0.152 m³/s, the head losses are calculated as follows:

$$h_L \text{ at max flow} = \frac{0.14}{2 \times 9.81 \text{ m/s}^2} \left[\frac{1.321 \text{ m}^3/\text{s}}{\pi/4 \times (0.46 \text{ m})^2} \right]^2$$
$$= 0.45 \text{ m}$$

$$h_L \text{ at min flow} = \frac{0.14}{2 \times 9.81 \text{ m/s}^2} \left[\frac{0.152 \text{ m}^3/\text{s}}{\pi/4 \times (0.46 \text{ m})^2} \right]^2$$
$$= 0.006 \text{ m}$$

These head loss values are 14.8 percent of the differential readings of the meter at respective flows.

D. Design Details. The design details of the flow-measuring system provided in the Design Example are given in Figure 10-4. The design details include universal Venturi tube, two pressure sensors, transmitter, and recorder.[10]

10-8 OPERATION AND MAINTENANCE

Before plant startup the primary Venturi tube flow meter shall be inspected for proper installation and freedom from debris. The pressure differential probe sensors shall be checked for proper connection. All tubings shall be checked frequently for clogging and slime growth. Periodic checks of indicated and recorded data against computed flows based on actual measured head conditions at the flow-measuring device or other backup system shall be made. The calibration curve shall also be checked frequently.

10-9 SPECIFICATIONS

Specifications of the Venturi tube flow meter, sensor, and transmitter are briefly presented below. The purpose of these specifications is to describe the various components that could not be fully covered in the Design Example. These specifications should be used as a tool to fully understanding the design. Detailed specifications should be prepared by the design engineer for a specific design in consultation with the equipment manufacturers. These specifications are partly adapted from Ref. 10.

10-9-1 General

The Venturi tube flow meter shall be capable of measuring flows of solids-bearing liquids. It shall consist of a primary flow element of a universal Venturi tube with deposit-resistant interior, two pressure sensors which operate on a null balance principle, and a differential pressure transmitter which shall translate differential pressure into an output signal. The Venturi tube meter, sensor, and transmitter shall be installed in a dry vault with ambient temperature range 0–38°C.

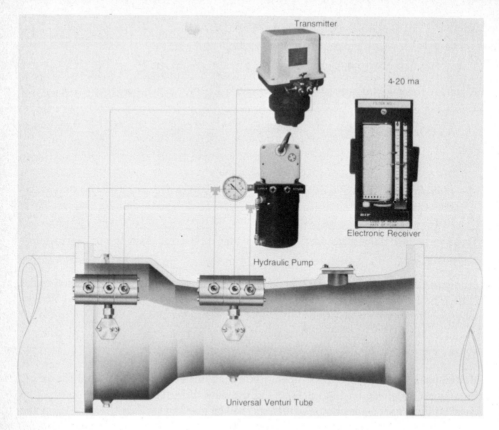

Figure 10-4 Installation Details of the Universal Venturi Tube. (Courtesy BIF.)

10-9-2 Primary Flow Element

1. The primary flow element shall be of the pressure differential producing type utilizing pure static pressure sensed at the inlet and throat. Devices amplifying the differential by causing change in the direction of the flow at the cross-sections where inlet and/or throat pressures are sensed shall not be acceptable.
2. The inlet section shall comprise cylindrical sections similar to the pipe (92 cm diameter). The inlet cone shall have an angle of reduction of 20°. The throat shall be 0.5 times the diameter of the pipe. The outlet cone shall have an incline angle of 5° and shall have a similar diameter as the pipe at its cylindrical section.
3. The metering element shall have no debris-collecting cavities, nor annular chambers, nor protrusions into the flow stream.
4. The tube, throat, and other primary elements shall be constructed of suitable material for intended service, meeting ASTM specifications for stem, valves, flanges, and pipe fillings.
5. The basic performance of the primary element shall be unaffected by line pressure and/or temperature changes.

6. The manufacturer shall furnish a certified calibration curve that shall have a plot of discharge versus differential pressure (see Figure 10-3).
7. The manufacturer shall furnish certified data substantiating tube proportions and performance. The data shall include (a) coefficient values, tolerance, and proof of their independence of pipe beta ratio, and Reynolds number of 50,000 and greater, (b) effects of convergence and divergence (c) head losses as a function of velocity head expanded, (d) error less than ± 0.75 percent of the actual rates of flow corresponding to the differential produced over the range of flow 10 : 1, and (e) 1 : 1 transfer of differential pressure at Venturi tube and the differential pressure output of the sensors.
8. Performance of the primary element shall be capable of being checked by a manometer without breaking line integrity or draining the line. It shall also be possible to inspect the interior of the primary element in place without breaking the line.

10-9-3 Sensor

1. The sensors shall be of a hydraulic type utilizing a pure liquid on one side of a diaphragm to oppose and sense the differential pressure of the solids-bearing fluids at the inlet and throat sections of the primary element.
2. The sensors shall operate on a pure null balance principle which will compensate automatically for changes in diaphragm effective area or spring rate. The actuation fluid shall be supplied by a miniature gear pump which shall be furnished with the unit and meets the power supply requirement.
3. The sensor shall not require purging system for primary element, nor electrical or electronic devices to prevent clogging or contaminant buildup which would affect sensor performance.
4. Sensors shall not protrude into the flowing liquid.
5. The sensors shall be removable from the system for inspection or repair without interruption of flow. Sensors shall also be designed to permit field checking and calibration of gauge independent of the electronic portion of the system.
6. The fluid system of the sensors shall be dynamic to compensate for line pressure (up to $700 \, kN/m^2$) and temperature variations. Static fluid systems will not be acceptable.

10-9-4 Differential Pressure Transmitter

1. The transmitter shall be of the null balance, differential pressure type employing a friction-free coaxial design for the orientation of the differential bellows, the linear differential transformer, and the force motor.
2. The transmitter shall have integral square root extraction and high precision power supply. The electronic circuit shall include reset action to eliminate droop, integrating, and driving the output until it equals the input differential.

10-10 PROBLEMS AND DISCUSSION TOPICS

10-1. A Venturi meter is 30 cm $\times$ 20 cm. The differential gauge is deflected 5.52 cm when flow is 0.04 m^3/s. The specific gravity of the gauge liquid is 1.25. Determine the meter coefficient K and discharge coefficient C_1.

10-2. A Venturi meter with a 10-cm throat is installed in a pipe that is inclined upward at an angle of 45° to the horizontal. The diameter of the pipe is 20 cm. The distance between the pressure taps along the pipe is 1.5 m, and the differential pressure is 69 kPa. Estimate the flow of water in the pipe.

10-3. A Venturi meter has a 30-cm throat. Calculate the head loss if $K = 0.12$ and the discharge is 0.1 m^3/s.

10-4. The design of a proportional weir for velocity-control is given in Chap. 14. Using this design, calculate the discharge through the channel if head over the weir crest is 0.7 m.

10-5. A Parshall flume is designed in Chap. 14. The throat width is 1.22 m (4 ft) and design dimensions are given in Figure 14-7. Calculate the discharge through the flume if the depth of flow H_a at the throat is 0.4 m.

10-6. Review the design of Palmer–Bowlus flume in Ref. 4. Study the Example 3-2 on page 91 of this reference. Determine the discharge if the depth of flow is 0.3 m.

10-7. Develop the theoretical formula for flow over a rectangular weir.

10-8. The Francis equation is generally used to calculate the discharge over a rectangular weir. This equation takes into account the contracted width of the weir

$$Q = \tfrac{2}{3}C_d L' \sqrt{2g}\, H^{3/2}$$

where L' is the contracted width $(L-0.1nH)$, n is the number of end contractions. A sharp-crested rectangular weir is used to measure flow in a rectangular channel that is 3 m wide. The weir is 1 m above the floor of the channel and extends across the channel width. Calculate the depth in the channel upstream of the weir if discharge in the channel is 12.0 m^3/s.

10-9. The discharge over a triangular weir is expressed by the equation

$$Q = \frac{8}{15}\, C_d \sqrt{2g}\, \tan\frac{\theta}{2}\, H^{5/2},$$

where

$$Q = \text{discharge in m}^3/\text{s}$$
$$H = \text{head over V-notch, m}$$
$$\theta = \text{angle of the notch}$$

An effluent weir plate contains 25 90° V-notches. Calculate the discharge if head over the notches is 20 cm. Assume $C_d = 0.6$.

10-10. Recorder and float system was used in a sewer manhole for recording flow. The depth of flow in the manhole is same as that in the 46-cm (18-in.) diameter sewer connected in the manhole. The slope of the sewer is 0.0012. Calculate the flow if the recorded depth at any time is 16 cm ($n = 0.013$).

REFERENCES

1. Shelley, Philip E., and G. A. Kirkpatrick, *Sewer Flow Measurement: A State of the Art Assessment,* U.S. Environmental Protection Agency, EPA-600/2-75-027, November 1975.

2. U.S. Environmental Protection Agency Technology Transfer, *Handbook for Monitoring Industrial Wastewater,* August 1973.

3. U.S. Department of the Interior, Bureau of Reclamation, *Water Measurement Manual,* U.S. Government Printing office, Washington, D.C., Stock No. 2403-00086, 1974.

4. Metcalf and Eddy, Inc., *Wastewater Engineering: Treatment, Disposal, Reuse,* McGraw-Hill Book Co., New York, 1979.
5. Wells, E. A., and H. B. Gotaas, "Design of Venturi Flumes in Circular Conduits," *Journal of the Sanitary Engineering Division,* American Society of Civil Engineers, Vol. 8, No. SA2, April 1956.
6. Chow, Ven T., *Open-Channel Hydraulics,* McGraw-Hill Book Co., New York, 1959.
7. Leupold and Stevens, Inc., *Stevens Water Resource Data Book,* Beaverton, Oregon, Second Edition (undated).
8. Albertson, M. L., J. R. Barton, and D. B. Simons, *Fluid Mechanics for Engineers,* Prentice-Hall, Englewood Cliffs, N.J., 1960.
9. Binder, R. C., *Fluid Mechanics,* Prentice-Hall, Englewood Cliffs, N.J., 1973.
10. BIF, *Solids Bearing Fluid Flowmetering System,* Ref. 185.20-1, 9/78, West Warwick, R.I., September 1978.

11

Grit Removal

11-1 INTRODUCTION

Grit includes sand, dust, cinder, bone chips, coffee grounds, seeds, eggshells, and other materials in wastewater that are nonputrescible and are heavier than organic matter. It is necessary to remove these materials in order to (1) protect moving mechanical equipment and pumps from unnecessary wear and abrasion, (2) prevent clogging in pipes, heavy deposits in channels, (3) prevent cementing effects on the bottom of sludge digesters and primary sedimentation tanks, and (4) reduce accumulation of inert material in aeration basins and sludge digesters which would result in loss of usable volume.

In this chapter the design information on various types of grit removal facilities is provided. Design procedure and equipment details, operation and maintenance, and equipment specifications for aerated grit removal facility are covered in the Design Example.

11-2 LOCATION OF GRIT REMOVAL FACILITY

With increasing mechanization of wastewater treatment plants, greater consideration is given to equipment protection. As a result, grit removal facilities are commonly provided at all treatment plants. It is desirable to locate grit removal facilities ahead of the raw sewage pumps. However, in many instances the incoming sewers are at such depths that location of grit removal facilities ahead of the pumping station becomes undesirable. In such instances, the grit removal facilities are provided after the pumping station or in conjunction with the primary clarifiers. The advantages and disadvantages of having a grit removal facility installed at different locations are given in Table 11-1.

11-3 TYPES OF GRIT REMOVAL FACILITIES

The quantity and quality of grit and the effect of the grit on treatment units are the important factors to be considered in selection of grit removal facilities. The choice of grit removal equipment may also be dictated by head loss and space requirements, and the type of equipment used in other parts of the plant. Grit removal facilities fall into two general categories: (1) selective removal from wastewater and (2) removal with organic matter followed by degritting.

TABLE 11-1 Comparison of Various Locations of Grit Removal Facility

Location	Advantages	Disadvantages
Ahead of lift station	Maximum protection of pumping equipment	Frequently deep in the ground, high construction cost, not easily accessible, and difficult raising the grit to ground level
After pumping station	Ground level structure. Accessible and easy to operate	Some abnormal wear to pumps
Degritter in conjunction with primary sludge	Usually low capital and operation and maintenance costs. Cleaner and drier grit	Pumping equipment not adequately protected

11-3-1 Selective Grit Removal from Wastewater

Grit is selectively removed from other organics in (1) a velocity-controlled grit channel or (2) an aerated grit chamber. Both unit operations are commonly used and are discussed below in detail.

Velocity-Controlled Grit Channel.   The grit in wastewater has specific gravity in the range of 1.5–2.7. The organic matter in the wastewater has specific gravity around 1.02. Therefore, differential sedimentation is a successful mechanism for separation of grit from organic matter. Also, the grit exhibits discrete settling whereas organic matter settles as flocculant solids.*

The velocity-controlled grit channel is a long narrow sedimentation basin with better control of flow through velocity. In some channels attempts have been made to control the velocity by use of multiple channels. A more economical arrangement and better velocity control are achieved by use of control sections on the downstream end of the channel. The control sections include proportional weir, Sutro weir, Parshall flume, parabolic flume, etc. These control sections maintain constant velocity in the channel at wide range of flows. The procedure for designing grit channel and discharge equations for various types of control sections are given in Refs. 5–9. Typical values of detention time, horizontal velocity, and settling velocity for a 65-mesh (0.21-mm diameter) material are, respective-

---

*Depending on the concentration and the tendency of particles to interact, four types of settling can occur in aqueous solutions: discrete, flocculant, hindered (also called zone), and compression. Discrete settling occurs when the particles settle as individual entities (for example, grit channel). Flocculant settling occurs when solids flocculate or agglomerate (for example, primary sedimentation). Hindered settling occurs when the concentration of solids is high, and solids settle as a "mass" or "blanket" (for example, upper zone of final clarifier). Compression settling occurs when solids remain supported on top of each other; further settling can occur only by compression of the structure (for example, thickener and lower zone of the final clarifier). References 1–4 provide excellent discussion on four types of settling.

ly, 60 s, 0.3 m/s, and 1.15 m/min.[1] The head loss through the velocity-controlled grit channel is 30–40 percent of the maximum water depth in the channel.

The grit channel may be manually cleaned or mechanically cleaned. Manually cleaned channels are used only at small plants. The channels have hoppers at the bottom for grit storage and are normally drained for manual removal of grit.

The mechanically cleaned grit channel utilizes a grit collection mechanism to move the grit to a sump. The grit is removed from the sump by mechanical means to a storage area. The grit collection and removal equipment is discussed in Sec. 11-4. Components of velocity-controlled grit channels are shown in Figure 11-1. Design calculations and details of a proportional weir are provided in Chapter 14.

Aerated Grit Chamber.   Aerated grit chambers are widely used for selective removal of grits. They are similar to standard spiral flow aeration tanks (Chapter 13). A spiral current within the basin is created by the use of diffused compressed air. The air rate is adjusted to create a velocity near the bottom, low enough to allow the grit to settle; whereas the lighter organic particles are carried with the roll and eventually out of the basin. The aerated grit chamber is normally designed to remove grit particles having a specific gravity of 2.5 and retained over a 65-mesh screen (0.21-mm diameter). However, smaller particles may also be effectively removed by reducing the air supply that also reduces the velocity of spiral flow.

Advantages of Aerated Grit Chamber.   The aerated grit chambers are extensively used at medium- and large-size treatment plants. They offer many advantages over the velocity-controlled grit channels. Some of the advantages are as follows:

1.   An aerated grit chamber can also be used for chemical addition, mixing, and flocculation ahead of primary treatment.
2.   Wastewater is freshened by the air, thus reduction in odors and additional BOD_5 removal may be achieved.
3.   Minimal head loss occurs through the chamber.
4.   Grease removal may be achieved if skimming is provided.
5.   By controlling the air supply, grit of relatively low putrescible organic content can be removed.
6.   By controlling the air supply, grit of any desired size can be removed. However, due to variable specific gravity and size and shape of the particles, there may be some limitations on removal.

Design Factors.   It is important that the design engineer must give consideration to those factors that control the design and performance of the process. Many of these factors include type of grit and other solids, detention time, air supply, inlet and outlet structures, dead spaces, tank geometry, and baffle arrangement. Different design factors, their significance, and their design values are given in Table 11-2. Many typical design configurations of aerated grit chambers are shown in Figure 11-2.

Grit collection and removal systems in aerated chambers are similar to those used for velocity-controlled channels. Often an air lift decanting arrangement is used with aerated

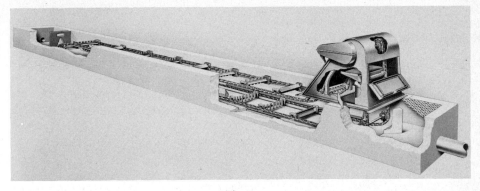

(a)

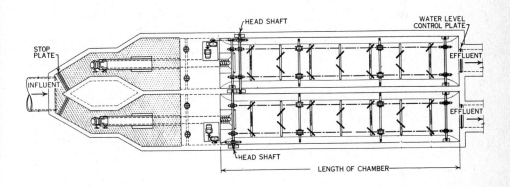

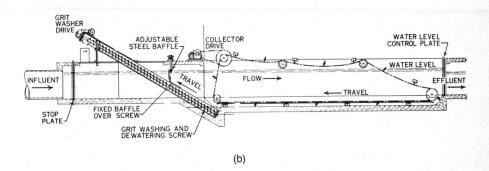

(b)

Figure 11-1   Details of Velocity Controlled Grit Channel. (a) Velocity controlled grit channel. (Courtesy Envirex Inc., a Rexnord company.) (b) Plan and longitudinal section of a double channel grit collector. [From Ref. 6. Used with permission of Water Pollution Control Federation, and American Society of Civil Engineers.]

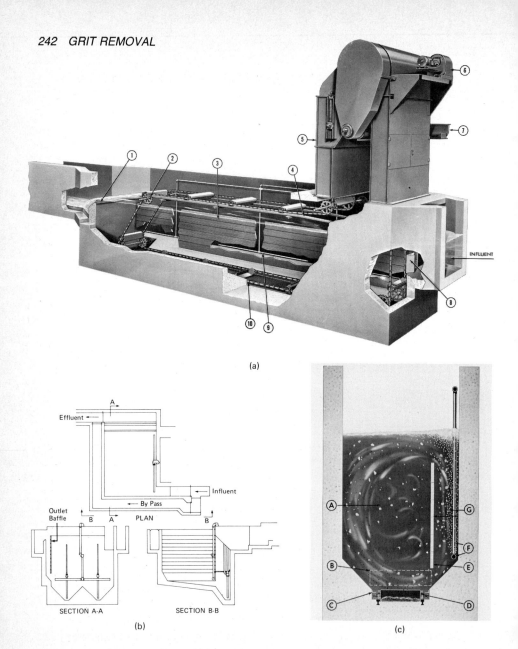

Figure 11-2 Details of Aerated Grit Chamber. (a) Components of aerated grit removal facility. 1. Effluent weir 2. Sprockets 3. Wooden circulation baffle 4. Guided chain support 5. Housing 6. Motorized drive unit 7. Screw conveyor 8. Inlet baffle 9. Air inlet pipe and headers 10. Chain and bucket collector mechanism. (Courtesy Envirex Inc., a Rexnord Company.) (b) Aerated grit chamber inlet and outlet arrangement. [From Ref. 6. Used with permission of Water Pollution Control Federation, and American Society of Civil Engineers.] (c) Simplified cross section of aerated grit removal facility. A. Flow through zone B. Grit and organic separation zone C. Collection chain guard D. Chain and bucket grit collector E. Circulation baffle F. Air inlet pipe and header G. Air circulation and grit conditioning zone. (Courtesy Envirex Inc., a Rexnord Company.)

TABLE 11-2 Design Factors and Typical Design Values for Aerated Grit Chambers

Design Factor	Range of Value	Comment
Dimensions		The width of the basin is limited to provide roll action in the tank
Depth, m	2–5	
Length, m	7.5–20	
Width, m	2.5–7.0	
Width/depth ratio	1 : 1–5 : 1	
Length/width ratio	2.5 : 1–5 : 1	
Transverse velocity at surface	0.6–0.8 m/s	
Detention time at peak flow, min	2 to 5	If grit chamber is used for preaeration or to remove grit less than 65-mesh (0.21 mm), longer detention time may be provided
Air supply	4.6 to 12.4 ℓ/s·m of tank length (3–8 cfm/ft)	Higher air rate should be used for wider and deeper tanks. Provision should be made to vary the air flow. An air flow rate of 4.6–8 ℓ/s·m in a 3.5- to 5-m-wide and 4.5-m-deep tank give surface velocity of approximately 0.5–0.7 m/s. The velocity at the floor of the tank is 75 percent of the surface velocity. A velocity of 0.23 m/s is required to move a 0.2-mm sand particle along the tank bottom
Inlet and outlet structures	—	Inlet and outlet structures must be such as to prevent short circuiting and turbulence. Inlet to the chamber should introduce the influent into circulation pattern. Outlet should be at a right angle to the inlet (Figure 11-2). The inlet and outlet are sized and built in such a way that the velocity exceeds 0.3 m/s under all flow conditions to minimize the deposits
Baffles	—	Longitudinal and transverse baffles improve grit removal efficiency. If grit chamber is much longer than the width, a transverse baffle should be considered

TABLE 11-2 Design Factors and Typical Design Values for Aerated Grit Chambers

Design Factor	Range of Value	Comment
Chamber geometry		Location of air diffusers, sloping tank bottom, grit hopper, and accommodation of grit collection and removal equipment should all be given consideration in chamber geometry. The diffusers are normally located approximately 0.6 m above the sloping tank bottom. Beneath the air diffusers would be the hopper for grit collection.

Source: Adapted in part from Refs. 1, 6–8, and 10.

grit channels. Different types of grit collection and removal systems are discussed in Sec. 11-4.

11-3-2 Combined Grit Removal with Organic Matter Followed by Degritting

Many systems are used to remove grit along with organic solids. Degritting of sludge is necessary to separate grit from the organic solids. Degritting devices include cyclone (hydrocyclone), centrifuge, and detritus tank. All these devices are discussed below.

Cyclone (Hydrocyclone). Cyclones are cone-shaped devices into which dilute primary sludge is pumped tangentially. As fluid spirals inward, centrifugal force pushes the grit toward the walls. The accumulated grit slide spirally down to the apex of the cone and then discharge into a chamber through an orifice. The effluent containing organic solids is discharged from the top.

The efficiency of the cyclone in removing grit depends on (1) effective size of the particles, (2) specific gravity of the particles, (3) differential pressure across the cone, and (4) diameter of the cone. For degritting purposes it is critical that the volume of pumped sludge and the resultant pressure across the cyclone be specified by the manufacturer. For sludge less than 1.5 percent solids, a pressure drop of $70-80 \text{ kN/m}^2$ (10–13 psi) across the cone has been suggested.[6]

Centrifuge. Centrifugal force is used to cause selective separation of grits from organic matter in various types of centrifuges. Discussion on centrifuges may be found in Chapters 16 and 18.

Detritus Tanks. Detritus tanks are square sedimentation tanks in which grit and organic solids are removed collectively. The solids are raked by rotating mechanism to a sump at the side of the tank, from which they are moved up an incline by a reciprocating rake mechanism. The organic solids are separated from the grit and fall back into the basin while passing up the incline. Cleaner and dryer grit is removed by this method. Design details and advantages and disadvantages of this device are given in Refs. 1 and 6.

11-4 GRIT COLLECTION AND REMOVAL

Mechanical grit collection in velocity-controlled channels and aerated grit chambers is achieved by conventional equipment with scrapers, screws, buckets, plows, or some combination of these. In some instances steep bottom slopes and artificial velocities (aerated chambers) are created that tend to move the settled grit to a central point for removal.

Grit removal is accomplished by tubular conveyors, bucket-type collector and elevator, screw conveyors, grit pumps, and clamshell buckets. In small aerated chambers, air pumps are also used. The tubular conveyors use a pipe in which there is an endless chain coupled with rubber flights. A portion of the pipe is mounted in the bottom of the hopper with the top of the pipe opened to allow grit to fall into a container between the flights.

The bucket-type collectors and elevators are equipped with chain-and-bucket conveyors. The buckets run the full length of the storage troughs, which move the grit to one end of the trough and then elevate it above the wastewater level in a continuous operation. Screw-type conveyors utilize helical blades that rotate and move the grits up an incline. Although wear is the general problem with all these grit collectors and elevators, tubular and screw conveyors produce grit that has less organics (due to additional washing in vertical rise) than that obtained from the bucket-type elevator or grit pump. Emptying of the basin may be necessary for the routine maintenance. For this reason many grit removal facilities utilize clamshell buckets that travel on monorails, and are centered over the grit collection and storage trough. These grab buckets operate on an intermittent basis. Multiple tanks are needed so that one unit can be bypassed during operation of the clamshell bucket and tank emptying may not be necessary. Many types of grit collection and removal systems are illustrated in Figure 11-3.

11-5 QUANTITY OF GRIT

The quantity of grit varies greatly depending on (1) type of collection system (separate or combined), (2) climatic conditions, (3) soil type, (4) condition of sewers and grades, (5) types of industrial wastes, (6) relative use of garbage grinders, and (7) proximity to the sandy bathing beaches. The grit and scum quantity may range from 5 to 200 $m^3/10^6$ m^3. Typical value is 30 $m^3/10^6$ m^3.

11-6 GRIT DISPOSAL

Various methods of grit disposal include sanitary landfill, land spreading, and incineration with sludge. For small- and medium-size plants it is best to bury and cover the grit because the residual organic contents can be a nuisance. Unwashed grit may contain 50 percent or more organic material. It attracts flies and rodents, and causes serious odor problems. Large treatment plants utilize grit washing in conjunction with grit-moving and

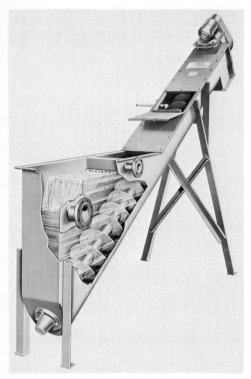

(a)

Figure 11-3   Grit Collector
and Removal System. (Courtesy Envirex, Inc., a Rexnord
Company.) (a) Screw collector and elevator. (b) Chain
and bucket grit collector at
Atlanta, Georgia South River
Plant.

(b)

raising equipment. Elevated grit storage facilities for direct truck loadings are also used at larger plants.

11-7 EQUIPMENT MANUFACTURERS OF GRIT REMOVAL FACILITIES

Several manufacturers of grit removal facilities are listed in Appendix D. Many design decisions must be made in advance prior to selection of equipment. Information such as type and number of grit collection and removal systems, flow conditions, space available, existing conditions, head loss constraints, etc., are necessary to make equipment selection. Important considerations for equipment selection and design engineer's responsibilities are covered in Sec. 2-10.

11-8 INFORMATION CHECKLIST FOR DESIGN OF GRIT REMOVAL FACILITY

The following information must be developed and decisions be made before a design engineer can start the design of a grit removal facility:

1. Design average, peak, and low initial flows
2. Information on existing facility if plant is being expanded, and site plan and topographic maps
3. Type of grit removal facility to be provided (velocity control grit channel, aerated grit chamber, or combined grit and organics removal system including degritting)
4. Influent pipe data, and static head, force main, and hydraulic grade line if grit removal is preceded by a pumping station
5. Head loss constraints for grit removal facility
6. Treatment plant design criteria prepared by the concerned regulatory agency
7. Equipment manufacturers and equipment selection guides (catalog)

11-9 DESIGN EXAMPLE

11-9-1 Design Criteria Used

The following design criteria are used for the Design Example:

1. Provide two grit chambers with spiral circulation, capable of removing grit particles of 0.21 mm (65 mesh) or larger. Each chamber shall be designed for half the peak flow. The influent and effluent structures shall be designed to handle the emergency flow conditions when one unit is out of service.
2. The grit removal facility shall be designed for maximum flow delivered by the pumping station. The maximum capacity of the pumping station under minimum static head is $1.56 \text{ m}^3/\text{s}$ (24,700 gpm). The station capacity is slightly more than the design peak flow of $1.321 \text{ m}^3/\text{s}$. Due to uncertainties in forecasting the friction head losses coupled with confidence in predicting flow rates, a slightly higher flow rate for the pump is often preferred. In actual practice, the variable-drive units are provided to

match the flow conditions. Therefore maximum design flow of 1.321 m³/s will be used for the design of the grit removal facility.

3. Provide detention time of at least 4 min at peak design flow when both chambers are in operation.
4. Provide air supply of at least 7.8 ℓ/s per meter of the tank length. Provide nozzle diffusers with coarse bubbles. Provision shall be made for 150 percent air capacity for peaking purposes.
5. The inlet and outlet shall be sized to provide a minimum velocity of 0.3 mps.
6. Provide width of the chamber 3.5 m.
7. Provide screw conveyor to move the grit to the hopper, and grab buckets for grit removal.

11-9-2 Design Calculations

A. Geometry of Grit Chamber

1. Calculate the dimensions of the grit chamber
   Provide two identical grit chambers for independent operation

Maximum design flow through each chamber	= (1.321 m³/s)/2
	= 0.661 m³/s
Volume of each chamber for 4-min detention period	= 0.661 m³/s × 4 min × 60 s/min
	= 158.6 m³
Provide average water depth at midwidth	= 3.65 m
Provide freeboard	= 0.8 m
Total depth of the grit chamber	= 4.45 m
Surface area of the chamber	= 158.6 m³/3.65 m
	= 43.5 m²
Provide length to width ratio	= 4 : 1
Length of the chamber	= 13.0 m
Width of the chamber	= 3.5 m
Surface area provided	= 45.5 m²

2. Select diffuser arrangement
   Locate diffusers along the length of the chamber on one side and place them 0.6 m above the bottom. The upward draft of the air will create a spiral roll action of the liquid in the chamber. The chamber bottom is sloped toward a collection channel located on the same side as the air diffusers. A screw conveyor is provided to move the grit along the channel length to a hopper at the down-stream end.

3. Check actual detention period

   Actual detention time at peak design flow when both chambers are in operation $= \dfrac{3.5 \text{ m} \times 3.65 \text{ m} \times 13 \text{ m}}{0.661 \text{ m}^3/\text{s} \times 60 \text{ s/min}} = 4.2$ min

Actual detention time at peak
design flow when one $= \dfrac{3.5 \text{ m} \times 3.65 \text{ m} \times 13 \text{ m}}{1.321 \text{ m}^3/\text{s} \times 60 \text{ min}} = 2.1 \text{ min}$
chamber is in operation

B. Design the Air Supply System

1. Determine air requirements

   Provide air supply at a rate of 7.8 ℓ/s per meter length of the chamber.

   Theoretical air required per chamber $= 7.8 \ \ell/\text{s·m} \times 13 \text{ m}$
   $= 101.4 \ \ell/\text{s}$

   Provide 150 percent capacity for peaking purposes

   Total capacity of the diffusers $\quad = 1.5 \times 101.4 \ \ell/\text{s}$
   $= 152.1 \ \ell/\text{s per chamber}$

   Blower capacity (both chambers) $\quad = 152.1 \ \ell/\text{s} \times 2 \times 60 \text{ s/min}$
   $\qquad\qquad \times \dfrac{1}{1000 \ \ell/\text{m}^3}$

   $= 18.3 \text{ standard m}^3/\text{min (s·m}^3/\text{min)}$

   Provide two blowers 20 sm^3/min each, with one blower used as a standby unit. Air piping shall deliver a minimum of 0.15 m^3/s air to each chamber. Control valves and flow meters shall be provided on all branch lines in each basin to balance the air flow.

2. Design the diffusers and blowers

   Provide coarse diffusers with air pipe headers and hanger feed pipes having swing joint assembly. The procedure for designing the diffusers, pipings, and blower is covered in detail in Chapter 13. Similar procedure should be used to design the air supply system for the aerated grit chambers.

C. Surface Rise Rate

1. Check overflow rate when both chambers are in operation

   Surface area of each chamber $\quad = 3.5 \text{ m} \times 13 \text{ m} = 45.5 \text{ m}^2$

   Overflow rate (surface rise rate) $= \dfrac{0.661 \text{ m}^3/\text{s} \times 86400 \text{ s/d}}{45.5 \text{ m}^2}$

   $= 1255.2 \text{ m}^3/\text{m}^2 \cdot \text{d} \ (30{,}805 \text{ gpd/ft}^2)$

2. Check overflow rate when one chamber is out of service

   Overflow rate (surface rise rate) $= 2 \times 1255.2 \text{ m}^3/\text{m}^2 \cdot \text{d}$
   $= 2510.4 \text{ m}^3/\text{m}^2 \cdot \text{d}$

D. Influent Structure

1. Select the arrangement of the influent structure

   Provide 1-m-wide submerged influent channel that diverts the flow to two grit chambers. The channel has two orifices 1.0×1.0 m that discharge the flow near the diffuser area. Provide a baffle at the influent to divert the flow transversally to follow the circulation pattern. Sluice gates are provided to remove one chamber from service for maintenance purposes. The details of the influent structure are shown in Figure 11-4.

2. Calculate the head losses through the influent structure
    The head loss through the influent structure is calculated by applying energy equation [Eq. (8-4)] at sections (1) and (2) as defined in Figure 11-5. Simplifying the energy equation

$$\Delta z = \frac{v_2^2}{2g} - \frac{v_1^2}{2g} + h_L \qquad (11\text{--}1)$$

where

$$v_1 = \text{average velocity into the influent channel}$$
$$v_2 = \text{average velocity into the grit chamber}$$
$$Z_1 - Z_2 = \Delta z, \text{ difference in elevation of free water surface into the channel and the chamber}$$
$$h_L = \text{head loss in the channel and the submerged orifice}$$

Generally, v_1 and v_2 are small and the factor $[(v_2^2/(2g) - (v_1^2/2g)]$ is ignored. In the present situation these velocities at peak design flow when only one chamber is in operation is calculated below:

$$v_1 = \frac{1.321 \text{ m}^3/\text{s}}{1 \text{ m (channel width} \times 4.06 \text{ m (assumed water depth in the channel)}}$$
$$= 0.33 \text{ m/s}$$

$$v_2 = \frac{1.321 \text{ m}^3/\text{s}}{3.5 \text{ m (chamber width)} \times 3.82 \text{ m (assumed water depth in the chamber)}}$$
$$= 0.10 \text{ m/s}$$

$$\frac{v_2^2}{2g} - \frac{v_1^2}{2g} = \frac{(0.10 \text{ m/s})^2}{2 \times 9.81 \text{ m/s}^2} - \frac{(0.33 \text{ m/s})^2}{2 \times 9.81 \text{ m/s}^2}$$
$$= -0.005 \text{ m (small)}$$

h_L equals the combined head losses into the influent channel and the exit loss through the influent port. Since the head loss in the influent channel and velocity head difference across sections (1) and (2) are small, h_L is approximately equal to the piezometric difference across the influent port, therefore from Eq. (11-1), h_L is substituted for Δz in Eq. (11-2).[11-13]

$$Q = C_d A \sqrt{2g\Delta z} \qquad (11\text{--}2)$$

where

$$A = \text{area of the orifice, m}^2$$
$$C_d = \text{coefficient of discharge} = 0.61 \text{ for square-edged entrance†}$$

*An alternative arrangement is to provide a division box with two weirs to divide flow into each chamber.
†A conservative value is selected since slime growth may restrict the orifice area.

At peak design flow of 1.321 m³/s, when one chamber is in operation,

$$\Delta z = \left[\frac{1.321 \text{ m}^3/\text{s}}{0.61 \times 1 \text{ m} \times 1 \text{ m} \times \sqrt{2 \times 9.81} \text{ m/s}^2} \right]^2 = 0.24 \text{ m}$$

If a larger influent port is selected, Δz will be smaller.

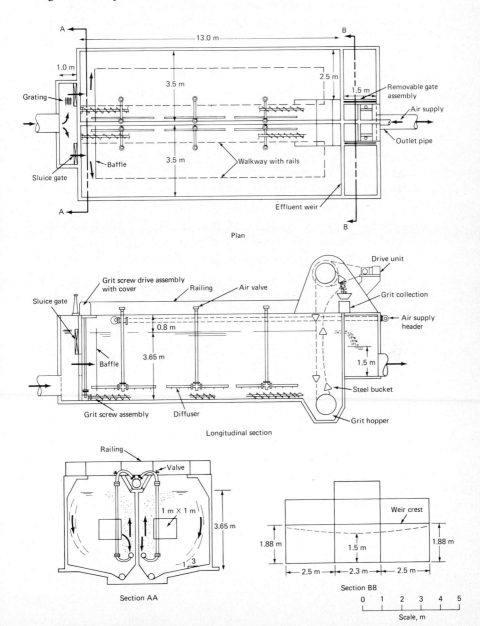

Figure 11-4   *Design Details of Aerated Grit Chamber.*

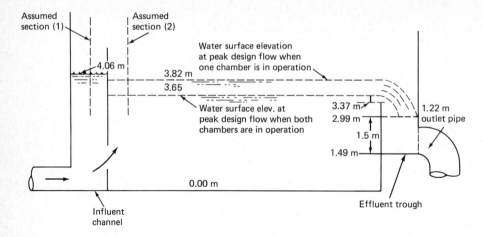

Figure 11-5   Hydraulic Profile Showing Differential Liquid Levels in the Influent Channel,
Grit Chamber, and Effluent Trough at Peak Design Flow When One Chamber Is in Operation. All elevations are with respect to the assumed datum at the midwidth of the chamber
bottom.

E. Effluent Structure

1.   Select an arrangement of the effluent structure.
     The effluent structure consists of a rectangular weir, an effluent trough, an
     effluent box, and an outlet pipe. The effluent weir is 2.5 m long, and the
     effluent trough is 2.5 m long × 1.5 m wide. The effluent box is common to both
     chambers and is 1.5 m × 2.3 m.* Removable gates are provided at the effluent
     box to drain the effluent trough when one chamber is removed from service.
     The carries flow to a collection-division box that precedes the primary sedi-
     mentation basin. Details of the effluent structure are given in Figure 11-4.
2.   Compute the head over the effluent weir at average design flow when both
     chambers are in operation.
     The head over the rectangular weir is calculated from the weir equation [Eq.
     (11-3)].[12]

$$Q = \frac{2}{3} C_d L' \sqrt{2gH^3} \qquad (11\text{--}3)$$

where

Q = flow over weir, m³/s
H = head over weir, m
C = coefficient of discharge, assume C = 0.624
L' = L − 0.2H
L = length of weir = 2.5 m

*2.3 m length of the effluent box includes 0.3 m thickness of the wall common in both chambers.

At peak design flow when both chambers are in operation, $Q = 0.661$ m³/s and H is calculated by trial and error.

Assume:

$L' = 2.44$ m

$$H = \left[\frac{0.661 \text{ m}^3/\text{s} \times 3/2}{0.624 \times 2.44 \text{ m} \times \sqrt{2} \times 9.81 \text{ m/s}^2} \right]^{2/3}$$

$= 0.28$ m

$L = 2.5$ m $- 0.2 \times 0.28$ m $= 2.44$ m (same as the initial assumption)

3. Compute the height of the weir crest above the bottom of the chamber.

   Height of the weir crest $= 3.65$ m $- 0.28$ m $= 3.37$ m.

4. Compute the head over the effluent weir at peak design flow when one chamber is out of service.

   Peak design flow $= 1.321$ m³/s

   Assume:

   $L' = 2.41$ m

   $$H = \left[\frac{1.321 \text{ m}^3/\text{s} \times 3/2}{0.624 \times 2.41 \text{ m} \times \sqrt{2} \times 9.81 \text{ m/s}^2} \right]^{2/3}$$

   $= 0.45$ m

   $L' = 2.5$ m $- 0.2 \times 0.45$ m $= 2.41$ m (same as the initial assumption)

5. Compute the water depth in the chamber at peak design flow when one chamber is out of service.

   $$\text{Water depth} = \left(\begin{array}{c} \text{height of weir crest above} \\ \text{the bottom of the chamber} \end{array} \right) + \text{head over weir}$$

   $= 3.37$ m $+ 0.45$ m $= 3.82$ m

6. Calculate the depth of the effluent trough.
   A free-falling weir discharging into a flume, trough or launder has varying flow throughout its entire length. At the discharge point the flow is maximum. The water surface profile is shown in Figure 11-6. For uniform velocity distribution, Chow expressed the drop in the water surface elevation between sections 1 and 2 by Eq. (11-4).[13]

   $$\Delta y' = \frac{q_1 \, v_{\text{avg}}}{g \, q_{\text{avg}}} \left[\Delta v + \frac{v_2}{q_1} \Delta q \right] + (S_E)_{\text{avg}} \Delta x \qquad (11\text{–}4)$$

   where

   $\Delta y' = $ drop in water surface elevation between sections 1 and 2, m (ft)

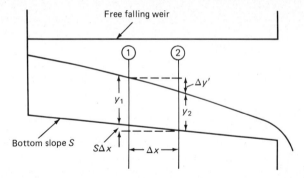

Figure 11-6 Water Surface Elevation in a Flume Receiving Weir Discharge.

y_1 and y_2 = depth of flow at sections 1 and 2 respectively, m (ft)
q_1 and q_2 = discharge at sections 1 and 2 respectively, m³/s (ft³/s)
v_1 and v_2 = velocity at sections 1 and 2 respectively, m/s (ft/s)
Δv = $(v_2 - v_1)$, m/s (ft/s)
Δq = $q_2 - q_1$, m³/s (ft³/s)
Δx = horizontal distance between sections 1 and 2, m (ft)
$(S_E)_{avg}$ = average slope of the energy line, m/m (ft/ft). It is obtained from Eq. (11–5) or (11–6).
v_{avg} = average velocity $(v_1 + v_2)/2$, m/s (ft/s)
q_{avg} = average discharge $(q_1 + q_2)/2$, m³/s (ft³/s)

$$(S_E)_{avg} = \frac{n^2 (v_{avg})^2}{(R_{avg})^{4/3}} \quad \text{(SI units)} \quad (11–5)$$

$$(S_E)_{avg} = \frac{n^2 (v_{avg})^2}{2.21 (R_{avg})^{4/3}} \quad \text{(U.S. customary units)} \quad (11–6)$$

where

$$R_{avg} = (R_1 + R_2)/2$$

The terms n and R are defined in Eq. (7-1).

The computational procedure for obtaining the depth of flow in the trough at the upstream section is given below:[14]

a. Determine the depth of flow at the lower end of the trough which is generally fixed by the downstream control condions.
b. Select an incremental distance Δx between two sections for computational purpose. Δx may be a constant or may be varied.
c. Take the first increment Δx at the lower end of the trough having water depths of y_2 and y_1 at lower and upper ends of the increment.
d. Assume $\Delta y'$ for the first increment.
e. Compute Δy from Eq. (11-7).

$$\Delta y = S\Delta x - \Delta y' \quad (11–7)$$

f. Compute the depth y_2 from Eq. (11-8).

$$y_1 = y_2 - \Delta y \qquad\qquad (11\text{--}8)$$

g. Determine discharge q_2 and q_1 below and above the selected incremental distance Δx.

h. Compute velocity v_2 and v_1.

i. Use Eq. (11-4) to compute $\Delta y'$.

j. If the computed value of $\Delta y'$ in step i is not within certain tolerance level set by the designer, repeat steps d through i.

k. After balancing the two sections, repeat steps c through j. At this time the computed depth y_1 at the upper end of the increment becomes the water depth y_2 at the lower end of the next selected increment.

Benefield et al. provided a computational scheme and solution of water surface profile for a lateral spillway channel receiving uniformly distributed flow along the entire channel length.[14] A similar solution using Eqs. (11-4)–(11-8) for point load discharges along the effluent channel is presented in Chapter 13 (Table 13-8). The computer program is given in Appendix B.

The use of above procedure is tedious and time consuming. Widely used practice by the designers is to utilize an approximate solution given by Eq. (11-9). This equation was originally developed for flumes with level inverts and parallel sides; channel friction is neglected; and the draw-down curve is assumed parabolic.[14,15]

$$y_1 = \sqrt{y_2^2 + \frac{2(q'L \cdot N)^2}{gb^2 y_2}} \qquad\qquad (11\text{--}9)$$

where

y_1 = water depth at the upstream end, m
y_2 = water depth in the trough at a distance L from the upstream end, m
q' = discharge per unit length of the weir, m³/s · m
b = width of the launder, m
N = number of sides the weir receives the flow (one or two)
g = acceleration due to gravity, m/s²

At peak design flow, when one chamber is out of service, the following conditions apply:

$q = 1.321 \text{ m}^3/\text{s}$

$$q' = \frac{q}{\text{length of weir}} = \frac{1.321 \text{ m}^3/\text{s}}{2.5 \text{ m}} = 0.5284 \text{ m}^3/\text{s} \cdot \text{m}$$

The water surface elevation in the effluent box is governed by the downstream conditions. Assume water depth in the effluent box at the exit point (center of the effluent pipe) is 1.5

m; therefore, the water depth in the trough at the effluent box y_2 is also 1.5 m* (see Chapter 21).

$$y_1 = \sqrt{(1.5 \text{ m})^2 + \frac{2 \times (0.5284 \text{ m}^3/\text{s m} \times 2.5 \text{ m} \times 1)^2}{9.81 \text{ m/s}^2 \times (1.5 \text{ m})^2 \times 1.5 \text{ m}}} = 1.54^\dagger \text{ m}$$

Allow 12 percent additional depth to account for friction losses, and add 15 cm to ensure a free fall. Total depth of the effluent trough = 1.54 m × 1.12 + 0.15 m = 1.88 m. The flow is conveyed from the effluent box to a collection box upstream of the sedimentation basin. The hydraulic calculations for pipe sections between the grit chambers and division box preceding the primary sedimentation basin are given in Chapter 21.

F. Head Loss Through the Grit Chamber

The total head loss through the grit chamber includes (1) head loss at the effluent structure, (2) head loss at the influent structure, (3) head loss in the basin, and (4) head loss due to baffles.

1. Compute head loss at the influent and effluent structures
    The head losses at the influent and effluent structures were calculated earlier in steps D and E.
2. Compute the head loss in the grit chamber
    The head loss in the basin is small due to short length and small velocity of flow. Therefore it can be neglected.
3. Compute the head loss due to influent and effluent baffles
    The influent and effluent baffles offer obstruction to the flow in the grit chamber. The momentum equation [Eq. (11-10)] is used to calculate the head loss due to the baffles.[16,17]

$$h_L = C_D \frac{v_2^2}{2g} \frac{A_b}{A} \tag{11-10}$$

where

h_L = head loss due to baffle
v_2 = horizontal component of the velocity in the chamber through the unobstructed area
A_b = vertical projection of the area of the baffle
A = cross-sectional area of the chamber
C_D = coefficient of drag. The value of C_D for flat plates are approximately 1.9[16]

If influent baffles occupy 50 percent of the cross-sectional area of the chamber, h_L at peak design flow when one chamber is out of service is calculated as follows:

*From Eq. (8-5) the critical depth is 0.43 m which is considerably less than 1.5 m depth provided at the control point. This presents a submerged outfall condition.

†The water depth at the upstream end of the effluent launder from Eqs. (11-4)–(11-8) is 1.528 m (see Problem 11-11).

$$\text{Velocity through the chamber} = \frac{Q}{\text{area}} = \frac{1.321 \text{ m}^3/\text{s}}{(3.5 \text{ m width}) (3.82 \text{ m water depth})}$$

$$= 0.099 \text{ m/s}$$

$$h_L = \frac{1.9 \ (0.099 \text{ m/s})^2}{2 \times 9.81 \text{ m/s}^2} \times \frac{0.5}{1} = 0.0005 \text{ m}$$

This is a small head loss and can be neglected. Similarly, the head loss due to the effluent baffle can also be calculated. The value will be small and can also be ignored.

G. Hydraulic Profile through the Chamber.   The hydraulic profile through the grit chamber at peak design flow when one chamber is out of service is shown in Figure 11-5. The hydraulic profile is with respect to the assumed datum at the bottom of the chamber. The velocity head due to the approach velocity at the effluent weir is ignored.

H. Effluent Quality.   Overall BOD_5 and total suspended solids removal in a grit chamber is small. It is therefore assumed that the concentrations of BOD_5 and total suspended solids in the effluent from the grit removal facility are the same as those in the raw wastewater ($BOD_5 = 250$ mg/ℓ and TSS $= 260$ mg/ℓ). However, the effluent from the aerated grit chamber is freshened by the air and dissolved oxygen, thus some reduction in odors may occur.

I. Quantity and Characteristics of Grit.   The quantity of grit that will be removed from the wastewater will vary greatly. Using the range and typical values given in Sec. 11-5, the following quantities are obtained:

Average quantity of grit $= 30 \text{ m}^3/10^6 \text{ m}^3 \times 0.44 \text{ m}^3/\text{s} \times 86400 \text{ s/d} = 1.14 \text{ m}^3/\text{d}$

The minimum and maximum quantities may range from 0.2 to 7.6 m^3/d.

The grit removal by the bucket elevator provides a certain degree of grit washing due to agitation in wastewater during the ascent. However, it is expected that the grit will contain 3–4 percent putrescible organic matter.

J. Grit Disposal.   Combined disposal of grit, screenings, and digested sludge will be achieved by sanitary landfilling. Environmental considerations (air, water, land, aesthetics) along with the design and operation of sanitary landfills are discussed in Chapter 19.

K. Design Details.   The design details of the aerated grit removal facility are provided in Figure 11-4.

11-10 OPERATION AND MAINTENANCE, AND TROUBLESHOOTING AT AERATED GRIT REMOVAL FACILITY

Operation and maintenance of aerated grit removal facility requires well-trained operators familiar with the peculiarities of sewer system and characteristics of wastewater.

It is important that grit removal is conducted at the highest efficiency possible so that the wear on downstream equipment is minimum. The operator must adjust the air flow to allow the grit to settle, but must also provide enough air to prevent organic material from settling. If short circuiting is noticed, submerged baffles must be installed. The design of a grit removal system allows one unit to be taken out of service routinely for maintenance without impairing the grit removal performance at peak design flow. Also, the swing-type diffusers will allow maintenance of aeration equipment without taking the units out of service. The following items should be considered as a troubleshooting guide:[18]

1. Rotten-egg odor is an indication of hydrogen sulfide formation. Increase the aeration, inspect the walls, channels, and the chamber for debris. Wash walls, weir, and influent and effluent channels with chlorine or hydrogen peroxide solution.
2. Corrosion or wear on equipment indicates inadequate ventilation and the production of hydrogen sulfide. Increase the air supply, and stop operation for routine maintenance and painting.
3. Increase air supply if the grit that is removed is grey in color, smells, and feels greasy.
4. If surface turbulence is reduced, diffusers may be covered by rags or grit. Cleaning of diffusers is needed.
5. Low recovery of grit is an indication of excessive aeration, and inadequate retention time. Reduce air supply.
6. Overflowing grit chamber indicates a pump surge problem. Adjust the pump controls.
7. Percent volatile matter should be determined regularly to obtain percent organics in the grit. High organics indicate low aeration.

The grit contains putrescible organics and therefore can cause serious odor problems if not properly handled. Daily disposal of removed grit will reduce odor and insect problem. The area must be washed daily with chlorine or hydrogen peroxide solution.

11-11 SPECIFICATIONS

The specifications for the grit removal facility designed in this chapter are briefly presented below. The purpose of these specifications is to describe many components that could not be fully covered in the Design Example. These specifications are neither complete nor definitive. Detailed specifications should be prepared in consultation with the equipment manufacturers. The following specifications should be used only as a guide.

11-11-1 General

Each aerated grit chamber shall comprise a complete assembly of concrete basin, influent and effluent structures, air diffusers and aeration equipment, grit collection by spiral conveyor, and bucket elevator for grit removal. The manufacturer shall furnish and deliver, ready for installation, a grit collection and removal mechanism suitable for installation in two identical rectangular grit chambers of the following dimensions:

Number of identical chambers with common wall	= 2
Length	= 13.0 m (42.7 ft)
Width	= 3.5 m (11.5 ft)
Water depth at midwidth	= 3.65 m (12.0 ft)
Bottom slope along width (toward spiral conveyor)	= 3 horizontal : 1 vertical
Freeboard	= 0.8 m (2.5 ft)

11-11-2 Materials and Fabrication

All structural steel shall conform to the "Standard Specifications for Structural Steel for Buildings" of the American Society for Testing and Materials. All iron castings shall be tough, closed-grained, free from blow holes, flaws, or excessive shrinkage.

11-11-3 Aeration Equipment

Air Diffusion Equipment. The air diffusion equipment in the aerated grit chamber includes air header piping, flow measurement, air control valve, riser pipe, air diffusers, and necessary supports and brackets. The air diffusers shall be nonmetallic, coarse bubble, fixed to a horizontal stainless steel tube, connected to a stainless steel tube downcomer. The diffusers shall be capable of delivering at least 150 ℓ/s air per chamber. The air diffusion system shall furnish a range of 3–12 liters per s per m of the basin measured along the longitudinal axis. The system shall be field-tested by filling each chamber with water and visually observing the operation of the air tubes.

Air Blowers.   The air blower shall have a capacity not less than 20 standard m^3/min when operating against a pressure of 27.6 kN/m^2 (gauge) (4.0 psig) measured at the blower outlet. There will be two identical blowers, one used as a standby. The blowers shall be of a heavy-duty, centrifugal type having end suction and top discharge. The blowers shall be directly connected to a 1.5-kW, (minimum), 440-V, 3-phase, 60-cycle motor through a flexible coupling. Motor and blower shall be mounted on an integral base plate. The blower's magnetic-motor starter shall be the full-voltage nonreversing combination circuit-breaker type. Each starter shall have 3-phase overload protection and a covermounted hand-off-automatic selector switch.

11-11-4 Conveyor and Elevator

The spiral conveyor shall be provided to move the grit into the hopper. The helix shall be made up from preformed heavy-steel flight sections welded to the shaft and fitted with replaceable wearing shoes. A plate and shroud shall be provided at the discharge end of the spiral to limit the feed to the bucket elevator and thus prevent overfeeding or stalling. The bearings shall be babbitted, and those above water line shall be provided with suitable lubrication fittings. The head shaft shall be equipped with take-up bearings with an adjustment of specified tolerance.

The drive unit shall consist of an all-motor-type helical gear reducer. Gears shall be made of steel with teeth cut to accurate shape, enclosed in a moisture- and oil-proof case. The driving mechanism shall be enclosed in a suitable guard. The motor shall be totally enclosed with moisture-proof impregnated windings, and shall be 0.5 kW, 440 V, 3 phase, and 60 cycle.

The bucket elevator shall consist of a single strand of chain passing over the top and bottom sprockets, with steel buckets attached to the chain, driven by a motor and speed reducer. The elevator shall be the continuous bucket type with the lower bottom shaft extended for driving the spiral conveyor. The speed shall be approximately 1.5 m/min. The elevator above the top of the grit chamber shall be completely enclosed and shall be complete with discharge spout. The capacity of the system shall be approximately 0.5 m^3/h. The motor starter for the grit elevator is similar to that of the blower as specified above.

11-11-5 Painting

All submerged metal parts (except bearing surfaces, shafts, and chain) shall be painted with one shop coat of metal primer (rust-resistant). The bearing surfaces and shaft shall be coated with grease. Chain shall be dipped in rust-inhibiting compound. All anchor bolts shall be hot-dipped galvanized steel.

11-12 PROBLEMS AND DISCUSSION TOPICS

11-1. Review Sec. 6-5 in Ref. 1. Describe four types of settling (discrete, flocculant, hindered or zone, and compression), and their occurrence in various wastewater treatment units.

11-2. Newton's and Stokes's equations for settling particles are

$$v_s = \left[\frac{4}{3} \frac{g(\rho_s - \rho)d}{C_D\rho} \right]^{1/2} = \left[\frac{4}{3} \frac{g}{C_D}(S_s - 1)d \right]^{1/2} \qquad \text{(Newton's law)}$$

$$v_s = \frac{g(\rho_s - \rho)d^2}{18\mu} = \frac{g(S_s - 1)d^2}{18\nu} \qquad \text{(Stokes's law)}$$

Develop these equations by equating gravitational force $(\rho_s - \rho)gV$ and frictional drag force

$$C_D A\rho v_s^2/2$$

where

$$
\begin{aligned}
v_s &= \text{settling velocity, m/s} \\
d &= \text{diameter of the particle, m} \\
\rho_s \text{ and } \rho &= \text{density of the particle and fluid} \\
C_D &= \text{coefficient of drag} \\
\mu &= \text{dynamic viscosity, N s/m}^2 \\
\nu &= \text{kinematic viscosity, m}^2/\text{s} \\
S_s &= \text{specific gravity of the particles} \\
A &= \text{surface area of the spherical particle} \\
V &= \text{volume of the spherical particle}
\end{aligned}
$$

11-3. A grit particle with $S_s = 2.65$ falls through water having kinematic viscosity $\nu = 1.1306 \times 10^{-6}$ m^2/s. The diameter of the particle is 0.3 mm. Calculate the settling velocity.

$$C_D = \frac{24}{N_R} + \frac{3}{\sqrt{N_R}} + 0.34$$

where

$$N_R = \text{Reynolds number} = v_s\, d/v$$

11-4. Design a velocity controlled grit channel for the Design Example. The maximum and average design flows are $1.321, 0.440 \text{ m}^3/\text{s}$. Detention time is 1 min, and velocity through the channel is $0.24 - 0.34$ m/s. Calculate the dimensions of the channel, and design the proportional weir using the procedure given in Chap. 14. The weir crest is 7.0 cm above the channel floor.

11-5. Calculate the dimensions of an aerated grit chamber treating wastewater flow of $0.5 \text{ m}^3/\text{s}$. Detention time is 4 min, depth $= 3$ m, and the $L : W$ ratio is 5 : 1. Also calculate total air supply if aeration rate is 12 ℓ/s per meter of the tank length. Also estimate average quantity of grit collected.

11-6. Develop the hydraulic profile through the grit chamber in the Design Example at average design flow of $0.44 \text{ m}^3/\text{s}$ when both chambers are in operation.

11-7. A weir trough is 10 m long and 1 m wide. The weir crest is on one side of the trough, and covers the entire length of the trough. Calculate the depth of the trough if discharge through the basin is $0.3 \text{ m}^3/\text{s}$. Depth of flow at the lower end of the trough is 0.9 m. Assume friction loss is 15 percent of the depth of water at the upper end, and freefall allowance is 6 cm. Use Eq. (11-9).

11-8. List the necessary information needed by a design engineer to start the calculations for design of a grit removal facility.

11-9. List the advantages of providing a grit removal facility. Under what conditions would you prefer a velocity controlled grit channel over an aerated grit chamber?

11-10. Determine the total head loss through a 4-m wide grit chamber. The details of the grit chamber are given below.

(a) The influent channel is 1.5 m wide and has one submerged orifice 1.5m × 1.5m. The invert of the influent channel is 0.5 m above the floor of the chamber. The depth of floor in the chamber is 3.0 m.

(b) The head loss in chamber is small and can be ignored.

(c) The influent and effluent baffles occupy 65 and 60 percent of the cross-sectional area of the chamber, respectively.

(d) The flow through the grit chamber is $1.6 \text{ m}^3/\text{s}$.

(e) There is 0.4 m head loss into the effluent structure. This head loss is the difference in water surface elevations in the chamber at the effluent weir and the outlet box of the grit chamber.

Draw the hydraulic profile through the grit chamber.

11-11. Study the computational scheme of determining water surface profile in a flume using Eqs. (11-4)–(11-8). A solution procedure is given in Chapter 13 (Sec. 13-7-4, step C4). Perform similar steps to determine the water depth in the effluent launder of grit removal facility in the Design Example. Use uniform increment of 0.5 m. Compare your result with that given in Sec. 11-9-2, step E6.

REFERENCES

1. Metcalf and Eddy, Inc., *Wastewater Engineering: Treatment, Disposal, Reuse,* McGraw-Hill Book Co., New York, 1979.

2. Ramalho, R. S., *Introduction to Wastewater Treatment Processes,* Academic Press, New York, 1977.

3. Schroeder, E. D., *Water and Wastewater Treatment,* McGraw-Hill Book Co., New York, 1977.

4. Sundstrom, D. W., and H. E. Klei, *Wastewater Treatment,* Prentice-Hall, Englewood Cliffs, N.J., 1979.

5. Metcalf and Eddy, Inc., *Wastewater Engineering: Collection, Treatment, Disposal,* McGraw-Hill Book Co., New York, 1972.

6. Joint Committee of the Water Pollution Control Federation and the American Society of Civil Engineers, *Wastewater Treatment Plant Design,* MOP/8, Water Pollution Control Federation, Washington, D.C., 1977.

7. Rex Chain Belt Co., *Weir of Special Design for Grit Channel Velocity Control,* Binder No. 315, Vol. 1, September 1963.

8. Rao, N. S. K., and D. Chandrasekaran, "Outlet Weirs for Trapazoidal Grit Chambers," *Journal Water Pollution Control Federation,* Vol. 44, No. 3, March 1972, p. 459.

9. Clark, J. W., W. Viessman, and M. J. Hammer, *Water Supply and Pollution Control,* IEP-A Dun-Donnelley, New York, 1977.

10. Neighbor, J. B., and T. W. Cooper, "Design and Operation Criteria for Aerated Grit Chamber," *Water and Sewage Works,* December 1965, p. 448.

11. Albertson, M. L., J. R. Barton, and D. B. Simons, *Fluid Mechanics for Engineers,* Prentice-Hall, Englewood Cliffs, N.J., 1960.

12. Morris, H. M., and J. K. Miggert, *Applied Hydraulics in Engineering,* Ronald Press, New York, 1972.

13. Chow, V. T., *Open-channel Hydraulics,* McGraw-Hill Book Co., New York, 1959.

14. Benefield, L. O., J. F. Judkins, and A. D. Parr, *Treatment Plant Hydraulics for Environmental Engineers,* Prentice-Hall, Englewood Cliffs, N.J., 1984.

15. Fair, G. M., J. C. Geyer, and J. C. Morris, *Water Supply and Wastewater Disposal,* McGraw-Hill Book Co., New York, 1954.

16. Rouse, Hunter, *Engineering Hydraulics,* Wiley & Sons, New York, 1950.

17. Hwang, N. H. C., *Fundamentals of Hydraulic Engineering Systems,* Prentice-Hall, Englewood Cliffs, N.J., 1981.

18. Culp, G. L., N. F. Heim, *Field Manual for Performance Evaluation and Troubleshooting at Municipal Wastewater Treatment Facilities,* U.S. Environmental Protection Agency, Washington, D.C., EPA-430/9-78-001, January 1978.

12

Primary Sedimentation

12-1 INTRODUCTION

The purpose of primary sedimentation (or clarification) is to remove the settleable organic solids. Normally, a primary clarification facility will remove 50–70 percent total suspended solids and 30–40 percent BOD_5. Primary clarification is achieved in large basins under relatively quiescent conditions. The settled solids are collected by mechanical scrapers into a hopper, from which they are pumped to a sludge-processing area. Oil, grease, and other floating materials are skimmed from the surface. The effluent is discharged over weirs into a collection trough.

The purpose of this chapter is to present the basic design considerations of primary sedimentation facility. The step-by-step design procedure, equipment selection, design details, facility operation and maintenance, and equipment specifications are covered in the Design Example.

12-2 TYPES OF CLARIFIERS

In general, the design of most of the clarifiers falls into three categories: (1) horizontal flow, (2) solids contact, and (3) inclined surface.

12-2-1 Horizontal Flow

In horizontal flow clarifiers, the velocity gradients are predominantly in horizontal direction. The common types of horizontal flow clarifiers are rectangular, square, or circular. The selection of any shape depends on:

1. Size of installation
2. Regulation and preference of regulatory authorities
3. Local site conditions, and
4. Preference, experience, and engineering judgment of the designer and plant personnel.

The advantages and disadvantages of rectangular clarifiers over circular clarifiers are summarized below.

Advantages

1. Occupy less land area when multiple units are used
2. Provide economy by using common walls for multiple units
3. Easier to cover the units for odor control
4. Provide longer travel distance for settling to occur
5. Less short circuiting
6. Lower inlet-outlet losses
7. Less power consumption for sludge collection and removal mechanisms

Disadvantages

1. Possible dead spaces
2. Sensitive to flow surges
3. Restricted in width by collection equipment
4. Require multiple weirs to maintain low weir loading rates
5. High upkeep and maintenance costs of sprockets, chain, and flights used for sludge removal

The details of rectangular and circular-horizontal flow clarifiers are illustrated in Figures 12-1 and 12-2.

(a)

Figure 12-1   Details of Horizontal Flow Rectangular Clarifiers. (a) Four primary settling tanks preceding aeration tanks at a municipal wastewater treatment plant. (b) Plan, longitudinal and cross sections of a double rectangular sedimentation basin with skimmer (standard dimensions may be found in Ref. 16). (Courtesy FMC Corporation, Material Handling Systems Division.)

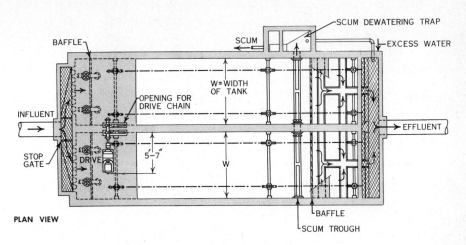

PLAN VIEW

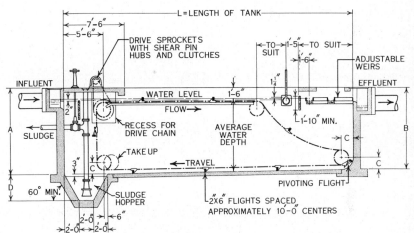

LONGITUDINAL SECTION WITH SKIMMER

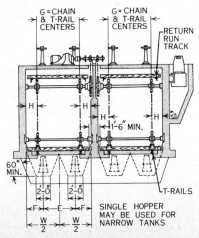

CROSS SECTION

(b)

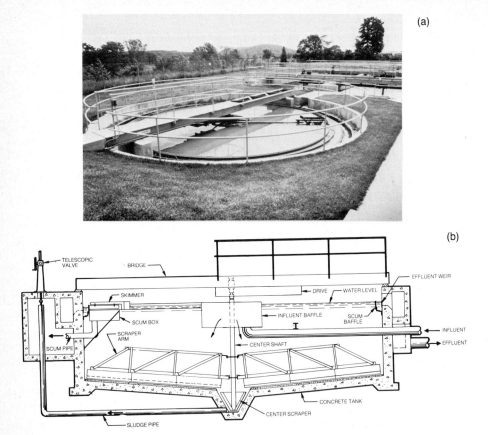

Figure 12-2   Details of Horizontal Flow Circular Clarifier. (a) One 9-meter diameter primary clarifier with skimmer. (b) Sectional view of a circular clarifier with central feed. (Courtesy FMC Corporation, Material Handling Systems Division.)

12-2-2 Solids Contact

The solids contact clarifiers utilize the principle of solids contact.[1-3] The incoming solids are brought in contact with a suspended sludge layer near the bottom. The incoming solids rise and come in contact with the solids in the sludge layer. This layer acts as a blanket, and the incoming solids agglomerate and remain enmeshed within this blanket. The liquid rises upward while a distinct interface retains the solids below. These clarifiers have better hydraulic performance and have reduced detention time for equivalent solids removal in horizontal flow clarifiers.

Both circular and rectangular units are used for the solids contact clarifier. Solids contact clarifiers are efficiently used for chemical flocculant suspensions. These units are not suitable for biological sludges because long sludge-holding times may create undesirable septic conditions. The operational principles of solids contact clarifiers are illustrated in Figure 12-3(a).

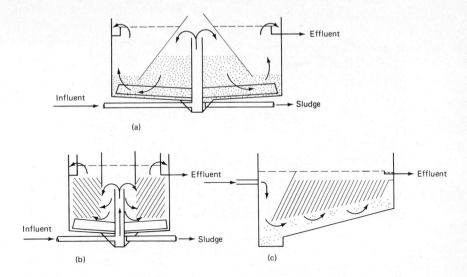

Figure 12-3   Types of Clarifiers. (a) Circular solids contact clarifier. (b) Parallel inclined plates in a circular clarifier. (c) Tube settlers in a rectangular clarifier.

12-2-3 Inclined Surface

The inclined surface basins utilize inclined trays to divide the depth into shallower sections. Thus, the depth of fall of particles (and therefore the settling time) is significantly reduced. This concept is frequently used to upgrade the existing overloaded primary and secondary clarifiers. There are two design variations to the inclined surface clarifiers: tube settlers and parallel plate separators.

Tube Settlers.   The inclined trays are constructed using thin-wall tubes. These tubes are circular, square, hexagonal, or any other geometric shape, and are installed in an inclined position within a basin. The incoming flow enters these tubes and flows upward. The solids settle on the inside of the tubes and slide down into a hopper.

Parallel Plate Separators.   The parallel inclined plate separators have parallel trays covering the entire tank. The operational principles for parallel plate separators are the same as those for the tube settlers.

The inclined surface clarifiers provide a large surface area, thereby reducing the clarifier size. There is no wind effect and the flow is laminar. Many overloaded horizontal flow clarifiers have been upgraded using this concept. The major drawbacks of inclined surface clarifiers include the following: (1) long periods of sludge deposits on inner walls may cause septic conditions, (2) effluent quality may deteriorate when sludge deposits slough off, (3) there may be clogging of the inner tubes and channels, and (4) serious short circuiting may occur when influent is warmer than the basin temperature. Some excellent references are available on this subject. [3-5] The operating principles of inclined surface clarifiers are illustrated in Figure 12-3(b) and (c).

12-3 DESIGN FACTORS

The design objective of a primary sedimentation facility is to provide sufficient time under quiescent conditions for maximum settling to occur. The solids removal efficiency of a clarifier is reduced due to the following conditions:

1. Eddy currents induced by the inertia of the incoming fluid
2. Surface currents produced by wind action
3. Vertical currents induced by outlet structure
4. Vertical convection currents induced by the temperature difference between the influent and the tank contents
5. Density currents causing cold or heavy water to underrun a basin, and warm or light water to flow across its surface
6. Currents induced due to the sludge scraper and sludge removal system.

Therefore, many factors, such as overflow rate, detention period, weir-loading rate, shape and dimensions of the basin, inlet and outlet structures, and sludge removal system, all enter into the design of the sedimentation basin. Many of the design factors are discussed in the following sections.[6-10]

12-3-1 Overflow Rate or Surface-Loading Rate

Overflow rate is expressed in cubic meters per day per square meter $(m^3/m^2 \cdot d)$* of the surface area of the tank. The effect of overflow rate on suspended solids removal varies widely depending on character of the wastewater, proportion of settleable solids, concentration of solids, and other factors. For a well-designed and operated basin, the typical removal of suspended solids and BOD_5 from municipal wastewater as a function of overflow rate is given in Figure 12-4. The overflow rate should be small enough to ensure satisfactory performance at peak flows.[8] The design overflow rates for various types of

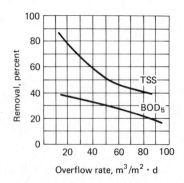

Figure 12-4   *Removal of Suspended Solids and BOD$_5$ with Respect to Overflow Rate in Primary Clarifiers.*

*1 $m^3/m^2 \cdot d$ = 24.5424 gpd/ft^2.

TABLE 12-1 Design Overflow Rates for Clarifiers

Condition	Range $(m^3/m^2 \cdot d)$	Typical $(m^3/m^2 \cdot d)$
Primary clarification prior to secondary treatment		
Average flow	30–50	40
Peak flow	70–130	100
Primary clarification with waste-activated-sludge return		
Average flow	25–35	30
Peak flow	45–80	60

Note: 1 $m^3/m^2 \cdot d$ = 24.5424 gpd/ft^2

clarification facilities are given in Table 12-1. Primary sedimentation facilities are generally designed for an overflow rate of 40 $m^3/m^2 \cdot d$ at average design flow.

12-3-2 Detention Period

For a given surface area, the detention time depends on the depth of a tank. The detention time for various overflow rates and tank depths are summarizd in Table 12-2. Most of the designs utilize 1–2 h detention period for primary tanks and 2–4 h for final clarifiers at average design flow. The removal of BOD_5 and TSS may vary greatly. Typical removals in a well-designed and operated basin as a function of detention time are shown in Figure 12-5.

TABLE 12-2 Detention Times for Various Overflow Rates and Tank Depths

Overflow Rate $(m^3/m^2 \cdot d)$	Detention Period (h)					
	2.0-m Depth	2.5-m Depth	3.0-m Depth	3.5-m Depth	4.0-m Depth	4.5-m Depth
30	1.6	2.0	2.4	2.8	3.2	3.6
40	1.2	1.5	1.8[a]	2.1	2.4	2.7
50	1.0	1.2	1.4	1.7	1.9	2.2
60	0.8	1.0	1.2	1.4	1.6	1.8
70	0.7	0.9	1.0	1.2	1.4	1.5
80	0.6	0.8	0.9	1.1	1.2	1.4

[a] A 3.0 m deep sedimentation basin having an overflow rate of 40 $m^3/m^2 \cdot d$ will provide a detention period of 1.8 h.

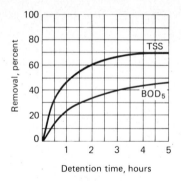

Figure 12-5   Removal of Suspended Solids and BOD$_5$ with Respect to Detention Time in Primary Clarifiers.

12-3-3 Weir-Loading Rate

Weir-loading rates have some effect on the removal efficiency of sedimentation basins. Sedimentation basins are generally designed for loading of less than 370 m^3 per d per m length of the weir.* The Ten-States Standards recommend the following weir loadings:[8]

- Weir-loading rate of 124 m^3/m·d for plants designed for average design flows of 44† ℓ/s or less
- Weir-loading rate of 186 m^3/m·d for plants designed for average flows in excess of 44 ℓ/s

12-3-4 Dimensions

The dimensions of a sedimentation basin are selected to accommodate the standard equipment supplied by the manufacturers. Also, consideration is given to the size of the installation; local site conditions; regulations of the local water pollution control agencies; the experience, judgment, and preference of the designer; and the economics of the system. Multiple units are essential to keeping the process in operation during the repair and servicing of one unit. Basic dimensions of rectangular and circular clarifiers is summarized in Table 12-3.

12-3-5 Solids Loading

Solids loading is not an important deciding factor for the primary sedimentation facility. It varies from 1.5 to 34 kg per m^2 per d.‡ In final clarifiers, the solids loading rate may vary greatly depending on the concentration of mixed liquor suspended solids and sludge volume index. A practical range of solids loading in final clarifier is 49–98 kg/m^2·d.[9,10]

*1 m^3/m^2·d = 80.52 gpd/ft
†1 mgd = 43.813 ℓ/s.
‡1 kg/m^2·d = 0.205 lb/ft^2·d

TABLE 12-3 Dimensions of Rectangular and Circular Basins

Clarifier	Range	Typical
Rectangular		
Length, m	10–100	25–60
Length-to-width ratio	1.0–7.5	4
Length-to-depth ratio	4.2–25.0	7–18
Sidewater depth, m	2.5–5.0	3.5
Width, m[a]	3–24	6–10
Circular		
Diameter, m[b]	3–60	10–40
Side depth, m	3–6	4

[a]Most manufacturers build equipment in width increments of 61 cm (2 ft). If width is greater than 6 m (20 ft), multiple bays may be necessary.
[b]Most manufacturers build equipment in 1.5-m (5-ft) increments of diameter.

12-3-6 Influent Structure

The influent structures are designed to serve many purposes. The objectives are:

1. To dissipate energy in incoming flow by means of baffles or stilling basin
2. To distribute flow equally along the width
3. To prevent short circuiting by disturbing thermal and density stratification, and
4. To provide small head loss

In the design of influent structure, provision is also made to control flows, remove scum, and facilitate maintenance. Normally, inlets are uncovered or have removable gratings. The velocity at inlet pipe is maintained at approximately 0.3 m/s. Many inlet details for rectangular tanks are shown in Figure 12-6.[10,11] Selection of any design may depend on tank size, flow conditions, and designer preference and experience.

Based on the influent structures the circular clarifiers are classified as center and peripheral feed. In center feed circular clarifiers the inlet is at the center and the outlet along the periphery. A concentric baffle distributes the flow equally in radial directions. The advantages of center feed clarifiers are low upkeep cost, and ease of design and construction. The disadvantages include short circuiting, low detention efficiency, lack of scum control, and loss of sludge into the effluent.

The center feed square clarifier is a modification of the circular clarifier. Both the inlet and outlet structures are similar to those of circular clarifiers. The inside corners are rounded to prevent solids accumulation. The benefit of square clarifiers is the opportunity to use the common walls in multiple units. The flow scheme of center feed circular clarifier is shown in Figure 12-7(a). In the peripheral feed clarifiers, the flow enters along the periphery. These clarifiers are considerably more efficient and have less short-

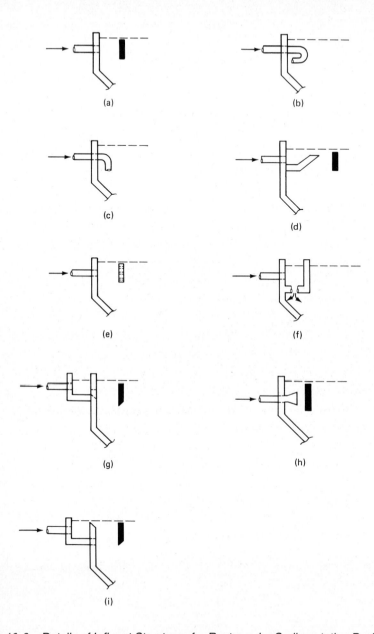

Figure 12-6   Details of Influent Structures for Rectangular Sedimentation Basin. (a) Inlet pipes discharging against a baffle. (b) U-shape elbow discharging against the wall. (c) A series of inlet pipes spaced across the width with turned elbow. (d) An inclined weir with baffle. (e) Perforated baffle. (f) A stilling basin with opening at the bottom. (g) Pipe discharging in a channel which has series of openings discharging against a baffle. (h) A bell shaped diverging pipe followed by a baffle. (i) An overflow weir followed by a baffle.

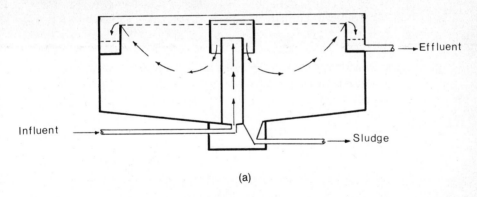

(a)

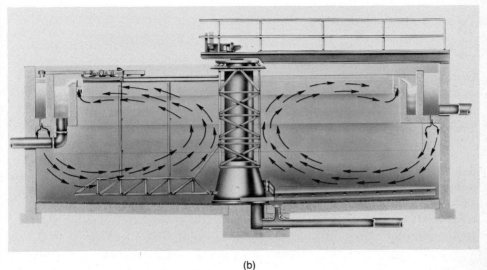

(b)

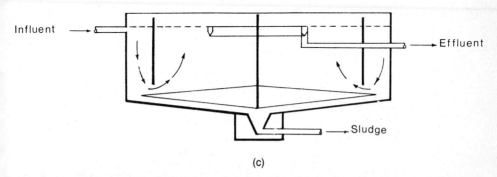

(c)

Figure 12-7 Influent and Effluent Structures for Circular Clarifiers. (a) Circular clarifier with center feed. (b) Peripheral feed circular clarifier with effluent and influent channels separated by a skirt. (Courtesy Envirex Inc., a Rexnord Company.) (c) Peripheral feed circular clarifier with effluent weirs near the center of the basin.

circuiting than the center feed clarifiers.[12-15] There are two major variations of peripheral feed clarifiers:

1.   The influent is distributed through the orifices in an influent raceway on the entire periphery. A skirt is put underneath the raceway. This skirt gives low initial velocity and minimizes the density currents and short-circuiting.
2.   A circular baffle is suspended a short distance from the tank wall to form an annular space into which the flow is discharged in a tangential direction. The wastewater flows spirally around the tank and underneath the baffle.

Both the variations of periphiral feed clarifiers are illustrated in Figure 12-7.

12-3-7 Effluent Structure

The effluent structures are designed to (1) provide a uniform distribution of flow over a large area, (2) minimize the lifting of the particles and their escape into the effluent, and (3) reduce the escape of floating matter to the effluent. Most common types of effluent structures for rectangular and circular tanks are weirs that are adjustable for leveling. These weir plates are sufficiently long to avoid high head that may result in updraft currents and lifting of the particles. A weir loading of 186 m^3 per d per linear m is used for plants larger than $44 \ \ell/s$.[8] Both straight edge and V notches either on one side of the trough or both sides have been used in rectangular and circular tanks. V notches provide uniform distribution at low flows. A baffle is provided in front of the weir to stop the floating matter from escaping into the effluent. Normally the weirs in rectangular tanks are provided on the opposite end of the influent structure. Different weir configurations in rectangular basins may be utilized in order to obtain the desired length of the weir. Some arrangements for rectangular basins are shown in Figure 12-8.

In circular clarifiers, the outlet weir can be near the center of the clarifier or along the periphery as shown in Figure 12-7. The center weir generally provides high velocity gradients which can result in solids carryover. Figure 12-9 shows the arrangement of weir notches, effluent launders, and outlet channel in rectangular and circular clarifiers.

12-3-8 Sludge Collection

Bottom Slope.   The floor of the rectangular and circular tanks are sloped toward the hopper. The slope is made to facilitate draining of the tank and to move the sludge toward the hopper. Rectangular tanks have a slope of 1–2 percent. In circular tanks, the slope is approximately 40–100 mm/m diameter.

Equipment.   In mechanized sedimentation tanks, the type of sludge collection equipment varies with size and shape of the tank. The sludge collection equipment for rectangular and circular clarifiers is discussed below.

Rectangular Tanks.   In rectangular tanks the sludge collection equipment may consist of (1) a pair of endless conveyor chains running over sprockets attached to the shafts or (2)

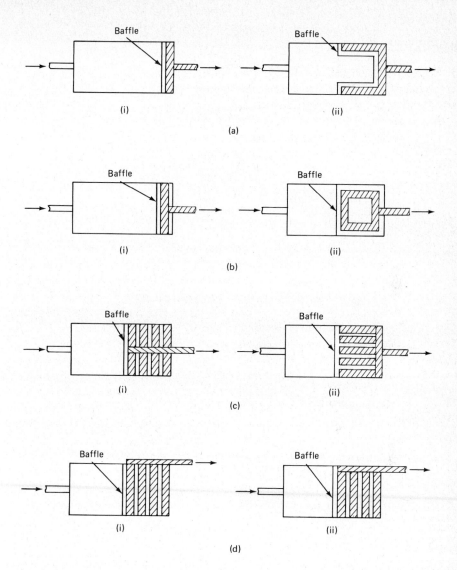

Figure 12-8 Various Configurations of Effluent Structure Used for Rectangular Clarifiers. (a) Single weir and trough (b) Double weirs and trough (c) Multiple weirs and troughs with outlet channel at the middle, and at the end (d) Multiple weirs and troughs with outlet channel at the side.

moving-bridge sludge collectors having a scraper to push the sludge into the hopper or suction-type arrangement to withdraw the sludge from the basin. Design details, and advantages and disadvantages of these types of sludge collection equipment are summarized in Table 12-4.[2,15,16] The details of endless-conveyor chain and moving-bridge sludge collector arrangements are shown in Figures 12-10 and 12-11.

(a)

(b)   (c)

Figure 12-9   Effluent Weir, Notches, Launder, and Outlet Channel in Rectangular and Circular Clarifiers. (a) V notches and suspended effluent launder in circular clarifier. (b) Effluent launders discharging into an outlet channel. (c) Effluent notches, launders, and outlet channel in a rectangular clarifier.

Figure 12-10 Endless Conveyor Chain Details. (a) Conveyor sludge collectors with skimmer. (Courtesy Envirex Inc., a Rexnord Company.) (b) Drive sprockets. (c) Chain drive with gear speed reducer. (d) Collector sprocket. (Courtesy FMC Corporation, Material Handling Systems Division.)

Circular Tanks. The circular tanks utilize two types of sludge collection equipment:

1. The scraping mechanism is installed with radial arms having plows set at an angle supported on center pier or on a beam spanning the tank. The clarifiers in excess of 10 m in diameter normally have a central pier, while the clarifiers with smaller diameter utilize beam support. The flight travel speed is 0.02–0.06 revolutions/min.
2. Suction-type units are used for handling light sludge. The suction mechanism is installed similar to scraping mechanism. The equipment details of circular clarifiers are shown in Figures 12-2 and 12-9(a).

TABLE 12-4 System Description and Advantages/Disadvantages of Sludge Collectors for Rectangular Tanks

Collector Type	Description	Advantages	Disadvantages
Conveyor chain	1) One endless chain is connected to a shaft and a drive unit	1) Arrangement is simple to install	1) High maintenance cost of chain and flight removal mechanism
	2) Linear conveyor speed is 0.3–1.0 m/min for primary and 0.3 m/min for final clarifier	2) Power consumption is small	2) Tank must be dewatered for gear or chain repair
	3) Cross-wood (flights) are attached to the chain at 3-m intervals and are up to 6 m in length. These flights scrape the sludge to the hopper. The cross-wood is 5 cm thick and 15–20 cm deep	3) Scum collection is efficient	3) Light sludge may resuspend
	4) For tanks greater than 6 m in width, multiple pairs of chains are used	4) Suitable for heavier sludge	
	5) The floating material is pushed in opposite direction of sludge, and is collected in a scum collection box		
Bridge drive scraper	1) Standard traveling beam bridges for spans up to 13 m (40 ft) and truss bridge for spans over 13 m span are used	1) All moving mechanisms are above water. No underwater bearings are used	1) High-power requirement to move the bridge

2) Bridge travel is accomplished by the use of a gear motor

3) The wheels run on rails which are attached to the footing wall along each side wall of the basin

4) Mechanical scrapers or rakes are hung from the top carriage that push the sludge to the hopper

5) Separate blades are provided on top to move the scum

2) Standard designs permit scraper repair or replacement without tank dewatering

3) No width restrictions as with chain type

4) Longer operation life

5) Lower maintenance cost in low-span bridges

2) Units will not operate with ice-covered tanks

3) In long-span bridges, the wheels may climb frequently over rails, causing break-downs.

Bridge drive sludge suction

1) The bridge design is similar to the above arrangement

2) The sludge removal mechanisms are attached to the bridge and provide continuous removal of sludge along the length of travel

3) Pump, siphon, or airlift arrangements are used to suck and remove the sludge

1) Better pick up of light sludge

2) Other advantages same as above

3) Used for biological and chemical sludges

1) Same as above

2) Not used in primary clarifier

Source: Adapted in part from Refs. 2, 15, and 16.

(a)

(b)

Figure 12-11   Moving Bridge Sludge Collector. (a) Traveling truss bridge with mechanical scraper hung from the top. The effluent weir is the type shown in Figure 12-8(c-ii). (b) Traveling truss bridge showing wheels that run on rails.

12-3-9 Sludge Removal

The sludge is removed from the hopper by means of a pump. The following design considerations are given to the sludge removal system:

1. Provision of continuous sludge pumping is desirable
2. Each sludge hopper should have individual sludge withdrawal line at least 15 cm in diameter
3. In rectangular tanks, cross-collectors are preferred over multiple hoppers. The multiple hoppers pose operational difficulties such as sludge accumulating in corners and slopes, and arching over the sludge-drawoff piping. Also, the cross collectors provide withdrawal of more uniform and concentrated sludge.
4. Screw conveyors for sludge removal are also used.
5. In new plants, a photocell-type or a sonic-type sludge blanket detector is often used to provide an indication of the depth of the sludge blanket. Therefore, an automatic control of sludge pump or siphon pipes can be achieved. This arrangement is particularly desirable for final clarifiers. The sludge pump or the siphon starts or stops automatically when the predetermined sludge blanket depth is reached.
6. The sludge pump used are self-priming centrifugal and normally discharge into a common manifold. One sludge pumping station can serve two rectangular clarifiers. The circular clarifiers are normally arranged in groups of two or four. The sludge is withdrawn to a control chamber located in the middle; from there it is pumped to other sludge-handling areas.

12-3-10 Scum Removal

Scum that forms on the surface of the primary clarifiers is generally pushed off the surface to a collection sump. In rectangular tanks the scum is normally pushed in the opposite direction by the flights of the sludge mechanism in its return travel. In circular clarifiers, the scum is moved by a radial arm which rotates on the surface with the sludge removal equipment. The scum can also be moved by water sprays.

The scum may be scraped manually or mechanically up an inclined apron. In small installations, a hand-tilt slotted pipe with a lever or screw is commonly used. This arrangement may result in a relatively large volume of scum liquor. Various types of scum removal arrangements used in rectangular and circular basins are shown in Figures 12-2, 12-9, and 12-12.

All effluent weirs have baffles to stop the loss of scum into the effluent. A scum sump is provided outside the tank. A scum pump transfers scum to the disposal facility.

The scum usually has a specific gravity of 0.95. Solids content may vary from 25 to 60 percent. The quantity of scum at a plant may vary from 2 to 13 kg per 10^3 m^3 (17–110 lb per million gallon). Pumping of scum is difficult as it may form clotty mass (called greaseballs) at lower temperatures. Pipings are often glass-lined and kept reasonably warm to minimize blockages. Scum has been digested in aerobic and anaerobic digesters, landfilled or incinerated. Heating value may range from 16,000 to 40,000 kJ/kg of dry solids (7000–17,000 Btu/lb).

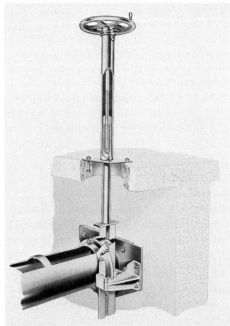

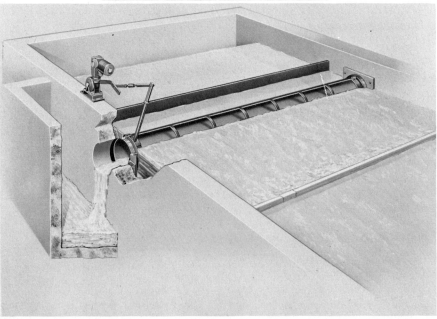

(a)

Figure 12-12   Scum Collection and Removal Arrangements. (a) Hand and motor operated system to tilt the open-top pipe to remove and discharge scum into dewatering trap. (b) Motor operated spiral skimmer that turns the blades to push and drop the scum into a scum trough. (Courtesy Envirex Inc., a Rexnord Company.)

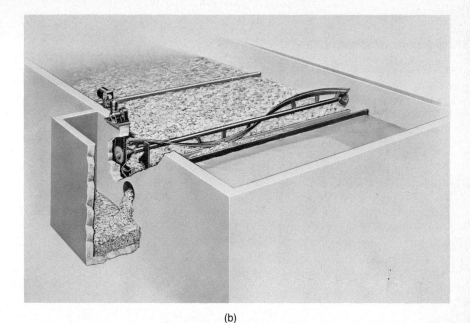

(b)

12-4   EQUIPMENT MANUFACTURERS OF SEDIMENTATION BASINS

Many manufacturers of pollution control equipment construct or assemble the hardware for sedimentation basins. These manufacturers specialize in different types of designs. The names and addresses of several manufacturers of sedimentation equipment are given in Appendix D. Basic considerations for equipment selection are covered in Sec. 2-10.

12-5   INFORMATION CHECKLIST FOR DESIGN OF SEDIMENTATION BASIN

The following information is necessary for designing a sedimentation facility:

1. Average and peak design flows including the returned flows from other treatment units
2. All sidestreams from thickener, digester, and dewatering facilities should be considered if flows are returned ahead of primary sedimentation.
3. Treatment plant design criteria prepared by the concerned regulatory agencies
4. Equipment manufacturers and equipment selection guide (catalog)
5. Information on the existing facility if the plant is being expanded
6. Available space and topographic map of the plant site
7. Shape of the tank (rectangular, square, or circular)
8. Influent pipe data, to include diameter, flow characteristics, and approximate water surface elevation or hydraulic grade line
9. Head loss constraints for sedimentation facility

12-6 DESIGN EXAMPLE

12-6-1 Design Criteria Used

The following design criteria shall be used for the Design Example:

1. Two rectangular units shall be designed for independent operation. A bypass to the aeration basin shall be provided for emergency conditions when one unit is out of service. Most regulatory agencies will allow such bypass.
2. Overflow rate and detention time shall be based on an average design flow of 0.44 m³/s (0.22 m³/s through each unit)
3. The overflow rate shall be less than 36 m³/m²·d (at average design flow).
4. The detention time shall be not less than 1.5 h (at average design flow).
5. The influent structure shall be designed to prevent short circuiting and reduce turbulence. The influent channel shall have a velocity less than 0.35 m/s at design peak flow (0.661 m³/s through each basin).
6. All sidestreams shall be returned to aeration basins.
7. The weir loading shall be less than 186 m³/m·d at average design flow.
8. The launder and outlet channels shall be designed at the peak design flow of 1.321 m³/s (0.661 m³/s through each unit).
9. The liquid depth in the basin shall be no less than 2 m.

12-6-2 Design Calculations

A. Basin Dimensions

1. Select basin geometry, and provide two rectangular basins with common wall*

- Average design flow through each basin $= 0.22$ m³/s

- Overflow rate at average design flow $= 36$ m³/m² · d

- Surface area $= \dfrac{0.22 \text{ m}^3/\text{s} \times 86{,}400 \text{ s/d}}{36 \text{ m}^3/\text{m}^2 \cdot \text{d}} = 528$ m²

- Use length-to-width ratio $L : W :: 4 : 1$ (see Table 12-3)

- $(4\ W)(W) = 528$ m²

  $W = 11.5$ m

Since most of the equipment manufacturers' assembled equipment in increments of 61 cm (2 ft), therefore:

Width	$= 11.58$ m (38 ft)
Length	$= 46.33$ m (152 ft)
Length-to-width ratio	$= 4 : 1$
Provide average water depth at mid-length of the tank	$= 3.1$ m (10 ft)

---

*For illustration purposes, the design procedure for the rectangular primary sedimentation basins is given in this chapter. The design procedure for circular clarifiers is provided in Chapter 13.

Length-to-depth ratio	$= 46.33 \text{ m}/3.1 \text{ m} = 15$
Freeboard	$= 0.6 \text{ m (2 ft)}$
Average depth of the basin	$= 3.7 \text{ m (12 ft)}$

2. Check overflow rate

$$\frac{\text{Overflow rate at}}{\text{average design flow}} = \frac{0.22 \text{ m}^3/\text{s} \times 86{,}400 \text{ s/d}}{11.58 \text{ m} \times 46.33 \text{ m}}$$

$$= 35.4 \text{ m}^3/\text{m}^2 \cdot \text{d}$$

$$\frac{\text{Overflow rate at}}{\text{peak design flow}} = \frac{0.661 \text{ m}^3/\text{s} \times 86{,}400 \text{ s/d}}{11.58 \text{ m} \times 46.33 \text{ m}} = 106.4 \text{ m}^3/\text{m}^2 \cdot \text{d}$$

3. Check detention time

Average volume of the basin $= 3.1 \text{ m} \times 11.58 \text{ m} \times 46.33 \text{ m} = 1663.2 \text{ m}^3$

$$\frac{\text{Detention time at}}{\text{average design flow}} = \frac{1663.2 \text{ m}^3}{0.22 \text{ m}^3/\text{s} \times 3600 \text{ s/h}} = 2.1 \text{ h}$$

$$\frac{\text{Detention time at}}{\text{peak design flow}} = \frac{1663.2 \text{ m}^3}{0.661 \text{ m}^3/\text{s} \times 3600 \text{ s/h}} = 0.7 \text{ h}$$

B. Influent Structure

1. Select the arrangement of the influent structure
    The influent structure includes a 1-m-wide influent channel that runs across the width of the tank. Eight submerged orifices 34 cm (13 in.) square each, are provided in the inside wall of the channel to discharge the flow into the basin. The purpose of these orifices is to distribute the flow over the entire width of the basin. A submerged influent baffle is provided 0.8 m in front, 1.0 m deep, and 5 cm below the liquid surface. The influent structure is shown in Figure 12-13.
2. Compute the head loss in the influent pipe connecting the junction box located downstream of the grit chamber and the influent structure of the sedimentation basin
    The influent pipe to the primary sedimentation basin connects the junction box below grit chamber with the influent channel. Normally, this pipe is a sewer or a pressure pipe. The elevation of the water surface in the influent channel of the basin is lower than the water surface elevation in the junction box downstream from the grit chamber. The difference (ΔH) being the sum of the head loss in the connecting pipe due to the entrance, friction, bends, and fittings; and exit loss into the influent channel of the sedimentation basin. These losses are calculated in Chapter 21 when the hydraulic profile is prepared through the entire treatment plant.
3. Compute the head losses at the influent structure
    Chao and Trussell have provided the design procedure of a distribution channel.[17] For a precise mathematical solution of losses at the influent structure the boundary layer theory and energy equations for manifolds are used.[18–20] In

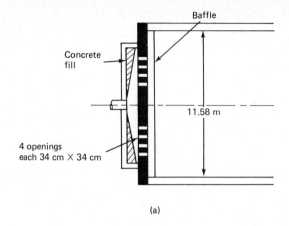

(a)

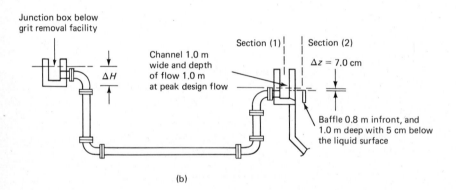

(b)

*Figure 12-13   Example of Influent Structure and Piping Details. (a) Plan of influent struc-
ture. (b) Connecting pipes from junction box below the grit removal facility to the influent
structure of the sedimentation basin.*

this design only an approximate solution using the energy equation is given.
Refer to Eq. (11-1) and sections (1) and (2) as defined in Figure 12-13(b). The
horizontal velocity in the sedimentation basin (v_2) is small and is ignored. The
average velocity in the influent channel (v_1) is calculated at peak design flow.
Half of the flow divides on each side of the basin.

$$\text{Discharge in each channel} = \frac{\text{peak design flow per basin}}{2} = \frac{0.661 \text{ m}^3/\text{s}}{2} = 0.331 \text{ m}^3/\text{s}$$

The depth of water into the influent channel is fixed by the designer. Suppose
the depth of water at the entrance of the influent channel is 1.0 m and the width
of the influent channel is 1.0 m.

$$\frac{\text{Velocity in the channel}}{\text{at peak design flow}} = \frac{0.331 \text{ m}^3/\text{s}}{1 \text{ m} \times 1 \text{ m}} = 0.331 \text{ m/s}$$

The velocity in the channel will change as the flow is successively reduced at various orifices. At the same time the width of the channel and depth of flow away from the entrance of the influent pipe are also reduced. Consequently, the discharge, and velocity in the channel and through the submerged orifices will vary.[12] Using approximation, the term Δz in Eq. (11-2) is equal to the sum of friction losses into the channel, and the head loss through the short pipes referred to as submerged orifices. The discharge through each orifice = (0.661 m³/s)/8 = 0.083 m³/s. Using Eq. (11-2) for a submerged orifice, Δz is calculated as follows:

$$\Delta z^* = \left[\frac{0.083 \text{ m}^3/\text{s}}{0.6 \times (0.34^{\dagger} \text{ m})^2 \times \sqrt{2 \times 9.81 \text{ m/s}^2}} \right]^2$$

$$= 0.07 \text{ m}$$

C. Effluent Structure

1. Select the arrangement of the effluent structure
    The effluent structure consists of weirs, launder, an outlet box, and an outlet pipe. The weirs are either sharp-crested, freefalling, or V notch. The total length of the weir is calculated from the weir-loading rate. The weir length is normally large and may cover a significant portion of the basin area. Normally, the weir plate is installed over concrete walls that form the launder. Often, metal weirs and launder are made as an integral part and then installed into the tank. The weir edge must be kept perfectly leveled to ensure uniform water depth at the notches or over the weir crest.
    If straight edge weirs are provided, the head calculations using the classical weir equations give extremely small heads over the weirs (1–2 mm). At such small heads the capillary clinging effects at the weir crest are significant and the use of the classical weir equations does not provide satisfactory head calculations.[18] At small heads the solids and slime tend to accumulate at the weir crest. Also, if the weir is not perfectly level, portions of the weir do not have any flow which creates nonuniform loading and solids accumulation problems. Due to these difficulties, generally V-notch weirs are preferred in the design of sedimentation basins.
2. Compute the length of the weir

    Weir loading   = 186 m³/m · d at average design flow

    $\dfrac{\text{Average design}}{\text{flow per basin*}}$ = 0.22 m³/s × 86,400 s/d = 19,008 m³/d

---

*Various terms are defined in Eq. (11-2).
†The orifice is 34 cm × 34 cm.

$$\text{Weir length} = \frac{19{,}008 \ \text{m}^3/\text{d}}{186 \ \text{m}^3/\text{m} \cdot \text{d}} = 102.2 \ \text{m}$$

Provide weir notches on both sides of the launder, as shown in Figure 12-14(b)

$$\begin{aligned}\text{Total length of the weir plate} &= 2(16.5 + 10.38)\text{m} + 2(15.3 + 9.18) \ \text{m} - 1.0 \ \text{m} \\ &= 101.72 \ \text{m}\end{aligned}$$

$$\text{Actual weir loading} = \frac{19{,}008 \ \text{m}^3/\text{d}}{101.72 \ \text{m}} = 186.9 \ \text{m}^3/\text{m} \cdot \text{d}$$

3. Compute the number of V notches
    Provide 90° standard V notches at a rate of 20 cm center to center on both sides of the launders. The weir arrangement and details of the notches are shown in Figure 12-14.

$$\text{Total number of notches} = 5 \ \text{notches per m} \times 101.72 \ \text{m} = 509$$

In order to leave sufficient space on the ends of the weir plate, provide a total of 505 notches. The arrangement of V notches is shown in Figure 12-14(a).

4. Compute the head over the V notches at the average design flow

$$\text{The average discharge per notch at average design flow} = \frac{0.22 \ \text{m}^3/\text{s}}{505 \ \text{notches}}$$

$$= 4.34 \times 10^{-4} \ \text{m}^3/\text{s per notch}$$

The discharge through a V notch is given by Eq. (12-1)[18–20]

$$Q = \frac{8}{15} \ C_\text{d} \ \sqrt{2g} \ \tan \frac{\theta}{2} \ H^{5/2} \tag{12–1}$$

where

$$\begin{aligned}Q &= \text{flow per notch, m}^3/\text{s} \\ C_\text{d} &= \text{coefficient of discharge} = 0.584 \\ H &= \text{head over notch, m} \\ \theta &= \text{angle of the V notch} = 90°\end{aligned}$$

The head over the notches is obtained as follows:

$$4.34 \times 10^{-4} = \frac{8}{15} \times 0.584 \times \sqrt{2 \times 9.81 \ \text{m/s}^2} \ \tan \frac{90}{2} \ H^{5/2}$$

where

$$H = 0.04 \ \text{m} = 4.0 \ \text{cm}$$

---

*Flow over weir = average design flow − sludge withdrawal rate. Sludge withdrawal rate is small and is ignored in these calculations.

5. Compute head over V notches at peak design flow

$$\text{Discharge per notch at peak design flow} = \frac{0.661 \text{ m}^3/\text{s}}{505} = 13.0 \times 10^{-4} \text{ m}^3/\text{s}$$

$$13.0 \times 10^{-4} \text{ m}^3/\text{s} = \frac{8}{15} \times 0.584 \sqrt{2 \times 9.81} \text{ m/s}^2 \tan\frac{90}{2} H^{5/2}$$

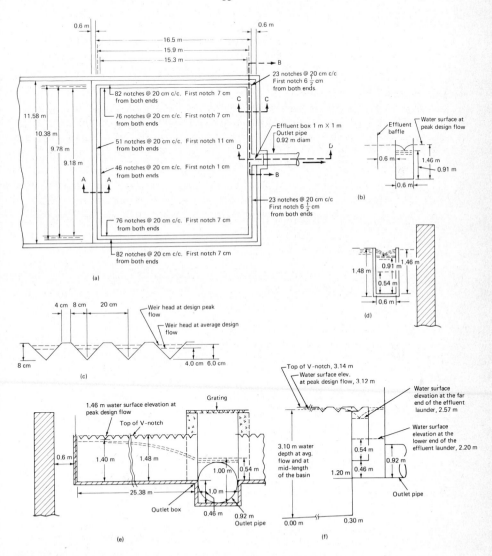

Figure 12-14   Weir Arrangement and Design Dimensions of the Effluent Structure. (a) Plan View. Weir arrangement. (b) Section AA. Weir trough (effluent launder) with weir notches on both sides. (c) Details of the V notches. (d) Section CC. Details of the effluent launder. (e) Section BB. Water surface profile in the effluent launder. (f) Section DD. Details of the outlet channel. All elevations are with respect to the floor level of the basin at mid length.

where

$$H = 0.06 \text{ m} = 6.0 \text{ cm}$$

6. Check the depth of the notch
The total depth of the notch is 8 cm. Maximum liquid head over the notch at peak design flow is 6.0 cm. This gives a safe allowance of 2.0 cm (0.8 in.) against submergence.

7. Compute the dimensions of the effluent launder
The approximate solution of the water profile in the effluent launder is obtained from the equation developed for flumes with level inverts and parallel sides. This equation was discussed in Chapter 11 [Eq. (11-9)]. A more accurate solution may be obtained by using Eqs. (11-4)–(11-8). A computer program for the solution of these equations is given in Appendix B.
The effluent launder discharges into the effluent box. An outlet pipe is connected to the effluent box that carries the primary treated effluent to the next junction-splitter box preceding the aeration basin. The dimensions of the effluent launder, effluent box, outlet pipe, and the water surface elevations in the effluent launder and effluent box are shown in Figure 12-14. The following dimensions may be noted:*

Width of the launder b	= 0.6 m
Width of the effluent box	= 1.0 m
Diameter of the outlet pipe	= 0.92 m
The depth of water in the effluent box[†]	= 1.00 m

Provide the invert of the effluent launder 0.46 m above the invert of the effluent box

$$\begin{array}{l} \text{Depth of water in the effluent launder at} \\ \text{exit point } y_2 \end{array} = 1.0 \text{ m} - 0.46 \text{ m}$$

$$= 0.54 \text{ m}[\ddagger]$$

Half of the flow divides on each side of the launder.

$$\begin{array}{l} \text{Flow on each side of the} \\ \text{launder} \end{array} = \frac{0.661 \text{ m}^3/\text{s}}{2} = 0.33 \text{ m}^3/\text{s}$$

$$\begin{array}{l} \text{Average half-length of the} \\ \text{launder } L \text{ (Figure 12-14)} \end{array} = \frac{9.78 \text{ m}}{2} + 15.90 \text{ m} + \frac{(9.78 - 1.00)\text{m}}{2}$$

$$= 25.18 \text{ m}$$

$$\begin{array}{l} \text{Total length of weir on} \\ \text{each side of the launder} \end{array} = 50.86$$

$$\begin{array}{l} \text{Discharge per unit length} \\ \text{of weir } q' \end{array} = \frac{0.33 \text{ m}^3/\text{s}}{50.86 \text{ m}} = 0.0065 \text{ m}^3/\text{s m}$$

*Various terms are defined in Eq. (11-9)

[†]The water depth in the effluent box is fixed by the designer. In this case the depth of water in the effluent box is equal to the diameter of the outlet pipe plus the entrance and other losses. Calculations for these losses are given in Chapter 21.

[‡]The critical depth of flow in the launder from Eq. (8-5) is 0.5 m. Since the actual depth at the lower end of the launder is greater than the critical depth, the outfall is submerged.

Number of sides that re-
ceive the flow N $= 2$

$$y_1 = \sqrt{(0.54 \text{ m})^2 + \frac{2(0.0065 \text{ m}^3/\text{s} \cdot \text{m} \times 25.18 \text{ m} \times 2)^2}{9.81 \text{ m/s}^2 (0.6 \text{ m})^2 \times 0.54 \text{ m}}} = 0.64 \text{ m}*$$

Provide a 42 percent allowance for losses due to friction, turbulence, and bends, and add 0.49 m for freefall

Water depth at the
far end of the trough $= 0.64 \text{ m} \times 1.42 \text{ m} = 0.91 \text{ m}$

Total depth of the
effluent launder $= 0.91 \text{ m} + 0.49 \text{ m} = 1.40 \text{ m}$

The details of the effluent structure and various dimensions are given in Figure 12-14.

D. Head Loss through Sedimentation Basin.   The total head loss through the primary sedimentation basin includes (1) head losses at the influent structure, (2) head losses at the effluent structure, (3) head loss in the basin, and (4) head losses at the influent and effluent baffles.

The head loss calculations for the influent and effluent structures are given above. The head loss through the basin is small and is ignored. The influent and effluent baffles offer little obstruction to the flow. The momentum equation can be used to determine the head losses due to the baffles.[18–20] The use of the momentum equation [Eq. (11-5)] for the calculation of head losses due to the baffles has been shown in Chapter 11. These losses are small and can be ignored.

E. Hydraulic Profile through the Basin.   Figure 12-15 illustrates the hydraulic profile through the sedimentation basin at peak design flow. A total head loss of 0.99 m (3.25 ft) occurs at the peak design flow.

F. Sludge Quantities.
  1.   Establish sludge characteristics
       It is desirable to produce a sludge that is as thick as possible. Normally, the primary sludge has a specific gravity of 1.03 and a solids content of 3–6 percent. The sludge produced from the facility designed in this example will have a typical solids content of about 4.5 percent.
  2.   Compute average quantity of sludge produced per day

Amount of solids produced per
basin per day at a removal rate of $=$ $260 \text{ g/m}^3 \times (0.63)$
63 percent[†] $\times 0.22 \text{ m}^3/\text{s} \times 86{,}400 \text{ s/d}$
              $\times \text{kg}/1000 \text{ g}$
              $= 3113.5 \text{ kg/d}$

---

*The water depth at the upstream end of the effluent launder from Eq. (11-4)–(11-8) is 0.644 m (see Problem 12-11).

[†]Influent TSS $= 260 \text{ mg/}\ell$ (see Table 6-9)

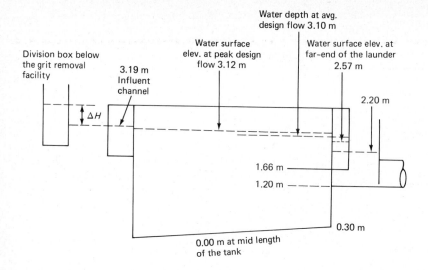

Figure 12-15   Hydraulic Profile through Primary Sedimentation Basin at Peak Design Flow. (The hydraulic profile is prepared with respect to the assumed datum at the bottom of the basin at mid-length.)

$$\begin{array}{l} \text{Average quantity of sludge} \\ \text{produced per day from both basins} \end{array} = 2 \times 3113.5 \text{ kg/d}$$
$$= 6227 \text{ kg/d}$$

3.   Compute the volume of sludge produced per minute per basin

Volume of sludge at specific gravity of 1.03 and 4.5 percent solids

$$= \frac{3113.5 \text{ kg/d per basin}}{1.03 \times \dfrac{1g}{cm^3} \times \dfrac{1}{1000 \text{ g/kg}} \times 0.045 \times (100 \text{ cm})^3/m^3 \times 1440 \text{ min/d}}$$

$$= 0.0467 \text{ m}^3/\text{min per basin}$$

4.   Determine sludge pump size and pumping cycle
     Provide separate sludge pumps for each basin. Arrange such that each pump will be able to serve both basins in case one pump is out of service. Operate each pump on a time cycle, at 16.5-min intervals with a 1.5-min pumping cycle per basin (total time 18 min per cycle).

$$\begin{array}{l} \text{The desired pumping capacity of the pump} \\ = \dfrac{0.0467 \text{ m}^3/\text{min per basin} \times 18 \text{ min per cycle}}{1.5 \text{ min pumping per cycle}} \end{array}$$

$$= 0.56 \text{ m}^3/\text{min per basin (150 gpm)}$$

When one pump is used to remove the sludge from two basins, the cycle time will be reduced.

Cycle interval in min for two basins

$$= \frac{0.56 \text{ m}^3/\text{min} \times 1.5 \text{ min pumping per cycle}}{0.0934 \text{ m}^3/\text{min for both basins}}$$

$$= 9 \text{ min per cycle}$$

G. Effluent Quality from Primary Sedimentation Basin

1. Establish BOD_5 and TSS removal
   The percent removal of BOD_5 and suspended solids in a well-designed and operated primary sedimentation basin may be estimated from Figures 12-4 or 12-5. Based on the actual overflow rate of 35.4 m³/m²·d, the following removal rates are expected:

   $$BOD_5 \text{ removal} \qquad = 34 \text{ percent}$$
   $$\text{Suspended solids removal} = 63 \text{ percent}$$

2. Compute BOD_5 and TSS in the effluent
   Based on the concentration of BOD_5 and suspended solids in the raw wastewater, the effluent quality from the primary basins is determined. It should be mentioned, however, that if the flows from the thickeners, digesters, and dewatering facilities are returned to the primary sedimentation basin, the load contributed by these return flows must be estimated properly by using material mass balance techniques. Many designers overlook this fact which may cause erroneous effluent quality estimates. An example of a material mass balance analysis may be found in Chapter 13. In this Design Example, the sidestream from the thickeners, digesters, and dewatering facilities are returned to the aeration basin. Thus, the primary treatment facilities do not receive any additional load.

   BOD_5 in the primary effluent

   $$= 250 \text{ g/m}^3 (1 - 0.34) \times 0.44 \text{ m}^3/\text{s} \times 86,400 \text{ s/d} \times \frac{\text{kg}}{1000 \text{ g}}$$

   $$= 6273 \text{ kg/d}$$

   TSS in the primary effluent

   $$= 260 \text{ g/m}^3 (1 - 0.63) \times 0.44 \text{ m}^3/\text{s} \times 86,400 \text{ s/d} \times \frac{\text{kg}}{1000 \text{ g}}$$

   $$= 3657 \text{ kg/d}$$

   Volume of primary effluent

   $$= \text{average flow to primary} - \text{sludge withdrawal}$$

   $$= \frac{0.44 \text{ m}^3/\text{s}}{\times 86,400 \text{ s/d}} - \frac{6227 \text{ kg/d} \times 1000 \text{ g/kg}}{0.045 \text{ g/g} \times 1.03 \times 1\text{g/cm}^3 \times 10^6\text{cm}^3/\text{m}^3}$$

   $$= 38,016 \text{ m}^3/\text{d} - 134 \text{ m}^3/\text{d} = 37,882 \text{ m}^3/\text{d}$$

BOD$_5$ concentration in effluent*

$$= 6273 \text{ kg/d} \times \frac{1}{37,882 \text{ m}^3/\text{d}} \times 1000 \text{ g/kg} = 165.6 \text{ g/m}^3 \text{ (mg/}\ell\text{)}$$

TSS concentration in effluent†

$$= 3657 \text{ kg/d} \times \frac{1}{37,882 \text{ m}^3/\text{d}} \times 1000 \text{ g/kg} = 96.5 \text{ g/m}^3 \text{ (mg/}\ell\text{)}$$

H. Scum Quantity

Average quantity of scum is 8 kg/1000 m^3

$$\begin{aligned}
\text{Average quantity of scum} \atop \text{having sp gr. of 0.95} &= \frac{8 \text{ kg}}{1000 \text{ m}^3} \times 38,016 \text{ m}^3/\text{d} = 304 \text{ kg/d} \\[2mm]
&= \frac{304 \text{ kg/d} \times 1000 \text{ g/kg}}{0.95 \times 1 \text{ g/cm}^3 \times 10^6 \text{ cm}^3/\text{m}^3} \\[2mm]
&= 0.32 \text{ m}^3/\text{d}
\end{aligned}$$

I. Design Details.   The design details of a primary sedimentation facility are given in Figure 12-16.

12-7 OPERATION AND MAINTENANCE, AND TROUBLESHOOTING AT PRIMARY SEDIMENTATION FACILITIES

The primary sedimentation facilities can create serious odors if the units are not carefully operated and maintained. When wastewater reaches the primary sedimentation basin, many odorous compounds are released into the atmosphere because these tanks offer a relatively large exposed surface area and there is a certain amount of turbulence at the influent and effluent structures. Furthermore, scum on the surface and at the collection devices offers highly favorable conditions for odor emission from the primary sedimentation basins. Settled sludge, if not pumped frequently, continues to undergo anaerobic breakdown thus intensifying the odor problems. Improper operation of the primary sedimentation tank may cause solids and BOD overloading in the secondary processes, and grease carryover may upset the biological processes in particular. Therefore, proper operation and maintenance of primary treatment facilities is essential for control of odors and smooth operation of downstream treatment units. The following items are considered important for the design and operation of the primary treatment facilities.

12-7-1 Common Operational Problems and Troubleshooting Guide

A number of operational problems may occur at a primary sedimentation facility. Some troubleshooting guides are given below:[22]

---

*Approximate concentration of BOD$_5$ in primary effluent = 250 mg/ℓ (1 − 0.34) = 165 mg/ℓ.
†Approximate concentration of TSS in primary effluent = 260 mg/ℓ (1 − 0.63) = 96 mg/ℓ. *Note:* 1 mg/ℓ = 1 g/m^3.

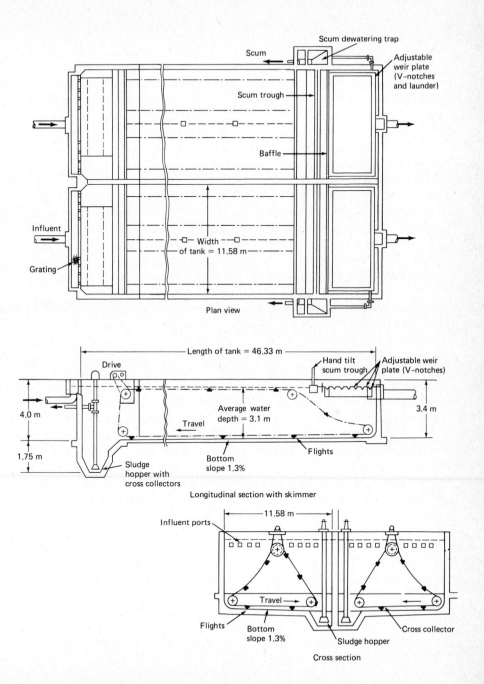

Figure 12-16   Layout of the Primary Sedimentation Basins in the Design Example (for equipment detail see Figure 12-1).

1. Black and odorous septic wastewater in the primary treatment facility is an indication of decomposing wastewater in the collection system, recycle of excessively strong digester supernatant, or inadequate pretreatment of organic discharges from the industries. Possible solution procedures include preaeration, chlorination or hydrogen peroxide treatment of wastes in the collection system, control of digester supernatant, and strict enforcement of industrial pretreatment regulations.

2. Floating sludge indicates excessive sludge accumulation in the basin, decomposing organics, or return of well-nitrified, waste-activated sludge. The scrapers may be worn or damaged, the sludge withdrawal line may be plugged, or the sludge withdrawal rate may be insufficient. Remove the sludge more frequently or at a higher rate, clean the sludge lines, or repair or replace sludge collection and pumping equipment.

3. Scum overflow is due to inadequate frequency of scum removal, excessive industrial contribution, worn or damaged scum wiper blades, or improper alignment of the skimmer. Remove scum more frequently, limit industrial waste contribution, clean or replace wiper blades, and adjust wiper blade alignment.

4. Sludge that is hard to remove from the sludge hopper may be due to excessive grit accumulation. Check the grit removal facility.

5. Low solids in the sludge may be due to excessive sludge withdrawal, short circuiting, or surging flow. Reduce sludge withdrawal, check and install baffles, and check and modify influent pumping rate.

6. Excessive sedimentation in the influent channel is due to low velocity. Agitate the influent channel with air or wastewater to resuspend solids and prevent decomposition.

7. Excessive slime growth on the surfaces and weirs may be due to accumulation of solids and scum. Inspect surfaces and clean them frequently.

8. Excessive corrosion of metals may be due to hydrogen sulfide gas. This may be due to septic sewage or sludge. Check items 1 and 2 above. Paint surfaces with corrosion-resistant paint.

9. Erratic operation of a sludge collection mechanism may be due to broken shear pin or damaged collection mechanism, excessive sludge accumulation, or rags or debris entangled around collector mechanism. Repair or replace damaged parts, remove debris, and increase sludge pumping rate.

10. Frequent broken scraper chain and shear pin failures are due to improper shear pin sizing and flight alignment, ice formation, or excessive loading on the sludge scraper. Realign flights and change shear pin size, break ice, or remove sludge more often.

11. A noisy chain drive, a chain that climbs sprockets, or a loose or stiff chain may be due to misalignment or improper assembly, worn out parts, faulty lubrication, or excessive rust or corrosion. Inspect and correctly align the entire drive mechanism; replace the chain, bearings, or sprockets and outer parts; remove dirt and rust; and lubricate properly.

12. A broken chain or broken sprockets in the chain drive system may be due to shock, overloading, wrong chain size, misalignment, excessive wear, or lack of lubrication. Replace parts, avoid shock and overload, correct corrosive condition, and lubricate properly.

13. Bearing or universal joint failure is due to excessive wear and lack of lubrication. Replace joints or bearing, and lubricate properly.

12-7-2 Operation and Maintenance

The following operation and maintenance items are necessary to keep the primary sedimentation facility in satisfactory operating condition:

1. Remove accumulations from influent and effluent baffles, weirs, and scum box. Clean all inside exposed walls and channels by squeegee on a regular basis.
2. Inspect all mechanical equipment at least once each shift.
3. Hose down and remove all wastewater and sludge spills as soon as possible.
4. Determine sludge level and underflow concentration and adjust primary sludge pumping rate accordingly. Observe scum pump operation and provide hosing as required. Both these items are considered regular duties.
5. Check daily the electrical motors for overall operation and bearing temperature. Also check the electrical motor overload detector daily.
6. Check oil levels in gear reducers and bearings on a regular basis.
7. Drain each primary basin annually and inspect the underwater portion of the concrete structure and all mechanical parts. Patch up defective concrete. Inspect all mechanical parts for wear, corrosion, and set proper clearance for flights at tank walls. Replace flights when necessary and supply protective coatings. Clean and paint the exposed metal surfaces as necessary.

12-8 SPECIFICATIONS

The specifications for the primary sedimentation facility designed in this chapter are briefly summarized here. The purpose of this section is to describe many components of the design that could not be fully covered in the Design Example. These specifications should be used only as a guide. Detailed specifications must be prepared for each design in consultation with the equipment manufacturers.

12-8-1 General

Each primary sedimentation basin shall include a complete assembly of a sludge collector mechanism with drive and collector chains with flights, access bridge and walkway, influent and effluent structures, pumping facilities, and overload alarm system. The manufacturer shall furnish and deliver ready for installation a conveyor-type sludge collector mechanism suitable for installation in two identical rectangular primary sedimentation basins of the following dimensions:

Number of basins	Two identical basins with common wall
Length	46.33 m (152 ft)
Width	11.58 m (38 ft)
SWD at effluent end	2.80 m (9.2 ft)
SWD at influent end	3.40 m (11.2 ft)

Number of basins	**Two identical basins with common wall**
SWD at mid-length	3.10 m (10.2 ft)
Bottom slope	1.3 percent
Freeboard	0.6 m (2 ft)

12-8-2 Materials and Fabrication

The structural steel shall conform to proper ASTM standards.* The minimum thickness of all submerged metal shall be not less than 6.4 mm ($\frac{1}{4}$ in.) and of all-above-water metal 4.8 mm ($\frac{3}{16}$ in.). All iron casting shall also conform to proper ASTM standards. Design and construction shall conform to all AISC[†] standards for structural steel buildings.

12-8-3 Collectors

Two longitudinal collectors and one cross-collector shall be provided in each basin.

Longitudinal Collectors. The longitudinal collectors shall consist of flights 7.6 × 20.3 cm (3 × 8 in.) nominal size, and select hard redwood scrapers spaced approximately 3.05 m (10 ft) apart on two strands of chain. Scrapers shall be equipped with wearing shoes to run on tee rails that will flush with tank bottom and on angle tracks for return run where required. The collector chain shall run over four sets of sprocket wheels at a speed of 0.61 m (2 ft) per min so as to clean the sludge from the entire tank bottom and skim the water surface in the basin.

Cross-Collectors. The cross-collector shall have flights 5.1 × 15.2 cm (2 × 6 in.) nominal size, select hard redwood, spaced 1.5 m (5 ft) apart, mounted on two strands of chain. These chains shall run over three sets of sprocket wheels at a speed double that of the longitudinal collectors.

12-8-4 Drive Unit

The drive assembly shall consist of a gear motor, gear reducer, drive base, shear pin coupling, overload alarm device, and drive sprocket and chain.

The motor for each drive shall be totally enclosed and shall have a sealed conduit box. The motor shall be for operation on 460 V, 3-phase, 60 Hz. It shall be industrial duty and shall have stainless steel hardware, a corrosion-resistant fan, epoxy paint, and cast iron end shields. All gears in the mechanism shall be heavy-duty type. All reduction gears shall be of the oil bath type, with an external oil level indicator. A shear pin for protecting the mechanism shall be provided at the gear motor output shaft and shall be set at the torque specified. A torque indicator scale shall be provided and shall be calibrated from 0 to the maximum stall torque of the individual mechanisms.

---

*American Society for Testing and Materials.
[†]American Institute of Steel Construction.
Note: All side water depths (SWD) are at average design flow when both basins are in operation.

Collector Chains. The collector chain shall have a specified ultimate strength. The links shall be made from a suitable cast iron steel which shall have a specified tensile strength and hardness.

Sprockets, Shafts, and Bearings. The sprockets for the drive and collector chains shall be semisteel, cast in a chill, and shall have a specified hardness. Sprocket teeth shall be accurately ground to fit the chain and shall have not less than the specified number of teeth. The driving sprockets shall be keyed firmly to the headshaft. All shafting shall be solid, cold finished steel, that is straight and true, and shall be held in alignment with collars and screws. The shafting shall contain keyways with fitted keys where necessary and shall be of sufficient size to transmit the power required. All shafting shall extend across the full width of the tank and shall turn in the bearings mounted on the tank walls. All bearings shall be bolted directly to the concrete wall in a manner that permits easy adjustment. All underwater bearings shall be babbitted and of the water-lubricated, ball-and-socket, self-aligning type, especially designed to prevent accumulation of settled solids on its surface.

12-8-5 Sludge Pump

The sludge pump shall be self-priming, centrifugal, and nonclogging. It shall be suitable for handling the maximum solids concentration anticipated in the primary sedimentation basin. The specified pumping rate is approximately 0.56 m^3/min (150 gpm) at an elevation difference from the sedimentation basin to the sludge-blending tank. Each basin shall have an individual pump assembly capable of servicing both basins in case one pump is out of service. The sludge line shall be sized to handle maximum pump capacity. The pump operation shall be programmed to provide operational flexibility.

12-8-6 Effluent Weir

The manufacturer shall provide an overflow weir box to be located on the outlet end of the basin as required. The weir box shall have 90° V-notch weirs 20 cm center-to-center as detailed. The effluent box shall have vertical adjustment for leveling. The weir box structure shall also provide baffle as shown in drawings to prevent escape of scum into the effluent.

12-8-7 Skimmer

Each sedimentation basin shall be equipped with a hand-operated scum trough for removal of floating oil, grease, and scum from the surface of the tank. The hand-operated scum trough shall have rotating steel pipes open at the top and running across the entire width of the tank. The skimming troughs shall remove the scum from the surface and deposit it into the scum-dewatering trap. The floating scum shall flow over an adjustable weir into the scum pit and be piped to the digester. The excess water shall return to the primary effluent.

12-8-8 Painting

All nonsubmerged ferrous materials shall be brushed cleaned, and submerged ferrous material shall be sandblasted to white metal. Cleaned surfaces shall be primed with approved epoxy primer and have a finished coat of approved paint or epoxy. All field welds shall be touched up with compatible paints.

12-9 PROBLEMS AND DISCUSSION TOPICS

12-1. A 30-m-diam sedimentation basin has average water depth of 3.0 m. It is treating 0.3 m^3/s wastewater flow. Compute overflow rate and detention time.

12-2. A sedimentation basin has an overflow rate of 80 m^3/m^2·d. What fraction of the particles that have velocity of 0.02 m/min will be removed in this tank?

12-3. Design a circular clarifier for the Design Example. The design data are given in Sec. 12-6-1. The effluent launder is 0.5 m wide and is installed around the circumference of the basin, 1 m away from the concrete wall. The weir notches are provided on both sides of the launder. The effluent box is 1 m × 1 m, and the depth of flow in the effluent box at peak design flow is 1 m. The invert of the effluent launder is 0.46 m above the invert of the effluent box.

12-4. Develop the hydraulic profile through the primary sedimentation basin in the Design Example at average design flow when both basins are in operation.

12-5. A primary sedimentation basin is designed for an average flow of 0.3 m^3/s. The TSS concentration in the influent is 240 mg/ℓ. The average solids removal efficiency of the basin is 60 percent. The sludge has average solids concentration of 4 percent, and specific gravity of 1.025. Calculate (a) the quantity and volume of sludge produced, (b) the effluent flow rate, and (c) the pump cycle time if the pumping rate is 570 ℓ/min.

12-6. A primary sedimentation basin is designed for an overflow rate of 30 m^3/m^2·d. Calculate different liquid depths if the basin is designed for the following detention times: 1.0, 1.5, 2.0, 2.5 and 3.0 hours. Assume flow rate of 0.5 m^3/s. Prepare a curve between detention times and calculated overflow rates.

12-7. A primary sedimentation facility was designed to treat an average flow of 0.6 m^3/s. The design overflow rate and detention time are 45 m^3/m^2·d and 2.5 h respectively. The length to width ratio of the rectangular basin is 4.3 : 1. Calculate the dimensions of the basin if 2, 3 or 4 basins are provided. Also compute the weir loading rate in each case if one weir trough is provided along the width in each basin as shown in Figure 12-8(b). The weir trough has weirs on both sides, and the outlet channel is 1 m wide.

12-8. What are the major differences in three types of clarifiers: horizontal flow, solids contact and inclined surface? Write the advantages and disadvantages, and major applications of each type.

12-9. Four primary sedimentation basins are designed for total average flow of 1.2 m^3/s. The TSS concentration in the primary treated effluent is 150 mg/ℓ. TSS removal is 63 percent. The sludge has an average solids concentration of 6 percent and specific gravity of 1.045. What is the capacity of the sludge pump in m^3/min per basin if pump is used on 15 min. pumping cycle per h?

12-10. 70,000 kg/d wet sludge (liquid and solids) is produced from a primary sedimentation basin. The sludge pump has a capacity of 0.2 m^3/min. If the sludge is pumped 50 times daily, how would you operate the pump? Assume that the specific gravity of wet sludge is 1.02.

12-11. Study the computational scheme of determining water surface profile in a flume using Eqs. (11-4)–(11-8). A solution procedure is given in Chapter 13 (Sec. 13-7-4, step 4). Perform similar steps to determine the depth of water in the effluent launder of the primary sedimentation facility given in the Design Example. Use a uniform increment of 1.0 m. Compare your result with that given in Sec. 12-6-2, step C7.

12-12. Study the effluent structure of the sedimentation facility given in Figure 12-1. How does this differ from that given in Figure 12-14? Prepare a computational scheme similar to that given in Sec. 11-9-2, step E6 for designing the effluent trough and central channel if the control conditions at the exit point of the central channel are known.

REFERENCES

1. Amirtharajah, A., "Design of Flocculation Systems," Chapter 11 in *Water Treatment Plant Design for Practicing Engineers,* (R. L. Sanks, Ed.), Ann Arbor Science, Michigan, 1979.
2. Richard of Rockford, Inc., Solids Separation Equipment Catalog, Sec. III, Design Engineering, February 1977.
3. American Water Works Association, Inc., *Water Quality and Treatment: A Handbook of Public Water Supplies,* Third Edition, McGraw-Hill Book Co., New York, 1971.
4. Weber, J. W., *Physicochemical Processes for Water Quality Control,* Wiley-Interscience, New York, 1972.
5. U.S. Environmental Protection Agency, *Process Design Manual for Suspended Solids Removal,* EPA 625/1-75-003N, January 1975.
6. Fair, G. M., J. C. Geyer, and D. A. Okum, *Water and Wastewater Engineering,* Vol. 2, Wiley & Sons, New York, 1968.
7. Clark, John W., W. Viessman, and M. J. Hammer, *Water Supply and Pollution Control,* IEP-A Dun-Donnelley, New York, 1977.
8. The Committee of the Great Lakes—Upper Mississippi River Board of State Sanitary Engineers (Ten-States), *Recommended Standards for Sewage Water,* Health Education Service, Albany, New York, 1973.
9. Metcalf and Eddy, Inc., *Wastewater Engineering: Treatment, Disposal, Reuse,* McGraw-Hill Book Co., New York, 1979.
10. Joint Committee of the Water Pollution Control Federation and the American Society of Civil Engineers, *Wastewater Treatment Plant Design,* MOP/8 Water Pollution Control Federation, Washington, D.C., 1977.
11. Steel, E. W., and T. J. McGhee, *Water Supply and Sewerage,* Fifth Edition, McGraw-Hill Book Co., New York, 1979.
12. Hammer, M. G., *Water and Wastewater Technology,* Wiley & Sons, New York, 1975.
13. Parker, H. W., *Wastewater Engineering,* Prentice-Hall, Englewood Cliffs, N.J., 1975.
14. Task Committee, Environmental Engineering Division, "Final Clarifiers for Activated Sludge Plants," *Journal of the Environmental Engineering Division,* American Society of Civil Engineers, Vol. 105, No. EES, October 1979, pp. 803–817.
15. Aqua-Aerobic Systems, Inc., *Clarifier,* Bulletin 302, 6306 North Alpine Road, Rockford, Ill., 1976.
16. Link-Belt Products, *Straightline Sludge Collectors,* Bulletin 16010, FMC Corporation, Environmental Equipment Division, Chicago, 1960.
17. Choa, J., and R. R. Trussell, "Hydraulic Design of Flow Distribution Channels," *Journal of the Environmental Engineering Division,* American Society of Civil Engineers, Vol. 106, No. EE2, April 1980, pp. 321–334.
18. Daugherty, R. L., and J. B. Franzini, *Fluid Mechanics with Engineering Applications,* McGraw-Hill Book Co., New York, 1977.
19. Morris, H. M., and J. J. Wiggert, *Applied Hydraulics in Engineering,* Ronald Press, New York, 1972.

20. Henderson, F. M., *Open Channel Flow*, Macmillan, New York, 1966.
21. Benefield, L. D., J. F. Judkins, and A. D. Parr, *Treatment Plant Hydraulics for Environmental Engineers*, Prentice-Hall, Inc., Englewood Cliffs, N.J., 1984.
22. Culp, G. L., N. F. Heim, *Field Manual for Performance Evaluation and Troubleshooting at Municipal Wastewater Treatment Facilities*, U.S. Environmental Protection Agency, Washington, D.C., ERA-430/9-78-001, January 1978.

Biological Waste Treatment

13-1 INTRODUCTION

The major goal of primary treatment is to remove those pollutants that can settle or float. The purpose of secondary treatment is to remove the soluble organics that escape the primary treatment and to provide further removal of suspended solids. These removals are typically achieved by using biological treatment processes. The biological treatment processes provide the similar biological reactions that would occur in the receiving waters if it had adequate capacity to assimilate the wastes. Although secondary treatment may remove more than 85 percent of the BOD_5 and suspended solids, it does not remove significant amounts of nitrogen, phosphorus, heavy metals, nonbiodegradable organics, bacteria, and viruses. These pollutants may require further removal where receiving waters are especially sensitive.

Secondary levels of treatment can also be achieved by physical-chemical or land treatment systems. These treatment systems are also applied to the existing biological treatment plants to upgrade the effluent quality or to achieve advanced wastewater treatment. Therefore, these treatment systems are presented in Chapter 24. The purpose of this chapter is to (1) present the fundamentals of biological waste treatment and (2) discuss the design aspects of important biological treatment processes. The major treatment processes discussed in this chapter include activated sludge process and various modifications, trickling filter and rotating biological contactor, and combined attached and suspended growth treatment. Since the activated sludge process is most commonly used for medium and large installations, step-by-step procedure for designing an activated sludge facility (aeration basin and clarifier) is presented separately in the Design Example.

13-2 FUNDAMENTALS OF BIOLOGICAL WASTE TREATMENT

13-2-1 Basic Requirements

Biological waste treatment involves bringing the active microbial growth in contact with wastewater so that they can consume the impurities as food. A great variety of micro-organisms come into play that include bacteria, protozoa, rotifers, nematodes, fungi, algae, and so forth. These organisms, in the presence of oxygen convert the biodegradable organics into carbon dioxide, water, more cell material, and other inert products. The basic ingredients needed for secondary biological treatment are the availability of (1) mixed populations of active microorganisms, (2) good contact between the microorganisms and waste material, (3) availability of oxygen, (4) availability of nutrients, and (5) maintenance of other favorable environmental conditions, such as temperature, pH, sufficient contact time, etc.

13-2-2 Growth

When organic waste is brought in contact with microorganisms, there is a rapid growth of biological mass. This phase, called the "log-growth phase," is typical of microorganisms when excess of food is around them. In this phase, the rate of growth of the cells and their subsequent division is limited by the ability of the microorganisms to process the substrate. At the end of the log-growth phase, the microorganisms grow at their maximum rate and consequently remove organic matter from solution at a maximum rate. As the food concentration becomes limited, a declining growth rate develops. Further decrease in food concentration inhibits microbial metabolism that results in a decrease in the biological mass. This is known as the "endogenous phase."

13-2-3 Role of Enzymes

Microorganisms grow and obtain their energy from substrates utilizing very complex and intricate biochemical reactions. Several enzymes are involved in a series of reactions forming a sequence of enzyme-substrate complexes, which are then converted to a product and the original enzyme. The enzymes are proteins that act as catalyst. Enzymes are also specific to each substrate and have a high degree of efficiency in converting the substrate to the end products. The enzymes are extracellular and intracellular. Extracellular enzymes convert the substrate to a form that can diffuse into the cell. The intracellular enzymes bring about oxidation, synthesis, and energy reactions within the cell. The enzyme activity, however, is substantially affected by pH, temperature, and substrate concentration.

When substrates are oxidized within the cell, energy is released. The energy thus released is stored by the phosphate enzyme system. Inorganic phosphate is added to adenosine diphosphate (ADP) to form adenosine triphosphate (ATP). In this way the energy is stored in the ATP rather than lost as heat. As the microorganisms require the energy, the ATP is reduced back to ADP with a transfer of energy to the chemical reactions needing it, or for growth and cell activity. The biochemistry of the metabolic pathways is very complex and beyond the scope of this book. Readers may consult Refs. 1–3 for more on this subject.

13-3 SUSPENDED GROWTH BIOLOGICAL TREATMENT

Suspended growth treatment systems are those in which the microorganisms remain in suspension. Common suspended growth processes used for secondary treatment include

(1) activated sludge and other modifications, (2) aerated lagoons, and (3) high-rate stabilization ponds. These processes are discussed below.

13-3-1 Activated Sludge

In the activated sludge process, microorganisms (MO) are mixed thoroughly with the organics so that they can grow and stabilize the organics. As the microorganisms grow and are mixed by the agitation of the air, the individual organisms clump together (flocculate) to form an active mass of microbial floc called "activated sludge." The mixture of the activated sludge and wastewater in the aeration basin is called "mixed liquor." The mixed liquor flows from the aeration basin to a secondary clarifier where the activated sludge is settled. A portion of the settled sludge is returned to the aeration basin to maintain the proper food-to-MO ratio to permit rapid breakdown of the organic matter. Because more activated sludge is produced than can be used in the process, some of it is wasted from the aeration basin or from the returned sludge line to the sludge-handling systems for treatment and disposal. Air is introduced into the aeration basin either by diffusers or by mechanical mixers. There are many modifications of the activated sludge process. These modifications differ in mixing and flow pattern in the aeration basin, and in the manner in which the microorganisms are mixed with the incoming wastewater. The basic design principles of the activated sludge processes are discussed below.

Types of Reactors.   The principal types of biological reactors (aeration basins) are plug flow, complete mix, and arbitrary flow. In a plug-flow reactor the particles pass through the tank and are discharged in the same sequence in which they enter. This type of flow is achieved in a long narrow basin. In a complete-mix reactor, the entering particles are dispersed immediately throughout the entire basin. Complete-mix flow is achieved in circular or square basins. Arbitrary-flow reactors exhibit partial mixing somewhere between the plug-flow and complete-mix reactors.

Biological Kinetic Equations.   In the past the designs of activated sludge plants were based on empirical parameters developed by experience. Many of these empirical parameters included organic loading, hydraulic loading, aeration period, etc. Today, however, the design utilizes empirical as well as rational parameters based on biological kinetic equations. These equations express biological (sludge) growth and substrate utilization rates in terms of biological kinetic coefficients, food-to-MO ratio, the mean cell residence time, etc. Using these equations the design parameters, such as volume of aeration basin, effluent quality, rates of return sludge and waste sludge, aeration period, and oxygen utilization rates, can be calculated. Many important design relationships for completely mixed reactors using sludge wasting from aeration basin or from the secondary clarifier (Figure 13-1) are expressed by Eqs. (13-1)–(13-17). Details on biological kinetic models and the derivation procedure for these equations may be found in several textbooks.[4-10] The use of these equations in the design of an activated sludge plant is shown in the Design Example.

$$E = \left[\frac{S_o - S}{S_o} \right] 100 \tag{13–1}$$

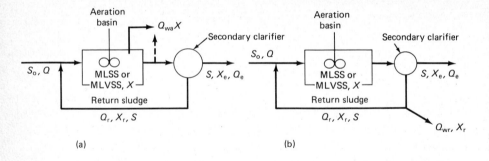

Figure 13-1   Completely Mixed Biological Reactor. (a) Sludge wasting from aeration basin. (b) Sludge wasting from return sludge line.

$$\theta = \frac{V}{Q} \tag{13-2}$$

$$X = \frac{\theta_c Y (S_o - S)}{\theta (1 + k_d \theta_c)} \tag{13-3}$$

$$\frac{\Delta X}{\Delta T} = \frac{X_o - X}{\Delta T} \tag{13-4}$$

$$V = \frac{\theta_c Q Y (S_o - S)}{X(1 + k_d \theta_c)} \tag{13-5}$$

$$U = \frac{Q(S_o - S)}{VX} \tag{13-6}$$

$$\frac{F}{M} = U = \frac{kS}{K_s + S} \tag{13-7}$$

$$\frac{1}{U} = \frac{K_s}{kS} + \frac{1}{k} \tag{13-8}$$

$$\theta_c = \frac{X}{\Delta X / \Delta T} \tag{13-9}$$

$$\theta_c = \frac{VX}{Q_{wa} X + Q_e X_e} \tag{13-10}$$

$$\frac{1}{\theta_c} = YU - k_d \tag{13-11}$$

$$Y_{obs} = \frac{Y}{1 + k_d \theta_c \text{ (or } \theta_{ct})} \tag{13-12}$$

$$P_x = Y_{obs} Q(S_o - S) \tag{13-13}$$

$$P_x = Q_{wr} X_r \text{ or } Q_{wa} X \tag{13-14}$$

$$Q_{wa} \approx \frac{V}{\theta_c} \tag{13-15}$$

$$Q_{wr} \approx \frac{VX}{\theta_c X_r} \tag{13-16}$$

$$O_2 = \frac{Q(S_o - S)}{BOD_5/BOD_L} - 1.42 \, P_x{}^* \tag{13-17}$$

where

E = process efficiency, percent
F/M = food-to-microorganism ratio, d^{-1}
k_d = endogenous decay coefficient, d^{-1}
K_s = substrate concentration at one-half the maximum growth rate, mg/ℓ (g/m^3)
k = maximum rate of substrate utilization per unit mass of microorganisms, d^{-1}
O_2 = oxygen utilization rate, kg/d
P_x = waste-activated sludge (VSS), kg/d
Q = influent wastewater flow rate, m^3/d
Q_e = treated effluent flow rate, m^3/d
Q_{wa} = waste sludge flow rate from aeration tank, m^3/d
Q_{wr} = waste sludge flow rate from the sludge return line, m^3/d
S_o = influent soluble BOD$_5$ concentration, mg/ℓ (g/m^3)
S = effluent soluble BOD$_5$ concentration, mg/ℓ (g/m^3)
θ = hydraulic detention time, d
θ_c = mean cell residence time based on solids in the aeration basin, d
θ_{ct} = mean cell residence time based on solids in the aeration basin and in the secondary clarifier, d
U = food-to-MO ratio, specific substrate utilization rate, d^{-1}
X = concentration of MLVSS maintained in the aeration basin, mg/ℓ (g/m^3)
$\Delta X/\Delta T$ = growth of biological sludge over time period ΔT, mg/ℓ (g/m^3)/d
X_o = concentration of MLVSS in the aeration basin at the end of time period ΔT, mg/ℓ (g/m^3). X_o includes the loss of VSS in the effluent
X_e = concentration of VSS in the treated effluent, mg/ℓ (g/m^3)
X_r = concentration of sludge in the return sludge line, mg/ℓ (g/m^3)
V = volume of aeration basin, m^3
Y = yield coefficient over finite period of log growth, g/g
Y_{obs} = observed yield, g/g

*Factor 1.42 is obtained as follows: $C_5H_7NO_2$ (cellular mass) + $5O_2 \longrightarrow 5CO_2 + 2H_2O + NH_3$. The molar ratios of cellular mass to $5O_2$ = 1.42. Thus BOD_L = 1.42 × cellular mass.

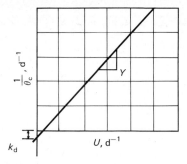

Figure 13-2   Procedure to Determine Y and k_d.

Determination of Kinetic Coefficients.   The values of kinetic coefficients Y, k, k_d, and K_s greatly influence the design of the activated sludge process. These values depend on the characteristics of the wastewater and therefore must be determined for every waste stream (especially if it contains industrial wastes) from bench or pilot plant studies. The procedure is to operate the experimental reactors at different MLVSS concentrations. Using the data collected under the steady-state conditions at each concentration of MLVSS, the mean values of Q, S_o, S, and X_o are determined. From these results the values of $\Delta X/\Delta T$, U, and θ_c are calculated from Eqs. (13-4), (13-6), and (13-9).

A plot of $1/\theta_c$ versus U gives a straight line [Eq. (13-11)]. The slope of the straight line is the value of Y and the intercept is k_d. Similarly, a plot of $1/U$ versus $1/S$ [Eq. (13-8)] gives a straight line. The slope of the straight line is K_s/k and the intercept is $1/k$. These plots are shown in Figures 13-2 and 13-3. The range and typical values of these kinetic coefficients for municipal wastewater are summarized in Table 13-1.

Process Modifications.   The major process modifications of activated sludge process are (1) conventional, (2) tapered aeration, (3) complete mix, (4) step aeration, (5) contact stabilization, (6) extended aeration, and (7) pure oxygen systems. Brief descriptions of

TABLE 13-1   Typical Values of Kinetic Coefficients for Activated Sludge Process

Coefficient	Basis	Values	
		Range	Typical
k	d^{-1}	2–8	4
k_d	d^{-1}	0.03–0.07	0.05
K_s	mg/ℓ, BOD_5	40–120	80
	mg/ℓ, COD	20–80	40
Y	VSS/BOD_5	0.3–0.7	0.5
	VSS/COD	0.2–0.5	0.4

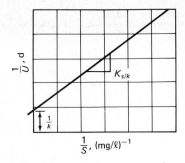

Figure 13-3   Procedure to Determine k and K_s.

these process modifications and principal design parameters are summarized in Table 13-2. Schematic flow schemes of different process modifications are illustrated in Figure 13-4.

Aeration System.   Two major types of aeration systems are used in the activated sludge process. These are (1) diffused aeration and (2) mechanical aeration. Each of these systems are discussed briefly below.

Diffused Aeration.   In diffused aeration, air is supplied through porous diffusers or through air nozzles near the bottom of the tank. Various components of the diffused aeration system include (1) diffuser or air nozzles, (2) pipings, and (3) blower or compressor. The factors affecting the oxygen transfer are bubble size, diffuser air rate, diffuser placement, and velocity of the surrounding medium.

The air diffusers are of various types. Common types include the bubble diffuser, tubular diffuser, and jet diffuser.[12,13] Brief descriptions and performance of many types of air diffusers are given in Table 13-3. Typical diffused air aeration devices are illustrated in Figure 13-5. Discussion on diffused aeration equipment for pure oxygen systems may be obtained in Refs. 4 and 12–14.

The air piping consists of header pipe, mains, valves, meters, and other fittings that transport compressed air from the blower to the diffusers. Different types of diffuser arrangements and pipings are shown in Figure 13-6. The basic design considerations for piping systems are listed below:

1. Piping is sized such that the head losses in the piping system are small in comparison to those in the diffusers.
2. Piping must be of corrosion resistant material (stainless steel, galvanized steel, Fiberglass, plastic, etc.).
3. Valves should be provided for flow regulation.
4. Piping losses should be calculated for maximum summer temperatures taking into account the theoretical adiabatic temperature rise during compression.
5. Friction loss in the piping is calculated using the Darcy–Weisbach or Hazen–Williams equation.

TABLE 13-2 Description and Design Parameters of Various Activated Sludge Process Modifications

Process Modification	Brief Description	Flow Regime	Sludge Retention Time, θ_c (d)	Food to MO Ratio[a] (d^{-1})	Aerator Loading[b] (kg/m^3d)	MLSS,[c] (mg/ℓ)	Aeration Period (h)	Recirculation Ratio Q_r/Q
Conventional	The influent and returned sludge enter the tank at the head end of the basin and are mixed by the aeration system (Figure 13-4a)	Plug	5–15	0.2–0.4	0.3–0.6	1500–3000	4–8	0.25–0.5
Tapered aeration	The tapered aeration system is similar to the conventional activated sludge process. The major difference is in arrangement of the diffusers. The diffusers are close together at the influent end where more oxygen is needed. The spacing of diffusers is increased toward the other end of the aeration basin.	Plug	5–15	0.2–0.4	0.3–0.6	1500–3000	4–8	0.25–0.5
Step aeration	The returned sludge is applied at several points in the aeration	Plug	5–15	0.2–0.4	0.6–1.0	2000–3500	3–5	0.25–0.75

basin. Generally, the tank is subdivided into three or more parallel channels with around-the-end baffles and the sludge is applied at separate channels or steps. The oxygen demand is uniformly distributed (Figure 13-4b)

Process	Flow regime	Description						
Complete mix aeration	Complete mix	The influent and the returned sludge are mixed and applied at several points along the length and width of the basin. The contents are mixed and the MLSS flows across the tank to the effluent channel. The oxygen demand and organic loading are uniform along the entire length of the basin (Figure 13-4c).	5–15	0.2–0.6	0.8–2.0	3000–6000	3–5	0.25–1.00
Extended aeration	Complete mix or plug	The extended aeration process utilizes large aeration basin where high population of MO is maintained. It is used for small	20–30	0.05–0.15	0.1–0.4	3000–6000	18–36	0.5–2.0

TABLE 13-2 Description and Design Parameters of Various Activated Sludge Process Modifications

Process Modification	Brief Description	Flow Regime	Sludge Retention Time, θ_c (d)	Food to MO Ratio[a] (d^{-1})	Aerator Loading[b] (kg/m^3d)	MLSS,[c] (mg/ℓ)	Aeration Period (hr)	Recirculation Ratio Q_r/Q
	flows from subdivisions, schools, etc. Prefabricated package plants utilize this process extensively. Oxidation ditch is a variation of extended aeration process. It has channel in shape of a race track. Rotors are used to supply oxygen and maintain circulation (Figure 13-4d)							
Contact stabilization	The activated sludge is mixed with influent in the contact tank in which the organics are absorbed by MO. The MLSS is settled in the clarifier. The returned sludge is aerated in the	Plug	5–15	0.2–0.6	1.0–1.2	1000–4000[d] 4000–10000[e]	0.5–1.0[d] 3.0–6.0[e]	0.5–1.0

reaeration basin to stabilize the organics. The process require approximately 50 percent less tank volume (Figure 13-4e)						
Pure oxygen	Oxygen is diffused into covered aeration tanks. A portion of gas is wasted from the tank to reduce the concentration of CO_2. The process is suitable for high-strength wastes where space may be limited. Special equipment for generation of oxygen is needed (Figure 13-4f)	Complete mix	8–20	0.25–1.0	1.6–3.3	6000–8000	2–5	0.25–0.5

[a] Food to microorganism ratio (F/M) is kg BOD_5 applied per day per kg of MLVSS in the aeration basin.
[b] Aerator loading is kg of BOD_5 applied per day per cubic meter of aeration capacity.
[c] Generally the ratio of MLVSS to MLSS is 0.75–0.85
[d] Contact tank.
[e] Reaeration or stabilization tank.
Source: Adapted in part from Refs. 4–11.

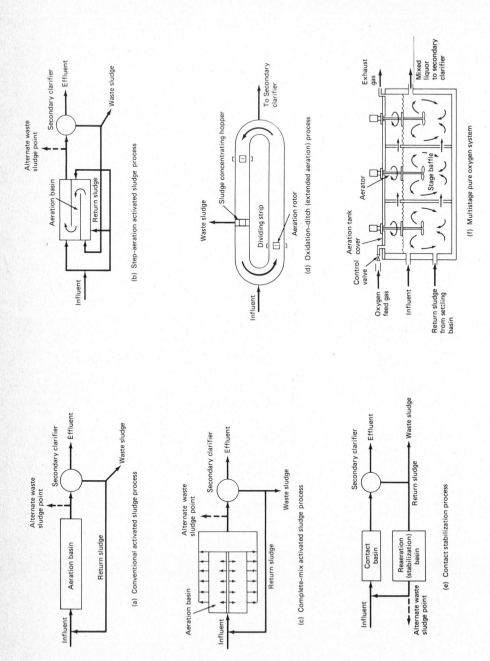

Figure 13-4 Modifications of Activated Sludge Process. [From Refs. 4–11.]

TABLE 13-3   Characteristics of Typical Aeration Devices

Aeration System	Description	Advantages	Disadvantages	Transfer Efficiency, (Percent)	Transfer Rate (Standard kgO$_2$/kW · h)[a]
Diffused air system	Air is introduced near the bottom of the tank through porous or nonporous diffusers. Oxygen transfer and mixing occurs as air bubbles rise to the surface				
Fine bubble	Porous plates, tube, or domes made of ceramic media such as bonded grains of fused crystalline aluminum oxide, vitreous-silicate-bonded grains, or resin-bonded grains of pure silica	Good mixing, varying air flow provides good operational flexibility, and good oxygen transfer	High initial and maintenance costs, air filter needed	10–30	1.2–2.0
Medium bubble	Made of a perforated stainless steel tube and coated with spiral winding of saran cord, or covering with woven fabric sock or sleeve to form a tube 7.5 cm diameter and 61 cm long.	Good mixing, lower maintenance cost as wrappers or sock could be changed, used to produce spiral flow	High initial cost, air filter may be needed.	6–15	1.0–1.6
Coarse bubble	Various nozzles or orifices with check-valve feature; sparger air escapes from perephery of a flexible disc that may lift over	Nonclogging; low maintenance; air filter not needed, used to produce spiral flow	High initial cost; low oxygen transfer, high power cost	4–8	0.6–1.2

TABLE 13-3 Characteristics of Typical Aeration Devices

Aeration System	Description	Advantages	Disadvantages	Transfer Efficiency, (Percent)	Transfer Rate (Standard kgO$_2$/kW · h)^a
	its seat under air pressure; slot orifice injector				
Tubular system	The air flows upward through a tortuous pathway within a tube. Mixing and oxygen transfer is accomplished because the tube aerator acts as air-lift pump.	Low initial cost, low maintenance, high transfer efficiency	Low mixing	7–10	1.2–1.6
Jet	Compressed air and liquid are mixed and discharged horizontally. The rising plume of fine bubbles produces mixing and oxygen transfer	Moderate costs, suited for deep tanks, high transfer efficiency	Requires blower and pumping equipment, nozzle clogging	10–25	1.2–2.4
Mechanical system	In mechanical aerators, the oxygen is entrained from the atmosphere. The aerators consist of submerged or partially submerged impellers that are attached to motors mounted on floats or on fixed structures. Surface aerators are classified according to the rotational speed of the impeller				

Radial flow, low-speed 20–60 rpm	Low speed, use large diameter impeller, floating or fixed base (bridge or platform), use gear reducer	Flexibility in tank shape and size, good mixing	Initial cost high, icing in cold climate, gear reducer may cause maintenance problem	—	1.2–2.4
Axial flow high-speed 300–1200 rpm	High speed, use smaller diameter propeller, and floating structure	Low initial cost, can be adjusted to varying water level, flexible operation	Icing in cold climate, poor accessibility for maintenance, mixing inadequate	—	1.2–2.4
Brush rotor	Used to provide aeration and circulation. Consists of a cylinder or drum with bristles of steel protruding from its perimeter into wastewater	Provide aeration and circulation, used in oxidation ditch, moderate initial cost, good maintenance accessibility	Tank geometry is limited, low efficiency	—	1.2–2.4
Submerged turbine	Provide violent agitation, compressed air may be introduced beneath the impeller by open pipe or a diffuser ring located beneath the impeller. Require fixed-bridge	Good mixing, high-capacity input per unit volume, suitable for deep tank, operational flexibility, no icing or splash	Require both gear reducer and blower, high total power requirement, high initial cost	—	1.0–1.5

[a]Standard conditions: tap water, 20°C, at 101.325 kN/m²(1 atm), and initial dissolved oxygen = 0 mg/ℓ
kg/kW·h × 1.644 = lb/HP·h
kN/m² × 0.145 = lb/in².
Source: Adapted in part from Refs. 11–14

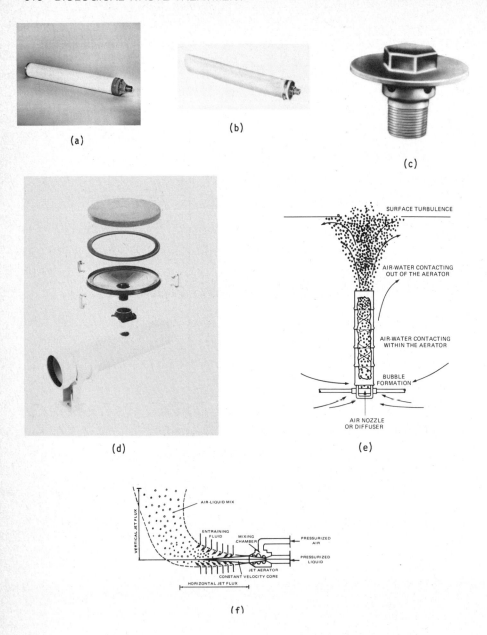

Figure 13-5   Diffused-air Aeration Devices. (a) Fine bubble diffuser cylindrical shape (Courtesy FMC Corporation, Material Handling Systems Division). (b) Fine bubble diffuser tube with fine sheath and holding clamp (Courtesy FMC Corporation, Material Handling Systems Division). (c) Coarse bubble diffuser. (Courtesy FMC Corporation, Material Handling Systems Division). (d) Fine bubble dome diffuser assembly (Courtesy Envirex Inc., a Rexnord Company). (e) Tubular diffuser [From Ref. 12. Used with permission of WPCF and ASCE.] (f) Jet diffuser [From Ref. 12. Used with permission of WPCF and ASCE.]

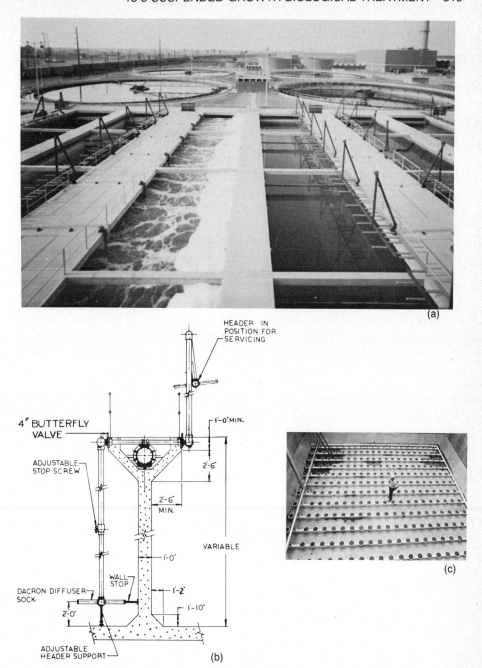

Figure 13-6   Diffused Aeration System Installation. (a) Diffused aeration system assembly, Big Blue Sewage Treatment Plant, Kansas City, Mo. (Courtesy FMC Corporation, Material Handling Systems Division). (b) Rotary lift diffuser showing header in position for service. (Courtesy Envirex Inc., a Rexnord Company). (c) Ceramic Grid Aeration System, Montpelier, Vermont (Courtesy Sanitaire Water Pollution Control Corp.).

6.   With porous diffusers producing fine-to-medium size bubbles, it is important to have swing-lift piping to permit maintenance without dewatering the tank.

A procedure for designing the air-piping system is covered in the Design Example.

The blower or compressor is designed to supply the required amount of air at the design pressure. Two types of blowers are in common use: (1) centrifugal and (2) rotary positive-displacement. A procedure for determining the power requirements and blower selection is presented in the Design Example.

Mechanical Aeration.   The mechanical aerators fall into two major groups: surface impeller and submerged turbine aerators. Brief descriptions and performance of various types of mechanical aerators are summarized in Table 13-3. Many types of mechanical aeration devices are illustrated in Figure 13-7.

Aeration Basin.   Aeration basins are generally rectangular tanks constructed of reinforced concrete. Important design factors of aeration basins are given below:

1.   The depth of aeration basin is 3–5 m (10–16 ft), with 0.3–0.6 m (1–2 ft) freeboard.
2.   For spiral flow mixing the width-to-depth ratio is 1.0:1 to 2.2:1. This limits the width of a tank by 3–11 m (10–36 ft).
3.   If the aeration tank volume exceeds 140 m^3 (5000 ft^3), two or more units should be provided. Each unit should be capable of independent operation.
4.   Common-wall construction should be used for multiple basins.
5.   Exceptionally long tanks should utilize multiple channels using around-the-end-flow baffles.
6.   Avoid dead spots by providing baffles and fillets in the corners.
7.   The foundation should be designed to prevent settlement and prevent flotation when tank is empty.
8.   The inlet and outlet structures should be designed to permit removal of an individual tank from service for routine maintenance.
9.   Suitable arrangement for draining the aeration basin should be made.
10.   Froth control system should be provided by installing effluent spray nozzle along the length on the opposite side of the diffuser. Provision for adding antifoaming agent into the spray water is often made.

Solids Removal System.   The mixed-liquor suspended solids must be settled in a sedimentation basin to produce well-clarified effluent. The design criteria and design procedure for solids removal systems have been presented in Chapter 12. The basic design considerations discussed in Chapter 12 include (1) overflow rate or surface-settling rate, (2) detention period, (3) weir-loading rate, (4) tank shape and dimensions, (5) solids-loading rate, (6) influent structure, (7) effluent structure, and (8) sludge collection and removal. The secondary clarifier in general must perform two functions: (a) provide clarification to produce high-quality effluent and (b) provide thickening of settled solids. Therefore, sufficient depth must be provided so that the solids are not lost in the effluent and at the same time there is storage for the settled solids for thickening and maintaining adequate sludge blanket. If sufficient sludge blanket is not maintained, unthickened

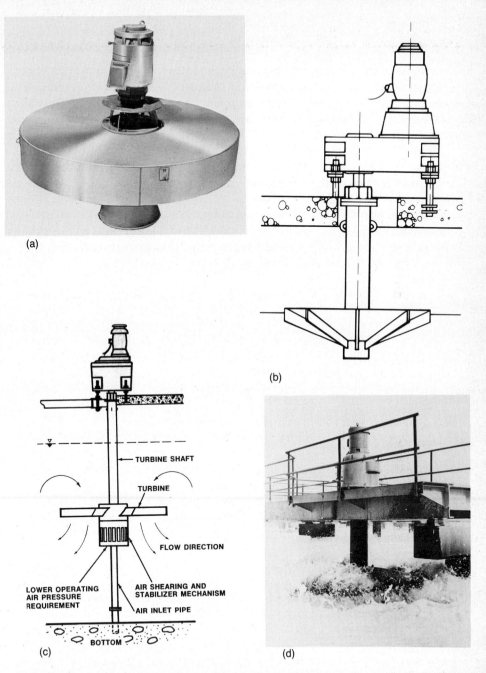

(a)

(b)

(c)

TURBINE SHAFT

TURBINE

FLOW DIRECTION

LOWER OPERATING
AIR PRESSURE
REQUIREMENT

AIR SHEARING AND
STABILIZER MECHANISM

AIR INLET PIPE

BOTTOM

(d)

Figure 13-7   Common Types of Mechanical Aeration Devices Used in Aeration Basins and Aerated Lagoons. (a) Floating aerator (Courtesy Aqua-Aerobic Systems, Inc.). (b) Impeller surface aerator Assembly (Courtesy EIMCO Process Equipment Company). (c) Turbine aerator (Courtesy EIMCO Process Equipment Company). (d) Impeller surface aerator installation (Courtesy EIMCO Process Equipment Company).

sludge will be returned to the aeration basin. The design factors are presented in detail in the Design Example. Some typical design values used for secondary clarifiers are listed below:

1. Overflow rate at average and peak design flows are 15–32 and 40–48 $m^3/m^2 \cdot d$.
2. Solids loading at average and peak design flows are 49–144 and 100–220 $kg/m^2 \cdot d$.
3. The tanks can be circular, rectangular, or square. Circular tanks are 10–60 m in diameter (preferably not to exceed 5 times the sidewater depth.) The desirable range of depth in circular and rectangular clarifiers is 3–6 m.
4. Influent and effluent structures, and sludge collection equipment for rectangular and circular basins are given in Chapter 12.

Return Sludge System.   The settled sludge is returned from the clarifier to the aeration basin to maintain the desired food-to-microorganism ratio. The return sludge system is designed for a total capacity of 50–150 percent of the average flow. Most common operational range is 20–30 percent. The return flow requirement is determined from some simple settling tests of the MLSS.*

Waste Activated Sludge.   The excess sludge is wasted either from the effluent line of the aeration basin or from the return sludge line. The waste sludge from the aeration basin is quite thin (0.2–0.6 percent solids). Thickening may be achieved by returning it into the primary sedimentation basin or by installing the proper thickening device. The waste sludge from secondary clarifier is considerably thicker (0.5–1.2 percent solids). The waste activated sludge may be thickened separately or may be mixed with the primary sludge and the combined sludge may be thickened. Various types of sludge thickening devices are discussed in Chapter 16.

13-3-2 Aerated Lagoon

The aerated lagoons are suspended growth reactors in earthen basins with no sludge recycle. Mechanical aerators are normally used for mixing and supplying oxygen demand. Since the aerated lagoons have a large detention period (2–6 days), a certain amount of nitrification is achieved. Higher temperatures and lower organic loadings generally encourage nitrification.

Design of aerated lagoons is similar to an activated sludge process with no recycle. The design equations for activated sludge derived from soluble substrate removal kinetics have been discussed in the preceding section. The procedure for designing an aerated lagoon may be obtained in Refs. 4, 14, and 15.

---

*Two simple settling tests are in common use. One method is to determine the volume of settled sludge in 30 min in a 1-ℓ graduated cylinder filled with MLSS. Qr/Q = [vol. of settled sludge, mℓ]/[1000 mℓ − vol. of settled sludge, mℓ]. The other method includes determination of sludge volume index (SVI). The sludge volume index is defined as the volume occupied (in mℓ) by 1g of settled sludge. It is also defined by the ratio of percent volume occupied by settled sludge in 30 min and MLSS concentration in percent.

$$Q_r/Q = \frac{1}{\left[\dfrac{100}{(\text{MLSS concentration percent}) \times \text{SVI}} - 1 \right]}$$

where Q_r/Q = ratio of return sludge to influent flow.

In the absence of a clarifier, the concentration of suspended solids in the effluent is high. Although the aerated lagoons are designed as completely mixed reactors, a certain amount of settling does occur in different parts of the basin. Aerated lagoons produce effluents that have suspended solids concentrations in the range of 80–250 mg/ℓ. To meet the secondary effluent standards, settling basins have been added to the existing aerated lagoons. Chemical coagulation and clarification of effluent from aerated lagoons will produce well-nitrified effluent that is also low in phosphorus. Design details of aerated lagoons are given in Figures 13-7 and 13-8.

13-3-3 Stabilization Pond*

A stabilization pond is a relatively shallow body of water contained in an earthen basin of certain shape, designed to treat wastewater. These ponds have become a popular means of wastewater treatment for small communities and industries that produce organic waste streams. The stabilization ponds have the advantage of low construction and operation costs. The major disadvantages are large land area required, odor and insect problems, possible groundwater contamination, and poor effluent quality.

Process Description. In a stabilization pond solids settle to the bottom. A wide variety of microscopic plants and animals find the environment a suitable habitat. Organic matter is metabolized by bacteria and protozoa as primary feeders. Secondary feeders include protozoa and higher animals such as rotifers and crustaceans. The nutrients released are utilized by algae and other aquatic plants. The main sources of oxygen are natural reaeration and photosynthesis. In the bottom layer, the accumulated solids are actively decomposed by anaerobic bacteria.

Types of Stabilization Ponds and Design Considerations. The stabilization ponds are usually classified as aerobic, facultative, and anaerobic. This classification is based on the nature of the biological activity taking place. Design factors such as depth, detention time, organic loading, and effluent quality also vary greatly for the three types of lagoons. Brief descriptions of the three types of lagoons, their design factors, and their effluent qualities are given in Table 13-4.

Effluent Quality. The effluent quality from a stabilization pond is poor and does not meet the EPA secondary treatment criteria. Although the effluent is low in soluble BOD_5, it is high in total suspended solids and total BOD_5. High values of total suspended solids and total BOD_5 are attributed to algae. If some type of solids removal system is used in conjunction with lagoon treatment, a high quality of effluent that is also low in nutrients can be obtained. Some solids removal techniques that have been used or are under investigation for removal of algae from pond effluent are as follows: (1) coagulation and clarification, (2) dissolved air flotation, (3) microscreening, (4) sand filtration, (5) rock filters, and (6) specialized algae-harvesting devices. A discussion of all these methods of algae removal from lagoon effluent, and design details may be found in Refs. 4, 16, and 17. Design details of stabilization ponds are shown in Figure 13-8.

*The terms *oxidation pond* and *lagoon* are also used for stabilization pond.

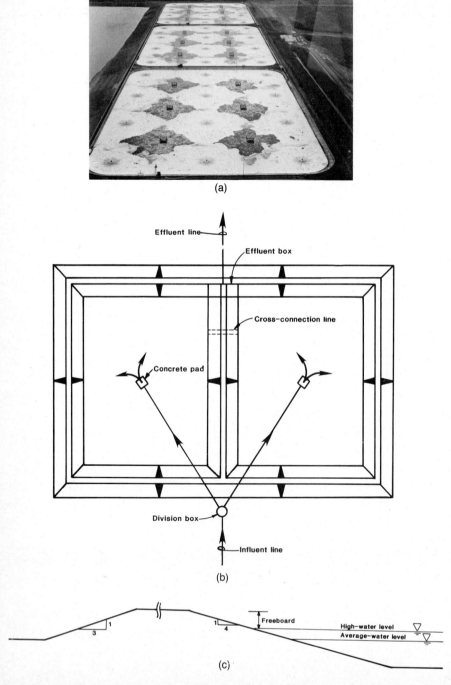

(a)

(b)

(c)

Figure 13-8   Design Details of Aerated Lagoon and Stabilization Ponds. (a) Aerated lagoons Mona Lake Project, Muskegon, Michigan (Courtesy Aqua-Aerobic Systems, Inc.) (b) Plan of stabilization pond. (c) Typical cross section of levees of stabilization pond.

TABLE 13-4 Process Description and Design Parameters for Aerobic, Facultative, and Anaerobic Stabilization Ponds

Parameter	Aerobic (High Rate)	Aerobic-anaerobic (Facultative)	Anaerobic
Process description	Aerobic condition prevails throughout the entire depth. Bacteria and algae remain in suspension. Principal sources of oxygen are natural reaeration and photosynthesis, which is governed by solar energy. Design is based on organic loading, first-order reaction kinetics, or by equating the oxygen resources of the pond to the applied organic loading. The BOD_5 removal rate constant is dependent on temperature, mixing, solar radiation, and type of wastewater	The upper layer is aerobic zone (maintained by algae and natural reaeration). Lower layers are facultative. Bottom layer of solids undergoes anaerobic decomposition. The design of facultative pond is based on organic loading and reaction kinetics. Accumulation of bottom sludge is also given consideration in oxygen utilization	Anaerobic conditions prevail throughout the pond. Design is based on the principles of anaerobic digesters with no mixing. Design details of anaerobic digesters are provided in Chapter 17. The ponds are associated with undesirable odors. Anaerobic ponds are normally used where concentrated industrial waste streams are treated. Examples are slaughterhouses, dairies, and canning and meat-processing plants
Detention time, days	5–20	10–30	20–50
Water depth, m	0.3–1[a]	1–2	2.5–5
BOD_5 loading kg/ha·d	40–120[b]	15–120	200–500
Soluble BOD_5 removal, percent	90–97	85–95	80–95
Overall BOD_5 removal, percent	40–80[c]	70–90	60–90
Algae concentration, mg/ℓ	100–200	20–80	0–5
Effluent TSS, mg/ℓ	100–250	40–100	70–120

Source: Adapted in part from Refs. (11), (14), (16)–(18).
[a] 1 m = 3.28 ft.
[b] 1 kg/ha·d = 0.8922 lb/acre·d.
[c] Overall BOD_5 removal is small because of high concentration of algae in the effluent.

13-4 ATTACHED GROWTH BIOLOGICAL TREATMENT

In attached growth biological treatment processes the population of active microorganisms is developed over a solid media (rock or plastic). The attached growths of microorganisms stabilize the organic matter as the wastewater passes over them. There are two major types of attached growth processes: (1) trickling filters and (2) rotating biological contactors. Both of these processes are discussed below.

13-4-1 Trickling Filter

The trickling filter consists of a shallow bed filled with crushed stones or synthetic media. Wastewater is applied on the surface by means of a self propelled rotary distribution system. The organics are removed by the attached layer of microorganism (slime layer) that develops over the media. The underdrain system collects the trickled liquid that also contains the biological solids detached from the media. The air circulates through the pores due to natural draft caused by thermal gradient. The trickled liquid and detached biological solids are settled in a clarifier. A portion of the flow is recycled to maintain a uniform hydraulic loading and to dilute the influent. The details of trickling filter and rotating biological contactor are given in Figure 13-9.

Types of Trickling Filters.   Based on the organic and hydraulic loading, the trickling filters are classified into low-rate, intermediate-rate, high-rate, and super-rate (roughing filters). Often two-stage trickling filters (two trickling filters in series) are used for treating high-strength wastes. Typical design information for different types of trickling filters are summarized in Table 13-5.

Design Methods.   Many design equations and procedures for designing trickling filters have been proposed over the years. Some of these equations were developed by the National Research Council (NRC),[19] Velz,[20] Eckenfelder,[21] Galler and Gotaas,[22] Schulze, Fairall, and Atkinson.[23,24] Discussions on these equations, design procedures and design examples may be obtained in several textbooks.[4,17,18,24–26]

13-4-2 Rotating Biological Contactor

A rotating biological contactor (also called bio-disc process) consists of a series of circular plastic plates (discs) mounted over a shaft that rotates slowly. These discs remain approximately 40 percent immersed in a contoured bottom tank. The discs are spaced so that wastewater and air can enter the space. The biological growth develops over the discs that receives alternating exposures to organics and the air. The excess growth of microorganisms becomes detached and therefore the effluent requires clarification. The rotating biological contactor has low power demand, greater process stability, higher organic loadings, and smaller quantity of waste sludge than the activated sludge process. The rotating biological contactor is shown in Figure 13-9. Technical discussion and design information may be obtained in Refs. 4, 5, 14, and 26.

(a)

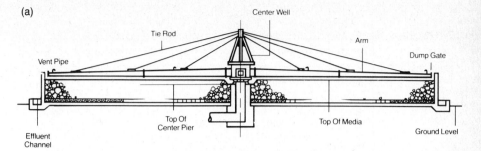

Machine Side View

(b)

(c)

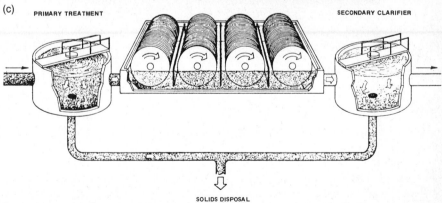

Figure 13-9   Trickling Filter and Rotating Biological Contractor. (a) Trickling filter assembly. (Courtesy EIMCO Process Equipment Company.) (b) Trickling filter installation. (Courtesy EIMCO Process Equipment Company.) (c) Rotating Biological Contactor flow scheme. [From Ref. 11].

TABLE 13-5 *Typical Design Information for Different Types of Trickling Filters*

Item	Low Rate	Intermediate Rate	High Rate	Super Rate or Roughing	Two-stage
Operation	Intermittent	Continuous	Continuous	Continuous	Continuous
Recirculation ratio	0	0–1.0	1.0–2.5	1.0–4.0	0.5–3.0
Depth, m	1.5–3.0	1.25–2.5	1.0–2.0	4.5–12	2.0–3.0
Hydraulic loading, $m^3/m^2 \cdot d$	1–4	4–10	10–40	40–200	10–40
BOD_5 loading, $kg/m^3 \cdot d$	0.08–0.32	0.24–0.48	0.32–1.0	0.8–6.0	1.0–2.0
Sloughing	Intermittent	Intermittent	Continuous	Continuous	Continuous
Media	Rock, slag	Rock, slag	Rock, slag, synthetic	Synthetic	Rock, slag, synthetic
Filter flies	Many	Medium	Small	None	None
Power $kW/10^3 m^3$	2–4	2–8	6–10	10–20	6–10
BOD_5 removal efficiency, percent	74–80	80–85	80–85	60–80	85–95
Effluent	Well nitrified	Well nitrified	Little nitrification	Little nitrification	Well nitrified

Source: Ref. 14.

13-4-3 Combined Attached and Suspended Growth Treatment

A combination of attached and suspended growth treatment is achieved in deep trickling filters 9–12 m (30–40 ft). These filters are circular or square, and utilize plastic media. Air is supplied at the bottom of the bed. The effluent from the filter is clarified and a portion of the sludge is returned to the filter to maintain hydraulic loading and to supply biological solids. The system differs from a conventional trickling filter because much of the active microbial growth is suspended in a manner similar to that of an activated sludge process. In many instances series arrangements of trickling filter and activated sludge processes are also used.[14]

13-5 EQUIPMENT MANUFACTURERS OF BIOLOGICAL WASTE TREATMENT PROCESSES

A list of the manufacturers of aeration equipment, clarifiers, and sludge pumps is provided in Appendix D. Basic considerations for equipment selection and responsibilities of the design engineer and equipment suppliers are given in Sec. 2-10. The selected equipment should provide the desired operational flexibility, maintenance requirements, and also meet the design criteria.

13-6 INFORMATION CHECKLIST FOR DESIGN OF BIOLOGICAL TREATMENT AND CLARIFICATION FACILITIES

The design engineer should make the following important decisions and develop the following data concerning the design of a biological treatment facility:

1. Select the type of biological treatment process. This includes a critical evaluation of fluid bed and attached bed reactors, and their modifications.
2. Develop the chemical characteristics of the wastewater reaching the biological treatment facility. A material mass balance analysis must be conducted to establish the effect of return flows to the biological treatment facility. The expected range of flows (minimum, average, and peak design values) and concentrations of soluble BOD_5, total suspended solid (TSS), nutrients, and toxic chemicals under average and sustained loading conditions should also be established.
3. Develop biological kinetic coefficients. Laboratory or pilot studies may be necessary if the kinetic coefficients cannot be estimated. Such studies may be particularly necessary for industrial wastes or combined municipal-industrial wastewaters.
4. Develop a preliminary site plan, piping layout, and location of collection boxes, return sludge pumps, etc., with respect to the biological reactor and solids removal system. Information on existing facility should also be obtained.
5. Obtain design criteria prepared by the concerned regulatory agencies (if any) for biological reactors and solids removal systems.
6. Obtain effluent quality criteria from the secondary treatment facility in terms of BOD_5, TSS, and nutrients.
7. Develop data on settling characteristics of the biological solids. If necessary, conduct

laboratory studies to determine the solids settling rate, solids flux values, solids loading rate, and overflow rate.
8.   Equipment manufacturers and equipment selection guide.

13-7 DESIGN EXAMPLE

13-7-1 Design Criteria Used

The following design criteria shall be used for the design of the biological reactor and solid separation facilities:

Biological Reactor

1.   Provide complete mix activated sludge process using diffused aeration system.
2.   The effluent shall have BOD_5 and TSS of 20 mg/ℓ or less.
3.   Provide *four* aeration basins with common walls. Each unit may be removed from operation for repairs and maintenance while other units shall continue to operate under normal operating procedures.
4.   Equipment for measuring raw wastewater flow, return sludge, waste sludge, and air supply shall be provided.
5.   Blowers shall be capable of delivering maximum air requirements considering the largest single unit out of service.
6.   Aeration equipment shall provide complete mixing of the mixed liquor suspended solids (MLSS) and shall be capable of maintaining a minimum of 1.5 mg/ℓ dissolved oxygen in the mixed liquor at all times.
7.   Diffusers and piping shall be capable of delivering 150 percent of the average air requirements. The diffusers and pipings shall be arranged in a manner that routine inspection and maintenance of the diffusers and piping can be accomplished without draining the aeration basin.
8.   The sludge pump and piping for return activated sludge shall be designed to provide capacity up to 150 percent of average design flow. The return activated sludge pump shall be capable of providing a variable capacity ranging from 0 to 150 percent of the average design flow.* A standby pumping unit equal in capacity to the largest single pump shall be provided. The sludge piping and channels shall be so arranged that flushing can be accomplished.
9.   All sidestreams from the sludge-handling facilities (thickeners, digesters, and dewatering units) shall be returned to the aeration basin. A material mass balance analysis shall be performed at the average design flow to determine the combined average flows, and concentrations of BOD_5 and total suspended solids.
10.   The sludge wasting shall be achieved from the common collection box containing the effluent MLSS from the aeration basins.
11.   The influent and effluent structures shall be designed and the basin hydraulics shall be checked at peak design flow plus the design return sludge flow when only three basins are in operation.

*The values given in Table 13-2 (Q_r/Q = 0.25–1.00) are the normal operation range.

12. The biological kinetic coefficients and operational parameters for the design purposes shall be determined from carefully controlled laboratory studies. The following kinetic coefficients and design parameters shall be used.

$$\theta_c = 10 \text{ d}$$
$$Y = 0.5 \text{ mg/mg}$$
$$MLVSS = 3000 \text{ mg/}\ell$$
$$k_d = 0.06/\text{d}$$

Return sludge concentration = 10,000 mg/ℓ (TSS)
Ratio of MLVSS to MLSS = 0.8
The value of BOD_5 = 68 percent of ultimate BOD (BOD_L)

The biological solids are 65 percent biodegradable

13. The following flow, BOD_5, and TSS values shall be used for the influent to the plant (Table 6-9):

Peak design flow = 1.321 m³/s
Average design flow = 0.440 m/s
BOD_5 = 250 mg/ℓ
TSS = 260 mg/ℓ

Secondary Clarifier.* The following design criteria shall be used for the solids separation facility:

1. Provided four circular clarifiers, each clarifier shall have independent operation with respect to the aeration basins.
2. Design the clarifiers for average design flow plus the recirculation.
3. Design the influent and effluent structures, and check the hydraulics at peak design flow when only three clarifiers are in operation.
4. Return sludge from each clarifier shall have an independent sludge withdrawal arrangement with flow measurement and control devices.
5. The design of the clarifier shall be based on the solids-settling rate obtained from laboratory results illustrated in Figure 13-10.
6. The surface area of the clarifier shall be large enough to meet the clarification as well as the thickening requirements for the effluent and the underflow, respectively.
7. The water depth of the clarifier shall be sufficient to provide an adequate clearwater zone, thickening zone, and sludge storage zone.
8. The overflow rates at average and peak flow conditions shall not exceed 15 and 40 m³/m²·d, respectively.
9. The solids-loading rates at average and peak design flows shall not exceed 50 and 150 kg/m²·d, respectively.
10. Scum baffles and scum collection system shall be provided.

*For comparison purposes see the design criteria for primary sedimentation facility given in Sec. 12-6-1.

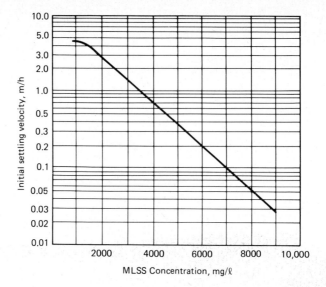

Figure 13-10 *Solids-Settling Rate Developed from Experimental Results for Design of Secondary Clarifier.*

11. The effluent weir shall be designed to prevent turbulence. The weir loading shall not exceed 124 m³/m·d (10,000 gpd/ft) at average design flow.

13-7-2 Unit Arrangement and Piping Layout

The preliminary layout of aeration basins, secondary clarifiers, return flow pumps and pipings, and aeration systems are shown schematically in Figure 13-11. The generalized flow scheme includes a junction splitter box that receives primary treated effluent, side streams from sludge-processing areas, and the return activated sludge. The combined flow is then discharged over four identical weirs for division into four pressure pipes that lead to the aeration basins. Each weir is adjustable,* has a stop gate,† and is equipped with a head measurement system over the weir.

In aeration basins, the swing-type diffusers are arranged in several rows perpendicular to the direction of flow. These diffusers provide complete mixing and also supply the required oxygen demand. The MLSS is discharged over a rectangular weir and then into a junction box. A portion of the MLSS is wasted from the junction box and pumped to the sludge-processing area. The MLSS is piped from the junction box to a flow splitter box that has four rectangular weirs similar to the junction splitter box ahead of the aeration basin. The MLSS from the division boxes is diverted to four secondary clarifiers. The effluent from the final clarifier is discharged into a common junction box and then conveyed to the chlorination facility. The return sludge from each clarifier is pumped to the junction box ahead of the aeration basin that also receives primary treated effluent.‡

*The adjustable weirs shall be used to divide the flows equally into the aeration basins.
†The stop gate shall be used to remove any aeration basin from operation.
‡Alternative arrangement provides gravity sludge flow from clarifiers into a common central well, pumping sludge from the well.

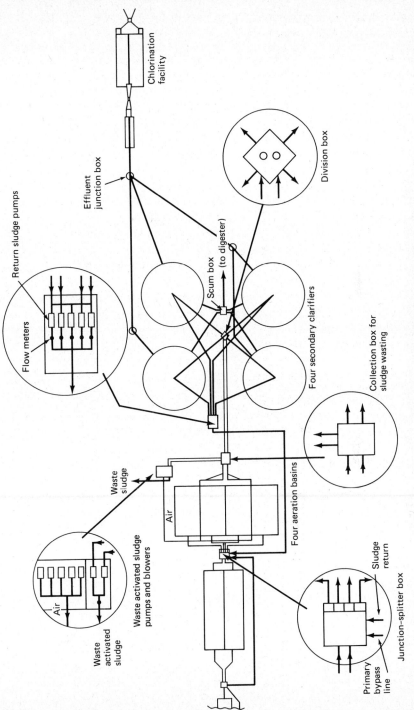

Figure 13-11   Schematic Flow and Piping Arrangement of Secondary Treatment Facility.

13-7-3 Design Calculations, Material Mass Balance, and Determination of Wastewater Quality

The sludge-processing facilities produce liquids that are normally returned to the plant. The return flows increase hydraulic, organic, and solids loadings to the treatment units below the point of discharge. Many designers prefer to return these flows to the wet well and blend them with the incoming wastewater. In many cases, however, these flows may also be returned to the primary sedimentation or secondary treatment units. Each location of returned flow must be carefully evaluated. Considerations should be given to (1) relative volume and incremental solids and organic loadings added by the returned flows, (2) geometry and location of treatment units, (3) treatment processes and available treatment capacity, and (4) odor problems that may result due to the return flows. A material mass balance analysis is needed to determine the combined flows and the concentrations of BOD_5 and TSS. If returned flows are produced intermittently such as in the case of filter back wash, filter presses, and the like, an equalization basin may be provided to avoid surge loadings.

In the Design Example, the flows from the thickener, digester, and dewatering facilities are returned to the aeration basin. The performance data of sludge-processing facilities are given in Chapters 4, 16–18. The material mass balance procedure is given in this section. An example of all return flows to the head of the plant may be found in Ref. 4.

A. First Iteration

1. Calculate influent flow, BOD_5, and TSS (see Chapter 12)

   Flow $= 0.440 \text{ m}^3/\text{s} \times 86,400 \text{ s/d} = 38,016 \text{ m}^3/\text{d}$
   $BOD_5 = 38,016 \text{ m}^3/\text{d} \times 250 \text{ g/m}^3 \times (1000 \text{ g/kg})^{-1} = 9504 \text{ kg/d}$
   TSS $= 38,016 \text{ m}^3/\text{d} \times 260 \text{ g/m}^3 \times (1000 \text{ g/kg})^{-1} = 9884 \text{ kg/d}$

2. Calculate primary sludge characteristics

BOD_5 (34 percent removal)	$= 9504 \text{ kg/d} \times 0.34 = 3231 \text{ kg/d}$
TSS (63 percent removal)	$= 9884 \text{ kg/d} \times 0.63 = 6227 \text{ kg/d}$
Solids concentration	$= 4.5$ percent
Specific gravity	$= 1.03$

   Sludge flow rate $= \dfrac{6227 \text{ kg/d} \times 1000 \text{ g/kg}}{0.045 \text{ g/g} \times 1.03 \times 1 \text{ g/cm}^3 \times 10^6 \text{ cm}^3/\text{m}^3}$

   $= 134 \text{ m}^3/\text{d}$

3. Calculate flow, BOD_5, and TSS in primary treated effluent

   $Q = 38,016 \text{ m}^3/\text{d} - 134 \text{ m}^3/\text{d} = 37,882 \text{ m}^3/\text{d}$

   $BOD_5 = 9504 \text{ kg/d} - 3231 \text{ kg/d} = 6273 \text{ kg/d}$

   $= \dfrac{6273 \text{ kg/d}}{37,882 \text{ m}^3/\text{d}} \times 1000 \text{ g/kg} = 165.6 \text{ g/m}^3$

$$\text{TSS} \quad = 9884 \text{ kg/d} - 6227 \text{ kg/d} = 3657 \text{ kg/d}$$

$$= \frac{3657 \text{ kg/d}}{37{,}882 \text{ m}^3/\text{d}} \times 1000 \text{ g/kg} = 96.5 \text{ g/m}^3$$

4. Calculate waste activated sludge

    Increase in TVSS^* $\quad = Y_{\text{obs}}\,(S_o - S)Q$

    $$= 0.3125\,(165.6 - 7.4) \text{ g/m}^3$$
    $$\times 37{,}882 \text{ m}^3/\text{d} \times (1000 \text{ g/kg})^{-1}$$

    $$= 1873 \text{ kg/d}$$

    Increase in TSS $\quad = 1873 \text{ kg/d}/0.8 = 2341 \text{ kg/d}$

    TSS in waste acti- $\quad$ = Increase in TSS–TSS lost in the effluent
    vated sludge

    $$= 2341 \text{ kg/d} - (20 \text{ g/m}^3 \times (37{,}882$$
    $$- 425^\dagger) \text{ m}^3/\text{d} \times (1000 \text{ g/kg})^{-1}$$

    $$= 1592 \text{ kg/d}$$

    MLSS concentration $\quad = \dfrac{3000 \text{ mg/}\ell}{0.8} = 3750 \text{ mg/}\ell$

    $$= 3.75 \text{ kg/m}^3$$

    Volume of waste acti- $\quad = 1592 \text{ kg/d} \times \dfrac{1}{3.75 \text{ g/m}^3}$
    vated sludge

    $$= 425^\dagger \text{ m}^3/\text{d}$$

    BOD_5 in waste acti- $\quad = 1592 \text{ kg/d} \times 0.65 \text{ g/g} \times 1.42 \text{ g/g} \times 0.68 \text{ g/g}$
    vated sludge*

    $$= 999 \text{ kg/d}$$

    Soluble BOD_5 $\quad = 7.4 \text{ g/m}^3 \times 425 \text{ m}^3/\text{d} \times (1000 \text{ g/kg})^{-1}$

    $$= 3 \text{ kg/d}$$

    Total BOD_5 in waste activated sludge $= 999 \text{ kg/d} + 3 \text{ kg/d} = 1002 \text{ kg/d}$

5. Calculate combined sludge

    $Q \quad = 134 \text{ m}^3/\text{d} + 425 \text{ m}^3/\text{d} = 559 \text{ m}^3/\text{d}$
    $\text{BOD}_5 = 3231 \text{ kg/d} + 1002 \text{ kg/d} = 4233 \text{ kg/d}$
    $\text{TSS} \quad = 6227 \text{ kg/d} + 1592 \text{ kg/d} = 7819 \text{ kg/d}$

---

*See Design Criteria (Sec. 13-7-1), and Sec. 13-7-4, steps A2 and A6 for the values of $Y_{\text{obs}} = 0.3125$, and $S = 7.4 \text{ g/m}^3$; TVSS/TSS $= 0.8$; biological solids are 65 percent biodegradable, 1 g of biodegradable solids $= 1.42$ g of ultimate BOD; and $\text{BOD}_5 = 0.68 \times$ ultimate BOD.
†Obtained by trial and error solution.

6. Calculate thickened sludge

$$TSS* = 0.85 \times 7819 \text{ kg/d} = 6646 \text{ kg/d}$$

$$Q^\dagger = \frac{6646 \text{ kg/d} \times 1000 \text{ g/kg}}{0.06 \text{ g/g} \times 1.03 \times 1 \text{ g/cm}^3 \times 10^6 \text{ cm/m}^3} = 108 \text{ m}^3/\text{d}$$

7. Calculate quantity of thickener return

$$Q = 559 \text{ m}^3/\text{d} - 108 \text{ m}^3/\text{d} = 451 \text{ m}^3/\text{d}$$
$$TSS = 0.15 \times 7819 \text{ kg/d} = 1173 \text{ kg/d}$$
$$BOD_5{}^\ddagger = 0.541 \times 1173 \text{ kg/d} = 635 \text{ kg/d}$$

8. Calculate supernatant from anaerobic digester
    Assume BOD_5 and TS in the supernatant = 3000 and 4000 mg/ℓ respectively;
    total solids in digested sludge = 5 percent (see Chapters 4 and 17).

$$TVS^\P = 0.752 \times 6646 \text{ kg/d} = 4998 \text{ kg/d}$$

$$TVS \text{ destroyed} = 0.52** \times 4998 \text{ kg/d} = 2599 \text{ kg/d}$$

$$TS \text{ after digestion} = \text{fixed solids} + \text{VS remaining}$$

$$= (6646 \text{ kg/d} - 4998 \text{ kg/d}) + (0.48 \times 4998 \text{ kg/d})$$

$$= 4047 \text{ kg/d}$$

Total mass in
digester
(liquid + solids)
$$= \frac{6646 \text{ kg/d}}{0.06} = 110{,}767 \text{ kg/d}$$

Gas produced††
$$= 0.936 \text{ m}^3/\text{kg} \times 0.86 \times 2599 \text{ kg/d} \times 1.162 \text{ kg/m}^3$$
$$= 2431 \text{ kg/d}$$

Mass out of
digester
$$= (110{,}767 - 2431) \text{ kg/d} = 108{,}336 \text{ kg/d}$$

If S = total solids (kg/d) in the supernatant from the sludge digester‡‡

---

*Solids recovery in thickened sludge = 85 percent (15 percent solids are lost in the side stream).
†Thickened sludge has 6 percent solids and specific gravity = 1.03.
‡Ratio of BOD_5 to TSS in the combined sludge $= \dfrac{4233 \text{ kg/d}}{7819 \text{ kg/d}} = 0.541$.

¶Organic portion of the combined sludge $= \dfrac{0.74 \times \text{primary sludge} + 0.8 \times \text{waste activated sludge}}{\text{total combined sludge}}$

$$= \frac{0.74 \times 6227 \text{ kg/d} + 0.8 \times 1592 \text{ kg/d}}{6227 \text{ kg/d} + 1592 \text{ kg/d}} = 0.752.$$

**TVS reduced in digester = 52 percent.
††Gas production = 0.936 m^3/kg of TVS reduced (15 ft^3/lb); gas is 86 percent lighter than air; and air weighs 1.162 kg/m^3 (0.07251 lb/ft^3).
‡‡TS in supernatant = 4000 mg/ℓ, and TS in digested sludge = 5 percent.
Note: A significant portion of TS in the digester supernatant will be in "soluble" form. If the ratio of TSS/TS is known, TSS concentration in the digester supernatant can be calculated. In this example it is assumed that TSS = TS.

$$\frac{S}{0.004} + \frac{4047 \text{ kg/d} - S}{0.05} = 108{,}336 \text{ kg/d}$$

$$S = 119 \text{ kg/d}$$

$$\text{Supernatant flow rate} = \frac{119 \text{ kg/d} \times 1000 \text{ g/kg}}{0.004 \text{ g/cm}^3 \times 10^6 \text{ cm}^3/\text{m}^3} = 30 \text{ m}^3/\text{d}$$

$$\text{BOD}_5 \text{ in the supernatant} = 3000 \text{ g/m}^3 \times 30 \text{ m}^3/\text{d} \, (1000 \text{ g/kg})^{-1}$$
$$= 90 \text{ kg/d}$$

9. Calculate quantity of digested sludge to dewatering facility

TS to the dewatering facility $= $ TS in digested sludge $-$ TS in supernatant

$$= 4047 \text{ kg/d} - 119 \text{ kg/d} = 3928 \text{ kg/d}$$

Flow of digested sludge* $=$ thickened sludge $-$ supernatant
$$= 108 \text{ m}^3/\text{d} - 30 \text{ m}^3/\text{d} = 78 \text{ m}^3/\text{d}$$

10. Calculate the characteristics of sludge cake and filtrate
    The sludge-dewatering facility consists of filter presses. It is assumed that the sludge cake has 25 percent solids; dewatering facility captures 95 percent solids; specific gravity of sludge cake is 1.06; inorganic and organic polymers added to the sludge for conditioning are 5 and 2 percent of sludge solids, respectively; and 75 percent added chemicals are incorporated into the sludge cake. The added chemicals are 25 percent recycled in the returned flow; and the BOD_5 of the filtrate is 1500 mg/ℓ.

$$\text{TS in sludge cake} = 0.95 \times 3928 \text{ kg/d} + 0.75 \times 3928 \text{ kg/d}$$
$$\times (0.05 + 0.02)$$
$$= 3938 \text{ kg/d}$$

$$\text{Volume of sludge cake} = \frac{3938 \text{ kg/d} \times 1000 \text{ g/kg}}{0.25 \times 1.06 \times 1 \text{g/cm}^3 \times 10^6 \text{ cm}^3/\text{m}^3}$$
$$= 15 \text{ m}^3/\text{d}$$

$$\text{Filtrate flow} = (78 - 15) \text{ m}^3/\text{d} = 63 \text{ m}^3/\text{d}$$

$$\text{BOD}_5 \text{ in the filtrate} = 1500 \text{ g/m}^3 \times 63 \text{ m}^3/\text{d} \times (1000 \text{ g/kg})^{-1}$$
$$= 95 \text{ kg/d}$$

$$\text{TSS in the filtrate}^\dagger = 0.05 \times 3928 \text{ kg/d} + 0.25 \times 3928 \text{ kg/d}$$
$$(0.05 + 0.02)$$
$$= 265 \text{ kg/d}$$

*Water lost in gas production is not considered.
[†]After addition of chemicals a large portion of dissolved solids will coagulate. It is assumed that TSS = TS.

$$\text{Concentration of TSS in the filtrate} = \frac{265 \text{ kg/d} \times 1000 \text{ g/kg}}{63 \text{ m}^3/\text{d}} = 4206 \text{ g/m}^3$$

$$\text{Concentration of BOD}_5 \text{ in the filtrate} = \frac{95 \text{ kg/d} \times 1000 \text{ g/kg}}{63 \text{ m}^3/\text{d}} = 1508 \text{ g/m}^3$$

11. Summary result of first iteration
     The results of material mass balance of the first iteration are summarized in Table 13-6.

B. Second and Third Iterations.   The above computational procedure was repeated for two more iterations (second and third). After the third iteration a stable value of influent quality to the aeration basin was obtained (the change from second iteration was less than 1 percent). The final results of the second iteration are summarized in Table 13-7. The final results of third iteration of material mass balance analysis are given in Figure 13-12.

Note: The effluent quantity discharged from the plant will be slightly less than the influent flow. This is due to the following factors: (1) evaporation losses, (2) loss of water in production of digester gases, and (3) moisture contained in the sludge cake. In the materials mass balance analysis the loss of water due to evaporation and gas production has not been included. The flow of effluent is approximately equal to the influent flow minus the volume of sludge cake (see Figure 13-12).

13-7-4 Design Calculations, Aeration Basin

A. Dimensions of Aeration Basin and Sludge Growth.   The complete mix biological reactor is designed using the kinetic coefficients given in Sec. 13-7-1. The design calculations are given below:

1. Estimate the flow and concentration of BOD_5 and TSS in the influent to the aeration basin
    BOD_5, TSS, and flow after the third iteration of material mass balance are 185 mg/ℓ, 137 mg/ℓ, and 38,511 m^3/d, respectively. Due to uncertainties associated with many assumptions used in the material mass balance analysis, it is a common practice to increase these values by 5–10 percent. Therefore, the following values are used for aeration basin design:

    BOD_5 = 200 mg/ℓ
    TSS = 150 mg/ℓ
    Average flow = 42,000 m^3/d = 0.486 m^3/s

2. Estimate the concentration of soluble BOD_5 in the effluent

    BOD_5 exerted by the solids
    in the effluent[*]   = 20 mg/ℓ × 0.65 × 1.42 × 0.68
                         = 12.6 mg/ℓ

[*]TSS in effluent = 20 mg/ℓ; the biological solids are 65 percent biodegradable, 1 g of biodegradable solids = 1.42 g BOD_L, and BOD_5 = 0.68 ultimate BOD_L.

TABLE 13-6   Summary Results of Material Mass Balance—First Iteration

Waste Stream	Flow (m³/d)	BOD₅		TSS	
		kg/d	mg/ℓ	kg/d	mg/ℓ
Primary sludge	134	3231	—	6227	—
Primary treated effluent	37,882	6273	165.6	3657	96.5
Thickener return	451	635	—	1173	—
Digester return	30	90	—	119	—
Filtrate return	63	95	—	265	—
Combined return flow	544	820	—	1557	—
Influent to aeration basin[a]	38,426	7093	185[b]	5214	136[c]

[a]Influent to aeration basin = primary treated effluent + combined return flow.
[b]7093 kg/d/38,426 m³/d × 1000 g/kg = 185 g/m³ (mg/ℓ)
[c]5214 kg/d/38,426 m³d × 1000 g/kg = 136 g/m³ (mg/ℓ)

$$\text{Soluble portion of the } BOD_5 \text{ in the effluent} = 20 \text{ mg/}\ell - 12.6 \text{ mg/}\ell$$
$$= 7.4 \text{ mg/}\ell$$

3.   Estimate treatment efficiency of biological treatment [Eq. (13-1)]

$$\text{Efficiency of biological treatment based on soluble } BOD_5 \text{ in the effluent} = \frac{200 \text{ mg/}\ell - 7.4 \text{ mg/}\ell}{200 \text{ mg/}\ell} \times 100$$
$$= 96 \text{ percent}$$

$$\text{Overall treatment efficiency of the plant including primary treatment} = \frac{250 \text{ mg/}\ell - 20 \text{ mg/}\ell}{250} \times 100$$
$$= 92 \text{ percent}$$

TABLE 13-7   Summary Results of Material Balance—Second Iteration

Waste Stream	Flow (m³/d)	BOD₅		TSS	
		kg/d	mg/ℓ	kg/d	mg/ℓ
Primary sludge	134	3231	—	6227	—
Primary treated effluent	37,882	6273	165.6	3657	96.5
Thickener return	529	664	—	1220	—
Digester return	31	93	—	124	—
Filtrate return	66	99	—	276	—
Combined return flow	626	856	—	1619	—
Influent to aeration basin	38,508	7129	185	5276	137

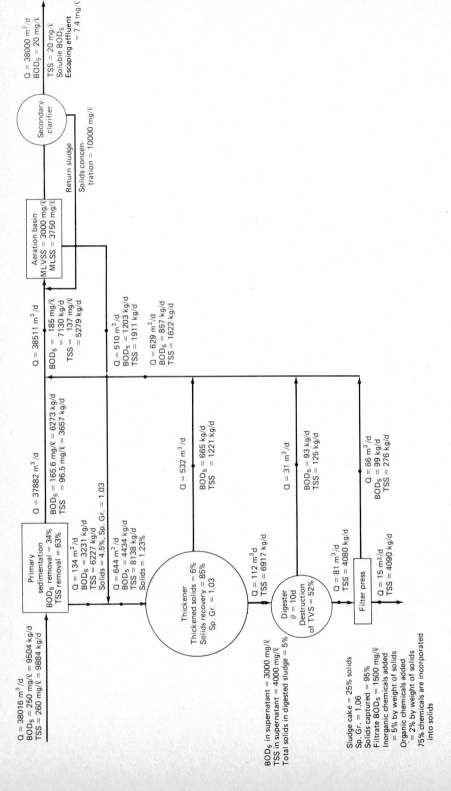

Figure 13-12 The Results of Material Mass Balance Analysis after Third Iteration.

4. Calculate the reactor volume [Eq. (13-5)]

$$V^* = \frac{Q\theta_c Y (S_o - S)}{X(1 + k_d\theta_c)} = \frac{42,000 \text{ m}^3/\text{d} \times 10 \text{ d} \times 0.5 (200 - 7.4) \text{ g/m}^3}{3000 \text{ g/m}^3 (1 + 0.06/\text{d} \times 10 \text{ d})}$$

$$= 8426 \text{ m}^3$$

5. Dimensions of aeration basin
   Provide four rectangular aeration basins with common walls. Use length-to-width ratio of 2 : 1. The basin dimensions are

Water depth	=	4.5 m (14.8 ft)
Length	=	31.0 m (102 ft)
Width	=	15.5 m (51 ft)
Freeboard	=	0.8 m (2.6 ft)

   The total volume of four aeration basins = 8649 m³. This is slightly larger than the required volume (8426 m³). These dimensions and other design details are given in Figure 13-13.

6. Estimate the quantity of waste activated sludge

   Observed Y_{obs} [Eq. (13-12)]
   $$= \frac{Y}{(1 + k_d\theta_c)} = \frac{0.5}{(1 + 0.06/\text{d} \ 10\text{d})}$$
   $$= 0.3125$$

   Increase in mixed liquor volatile suspended solids (MLVSS) [Eq. (13-13)]
   $$= Y_{obs} \ Q(S_o - S)$$

   $$= \frac{0.3125 \times 42,000 \text{ m}^3/\text{d} (200 - 7.4) \text{ g/m}^3}{1000 \text{ g/kg}}$$

   $$= 2528 \text{ kg/d}$$

   Increase in MLSS = 2528 kg/d $\times$ 1/0.8 = 3160[†] kg/d

   Since a portion of the solids is lost in the effluent, the remaining solids must be wasted from the aeration basin.

   Total solids wasted from the MLSS collection box below the aeration basin
   $$= 3160 \text{ kg/d}$$

   $$- \frac{(42,000 \text{ m}^3/\text{d} - 622[‡] \text{ m}^3/\text{d}) \times 20 \text{ g/m}^3}{1000 \text{ g/kg}}$$

   $$= 3160 \text{ kg/d} - 828 \text{ kg/d} = 2332 \text{ kg/d}$$

---

*Values of constants Y, θ_c, and k_d are the design criteria (Sec. 13-7-1).
[†]MLVSS/MLSS = 0.8
[‡]Effluent quantity = influent to aeration basin − volume of activated sludge wasted. The value of 622 m³/d of waste sludge is obtained by trial and error.

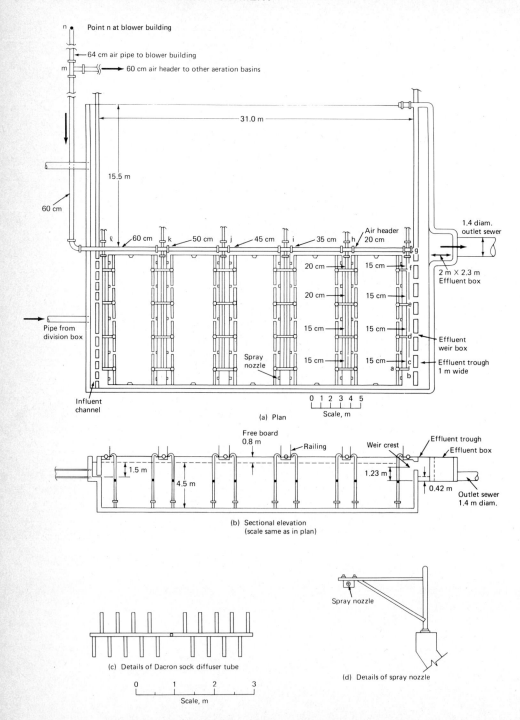

Figure 13-13   Design Details of Aeration Basin Diffusers and Pipings.

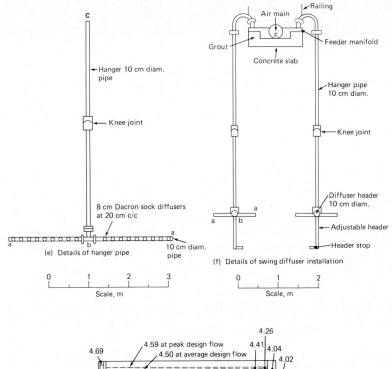

(e) Details of hanger pipe

8 cm Dacron sock diffusers
at 20 cm c/c

Hanger 10 cm diam.
pipe

Knee joint

10 cm diam.
pipe

Scale, m

(f) Details of swing diffuser installation

Air main
Railing
Grout
Feeder manifold
Concrete slab
Hanger pipe
10 cm diam.
Knee joint
Diffuser header
10 cm diam.
Adjustable header
Header stop

Scale, m

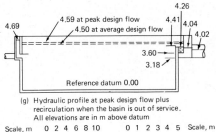

4.26
4.59 at peak design flow
4.41
4.50 at average design flow
4.04
4.69
4.02
3.60
3.18

Reference datum 0.00

(g) Hydraulic profile at peak design flow plus
recirculation when the basin is out of service.
All elevations are in m above datum

Scale, m   0 2 4 6 8 10   0 1 2 3 4 5   Scale, m
Horizontal   Vertical

$$\text{Sludge wasting rate from the MLSS collection box} = \frac{2332 \text{ kg/d}}{3.75 \text{ kg/m}^3} = 622 \text{ m}^3/\text{d}$$

7. Estimate the return sludge rate

The return sludge rate is calculated based on the concentration of MLSS in the aeration basin and TSS in the return sludge. Assume that the TSS in the influent is small and Q_r is the flow of return sludge in m^3/s, using mass balance.

$$\text{MLSS } (Q + Q_r) = \text{TSS in sludge} \times Q_r$$

$$3750 \text{ mg/}\ell \text{ } (0.486 + Q_r) \text{ m}^3/\text{s} = 10,000 \text{ mg/}\ell \times Q_r$$

$$Q_r = 0.292 \text{ m}^3/\text{s}$$

$$\frac{Q_r}{Q} = \frac{0.292 \text{ m}^3/\text{s}}{0.486 \text{ m}^3/\text{s}} = 0.6$$

8. Check aeration period (θ) against the design criteria* [Eq. (13-2)]

$$\text{Aeration period in hours} = \frac{\text{volume of basin, m}^3}{\text{flow, m}^3/\text{d}} \times 24 \text{ h/d}$$

$$= \frac{8649 \text{ m}^3 \times 24 \text{ h/d}}{42,000 \text{ m}^3/\text{d}} = 4.94 \text{ h}$$

9. Check food-to-MO ratio, U [Eq. (13-6)]

$$U = \frac{Q(S_o - S)}{VX} = \frac{42,000 \text{ m}^3/\text{d} \,(200 - 7.4) \text{ g/m}^3}{8649 \text{ m}^3 \times 3000 \text{ g/m}^3} = 0.312/\text{d}$$

10. Check organic loading

$$\frac{\text{Organic}}{\text{loading}} = \frac{S_o Q}{V} = \frac{200 \text{ g/m}^3 \times 42,000 \text{ m}^3/\text{d}}{8649 \text{ m}^3 \times 1000 \text{ g/kg}} = 0.97 \text{ kg BOD}_5/\text{m}^3 \cdot \text{d}$$

B. Influent Structure

1. Select the arrangement of the influent structure
    The influent structure consists of a rectangular channel constructed along the width of the aeration basin. The influent enters the channel in the center and divides equally along its length. The channel has 10 square submerged orifices (five on each side) that distribute the influent (a mixture of primary treated wastewater and the return sludge) along the width of the aeration basin.
2. Compute the head loss

    Maximum flow to each aeration basin† $= 0.538 \text{ m}^3/\text{s}$ ($0.269 \text{ m}^3/\text{s}$ on each side)

    Width of the influent channel $= 0.5$ m (Figure 13-13)

    Submergence of the influent channel $= 1.5$ m (Figure 13-13)

    Velocity in the channel $= \dfrac{0.269 \text{ m}^3/\text{s}}{1.5 \text{ m} \times 0.5 \text{ m}} = 0.36$ m/s

    Discharge through each orifice $= \dfrac{0.269 \text{ m}^3/\text{s}}{5} = 0.054 \text{ m}^3/\text{s}$ per orifice

    The differential elevation of water surface in the channel and the aeration basin is calculated from Eqs. (11-1) and (11-2).

---

*Many agencies specify aeration period, food to MO ratio, and organic loading values in their design criteria.
†Maximum flow to the aeration basin will occur at peak design flow plus recirculation when one basin is out of service. Flow = $(1.321 \text{ m}^3/\text{s} + 0.292 \text{ m}^3/\text{s})/3 = 0.538 \text{ m}^3/\text{s}$.

$$\Delta z = \left[\frac{Q}{C_d \times A \sqrt{2g}} \right]^2$$

If the orifices are 25 cm × 25 cm each, $C_d = 0.61$

$$\Delta z = \left[\frac{0.054 \ m^3/s}{0.61 \times 0.25 \ m \times 0.25 \ m \ \sqrt{2 \times 9.81 \ m/s^2}} \right]^2$$

$$= 0.10 \ m$$

C. Effluent Structure

1. Select arrangement of the effluent structure

    The effluent structure consists of effluent weir boxes, 1-m-wide effluent trough (launder), 2 m × 2.3 m effluent box,* and 1.4-m diameter outlet sewer. Provide eight effluent weir boxes, each box having an adjustable rectangular weir 0.5 m in length. Provide stop-gates at each weir box for the flexibility to close some openings if a minimum head over weirs is desirable under initial flow conditions. The details of effluent structure are shown in Figure 13-13.

2. Compute head over weir at average design flow

    The head over the weir under maximum flow is calculated from the weir equation [Eq. (11-3)].

    Average flow plus recirculation $= 0.486 \ m^3/s + 0.292 \ m^3/s$

    $= 0.778 \ m^3/s$

    Flow per basin $= 0.778 \ m^3/s \times \frac{1}{4} = 0.195 \ m^3/s$

    Flow per weir box $= \frac{0.195}{8} \ m^3/s = 0.024 \ m^3/s$

    Using length of weir $L = 0.5$ m and $C_d = 0.624$, the head over the weir at average design flow is calculated by trial and error as follows. Using $L' = 0.48$ m,

    $$H = \left[\frac{3}{2} \times \frac{0.024 \ m^3/s}{0.624 \times 0.48 \ m \ \sqrt{2 \times 9.81 \ m/s^2}} \right]^{2/3}$$

    $$= 0.09 \ m \ (3.5 \ in.)$$

    $$L' = 0.5 \ m - 0.2 \times 0.09 \ m = 0.48 \ m$$

3. Compute the head over the weir at peak design flow plus recirculation when one basin is out of service.

---

*2.3 m length of the effluent box includes 0.3 m thickness of the common wall in two aeration basins.

The maximum flow occurs when one aeration basin is out of service and the peak design flow plus average return sludge is reaching the basins.

$$\text{Maximum } Q = (1.321 \text{ m}^3/\text{s} + 0.292 \text{ m}^3/\text{s}) \times \frac{1}{3}$$

$$= 0.538 \text{ m}^3/\text{s per basin}$$

$$\text{Maximum } Q \text{ per weir box} = 0.067 \text{ m}^3/\text{s (eight boxes)}$$

Using $L' = 0.463$ m and $C_d = 0.624$, the head over the weir is calculated by trial and error as follows:

$$H = \left[\frac{3 \times 0.067 \text{ m}^3/\text{s}}{2 \times 0.624 \times 0.463 \text{ m } \sqrt{2 \times 9.81 \text{ m/s}^2}} \right]^{2/3}$$

$$= 0.183 \text{ m} \approx 0.18 \text{ m (7.1 in)}.$$

$$L' = 0.5 \text{ m} - (0.2 \times 0.18 \text{ m}) = 0.463 \text{ m}$$

4. Design the effluent trough (launder)

The design procedure for the effluent trough (launder) is given in Chapters 11 and 12. Eq. (11-9) is used to obtain an approximate solution, while Eqs. (11-4)–(11-8) provide a more exact solution. The computational procedure based on Eqs. (11-4)–(11-8) is tedious and seldom used by the designers. However, in this design both procedures are given to compare the results. Using the notations defined in Eq. (11-9):

$$L = 15.5 \text{ m} - 1.0 \text{ m} = 14.5 \text{ m (Figure 13-13)}$$
$$y_2 = 0.44^* \text{ m (assume)}$$
$$N = 1$$

$$q' = \frac{\text{max. flow from eight weir boxes}}{\text{length of the effluent trough}}$$

$$= \frac{0.538 \text{ m}^3/\text{s per basin when one basin is out of service}}{14.5 \text{ m}}$$

$$= 0.0371 \text{ m}^3/\text{s per m length}$$

$$y_1 = \sqrt{(0.44 \text{ m})^2 + \frac{2(0.0371 \text{ m}^3/\text{s} \cdot \text{m} \times 14.5 \text{ m} \times 1)^2}{9.81 \text{ m/s}^2 \times (1 \text{ m})^2 \times 0.44 \text{ m}}} = 0.57 \text{ m}$$

Allow 16 percent for friction losses, turbulence, and bend, and add 0.15 m drop to ensure free fall.

$$\text{Total depth of the effluent trough} = 0.57 \text{ m} \times 1.16 + 0.15 \text{ m}$$
$$= 0.81 \text{ m}$$

*Critical depth = 0.31 m. (submerged outfall).

Computational scheme for using Eqs. (11-4)–(11-8) has been presented in Chapter 11. A computer program (given in Appendix B), is used to determine the water surface profile in the effluent trough receiving eight point discharges distributed over the entire length of the weir. The design details of the effluent trough are given in Figure 13-13(a). The results of the final iteration of the computer solution are summarized in Table 13-8. Note that water depth at the upper end of the effluent trough from Eqs. (11-4)–(11-8) is 0.577 m, compared to 0.57 m from Eq. (11-9).

5. Design the effluent box and the outlet sewer
The effluent box is 2 m × 2.3 m and the outlet sewer is 1.4 m in diameter. The flow in the effluent box and outlet sewer comes from two basins. Peak flow to the outlet sewer $q' = 2 \times 0.538$ m³/s $= 1.08$ m³/s. The depth of flow and velocity in the outlet sewer is calculated from the Manning equation [Eqs. (7-1) and (7-2)] and Figure 7-5. Using $n = 0.013$, and providing slope of the outlet sewer at 0.00075 m/m

$$Q_{full} = 1.61 \text{ m}^3/\text{s}$$

$$V_{full} = 1.05 \text{ m/s}$$

$$\frac{q}{Q} = \frac{1.08 \text{ m}^3/\text{s}}{1.61 \text{ m}^3/\text{s}} = 0.67$$

From Figure 7-6, $d/D = 0.60$ and $v/V = 1.09$

The depth of flow in the outlet sewer $= 0.6 \times 1.4$ m $= 0.84$ m

The velocity in the outlet sewer at peak flow $= 1.09 \times 1.05$ m $= 1.14$ m/s

Head loss due to entrance (from Eq. (7-5)) $= \dfrac{0.3 \text{ (assume)} \cdot (1.14 \text{ m/s})^2}{2 \times 9.81 \text{ m/s}^2}$

$= 0.02$ m

Depth of flow in the effluent trough $= 0.84$ m $+ 0.02$ m $= 0.86$ m

The depth of water in the effluent trough is equal to the depth of water in the effluent trough at the lower end plus the difference in invert elevations of the trough and the box.

Differential invert elevation of effluent trough and effluent box

$= 0.86$ m $- 0.44$ m $= 0.42$ m

D. Hydraulic Profile. The hydraulic profile through the aeration basin at peak design flow when one basin is out of service is given in Figure 13-13(g). The hydraulic profile is prepared from the head loss calculations given for influent and effluent structures.

TABLE 13-8 Computational Scheme and Results of Final Iteration of Solution of Water Surface Profile in Effluent Launder Using Eqs. (11-4)–(11-8)

Length from Upper End (m)	y_2 (m)	Assumed $\Delta y'$ (m)	y_1 (m)	q_2 (m³/s)	q_1 (m³/s)	v_2 (m/s)	v_1 (m/s)	$\Delta y'$ Calculated from Eq. (11-4) (m)
14.500	0.440[a]	0.045[b]	0.485[c]	0.538[d]	0.471[e]	1.223[f]	0.970[g]	0.045[h]
14.000	0.485	0.002	0.487	0.471	0.471	0.970	0.967	0.002
12.500	0.487	0.029	0.516	0.471	0.404	0.967	0.782	0.029
12.000	0.516	0.001	0.517	0.404	0.404	0.782	0.781	0.001
10.500	0.517	0.020	0.537	0.404	0.336	0.782	0.626	0.021
10.000	0.537	0.001	0.538	0.336	0.336	0.781	0.625	0.001
8.500	0.538	0.015	0.553	0.336	0.269	0.626	0.487	0.015
8.000	0.553	0.000	0.553	0.269	0.269	0.625	0.487	0.000
6.500	0.553	0.011	0.564	0.269	0.202	0.487	0.358	0.011
6.000	0.564	0.000	0.564	0.202	0.202	0.487	0.358	0.000
4.500	0.564	0.007	0.571	0.202	0.135	0.358	0.236	0.007
4.000	0.571	0.000	0.571	0.135	0.135	0.358	0.236	0.000
2.500	0.571	0.004	0.575	0.135	0.067	0.236	0.117	0.004
2.000	0.575	0.000	0.575	0.067	0.067	0.236	0.117	0.000
0.500	0.575	0.001	0.577	0.067	0.067	0.117	0.000	0.001

[a] $y_2 = 0.440$ m based on the lower end conditions

[b] $\Delta y' =$ assumed value (0.045 m is the value obtained from the final iteration)

[c] $y_1 = y_2 - \Delta y$ (Eq. (11-8)); $\Delta y = S \Delta x - \Delta y'$ (Eq. (11-7)); $S = 0$, $\Delta x = 0.5$ m, and $\Delta y = -\Delta y'$; $y_1 = 0.440\text{m} + 0.045\text{m} = 0.485\text{m}$

[d] $q_2 = 0.538$ m³/s (total flow at the lower end)

[e] $q_1 = 0.538$ m³/s $- 0.067$ m³/s (discharge from each weir box) $= 0.471$ m³/s

[f] $v_2 = \dfrac{0.538 \text{ m}^3/\text{s}}{0.440 \text{ m} \times 1.0 \text{ m (width of the trough)}} = 1.223$ m/s

[g] $v_1 = \dfrac{0.471 \text{ m}^3/\text{s}}{0.485 \text{ m} \times 1.0 \text{ m}} = 0.971$ m/s

[h] $\Delta y'$ is calculated from Eq. (11-4) as follows:

$$R_{avg} = \frac{R_2 + R_1}{2} = \frac{1}{2}\left[\frac{1.0\text{m} \times 0.440\text{m}}{1.0\text{m} + 2 \times 0.440\text{m}} + \frac{(1.0\text{m} \times 0.485\text{m})}{(1.0\text{m} + 2 \times 0.485\text{m})}\right] = 0.24 \text{ m}$$

For Manning's $n = 0.013$, $(S_E)_{avg}$ is calculated from Eq. (11-5).

$$(S_E)_{avg} = \frac{0.013^2 (1.097)^2}{(0.24)^{4/3}} = 0.001369 \text{ m/m}.$$

$$\Delta y' = \left[\frac{0.471 \text{ m}^3/\text{s} \times 1.097 \text{ m/s}}{9.81 \text{ m/s}^2 \times 0.505 \text{ m}^3/\text{s}}\left(0.252 \text{ m/s} + \frac{1.223 \text{ m/s} \times 0.067 \text{ m}^3/\text{s}}{0.471 \text{ m}^3/\text{s}}\right)\right.$$

$$\left. + 0.001369 \text{ m/m} \times 0.5\text{m (length of the assumed increment)}\right] = 0.045 \text{ m}$$

Note: (1) The iteration must be continued until the assumed value of $\Delta y'$ and that calculated from Eq. (11-4) are within a certain allowable tolerance. In this solution the values are from the final iteration.

(2) The final value of y_1, obtained for the first increment Δx, is used as y_2 for the next increment, and the process is repeated.

E. Oxygen Requirements

1. Compute theoretical oxygen requirement
    The theoretical oxygen requirements are calculated from Eq. (13-17).

$$O_2 \text{ kg/d} = \frac{Q(S_o - S)}{BOD_5/BOD_L} - 1.42P_x$$

$$= \frac{42,000 \text{ m}^3/\text{d}(200 - 7.4) \text{ g/m}^3 \times (1000 \text{ g/kg})^{-1}}{0.68}$$

$$- 1.42 \times 2528 \text{ kg/d}$$

$$= 8306 \text{ kg/d}$$

2. Compute standard oxygen requirement (SOR) under field conditions.
    SOR under field condition is calculated from Eq. (13-18).[25]

$$\text{SOR kg/d} = \frac{N}{[(C'_{sw} \beta F_a - C)/C_{sw}] (1.024)^{T-20} \alpha} \qquad (13\text{--}18)$$

where

N = theoretical oxygen required, kg/d

C_{sw} = solubility of oxygen in tap water at standard 20°C = 9.15 mg/ℓ (Appendix A)

C'_{sw} = solubility of oxygen in tap water at field temperature, mg/ℓ

C = minimum dissolved oxygen maintained in the aeration basin, mg/ℓ

β = salinity surface tension factor, usually 0.9 for wastewater (DO saturation wastewater/DO saturation tap water)

α = oxygen transfer correction factor for wastewater usually 0.8–0.9 (oxygen transfer wastewater/oxygen transfer tap water)

F_a = oxygen solubility correction factor for elevation. The value of F_a may be calculated from Eq. (13-19)[25]

T = average temperature of wastewater in the basin (°C) under field conditions. This is dependent on the ambient average air temperature and the influent temperature. Equation (13-20) is normally used to estimate the temperature in the aerated lagoons[4]

$$F_a = \left(1 - \frac{\text{altitude, m}}{9450}\right) \qquad (13\text{--}19)$$

$$T = \frac{AfT_a + QT_i}{Af + Q} \qquad (13\text{--}20)$$

where

A = surface area, m^2

T = same as in Eq. (13-18)

T_a = average ambient air temp, °C

T_i = average influent temperature, °C

f = proportionality factor, 0.5 (m/d)

Q = flow rate, m^3/d

Assume average operating temperature in aeration basin = 24°C

C'_{sw} at 24°C = 8.5 mg/ℓ (Appendix A)

Using Eq. (13-18), SOR is calculated from the following data. $C = 1.5$ mg/ℓ, α = 0.95, $\beta = 0.9$, $C'_{sw} = 9.1$ mg/ℓ, $F_a = 0.95$ for an altitude correction of 500 m [Eq. (13-19)]:

$$SOR = \frac{8306 \text{ kg/d}}{\dfrac{(8.5 \text{ mg/ℓ} \times 0.9 \times 0.95 - 1.5 \text{ mg/ℓ})(1.024)^{24-20} \times 0.95}{9.15 \text{ mg/ℓ}}}$$

$$= 12,615 \text{ kg/d}$$

3. Compute the volume of air required
   Assuming that air weighs 1.201 kg/m^3 and contains 23.2 percent oxygen by weight

   Theoretical air required under field condition $= \dfrac{12,615 \text{ kg/d}}{1.201 \text{ kg/}m^3 \times 0.232\text{g } O_2/\text{g air}} \approx 45,500 \text{ } m^3/d \text{ air}$

   Assume that the efficiency of air diffusers = 8 percent

   Theoretical air required $= \dfrac{45,500 \text{ } m^3/d}{0.08} = 568,750 \text{ } m^3/d$

   Provide design air at 150 percent of the theoretical air,

   Total design air = 568,750 $m^3/d \times 1.5$ = 853,125 m^3/d
   = 592 m^3/min for four basins
   = 148 m^3/min per basin

4. Check the volume of air per kg BOD_5 removed, per m^3 of wastewater treated and per m^3 of aeration tank volume

$$\text{Volume of air supplied per} \atop \text{kg of BOD}_5 \text{ removed} = \frac{853,125 \text{ m}^3/\text{d} \times 1000 \text{ g/kg}}{(200 - 7.4) \text{ g/m}^3 \times 42,000 \text{ m}^3/\text{d}}$$

$$= 106 \text{ m}^3/\text{kg} \ (1700 \text{ ft}^3/\text{lb})$$

$$\text{Volume of air supplied (m}^3) \atop \text{per m}^3 \text{ of wastewater treated} = \frac{853,125 \text{ m}^3/\text{d}}{42,000 \text{ m}^3/\text{d}}$$

$$= 20.3 \text{ m}^3/\text{m}^3 \ (2.7 \text{ ft}^3/\text{gal})$$

$$\text{Volume of air supplied, (m}^3 \atop \text{per day per m}^3 \text{ of the aera-} \atop \text{tion tank volume)} = \frac{853,125 \text{ m}^3/\text{d}}{8649 \text{ m}^3}$$

$$= 99 \text{ m}^3/\text{m}^3 \cdot \text{d} \ (99 \text{ ft}^3/\text{ft}^3 \cdot \text{d})$$

F. Design of Diffused Aeration System

1. Select diffuser tube

    Provide Dacron sock diffusers, standard tube dimension 61 cm $\times$ 7.5 cm (ID), discharging 0.21 m^3 standard air per minute per tube (7.4 cfm).

2. Calculate the number of diffuser tubes and arrangement

    $$\text{Total number of diffuser tubes} = \frac{592 \text{ m}^3/\text{min}}{0.21 \text{ m}^3/\text{min/tube}}$$

    $$= 2819 \text{ tubes}$$

    Provide 2880 diffusers.[*]

    Number of diffuser tubes per basin = 2880/4 = 720

    Provide 10 rows of diffuser tubes along the width of the aeration basin.

    Number of diffuser tubes per row = 720/10 = 72

    Provide four knee and swing joint vertical hanger pipes per row.

    $$\text{Number of diffuser tubes per} \atop \text{hanger pipe} = 72/4 = 18$$

    The arrangement of the diffuser tubes and the piping layout is illustrated in Figure 13-13.[†]

3. Calculate the head losses in pipings and diffusers

    The power required to supply air to the diffusers depends on the supply pressure at the blowers. The supply pressure at the blowers depends on the head losses in

---

[*]Actual number of diffuser tubes provided is 2880. This number is based on the design symmetry. There are 10 rows of diffusers in each basin, and each row has four hanger pipes, with 18 diffuser tubes in each hanger pipe. Total number of diffuser tubes = 4 basins $\times$ 10 rows $\times$ 4 hanger pipes/row $\times$ 18 diffuser tubes/hanger pipe = 2880 diffuser tubes in four basins. If 17 diffuser tubes are provided in each hanger pipe, then total number of diffusers will be 4 $\times$ 10 $\times$ 4 $\times$ 17 = 2720 (less than required). For details, study Figure 13-13 and Table 13-10.
[†]An alternative arrangement is to provide a grid system of headers.

pipings, valves, and fittings, and the depth of submergence. The technique for determining the head losses through the air-piping system is similar to that for pumping equipment. Basic rules are summarized below.

a. The normal range of air velocities used to design the pipings are given in Table 13-9.

b. The total head loss in pipe headers is generally 5–20 cm of water (2–8 in.).

c. Head losses through diffusers generally range from 40 to 50 cm of water (16–20 in.). Most manufacturers provide rating curves for the diffusers. Allowance for clogging is also recommended by the manufacturers.

d. As an approximation, the friction factor for steel pipes carrying air may be obtained from Eq. (13-21).[25]

$$f = \frac{0.029 \ (D \ \text{in m})^{0.027}}{(Q \ \text{in m}^3/\text{min})^{0.148}} \tag{13-21}$$

e. The head loss in the straight pipe may be calculated from Eqs. (13-22) or (13-23).[25]

$$h_L = f \frac{L}{D} h_v \tag{13-22}$$

$$h_L = 9.82 \times 10^{-8} \frac{fLTQ^2}{PD^5} \tag{13-23}$$

where

$$
\begin{aligned}
h_L &= \text{head loss, mm H}_2\text{O} \\
L &= \text{equivalent length, m} \\
D &= \text{pipe diameter, m} \\
h_v &= \text{velocity head, mm H}_2\text{O} \\
Q &= \text{air flow, m}^3/\text{min} \\
P &= \text{air supply pressure, atmospheres} \\
T &= \text{temperature in pipe, °K, obtained from Eq. (13-24)}
\end{aligned}
$$

$$T = T_o \ (P/P_o)^{0.283} \tag{13-24}$$

TABLE 13-9 *Design Range of Air Velocities in Header Pipes*

Pipe Diameter (cm)	Velocity, Standard Air (m/min)
2–8	360–500
8–25	500–900
25–40	900–1050
40–60	1050–1200
60–80	1200–1350
80–150	1350–1950

where

T_o = ambient air temperature, °K
P_o = ambient barometric pressure, atmospheres
P = air supply pressure, atmospheres

f. The losses in pipe fittings (elbows, tees, valves, meters, etc.) may be computed using equivalent pipe length [Eq. (13-25)].[25,26]

$$L = 55.4\ C\ D^{1.2} \tag{13–25}$$

where

L = equivalent length of pipe fitting in meters for pipe diameter of D,
C = factor for equivalent length of pipe given in Table 13-10

Another method of calculating the pressure drop through valves and fittings is to apply appropriate multiplier K to the velocity head, similar to those for pumps. The values of K may be obtained in standard hydraulics textbooks.

g. The losses through air filters, blower, silencer, check valves etc., should be obtained from the equipment manufacturers. The following values may be used as a guide:

- Air filter losses   13–76 mm (0.5–3.0 in.)
- Silencer losses:
  - Centrifugal blower   13–38 mm (0.5–1.5 in.)
  - Positive displacement blower   152–216 mm (6.0–8.5 in.)
- Check valve losses   20–203 mm (0.8–8.0 in.)

h. The overall pressure drop in pipings, diffusers and other accessories is usually 20–40 percent of the submergence over the diffusers.
i. Air pipings with fine bubble diffusers must be of noncorroding and

TABLE 13-10 Factors to Convert Fitting Losses into Equivalent Lengths of Pipe

Fittings	C Values for Equivalent Pipe Lengths
Gate valve	0.25
Long-radius ell or run of standard tee	0.33
Medium-radius ell or run of tee reduced in size 25 percent	0.42
Standard ell or run of tee reduced in size 50 percent	0.67
Angle valve	0.90
Tee through side outlet	1.33
Globe valve	2.00

Source: From Ref. 26. Used with permission of Water Pollution Control Federation, and American Society of Civil Engineers.

nonscaling material. Steel, galvanized, or plastic pipes are commonly used.

j.   The head losses in the air pipings are calculated for the line that gives the maximum head loss. The sum of all the head losses plus the submergence of the diffusers gives the discharge pressure at the blower. Since the head losses in the pipings and the diffusers depends on the supply pressure and temperature of the air, an iterative procedure is required. The calculation steps for the final iteration are summarized in Table 13-11.*

G. Design of Blower.   Blowers develop a pressure differential between the inlet and discharge points. They move air or gases under pressure. There are two types of blowers: centrifugal and rotary positive displacement. The centrifugal blowers are commonly used for pressures 50–70 kPa (7–10 psi) and air flows above 15 m^3/min (5000 cfm). These blowers have head capacity curves similar to low specific speed centrifugal pumps (Chapter 9). The operating point is determined by the intersection of the head capacity curve and the system curve. The flow may be adjusted by throttling the inlet. Throttling the outlet of centrifugal blowers is not recommended because these machines will surge† if throttled close to the shutoff head.

Rotary positive displacement blowers are used for smaller installations (less than 45 m^3/min). Select centrifugal blowers in this example. The blowers should not be throttled as damage to the machine may occur due to overheating. The high-pitched whine emitted by blowers could be very disagreeable and silencers should be installed.

1.   Calculate supply pressure at the blower
     Supply pressure at the blower (in terms of head of water) is calculated as follows:

Total losses in pipings (Table 13-11)	= 237.40 mm
Losses in air filter (manufacturers' data)	= 50.00 mm
Losses in silencers (centrifugal)	= 30.00 mm
Losses in compressor pipings and valvings used for parallel combinations (check valve, butterfly valves, relief valve, piping and connections, etc.)	= 305.40 mm
Submergence (water depth above the diffusers)	= 4000.00 mm
Diffuser losses, fine bubble diffuser tubes (manufacturers' data)	= 300.00 mm
Allowance for clogging of diffusers and miscellaneous head losses under emergency conditions	= 880.20
Total	= 5803 mm
	= 5.803 m (19.05 ft)

The absolute supply pressure = (5.80 m + 10.34 m)/10.34 m‡ = 1.56 atm.

---

*An alternative arrangement would be to provide a pipeheader loop around the tank with feed crossheaders.
†Surging is a phenomenon in which a blower starts to operate alternately at zero and full capacity. This follows with vibration and overheating.
‡1 std atmosphere = 10.34 m of water.

TABLE 13-11 Head Loss Calculations in Air Pipings Designed for Aeration Basin

Line[a]	Description	Diam. (cm)	Air Flow[b] (m³/min)	Vel.[c] (m/min)	L[d] (m)	f[e]	$h_L{}^f$ (mm)
ab	Horizontal diffuser header contains 9 diffuser tubes	10	0.21–1.89 Avg-1.05	134	3.0	0.027	0.20
bc	Hanger pipe contains horizontal diffuser header with 18 tubes	10	3.78	481	12.0	0.022	8.28
cd	Pipe header along the width of the basin supplying air to one hanger pipe	15	3.78	214	8.0	0.023	0.76
de	Pipe header along the width of the basin supplying air to two hanger pipes	15	7.56	428	8.0	0.020	2.64
ef	Pipe header along the width of the basin supplying air to three hanger pipes	15	11.34	642	8.0	0.019	5.65
fg	Pipe header along the width of the basin supplying air to four hanger pipes (one row of diffusers)	15	15.12	856	18.0	0.018	21.41
gh	Pipe header (along the length of the basin) supplying air to two rows of pipe headers (one row in each basin)	20	30.24	963	10.0	0.017	10.66

hi	Pipe header supplying air to 6 rows of pipe headers (3 rows in each basin)	35	90.72	943	10.0	0.015	5.16
ij	Pipe header supplying air to 10 rows of pipe headers (5 rows in each basin)	45	151.20	951	10.0	0.014	3.81
jk	Pipe header supplying air to 14 rows of pipe headers (7 rows in each basin)	50	211.68	1078	10.0	0.013	4.09
kl	Pipe header supplying air to 18 rows of pipe headers (9 rows in each basin)	60	272.16	963	10.0	0.013	2.72
lm	Pipe header supplying air to 20 rows of pipe headers (10 rows in each basin)	60	302.40	1070	80.0	0.012	24.77
mn	Air main supplying air from the compressor to the aeration basins	64	604.80	1880	179	0.011	147.22
	TOTAL						237.37

[a]Air lines are marked in Figure 13-13.
[b]Air flow is based on standard oxygen requirement under field conditions.
[c]Velocity = air flow/area.
[d]Estimated equivalent length of pipe.
[e]Calculated from Eq. (13-21).
[f]Calculated from Eqs. (13-23) and (13-24) using $P = 1.56$ atm, $P_o = 0.95$ atm [from Eq. (13-19) for 500-m altitude], $T_o = 273 + 30°C$ (summer temp.) $= 303°K$.

2.   Compute the volume of air
   The volume of air to be supplied by the blower is determined as follows:

   Total standard air (150 percent of theoretical air)
   for two basins (Table 13-11).   = 302.40 m³/min

   Total air for four basins (Table 13-11)   = 604.80 m³/min

3.   Select number of blowers
   The number of centrifugal blowers chosen to deliver the required air volume should be an integer divisor of the total flow plus one extra machine for standby service.
   Provide a total of five centrifugal blowers each of 155 m³/min (5500 cfm) design capacity. These blowers shall be arranged in parallel. Each blower shall have a surge point of approximately 50 percent below the total design flow. The parallel arrangement shall provide the following operational flexibility.

- Four blowers will meet the 150 percent average air requirements (design period) = 620 m³/min.
- Three blowers will meet the average air requirements (design period)* = 465 m³/min.
- Two blowers will meet the average air requirement (initial period)† = 310 m³/min.
- Five blowers will provide one unit for standby service.

   All blowers shall be provided in a blower building as discussed in Chapter 20 and shown in Figure 13-11. Proper suction and discharge silencers and suitable foundations for compressors shall be provided.
4.   Calculate power requirements
   Blower power requirements are estimated from air flow, discharge and inlet pressures, and air temperature by using Eq. (13-26).[25] This equation is based on an assumption of adiabatic conditions.

$$P_w = \frac{wRT_o}{8.41e}\left[\left(\frac{P}{P_o}\right)^{0.283} - 1\right]$$ (13–26)

   where

   P_w = the power requirement of each blower, kW
   w = the air mass flow, kg/s
   R = gas constant, 8.314, kJ/k mole °K
   8.41 = constant for air, kg/k mole
   T_o = inlet temperature, °K
   P_o = absolute inlet pressure, atm

*Average air requirement = $\frac{604.8 \text{ m}^3/\text{min}}{1.5}$ = 403 m³/min. Three blowers will provide 465 m³/min.

†Based on volume of wastewater treated, average air requirement for the initial year = $\frac{0.304 \text{ m}^3/\text{s}}{0.440 \text{ m}^3/\text{s}}$ × 403 m³/min = 278 m³/min (see Table 6-9 for flows). Two blowers will provide 310 m³/min.

$$P = \text{absolute outlet pressure, atm}$$
$$e = \text{efficiency of the machine (usually 70–80 percent)}$$

At 75 percent efficiency,

$$P_w = \frac{155 \text{ m}^3/\text{min} \times 1.201 \text{ kg/m}^3 \times 8.314 \text{ kJ/k mole } ^\circ\text{K} \times 303 \,^\circ\text{K}}{8.41 \text{ kg/kmole} \times 0.75 \times 60 \text{ s/min}}$$

$$\times \left[\left(\frac{1.56}{0.95} \right)^{0.283} - 1 \right]$$

$$= 187 \text{ kW (250 HP)}$$

The procedure presented above permits tentative selection of diffused aeration equipment based on many manufacturers' claims. The design engineer must develop similar data for his design using the details supplied by the equipment manufacturers.

H. Design of Waste Sludge System

1. Select arrangement for waste sludge withdrawal

   The MLSS from four aeration basins is discharged into a collection box. The excess solids are pumped from the collection box to the sludge thickener. The amount of waste solids is 2332 kg/d or 622 m³/d at a solids concentration of 3750 mg/ℓ (g/m³).

2. Select pumps, piping, and pumping cycle

   Provide two identical constant-speed waste sludge pumps, each pump capable of independent operation and each having a design pumping capacity of 1.5 m³/min (400 gpm).* The pumping duration and frequency of operation shall be controlled by the continuous solids monitoring system in the collection box. In addition, the pumping operation shall also be controlled by an automatic time-controlled clock where both pumping duration and frequency of operation may be controlled in the event the solids-monitoring system does not function. Only one pump shall operate while the other shall serve as a standby unit. The pipings and arrangements of the waste sludge pumps are presented in Chapter 20 and shown in Figure 13-11. The procedure for pump selection and design of pumping station are given in Chapter 9.

I. Design of Spray Nozzles.

The design of spray nozzles and their spacings are shown in Figure 13-13. Two nozzles shall be capable of producing a hard, flat spray about 10 ℓ/min at 103 kN/m². The pump shall be capable of pumping the total flow of all nozzles at the required nozzle pressure.

13-7-5 Design Calculations, Solids Separation Facility

A. Surface Area of Secondary Clarifier

1. Establish design flow

*1.5 m³/min (400 gpm) pumping rate is large enough to select a nonclog pump for sludge wasting.

$$\text{Design flow to the secondary clarifier} = \text{average design flow + return sludge flow} - \text{MLSS wasted}$$

$$= 0.486 \text{ m}^3/\text{s} + 0.292 \text{ m}^3/\text{s} - 622^* \text{ m}^3/\text{d}$$
$$\times (86,400 \text{ s/d})^{-1} = 0.771 \text{ m}^3/\text{s}$$

$$\text{Design flow to each secondary clarifier} = 1/4 \times 0.771 \text{ m}^3/\text{s} = 0.193 \text{ m}^3/\text{s}$$

2. Prepare solids flux curves
   From the MLSS settling curve given in Figure 13-10, prepare the solids flux curve. The computation data is summarized in Table 13-12. Solids flux curve is shown in Figure 13-14.
3. Determine limiting solids-loading rate
   From Figure 13-14 determine the limiting solids loading (SF) for an underflow concentration of 10,000 mg/ℓ. This is obtained by drawing a tangent to the solids flux curve from 10,000 mg/ℓ solids concentration in return sludge. The solids flux value is 2.0 kg/m²·h = 2.0 × 24 = 48 kg/m²·d (9.81 lb/ft²·d)
4. Calculate area and diameter of the secondary clarifier
   The area of the clarifier is obtained from Eq. (13-27):

$$A = \frac{QX}{SF} \tag{13–27}$$

where

A = area of the clarifier, m²
Q = total flow to the clarifier including recirculation, m³/h
X = MLSS, kg/m³
SF = limiting solids value obtained from Figure 13-14, kg/m² · h

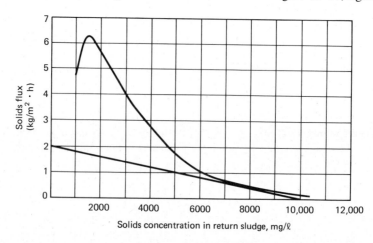

Figure 13-14 Solids Flux Curve Prepared from the Settling Data Given in Figure 13-10.

*From Sec. 13-7-4, step A6.

TABLE 13-12 Computation of Solids Flux Rate

Solids Concentration X g/m³	1000	1500	2000	3000	4000	5000	6000	7000	8000	9000
Initial settling rate V_i, m/h	4.40	4.20	2.80	1.30	0.67	0.34	0.20	0.10	0.05	0.03
Solids flux rate, $X V_i$, kg/m²·h	4.4	6.3	5.6	3.9	2.7	1.7	1.2	0.7	0.4	0.3

The flow to each clarifier including return sludge flow is 0.193 m³/s × 3600 s/h = 695 m³/h.

$$A = \frac{695 \text{ m}^3/\text{h} \times 3.750 \text{ kg/m}^3}{2.0 \text{ kg/m}^2 \cdot \text{h}} = 1303 \text{ m}^2$$

$$\text{Diameter of the secondary clarifier} = \sqrt{\frac{1303 \text{ m}^2 \times 4}{\pi}}$$

$$= 40.7 \text{ m (134 ft)}$$

Provide four clarifiers each of 40.7 m diameter

$$\text{Actual area} = \frac{\pi}{4} (40.7 \text{ m})^2 = 1301 \text{ m}^2$$

5. Check the overflow rate at average design flow

$$\text{The overflow rate} = \frac{Q}{A} = \frac{0.193 \text{ m}^3/\text{s} \times 86,400 \text{ s/d}}{1301 \text{ m}^2}$$

$$= 12.8 \text{ m}^3/\text{m}^2 \cdot \text{d} (314 \text{ gal/ft}^2 \cdot \text{d})$$

(this is less than the design criteria of 15 m³/m² · d)

6. Check the clarifier area for clarification requirement

Calculated overflow rate = 12.8 m³/m² · d = 0.533 m/h

From Figure 13-10 the MLSS concentration corresponding to a 0.533 m/h settling rate is about 4400 mg/ℓ. Since the design value of MLSS concentration is 3750 mg/ℓ, the area for clarification will be sufficient.

7. Check the overflow rate at peak design flow

$$\text{At peak design flow plus recirculation} \atop \text{the flow to each clarifier} = \frac{1.321 \text{ m}^3/\text{s} + 0.292 \text{ m}^3/\text{s}}{4}$$

$$= 0.403 \text{ m}^3/\text{s}$$

$$\text{Overflow rate} = \frac{0.403 \text{ m}^3/\text{s} \times 86{,}400 \text{ s/d}}{1301 \text{ m}^2}$$

$$= 26.8 \text{ m}^3/\text{m}^2 \cdot \text{d}$$
$$(\text{satisfactory})$$

At peak design flow plus recirculation when three clarifiers are in operation, the flow to each clarifier
$$= \frac{1.321 \text{ m}^3/\text{s} + 0.292 \text{ m}^3/\text{s}}{3}$$

$$= 0.538 \text{ m}^3/\text{s}$$

Overflow rate when three clarifiers are in operation
$$= \frac{0.538 \text{ m}^3/\text{s} \times 86{,}400 \text{ s/d}}{1301 \text{ m}^2}$$

$$= 35.7 \text{ m}^3/\text{m}^2 \cdot \text{d}$$

(this is satisfactory being less than the design criteria of 40 m³/m² · d)

8. Calculate the solids loadings

The limiting solids loading at average design flow
$$= \frac{0.193 \text{ m}^3/\text{s} \times 3750 \text{ g/m}^3 \times 86{,}400 \text{ s/d}}{1000 \text{ g/kg} \times 1301 \text{ m}^2}$$

$$= 48.1 \text{ kg/m}^2 \cdot \text{d} < 50 \text{ kg/m}^2 \cdot \text{d}$$
$$(\text{satisfactory})$$

Solids loading at peak design flow
$$= \frac{0.403 \text{ m}^3/\text{s} \times 3750 \text{ g/m}^3 \times 86{,}400 \text{ s/d}}{1000 \text{ g/kg} \times 1301 \text{ m}^2}$$

$$= 100.4 \text{ kg/m}^2 \cdot \text{d}$$

Solids loading at peak design flow when three clarifiers are in operation
$$= \frac{0.538 \text{ m}^3/\text{s} \times 3750 \text{ g/m}^3 \times 86{,}400 \text{ s/d}}{1000 \text{ g/kg} \times 1301 \text{m}^2}$$

$$= 134.0 \text{ kg/m}^2 \cdot \text{d} < 150 \text{ kg/m}^2 \cdot \text{d}$$
$$(\text{satisfactory})$$

B. Depth of Secondary Clarifier.   The liquid depth of the secondary clarifier = depth of clear water zone + depth of thickening zone + depth of sludge storage zone.
  1. Determine clearwater and settling zones
      The clearwater and settling zones are generally 1.5–2.0 m. Provide 2.0-m clearwater and settling zones.
  2. Compute the depth of thickening zone
      The depth of thickening and sludge storage zones are calculated using the procedure and assumptions given in Ref. 4.
      It is assumed that under normal conditions, the mass of sludge retained in the clarifier is 30 percent of the mass of solids in the aeration basin,* and the average concentration of sludge in the clarifier is 7000 mg/ℓ.

---

*The aeration basin is 31 m long × 15.5 m wide and 4.5 m deep.

$$\text{Total mass of solids in each aeration basin} = \frac{3750 \text{ g/m}^3 \times 4.5 \text{ m} \times 31 \text{ m} \times 15.5 \text{ m}}{1000 \text{ g/kg}}$$

$$= 8108 \text{ kg}$$

$$\text{Total mass of solids in each clarifier} = 0.3 \times 8108 = 2432 \text{ kg}$$

$$\text{Depth of thickening zone} = \frac{\text{total solids in the clarifier}}{\text{concentration} \times \text{area}}$$

$$= \frac{2432 \text{ kg} \times 1000 \text{ g/kg}}{7000 \text{ g/m}^3 \times 1301 \text{ m}^2} = 0.27 \text{ m} \approx 0.3 \text{ m}$$

3. Compute the depth of sludge storage zone

   The sludge storage zone is provided to store the sludge in the clarifier. This may be necessary if the sludge-processing facilities are unable to handle the solids for such reasons as equipment breakdown or unusual sustained waste-loading conditions.

   Provide the sludge storage capacity for two days under sustained peak flow rate and BOD_5 loadings. Assume that the sustained flow rate and sustained BOD_5 factors are 2.5 and 1.5, respectively.[4]

   $$\text{Total volatile solids produced under sustained loadings} = Y_{obs} \, Q(S_o - S) \times (10^3 \text{ g/kg})^{-1}$$

   $$= (0.3125 \times 42,000 \text{ m}^3\text{/d}) (200 - 7.4) \text{ g/m}^3$$
   $$\times 1.5 \text{ (sustained BOD}_5 \text{ factor)} \times 2.5$$
   $$\text{(sustained flow)} \times (10^3 \text{ g/kg})^{-1}$$

   $$= 9480 \text{ kg/d}$$

   If two days storage for solids is provided and TSS = TVSS/0.8,

   $$\text{Total solids stored} = 2 \times 9480 \text{ kg/d}/0.8 = 23,700 \text{ kg}$$

   $$\text{Solids stored in each clarifier} = \frac{23,700 \text{ kg}}{4} = 5925 \text{ kg}$$

   $$\text{Total solids in the clarifier} = 5925 \text{ kg} + 2432 \text{ kg} = 8357 \text{ kg}$$

   $$\text{Clarifier depth for solids storage} = \frac{8357 \text{ kg} \times 1000 \text{ g/kg}}{7000 \text{ g/m}^3 \times 1301 \text{ m}^2}$$

   $$= 0.92 \text{ m} \approx 1.0 \text{ m}$$

4. Compute total depth of clarifier

   $$\text{Total depth of clarifier} = 2.0 \text{ m} + 0.3 \text{ m} + 1.0 \text{ m} = 3.3 \text{ m}$$

   $$\text{Provide average side water depth in the clarifier} = 3.5 \text{ m (11.5 ft)}$$

For additional safety provide a free board of 0.5 m

Total depth of clarifier   = 4.0 m

C. Detention Time

1. Calculate the volume of the clarifier

$$\text{Average volume of the clarifier} = \frac{\pi}{4} \times (40.7)^2 \text{ m} \times 3.5 \text{ m}$$

$$= 4554 \text{ m}^3$$

2. Calculate detention time under different flow conditions

Detention time under average design flow plus recirculation $= \dfrac{4554 \text{ m}^3}{0.193 \text{ m}^3/\text{s} \times 3600 \text{ s/h}}$

$$= 6.6 \text{ h}$$

Detention time at peak design flow plus recirculation $= \dfrac{4554 \text{ m}^3}{0.403 \text{ m}^3/\text{s} \times 3600 \text{ s/h}}$

$$= 3.1 \text{ h}$$

Detention time under emergency condition (peak design flow plus recirculation when one clarifier is out of service) $= \dfrac{4554 \text{ m}^3}{0.538 \text{ m}^3/\text{s} \times 3600 \text{ s/h}}$

$$= 2.4 \text{ h}$$

D. Summary of Clarifier Design Information.   The design information for the secondary clarifier is summarized in Table 13-13. The layout plan of the clarifier is provided in Figure 13-15.

E. Influent Structure.   The influent structure consists of a central feed well. An influent pipe is installed across the clarifier that will discharge into the central feed well. The influent will pass under the baffle and then distribute uniformly throughout the tank. The details of the influent structure are shown in Figure 13-15.

F. Effluent Structure.   The effluent structure consists of effluent baffle, V notches, effluent launder, effluent box, and a pressure outlet pipe. The details of the effluent structure are shown in Figure 13-15. General discussion on weir types as well as configurations of rectangular and circular basins may be found in Chapter 12.

1. Select weir arrangement, and dimensions of effluent launder, effluent box, and outlet sewer
   Provide 90° standard V notches on the weir plate that shall be installed on one side of the effluent launder

Provide width of launder   = 0.5 m
Length of effluent weir plate* = $\pi(40.7 \text{ m} - 1.0 \text{ m})$
                              = 124.7 m

*Diameter of the clarifier is 40.7 m.

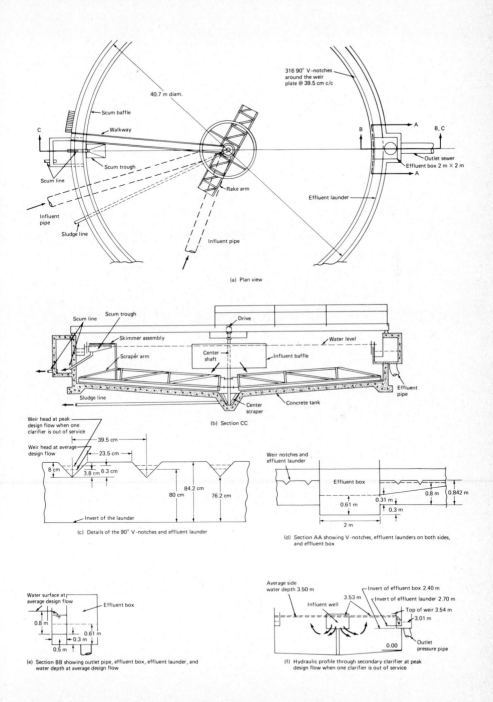

Figure 13-15   Design Details and Layout of Final Clarifier.

TABLE 13-13   Summary Design Information of the Secondary Clarifier

Item no.	Design Conditions	Design Value[a]	
		SI Units	U.S. Customary Units
1	Number of clarifiers	4	4
2	Surface area, each clarifier	1301 m^2	14,004 ft^2
3	Diameter	40.7 m	134 ft
4	Water depth	3.5 m	11.5 ft
5	Freeboard	0.5 m	1.6 ft
6	Average design flow plus recirculation	0.193 m^3/s	4.4 mgd
7	Peak design flow plus recirculation	0.403 m^3/s	9.2 mgd
8	Emergency flow, peak design flow plus recirculation	0.538 m^3/s	12.3 mgd
9	Detention period		
	At average design flow	6.6 h	6.6 h
	At peak design flow	3.1 h	3.1 h
	At emergency flow	2.4 h	2.4 h
10	Overflow rate		
	At average design flow	12.8 m^3/m^2·d	314 gpd/ft^2
	At peak design flow	26.8 m^3/m^2·d	658 gpd/ft^2
	At emergency flow	35.7 m^3/m^2·d	876 gpd/ft^2
11	Limiting solids flux (solids-loading rate)		
	At average design flow	48.1 kg/m^2·d	9.9 lb/ft^2·d
	At peak design flow	100.4 kg/m^2·d	20.61 lb/ft^2·d
	At emergency flow	134.0 kg/m^2·d	27.4 lb/ft^2·d

[a]1 m^3/m^2·d = 24.5424 gpd/ft^2. 1 kg/m^2·d = 0.2048 lb/ft^2·d.

Provide 8 cm deep 90° V notches at 39.5 cm center-to-center

$$\text{Total number of notches} = \frac{124.7 \text{ m}}{39.5 \text{ cm/notch} \times (100 \text{ cm/m})^{-1}}$$

$$= 316$$

2.   Compute head over V notch at average design flow

$$\begin{array}{l} \text{Average design flow} \\ \text{from the clarifier}^\dagger \end{array} = \begin{array}{l} \text{Average design flow to aeration basin} \\ - \text{ MLSS wasted} \end{array}$$

$$= 0.486 \text{ m}^3/\text{s} - [622 \text{ m}^3/\text{d} \times (86,400 \text{ s/d})^{-1}]$$
$$= 0.479 \text{ m}^3/\text{s}$$

†Effluent over weir will not include recirculation, as the sludge is returned from the bottom of the clarifier.

Flow per clarifier $= \dfrac{0.479 \text{ m}^3/\text{s}}{4} = 0.12 \text{ m}^3/\text{s per basin}$

Flow per notch at avg. design flow $= \dfrac{0.12 \text{ m}^3/\text{s}}{316 \text{ notch}} = 0.00038 \text{ m}^3/\text{s/notch}$

The head over V notch is calculated from Eq. (12-1):

Head over V-notch $H = \left[\dfrac{15}{8}\left(\dfrac{0.00038 \text{ m}^3/\text{s}}{0.584 \sqrt{2 \times 9.81} \text{ m/s}^2 \times \tan 45}\right)\right]^{2/5}$

$= 0.038 \text{ m} = 3.8 \text{ cm}$

3. Compute head over V notch at peak design flow when one unit is out of service

Flow per notch at peak design flow when one unit is out of service* $= \dfrac{1.321 \text{ m}^3/\text{s}}{3 \text{ clarifiers} \times 316 \text{ notch}}$

$= 0.00139 \text{ m}^3/\text{s per notch}$

Head over V notch $H = \left[\dfrac{0.00139 \text{ m}^3/\text{s}}{8/15 \times 0.584 \sqrt{2 \times 9.81} \text{ m/s}^2 \times \tan 45}\right]^{2/5}$

$= 0.063 \text{ m} = 6.3 \text{ cm}$

4. Compute actual weir loading

Weir loading at average design flow $= \dfrac{0.12 \text{ m}^3/\text{s} \times 86,400 \text{ s/d}}{124.7 \text{ m}}$

$= 83.1 \text{ m}^3/\text{m} \cdot \text{d}$ (less than the design loading of 124 $\text{m}^3/\text{m} \cdot \text{d}$)

Weir loading at peak design flow when one clarifier is out of service $= \dfrac{1.321 \text{ m}^3/\text{s} \times 86,400 \text{ s/d}}{3 \times 124.7 \text{ m}}$

$= 305 \text{ m}^3/\text{m} \cdot \text{d}$

5. Compute the depth of the effluent launder

Width of effluent launder $= 0.5 \text{ m}$

Provide effluent box 2 m × 2 m
Provide 0.8-m-diameter outlet pressure pipe. The pipe is an inverted syphon connected to a common junction box. The water surface elevation in the junction box is kept such that the depth of flow in the effluent box at peak design flow is maintained at 0.61 m. Provide invert of the effluent launder 0.3 m above the invert of the effluent box. The depth of the effluent launder is calculated from Eq. (11-9). (Various terms in this equation have been defined in Chapter 11):

*Does not include return flow, and the volume of waste sludge is ignored.

y_2 = depth of water in the effluent box − invert height of effluent launder above the invert of effluent box

= 0.61 m − 0.30 m = 0.31 m

b = 0.5 m

N = 1

Half of the flow divides on each side of the launder, therefore flow on each side of the launder* $= \dfrac{1.321 \text{ m}^3/\text{s}}{2 \times 3 \text{ clarifiers in operation}}$

= 0.22 m³/s

Average half length of the launder, L $= \dfrac{1}{2} [\pi(40.7 \text{ m} - 0.50 \text{ m}) - 2 \text{ m}]$

= 62.15 m

Flow per m length of the weir, q' $= \dfrac{0.22 \text{ m}^3/\text{s}}{62.4 \text{ m}}$

= 0.00353 m³/m · s

$$y_1 = \sqrt{(0.31 \text{ m})^2 + \frac{2(0.00353 \text{ m}^3/\text{m} \cdot \text{s} \times 62.15 \text{ m} \times 1)^2}{9.81 \text{ m/s}^2 \times (0.5 \text{ m})^2 \times 0.31 \text{ m}}} = 0.47 \text{ m}$$

Provide 16 percent losses for friction, turbulence, and bends, and provide 25 cm additional depth to ensure free fall

Total depth of the effluent launder = (0.47 m × 1.16) + 0.25 m
= 0.80 m

The water surface elevation in the clarifier at average design flow is kept 0.80 m above the invert of the effluent launder. See the details of effluent structure in Figure 13-15.

G. Sludge Collection System and Skimmer.   The sludge collection system shall consist of a rotating rake structure with scraper blades that will scrape the settled sludge from the tank bottom to a sludge pocket located near the center of the basin. The fixed access bridge shall house the drive machinery and shall be supported by a column at the center of the tank. The skimmer shall remove the scum and deposit it into the scum trough.

H. Return Sludge Pumps.   Four return sludge pumps each having a rated pumping capacity of 0.182† m³/s (150 percent of design average flow per basin) shall be provided.

---

*Recirculation does not reach the effluent launder.
†Design flow to the aeration basin = 0.486 m³/s (see Sec. 13-7-4, step A1).
Rated pumping capacity of each pump $= \dfrac{0.486 \text{ m}^3/\text{s}}{4} \times 1.5 = 0.182 \text{ m}^3/\text{s}$.

Each pump shall be variable speed and provide independent operation of one clarifier. An identical pump (fifth pump) shall serve as a standby unit and cross-connected to serve all four clarifiers. A magnetic flow meter shall be provided in each force main to control the flow of the return sludge from each clarifier. A sonic sludge blanket meter shall be provided in each clarifier to control the pump speed.

The design and selection of pumping equipment has been presented in Chapter 9. The flow-measuring systems are discussed in Chapter 10. The design engineers should consult these chapters to develop the specific design of return sludge and flow measurement devices mentioned above. The design is similar to that of a dry-well centrifugal pump with flooded suction pipe.

I. Hydraulic Profile. The hydraulic profile for the secondary clarifier is shown in Figure 13-15. The profile is prepared for the peak design flow plus the recirculation when one unit is out of service. The head loss in the influent structure is small as the baffle at the central well offers little obstruction to the flow. The momentum equation [Eq. (11-5)] should be used to calculate the head loss due to the baffle (the calculation procedure is presented in Chapter 11). The head loss calculations for the connecting pipings between the division box and the influent well are given in Chapter 21.

13-8 OPERATION AND MAINTENANCE, AND TROUBLESHOOTING AT ACTIVATED SLUDGE TREATMENT FACILITY

Many articles, reports, manuals, and books have been written on the control of activated sludge systems. Proper control of these facilities include a balance between air supply, mixing, influent quality, return sludge, waste sludge, and sludge blanket maintained in the secondary clarifier. Common operation and maintenance problems that may develop at the activated sludge treatment facility, and procedures to correct them are summarized below.

13-8-1 Starting a New Plant

Before starting the plant, check to see that all mechanical equipment is in good working order, properly lubricated, filters clean, and all pipings free of debris. The following procedure should be followed:

1. Start treatment with a portion of incoming wastewater (one-third to one-fourth).
2. Use only few units of the plant needed to handle the flow (one to two aeration basins and clarifiers).
3. Provide sufficient air to maintain a dissolved oxygen in the mixed liquor between 2–4 mg/ℓ.
4. Continue returning the entire sludge until MLSS concentration reaches 400–800 mg/ℓ.
5. Slowly increase the influent and bring other units in operation.
6. The normal plant operation shall be reached in 2–4 weeks. In cold climates the sludge growth may be slower than that in the warmer climates.

13-8-2 Common Operating Problems and Suggested Solutions

There are many common problems that may occur in the aeration basin and the clarifier. Important problems and corrective procedures are listed below.

1. Sludge floating to the surface of the clarifiers (bulking of sludge) may be due to growth of filamentous organisms. Generally, SVI is greater than 150, and microscopic examination of MLSS shows the presence of filamentous organisms. The following control measures may be taken: Increase DO in the aeration basin if it is less than 1 mg/ℓ; increase pH to 7; add nutrients (nitrogen, phosphorus, iron, etc.); add 5–6 mg/ℓ of chlorine to the return sludge until SVI is less than 150; add 5–6 mg/ℓ hydrogen peroxide in the aeration basin until SVI is less than 120; or decrease food to MO ratio. This is achieved by (1) increasing the sludge age, (2) reducing sludge wasting, or (3) decreasing the food supply.

    Often denitrification occurring in the secondary clarifier may be the cause of nitrogen bubbles attaching to sludge particles and sludge rising in clumps. The problem may be overcome by increasing the sludge return rate and DO in the aeration basin, and reducing the sludge age.

2. Turbid effluent (pin floc in effluent) but good SVI may be due to excessive turbulence in the aeration basin or overoxidized sludge. The problem is easily overcome by reduced aeration or agitation, increased sludge wasting, or decreased sludge age. Turbid effluent may also be the result of anaerobic conditions in the aeration basin. The problem is easily solved by increasing aeration. Often toxic shock loading may cause turbid effluent. Microscopic examination of MLSS will certainly show inactive protozoa. Reseeding from another plant may be necessary. Enforcement of industrial waste pretreatment must be done.

3. Very stable, dark tan foam on aeration tanks (which may not be broken by the sprays) may result from too long sludge age. Increase the sludge wasting.

4. Thick billows of white sudsy foam on aeration basin may be due to low MLSS. A reduction in sludge wasting may help the problem.

5. Different concentrations of MLSS in different aeration basins may be the result of unequal flow distribution to each basin. Valves and gates should be adjusted to equally distribute the influent and return sludge.

6. Sludge blanket uniformly overflowing the weir may be due to excessively high solids loading, peak flows overloading the clarifiers, unequal flow distribution on clarifiers, excessively high MLSS, and inadequate return sludge. Check the solids loading and overflow rates; distribute flows equally in all clarifiers; check return sludge pumps and pipings for malfunctioning or blockage; increase rate of return sludge to maintain at least 1 m clearwater zone in the clarifier; or increase sludge wasting.

7. Sludge blanket discharging over the weir in one portion of the clarifier may be the result of unequal flow distribution. Level the effluent weirs.

8. In diffused aeration systems air rising in large bubbles or clumps in some areas is an indication of faulty diffusers. Clean or replace diffusers, check air supply, and install filter ahead of blowers.

9. Low pH of MLSS (pH below 6.7) is an indication of nitrification or acid wastes reaching the plant. Check the ammonia and nitrate concentrations in effluent as well as the pH of the influent. Proper control measures include decreasing the sludge age by wasting, lime addition, or proper control of influent.
10. Low solids concentration in the return sludge (less than 8000 mg/ℓ) may be due to filamentous growth, high return sludge rate, or excessive sludge wasting. Proper control measures include reducing sludge return rate, reducing sludge wasting, and raising DO. If microscopic examination indicates filamentous growth, follow the procedure given above.
11. Dead spots in the aeration basin may be due to plugged diffusers or under aeration. Check air supply and clean or replace diffusers.

13-8-3 Routine Operation and Maintenance

The following procedures are necessary for routine operation and maintenance of aeration basins and secondary clarifiers:

Aeration Basins

1. Inspect distribution box daily and clean weirs, gates, to remove solids.
2. Remove accumulations of debris from inlet channels and gates, and outlet weir each day.
3. Keep daily record of DO in the aeration basin, MLSS concentration, SVI, and sludge age. If any unusually high or low values are found, take corrective measures.
4. Clean all vertical walls and channels by squeegee on daily basis.
5. Hose down and remove wastewater spills without delay.
6. Inspect gratings and exposed metal during daily cleanup for signs of corrosion and deterioration of paint.
7. Prepare lubrication chart for mechanical equipment according to manufacturers' recommendations.
8. Drain each aeration basin annually to inspect underwater portions of the concrete structure, pipings, etc. Replace or repair all defective parts. Patch defective concrete and repaint all clean metal surfaces as required.

Secondary Clarifier

1. Remove accumulations from the influent baffles, effluent weirs, scum baffles, and scum box each day.
2. Observe sludge return from individual clarifier, and adjust the flow rate as required from laboratory tests.
3. Determine sludge level and adjust waste sludge pump as necessary.
4. Observe operation of scum pump and provide hosing as necessary.
5. Clean daily all inside exposed vertical walls and channels by a squeegee.
6. Inspect distribution box and clean weirs, gates, and walls as necessary and remove all settled solids. Also check flow to all clarifiers.
7. Inspect effluent box, and clean weir and walls as necessary. Measure the head over the weir daily.

8. Hose down and remove wastewater sludge and spills without delay.
9. Check electrical motors for overall operation, bearing temperature, and overload detector twice each day.
10. Check oil level, grease reducer, and rollers on skimmer each week.
11. Check oil level for turntable bearings each week and refill as required.
12. Grease the main bearings each week.
13. Change the oil in the gear reducer quarterly.
14. Drain each clarifier annually to inspect underwater portion of concrete structure and mechanism. Inspect mechanical equipment for wear and corrosion, and apply protective coating. Inspect concrete structure and patch defective areas.
15. Inspect sludge collection and other equipment annually for indication of corrosion. Clean and paint all metal works as necessary.

13-9 SPECIFICATIONS

The general specifications of pumping equipment, flow-measuring devices, and sedimentation facility have already been presented in Chapters 9, 10, and 12. Therefore, specifications for return sludge, waste sludge and scum pumps, flow meters, and secondary clarifiers are not given in this section. The design engineer should consult Chapters 9, 10, and 12 to develop the specifications for these facilities.

The general specifications for the aeration equipment are presented in this section. These specifications should be used only as a guide. Detailed specifications for each unit or components shall be prepared in consultation with the equipment manufacturers.

13-9-1 General

This item governs the furnishing and installation of equipment required in the aeration basin. The aeration equipment shall be composed of Dacron sock diffuser. Each diffuser assembly shall be arranged for raising the entire assembly to an accessible position for servicing without dewatering the tank and shall be provided with positive means for both regulation and complete shutoff of the air supply.

The equipment manufacturer shall be experienced in the design, fabrication, installation, and operation of aeration equipment, and shall submit a list of installations of this type in operation over a reasonable period.

13-9-2 Aeration System

Diffusers.   Each Dacron sock diffuser shall consist of a tubular flexible Dacron cloth sheath having an internal dimension of 7.5 cm diameter, closed on one end and open at the other to fit over a formed rod, and clamped by a stainless steel clamp to an adapter fitting. Formed rod shall be approximately 61 cm long, 0.63 cm mild carbon steel attached to the adapter and the whole assembly dipped in vinyl. The differential pressure across the diffuser media when the tube is discharging air at the rate of 0.2 m^3/min with 200 mm of water over the center line of the tube shall not be greater than 300 mm.

Diffuser Tube Headers. Each diffuser tube header shall comprise two lengths of galvanized fabricated steel pipe, flange connected to a T fitting which is screw-connected to the hanger pipe. Each diffuser tube header shall have bosses drilled and tapped to accommodate the diffuser tube assemblies as detailed on the drawings.

Hanger Feed Pipes. Each hanger pipe shall comprise an upper section and a lower section of galvanized iron pipe jointed by a knee joint. The knee joint shall be equipped with wearing rings to provide an air-tight connection. The hanger pipe shall serve as a support and air feed pipe to the diffuser tube header.

Swing Joint Assembly. Each swing joint assembly shall comprise a galvanized steel feeder manifold and one duplex elbow fitting swing joint with built-in valve. One feeder manifold shall serve one or two air diffusion units as detailed on the plans. The feeder manifold shall be attached to the air main by means of a flanged connection. The movable member of the swing joint shall be connected to the hanger pipe by a cast iron screwed connection. Elbow fittings, anchor, interlock, wearing rings, etc., shall be provided for stationary and swing joints and airtight connections.

Portable Hoist. One portable hoist for raising the air diffusion units from the aeration basin shall be provided.

Blowers Assembly. The contractor shall furnish and install five centrifugal blowers with motors and accessories as indicated in plans and as covered herein.

Blower. Each blower will be of the multistage centrifugal type capable of compressing 155 m^3/min of air to a discharge pressure of 57 kPa* (8.25 lb/in.^{2} or 0.56 atm) when operating at an elevation of 500 m, and an air temperature of 30°C. When volumetric capacity is reduced by at least 30 percent, each blower under specified inlet conditions shall develop at least 3.5 kN/m^2 (0.5 lb/in.^{2}) above specified discharge pressure and shall not be in surge. The manufacturer shall submit a certified test report attesting to the date and place of testing and accuracy achieved.

Motor. Five electric motors for the driving blower shall be each horizontal squirrel cage induction type 187 kW, 460 V, three phase, 60Hz, 1.15 service factor, and shall be designed for full-voltage starting. Motor shall operate at a speed not to exceed 1720 rpm. Motor shall be subjected to standard commercial tests and certified copies of test data together with a certified statement of compliance with minimum specified efficiency shall be furnished.

Each blower and motor unit shall be mounted on a single heavy-strength steel or cast iron frame, properly cross-braced to form a rigid support for the entire unit.

Air Filter. The contractor shall furnish and install an air filter of the two-stage, high-efficiency type for an air flow of 620 m^3/min. The filter shall consist of an approved

*kPa = kN/m^2.

renewable media filter. The overall efficiency of the filter shall average not less than 97 percent on atmospheric dust. The initial resistance to rated air flow shall not exceed 50 mm water.

Accessories.   The following accessories shall be provided for each blower: a vibration-sensing device; alarm circuits, and control panel, check valve; butterfly valves; manometer (calibrated to read 0–70 kN/m^2, and suitable for wall mounting), thermobarometer; and intake silencer of sufficient capacity to handle the air requirements without unreasonable pressure drop.

13-10 PROBLEMS AND DISCUSSION TOPICS

13-1.   Using the results of the first iteration of material mass balance analysis given in Table 13-6, perform second iteration. Compare your results with those given in Table 13-7.

13-2.   Using the results of the second iteration of the material mass balance analysis, perform the third iteration. Check your results with those given in Figure 13-12.

13-3.   Four completely mixed bench scale reactors were operated. The reactors were identical in design and operation. The only difference was in the mixed liquor suspended solid (MLSS) concentration that was maintained in each reactor. When the reactors reached a steady-state condition the data collection began. Each day the following measurements were made:

(a)   total volume of the effluent
(b)   TVSS in the effluent
(c)   soluble COD and BOD$_5$ of the feed
(d)   soluble COD and BOD$_5$ of the effluent
(e)   MLSS and MLVSS in the reactor
(f)   certain volume of MLSS from each reactor was wasted to give the desired MLSS concentration that was maintained into each reactor. Equal volume of distilled water was added in each reactor.

From the results, ratios of MLVSS/MLSS, BOD$_5$ influent/COD influent, and BOD$_5$ effluent/COD effluent were calculated. The seven-day average experimental results at each MLSS concentration are summarized in Table 13-14. Volume of each reactor including the portion partitioned for settling is 10.9 l. Determine the biological kinetic coefficient, k, Y, k_d, and K_s based on MLVSS and BOD$_5$ results.

13-4.   A completely mixed aerated lagoon is treating 4500 m^3/d municipal wastewater. The aeration period is 4 d. Estimate the soluble BOD$_5$, and total BOD$_5$ by considering the biological solids. Use the following data:

$$\text{Total BOD}_5 \text{ influent} = 200 \text{ mg/}\ell$$
$$\text{Soluble BOD}_5 \text{ influent} = 150 \text{ mg/}\ell$$
$$\text{TSS influent} = 240 \text{ mg/}\ell$$
$$k = 4.0 \text{ d}^{-1}$$
$$K_s = 80 \text{ mg/}\ell$$
$$Y = 0.45$$
$$k_d = 0.05 \text{ d}^{-1}$$

$$S = \frac{K_s (1 + \theta k_d)}{\theta (Yk - k_d) - 1} \qquad (13\text{--}28)$$

$$X = \frac{Y (S_o - S)}{(1 + k_d\theta)} \qquad (13\text{--}29)$$

TABLE 13-14 Results of Experimental Data for Example 13-3

Reactor	MLSS Maintained, mg/ℓ	MLVSS/ MLSS %	Influent COD, mg/ℓ	Influent BOD$_5$/COD, %	Effluent COD, mg/ℓ	Effluent BOD$_5$/COD, %	Average Effluent Volume Each Day, ℓ	Total Average MLVSS at the End of Each Day,[a] mg/ℓ
1	2250	73.0	138.4	70.1	25.0	0.20	37.0	1692.3
2	1850	73.0	149.7	70.8	25.0	0.27	37.0	1422.5
3	1350	72.5	153.3	71.1	32.3	0.28	36.4	1067.2
4	1100	77.5	161.6	71.8	31.7	0.35	35.6	951.0

[a]Includes VSS lost in the effluent

13-5. Determine the dimensions of an aerated lagoon to treat 3000 m³/d having soluble BOD_5 of 150 mg/ℓ. Also calculate the power requirements of the aerators under field condition (average temperature 25°C, and pressure 1 atm). Effluent soluble BOD_5 not to exceed 15 mg/ℓ. Use Eqs. (13-28), and (13-29), and the following data and the following equations:

$$k = 5.0 \text{ d}^{-1}$$
$$K_s = 100 \text{ mg/}\ell$$
$$Y = 0.6$$
$$k_d = 0.05 \text{ d}^{-1}$$

Theoretical oxygen required $= 1.47 \times BOD_5$ removed

$$N = N_o \left[\frac{(C'_{sw}\beta \, F_a - C)}{C_{sw}} \right] 1.024^{(T-20)} \, \alpha$$

where

N_o = kgO₂/kW · h transferred in water at 20°C, and at zero dissolved oxygen
N = kgO₂/kW · h transferred under field condition

Other constants are defined in Eq. (13-18). (Use $C = 2.5$ mg/ℓ, $\alpha = 0.9$, $\beta = 0.9$, and N_o from the manufacturer's data for selected aerator is 1.2 kgO₂/kW · h.

13-6. MLSS concentration in an aeration basin is 3000 mg/ℓ. After settling it in a 1-ℓ graduated cylinder, the settled volume is 300 mℓ. Calculate SVI.

13-7. 2000 m³/d flow from a dairy is treated in an activated sludge plant. The BOD_5 is 300 mg/ℓ, and most of it is in soluble form. Calculate the volume of aeration basin. Also calculate aeration period, return sludge, waste sludge, food to MO ratio, and oxygen requirement. Use the following data:

$$\theta_c = 10 \text{ d}$$
$$Y = 0.55$$
$$k = 4.0 \text{ d}^{-1}$$
$$K_s = 95 \text{ mg/}\ell$$
$$k_d = 0.04/\text{d}$$
$$BOD_5 = 0.7 \, BOD_L$$
$$\text{Effluent } BOD_5 = 30 \text{ mg/}\ell$$
$$\text{TSS in effluent} = 30 \text{ mg/}\ell$$
$$\text{Biological solids are} \quad 72 \text{ percent biodegradable}$$
$$1 \text{ gram of biodegradable solids} = 1.42 \text{ g ultimate BOD}$$
$$\text{MLVSS maintained} = 2800 \text{ mg/}\ell$$
$$\text{MLVSS/MLSS} = 0.68$$
$$\text{Return sludge has} \quad 12,000 \text{ mg/}\ell \text{ TSS}$$

13-8. An aerobic stabilization pond is designed to treat 500 m³/d wastewater from a community. The BOD_5 loading is 60 kg/ha·d. Calculate the area, volume, and detention time. Average depth of the stabilization basin is 1 m. Also calculate TSS and total BOD_5 of the effluent. Assume 80% TSS in effluent is due to algae. Use the following data:

$$\text{Average algae concentration} = 100 \text{ mg/}\ell$$
$$\text{Biodegradable portion of algae cell} = 65 \text{ percent}$$
$$\text{Soluble } BOD_5 \text{ in effluent} = 10 \text{ mg/}\ell$$

$$BOD_5/BOD_L = 0.68$$
Influent BOD_5 = 250 mg/ℓ
1 g of biodegradable solids = 1.42 g BOD_L

13-9. A trickling filter is designed to treat 500 m³/d primary settled wastewater having a BOD_5 of 150 mg/ℓ. The BOD_5 loading is 0.5 kg/m³·d. Calculate the area and depth of the trickling filter, and the recirculation ratio. Use the following data:

Total BOD_5 effluent = 20 mg/ℓ
TSS, effluent = 20 mg/ℓ
TSS in effluent is 60 percent biodegradable, and each gram of biodegradable solids = 1.42 g BOD_L

BOD_5/BOD_L = 0.68
L_D/L_o = 10^{-KD}

where

L_D = soluble BOD_5 at depth D in meters, mg/ℓ
L_o = applied soluble BOD_5 at the filter after dilution, mg/ℓ
K = rate of BOD_5 removal. Use K = 0.49/m in this problem
Filter efficiency = 68 percent

13-10. A diffused aeration basin requires 1200 kg of O_2 per day. The oxygen transfer efficiency of the diffusers is 8 percent and water depth over diffusers is 4.5 m. Determine the power requirement of the compressor. Assume inlet temperature and pressure are 30°C and 1 atm. The efficiency of the compressor is 75 percent. Head loss in piping, air filter, silencer, diffuser losses and allowance for diffuser clogging, etc., add up to 1.7 m of water.

13-11. The dimensions of the final clarifier in the Design Example were developed using solids concentrations and settling rate data given in Figure 13-10 and Table 13-11. The return sludge rate of 0.292 m³/s was obtained from the design of the aeration basin. Calculate from the clarifier data the underflow concentration at this return sludge rate. (Use Fig. 13-14 to calculate the limiting solids flux at different underflow concentrations.)

13-12. The hydraulic profile through the aeration basin in the Design Example was developed at peak design flow when one basin was out of service. Develop the hydraulic profile at average design flow plus recirculation when all four basins are in operation.

13-13. The hydraulic profile through the final clarifier in the Design Example was developed at peak design flow when one unit was out of service. Develop the hydraulic profile at average design flow plus recirculation when all four clarifiers are in operation.

13-14. Calculate the maximum depth of flow in the effluent launder of the final clarifier using Eqs. (11-4)–(11-8). Use design peak flow condition when one clarifier is out of service.

13-15. Describe various modifications of activated sludge process. List the advantages and disadvantages of each process.

REFERENCES

1. Lehninger, A. L., *Principles of Biochemistry*. Worth, New York, 1982.
2. Gaudy, A. F., and E. T. Gaudy, *Microbiology for Environmental Scientists and Engineers*, McGraw-Hill Book Co., New York, 1980.
3. Mazur, Abraham, and B. Harrow, *Textbook of Biochemistry*, 10th Edition, W. B. Saunders, Philadelphia, 1971.

4. Metcalf and Eddy, Inc., *Wastewater Engineering: Treatment, Disposal and Reuse,* McGraw-Hill Book Co., New York, 1979.
5. Benefield, L. D., and C. W. Randall, *Biological Process Design for Wastewater Treatment,* Prentice-Hall, Englewood Cliffs, N.J. 1980.
6. Sundstrom, D. L., and H. E. Klei, *Wastewater Treatment,* Prentice-Hall, Englewood Cliffs, N.J., 1979.
7. Schroeder, E. D., *Water and Wastewater Treatment,* McGraw-Hill Book Co., New York, 1977.
8. Reynolds, T. D., *Unit Operations and Processes in Environmental Engineering,* Brooks/Cole, Engineering Division, Monterey, Cal., 1982.
9. Qasim, S. R., and M. L. Stinehelfer, "Effect of a Bacterial Culture Product on Biological Kinetics," *Journal Water Pollution Control Federation,* Vol. 54, No. 3, March 1982.
10. Palit, T., and S. R. Qasim, "Biological Treatment Kinetics of Landfill Leachate," *Journal of the Environmental Engineering Division,* American Society of Civil Engineers, Vol. 103, No. EE2, April 1977.
11. Culp, G. L., and N. F. Heim, *Performance Evaluation and Troubleshooting at Municipal Wastewater Treatment Facilities,* U.S. Environmental Protection Agency, NO-16 EPA-430/9-78-001, Washington, D.C., January 1978.
12. Joint Committee of the WPCF and ASCE, *Wastewater Treatment Plant Design,* MOP/8, Water Pollution Control Federation, Washington, D.C., 1977.
13. Gibbon, D. L. (Editor), *Aeration of Activated Sludge in Sewage Treatment,* Pergamon Press, Fairview Park, N.Y., 1974.
14. U.S. Environmental Protection Agency, *Process Design Manual for Upgrading Existing Wastewater Treatment Plants,* Technology Transfer, EPA 625/1-71-004a, Washington, D.C., October 1974.
15. Boulier, G. A., and T. J. Atchison, *Practical Design and Application of the Aerated-Facultative Lagoon Process,* Hinde Engineering Company, Highland Park, Ill., April 1975.
16. Middlebrooks, E. J., et al., *Design Manual, Municipal Wastewater Stabilization Ponds,* U.S. Environmental Protection Agency, Technology Transfer, EPA-625/1-83-015, October 1983.
17. Fair, G. M., J. C. Geyer, and D. A. Okun, *Water and Wastewater Engineering, Volume 2, Water Purification and Wastewater Treatment,* Wiley & Sons, New York, 1968.
18. Clark, J. W., W. Viessman, and M. J. Hammer, *Water Supply and Pollution Control,* 3rd Edition, IEP-A Dun-Donnelley, New York, 1977.
19. National Research Council, "Sewage Treatment at Military Installations," *Sewage Works Journal,* Vol. 18, No. 5, 1946, pp. 787–1,028.
20. Velz, C. J., "A Basic Law for the Performance of Biological Filters," *Sewage Works Journal,* Vol. 20, No. 4, 1948, p. 607.
21. Eckenfelder, W. W., "Trickling Filter Design and Performance," *Transactions American Society of Civil Engineers,* Vol. 128, Part III, 1963, p. 37.
22. Galler, W. S., and H. B. Gotaas, "Analysis of Biological Filter Variables," *Journal of the Sanitary Engineering Division,* ASCE, Vol. 90, No. 6, 1964, pp. 59–79.
23. Schulze, K. L., "Load and Efficiency of Trickling Filters," *Journal Water Pollution Control Federation,* Vol. 32, No. 3, 1960, pp. 245–261.
24. Metcalf and Eddy, Inc., *Wastewater Engineering: Collection, Treatment and Disposal,* McGraw-Hill Book Co., New York, 1972.
25. Steel, E. W., and T. J. McGhee, *Water Supply and Sewerage,* Fifth Edition, McGraw-Hill Book Co., New York, 1979.
26. Joint Committee of the ASCE and WPCF, *Sewage Treatment Plant Design,* American Society of Civil Engineers and Water Pollution Control Federation MOP8, 1961.

14

Disinfection

14-1 INTRODUCTION

Wastewater contains many types of human enteric organisms that are associated with various waterborne diseases. Typhoid, cholera, paratyphoid, and bacillary dysentery are caused by bacteria, and amebic dysentery is caused by protozoa. Common viral diseases include poliomyelitis and infectious hepatitis. Disinfection refers to selective destruction of disease-causing organisms in water supply or wastewater effluent.* Chlorination of the water supply has been practiced since about 1850. Presently, chlorination of both water supply and wastewater effluent is an extremely widespread practice for the control of waterborne diseases. However, chlorination may result in the formation of chlorinated hydrocarbons, some of which are known to be carcinogenic. Therefore, alternate methods of disinfection are receiving a great deal of attention.

In this chapter various alternative methods of disinfection, the advantages and disadvantages of chlorination, chlorine chemistry, and the design details of a chlorination facility are presented.

14-2 METHODS OF DISINFECTION

Methods of disinfection broadly fall into three major categories: (1) physical, (2) radiation, and (3) chemical.

14-2-1 Physical Methods

Physical methods of disinfection include heat (pasteurization) and ultraviolet radiation. Pasteurization of milk, dairy products, and beverages has been used for decades. Pasteurization of sludge is also used. However, pasteurization is not applicable to water and wastewater because of the high energy cost in heating large volumes of liquid. Ultraviolet light is an excellent disinfectant and does not leave any known residual chemicals. Special mercury lamps have been developed that emit ultraviolet rays over thin films of water. The efficiency of ultraviolet radiation depends on the (1) depth of penetration, (2) time of

---

*The term *sterilization* is applied to the complete destruction of all organisms. Pasteurization is selective destruction of undesired organisms by heat.

contact, and (3) turbidity or suspended solids that may reduce the effective depth of penetration. Sunlight has also been effectively used for disinfection.[1]

14-2-2 Gamma Radiation

Gamma rays are emitted from a radioactive source, such as cobalt 60, and are effectively used to disinfect or sterilize water, wastewater, and sludge. These rays have high penetrating power. The method is reliable and does not leave residual chemicals in the effluent. In sludges the gamma rays alter the makeup of sludge to benefit dewaterability and stability. The disadvantages of the system include the radiation emitted as a threat to safety.

14-2-3 Chemical Agents

Many types of chemical disinfectants are used for different applications. These include (1) oxidizing agents, such as halogens (chlorine, bromine, and iodine), ozone, hydrogen peroxides, potassium permanganate, (2) alcohols, (3) phenol and phenolic compounds, (4) heavy metals, (5) quarternary ammonium compounds, (6) soaps and synthetic detergents, and (7) alkalies and acids. Among all these chemicals, oxidizing agents are commonly used for disinfection of water and wastewater. Chlorine and its compounds are most universally used for disinfection of water supply and wastewater effluents. Bromine and iodine occasionally are used for swimming pool water but have not been used for treated wastewater.[2] Ozone is a highly effective disinfectant and is gaining popularity in spite of the fact that it does not build any residual and must be generated at the site. However, it does not produce any undesirable reaction products and is very effective in destroying odors and color-causing compounds in water. Hydrogen peroxide is used for odor control, and to inhibit the growth of microorganisms in the collection system. It is a strong oxidant but a poor disinfectant. It neither builds residual nor produces undesirable reaction products in water. Potassium permanganate ($KMnO_4$) is also a powerful oxidizing agent; however, its use in effluent treatment is very limited.

14-3 DISINFECTION WITH CHLORINE

Chlorination has received wide application in water and wastewater treatment. It is used for many applications including disinfection, taste and odor control, and color removal. Chlorine is also used for oxidation of ammonia, iron, manganese, and sulfide as well as for BOD removal. Chlorine is cheap, effective, available in large quantities, nontoxic in low concentrations to higher forms of life, and builds up a residual. The basic disadvantages of chlorine include acid generation (HCl), buildup of total dissolved salts, and formation of potentially carcinogenic, halogenated organic compounds.

14-3-1 Chlorine Compounds

The most common chlorine compounds used in water and wastewater treatment plants are chlorine gas (Cl_2), calcium hypochlorite [$Ca(OCl)_2$], sodium hypochlorite (NaOCl), and chlorine dioxide (ClO_2). Important properties and common applications of chlorine compounds and of ozone and hydrogen peroxide are compared in Table 14-1. Selection of any of these compounds depends on the size of the treatment facility, objectives, economics, and safety considerations.

14-3-2 Chlorine Chemistry

Free Chlorine Residual.   A number of reactions can occur when chlorine is added to the wastewater. The rate and efficiencies are dependent on temperature, pH, buffering capacity, and the form in which the chlorine is supplied. Chlorine in aqueous solution produces hypochlorus acid and hypochlorite ion [Eqs. (14-1) and (14-2)]:

$$Cl_2 + H_2O \rightleftharpoons \begin{matrix} HOCl + H^+ + Cl^- \\ \text{Hypochlorous acid} \end{matrix} \tag{14-1}$$

$$HOCl \rightleftharpoons \begin{matrix} H^+ + OCl^- \\ \text{Hypochlorite ion} \end{matrix} \tag{14-2}$$

The quantity of HOCl and OCl^- that is present in water is called free residual chlorine. The relative distribution of HOCl and OCl^- is important in chlorination as disinfecting power of HOCl is about 40–80 times greater than that of OCl^-. This is why wastewaters with lower pH are easier to disinfect by chlorination. The relative distribution of HOCl and OCl^- varies with temperature and pH. At 20°C, the relative distributions with temperature are reported in Table 14-2.

Free chlorine is also added in water from hypochlorite salts:

$$Ca(OCl)_2 + 2H_2O \rightleftharpoons 2HOCl + Ca(OH)_2 \tag{14-3}$$
$$NaOCl + H_2O \rightleftharpoons HOCl + NaOH \tag{14-4}$$

As seen from the chemical Eqs. (14-1)–(14-4), chlorine gas lowers the pH while hypochlorite in solution raises the pH. High pH favors the formation of OCl^- which is much less effective than HOCl. Therefore, for equal amount of chlorine added to poorly buffered effluents, a higher disinfection is achieved with chlorine gas than hypochlorite solutions. Hypochlorite solutions will also add additional dissolved solids to the effluent.

Combined Chlorine Residual.   Chlorine reacts readily with ammonia to form three types of chloramines: mono-, di-, and trichloramines:

$$NH_3 + HOCl \rightarrow NH_2Cl \text{ (monochloramine)} + H_2O \tag{14-5}$$
$$NH_2Cl + HOCl \rightarrow NHCl_2 \text{ (dichloramine)} + H_2O \tag{14-6}$$
$$NHCl_2 + HOCl \rightarrow NCl_3 \text{ (trichloramine)} + H_2O \tag{14-7}$$

Combined residual chlorine is due to chloramines. In this form chlorine has a lower disinfecting property.

Chlorine dioxide (ClO_2) has some unusual properties. It does not react with ammonia and its disinfection properties are not fully known yet. A discussion of chlorine dioxide and its uses in water and wastewater may be found in Ref. 3.

TABLE 14-1 Comparison of Important Properties and Common Applications of Chlorine Compounds, Ozone and Hydrogen Peroxide

Characteristics	Chlorine	Sodium Hypochlorite	Calcium Hypochlorite	Chlorine Dioxide	Ozone	Hydrogen Peroxide
Chemical formula	Cl_2	NaOCl	$Ca(OCl)_2$	ClO_2	O_3	H_2O_2
Form	Liquid, gas	Solution	Powder, pellets, or 1 percent solution	Gas	Gas	Liquid
Containers for shipping	45.4-, 68- and 907-kg cylinders. Tank cars	4.9- to 7.6-m^3 tanks, and tank cars	45- to 360-kg drums	On-site generation	On-site generation	Fiberglass drums 114 to 208 ℓ. Tank cars
Commercial strength, percent	100	12–15	70	Up to 0.35	2	35–70
Stability	Stable	Light yellow liquid, unstable	Stable	Greenish yellow gas, explosive	Unstable	Stable
Toxicity to microorganisms	High	High	High	High	High	Medium

Hazards associated with handling and use	High	Medium	Medium	High	High	Medium
Corrosion	High	Medium	Medium	High	High	Low
Deodorizing	High	Medium	Medium	High	High	High
Cost	Low	Medium	Medium	Medium	High	High
Common applications	Control of slime growth, H$_2$S, and odor, BOD reduction, fly control, sludge bulking and foaming control, ammonia oxidation, disinfection	Control of slime growth, disinfection	Control of slime growth, disinfection	Control of slime growth and odor, disinfection	Odor control, BOD reduction, oxidation of refractory organic compounds, disinfection	Control of slime growth, H$_2$S, and odor, sludge bulking

TABLE 14-2   Relative Distribution of HOCl and OCl⁻ with pH at 20°C

pH	HOCl (Percent)	OCl⁻ (Percent)
6.0	96.8	3.2
7.0	75.2	24.8
7.5	49.1	50.9
8.0	23.2	76.8
9.0	2.9	97.1

Breakpoint Chlorination.   When chlorine is added to water it is consumed in oxidizing a wide variety of compounds present in the water. No chlorine residual can be measured until the chlorine demand is satisfied. Then chlorine reacts with ammonia producing combined residual. Combined chlorine residual increases with additional dosage until a maximum combined residual is reached. Further addition of chlorine causes a decrease in combined residual. This is called *breakpoint chlorination* [Eqs. (14-8)–(14-11)]. At this point the choramines are oxidized to oxides of nitrogen or other gases.

$$NH_2Cl + NHCl_2 + HOCl \rightarrow N_2O + 4HCl \tag{14-8}$$
$$4NH_2Cl + 3Cl_2 + H_2O \rightarrow N_2 + N_2O + 10HCl \tag{14-9}$$
$$2NH_2Cl + HOCl \rightarrow N_2 + H_2O + 3HCl \tag{14-10}$$
$$NH_2Cl + NHCl_2 \rightarrow N_2 + 3HCl \tag{14-11}$$

After breakpoint chlorination is reached, free-chlorine residual develops at the same rate as applied dosage. A typical breakpoint chlorination curve is shown in Figure 14-1.

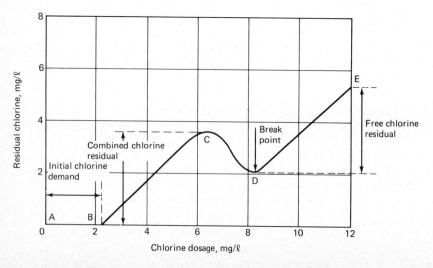

Figure 14-1   *Chlorine Residuals and Breakpoint Chlorination Curve.*

TABLE 14-3 Chlorine Dosages for Proper Disinfection
of Wastewater Effluents

Effluent from	Dosage Range (mg/ℓ)
Untreated wastewater (prechlorination)	6–25
Primary sedimentation	5–20
Chemical precipitation	3–10
Trickling filter	3–10
Activated sludge	2–8
Multimedia filter following activated sludge plant	1–5

Source: Adapted in part from Ref. 4.

Disinfection Efficiency. The disinfection efficiency of chlorine depends on the following factors: (1) contact time, (2) chlorine dosage, (3) temperature, (4) pH, (5) nature of liquid and suspended matter, and (6) type and number of organisms. For disinfection of wastewater a chlorine residual of 0.5 mg/ℓ after 20–30 min of contact period is required. The effluent quality to be disinfected must be evaluated for chemical dosage and contact period. Organic matter will react with chlorine thus reducing its effectiveness. Furthermore, the turbidity will also reduce effectiveness by adsorption and by protecting entrapped organisms. The typical chlorine dosages for proper disinfection of wastewater effluents are summarized in Table 14-3.

The type and number of organisms also affect the effectiveness of the disinfectants. The spores of organisms are much more resistant than organisms themselves. Also, the larger the number of organisms, the longer the time required for a given kill. Total coliform remaining in primary and secondary effluents at different chlorine residuals are summarized in Table 14-4. These results are obtained after 30 min of contact time, and assuming primary and secondary effluents contain 35×10^6 and 1×10^6 total coliform per 100 ml samples.[4] Several relationships have been developed that give the number of organisms remaining at different contact time, disinfectant concentration, temperature, etc. These relationships may be found in Refs. 2, 5, and 6.

TABLE 14-4 Total Coliform Remaining in the Effluent

Total Chlorine Residual (mg/ℓ)	Total Coliform (Numbers/100 mℓ)	
	Primary Effluent	Secondary Effluent
0.5–1.5	24,000–400,000	1,000–12,000
1.5–2.5	6,000–24,000	200–1,000
2.5–3.5	2,000–6,000	60–200
3.5–4.5	1,000–2,000	30–60

Source: Adapted in part from Ref. 4.

14-3-3 Design of Chlorination System

A chlorination system for disinfection of wastewater effluent consists of four separate subsystems: (1) chlorine supply, (2) chlorine feed, (3) mixing and contact, and (4) control systems. Design considerations of each system are discussed below.

Chlorine Supply

Chlorine. Gaseous or liquid chlorine can be supplied in 45.4- to 68-kg (100- to 150-lb) cylinders, 907-kg (1-ton) containers, and in tank cars.* Selection of the size of chlorine containers depends on transportation and handling costs, and available space, and quantities used. The use of a 907-kg cylinder is generally desirable for moderate-size users.[†] Some elements of the supply system to handle 907-kg containers include scale, pipe headers for delivering chlorine at the point of use, gauges for checking pressures, and an overhead crane for handling the cylinders.

Chlorine storage and handling systems must be designed with full safety considerations as chlorine gas is very poisonous and corrosive. Many important design and safety considerations are summarized below.[2,3]

- Chlorination room should be near the point of application.
- Chlorine storage and chlorinator equipment must be housed in a separate building. If not, it should be accessible only from the outdoors.
- Adequate exhaust ventilation at floor level should be provided because chlorine gas is heavier than air.
- Chlorine storage should be separate from the chlorine feeders and accessories.
- The chlorinator room should have temperature control. A minimum temperature of 21°C is recommended. The chlorine supply area should be kept cooler than the chlorinator; however, the temperature in the chlorine supply area should not be allowed to drop below 10°C.
- The sun should not be permitted to shine directly on the cylinders, and heat should never be applied directly to the cylinders.
- The chlorine storage and feed system should be protected from fire hazards. Water must be available for cooling the cylinders in case of fire.
- A clear viewing window should be provided for viewing the chlorination equipment. Blower control and gas masks should be located at the room entrance.
- Wrought iron piping should be provided for liquid chlorine and chlorine gas. Tough plastic piping should be provided for chlorine solutions. Valves and pipe fittings should be specifically designed for chlorine use. Liquid chlorine has a very high coefficient of volume expansion, therefore sufficient air cushion or expansion chambers should be provided in liquid chlorine lines.
- Chlorine cylinders in use should be set on platform scale and loss of weight should be used for record keeping of chlorine dosages.

Many other design guides and safety considerations for chlorine-handling equipment are provided in Ref. 3. The Chlorine Institute provides standards for chlorine-handling

---

*Tank cars have capacities 14,500, 27,200 and 49,900 kg (16, 30, and 55 tons).
[†]Withdrawal rates vary from 180 kg/d for gaseous to over 900 kg/d for liquid withdrawal system.

equipment and safety procedures.[7] All chlorine systems should conform to these standards.

Hypochlorite. The potential hazard associated with transportation, storage, and handling of chlorine has resulted in the use of hypochlorite solution. Hypochlorite is somewhat more expensive, loses strength in storage, and may be difficult to feed. However, for safety reasons alone, many larger plants in urban areas use hypochlorites.

Sodium hypochlorite solution is available in 1.5- to 15-percent strength, in 4.9- to 7.6-m^3 tanks, and tank cars. Stronger solutions decompose readily by exposure to light and heat. High-test calcium hypochlorite contains at least 70 percent available chlorine. It is available in 45- to 360-kg drums as powder, granules, or compressed tablets or pellets. Hypochlorites and solutions must be stored at cool and dry places for a better shelf life.

Chlorine Feed. The chlorine feed or injector system is very essential because it provides the required dosage at the point of application. Chlorine feed systems are of two types: (1) pressure injection of gas and (2) vacuum feed. The pressure injection of gas may pose risks of gas escaping. It is normally used in small plants or in large facilities where safety precautions are rigidly followed. In vacuum feed systems, a specified vacuum is applied to evaporate and move chlorine gas from the supply source to the chlorinator, where it is mixed with water and carried to the point of application. The amount of water must be enough to (1) maintain chlorine concentration in the solution below a given concentration (3500 mg/ℓ),[3] and (2) create required amount of vacuum in the line to the chlorinator and in all of the components of the chlorinator system. Manufacturers of chlorinator systems provide injector operating curves that specify the amount of water and pressure required for a given amount of chlorine to be applied against a given back pressure. From the injector the chlorine solution (in form of hypochlorous acid) flows to the point where it is applied into the effluent. Large installations using chlorine in excess of 180 kg/d utilize evaporators for chlorine gas withdrawal from single or multiple 907-kg (1-ton) containers.

Mixing and Contact. Rapid mixing of chlorine solution into wastewater followed by a quiescent contact period is essential for effective disinfection. The chlorine solution is provided through a diffuser system. It is then mixed rapidly by (1) mechanical means, (2) baffle arrangement, and (3) hydraulic jump cleated downstream of a weir, Venturi flume, or Parshall flume. Various types of chlorine diffusers and mixing arrangements are illustrated in Figures 14-2 and 14-3.

The purpose of chlorine contact chamber is to provide the contact time necessary for the disinfecting compound to reduce the number of organisms to acceptable levels. Regulatory agencies normally specify the contact time. It may range from 15 to 30 min; periods of 15 min at peak flow are common.[2,3] Irregularly shaped, circular, and rectangular basins have been used. The design objectives are to (1) minimize short-circuiting and dead spaces, (2) maximize mixing for better disinfection, and (3) reduce settling of solids in the basin. Chlorination improves the settling characteristics of the solids, and accumulation of solids in chlorine contact basin has often caused serious problems. Deposited solids exert greater chlorine demand. In addition, the deposited solids permit the growth of anaerobic organisms. The gases produced from the anaerobic areas rise and carry the solids, causing occasional high solids in the effluent.[7] Design changes were necessary in

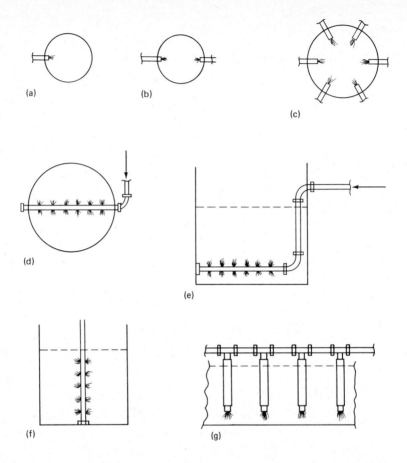

Figure 14-2   Typical Chlorine Diffusers. (a) Single injector for small pipes. (b) Dual injector for small pipes. (c) Multiple injectors for medium-size pipes. (d) Injector system for large pipes. (e) Single horizontal diffuser in open channel. (f) Single vertical diffuser for open channel. For wide channels, multiple diffusers along the width may be used. (g) Typical hanging-nozzle-type-multiple diffusers along the length of an open channel.

many cases to reduce this problem. Air agitation, maintaining high horizontal velocity through the basin, mechanical solids collection equipment (similar to sedimentation basin), and multiple units to take one chamber out of service for solids removal have all been attempted with some success. Many states require mechanical solids removal equipment in conjunction with chlorine contact chambers. Thalhmer proposed a solids flushing arrangement in an outfall sewer that was used as a contact basin.[7]

Most common design of contact chamber utilize longitudinal baffles with two to four passes around the ends to simulate a long, narrow channel (length-to-width ratio 10 : 1 or more). This design gives plug flow regime. However, lack of mixing and dead spaces

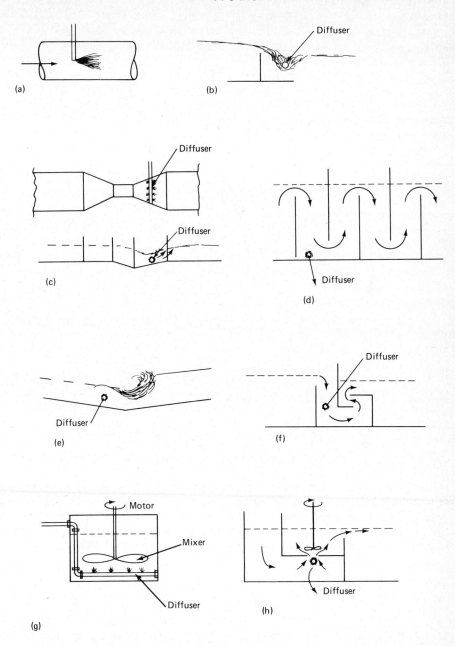

Figure 14-3   Different Types of Mixing Arrangements Used for Dispersing Chlorine Solution in Wastewater. (a) Jet nozzle in a closed conduit, and mixing by natural turbulence and diffusion. (b) Downstream of a weir, natural mixing due to turbulence. (c) Natural mixing in a Parshall flume. (d) Mixing in a tank using over-and-under baffles. (e) Mixing at the hydraulic jump in a channel. (f) Mixing by baffle arrangement. (g) Mechanical mixer in open channel. (h) Mechanical mixer with baffles in open channel.

around the corners result in solids accumulation. Rounded corners and additional baffles have been suggested to improve the design.[2,3] Cross-baffles having flow over and under the baffles are also used. This arrangement gives good mixing; however, solids accumulation between the baffles is a serious problem.

Control Systems.   The chlorination system must maintain a given chlorine residual at the end of the specified contact time. Since wastewater flow is variable and the effluent quality to be chlorinated may also change, chlorine dosage must be adjusted frequently to provide the given residual. At small installations manual control is used. The operator determines the chlorine residual and then adjusts the feed rate of chlorine solution. Simple orifice-controlled, constant-head arrangement or low-capacity proportioning pumps are used to feed the chlorine solutions. Often constant-speed feed pumps are programmed by time clock arrangement to start the pump at desired intervals.

At large facilities, complex automatic proportional control systems with recorders are used. Signals from a flow meter transmitter (Parshall flume) and/or chlorine residual analyzes (measuring the residual at one-quarter to one-half contact time) are transmitted to the chlorine feeder to adjust the feed rate to maintain a constant preset chlorine residual at all flow rates as well as effluent quality.

Several alarms are also considered an essential part of the control system. These include high and low pressures in chlorine containers, chlorine leaks, high and low injector vacuum, high and low temperatures for evaporator water bath, and high and low chlorine residuals.

14-4   EQUIPMENT MANUFACTURERS OF A CHLORINATION FACILITY

A number of equipment manufacturers supply chlorination systems. Names and addresses of many manufacturers are given in Appendix D. The design engineer should work closely with the local representatives of the equipment supplier. A good performance record of chlorination equipment at other installations should be considered for equipment selection. Responsibilities of the design engineer for proper equipment selection are given in Sec. 2-10.

14-5   INFORMATION CHECKLIST FOR DESIGN OF A CHLORINATION FACILITY

The following information must be obtained and decisions made before the design engineer should proceed with the design of a chlorination facility:

1.   Peak wet weather, peak dry weather, average design, and minimum initial flows
2.   Treatment plant design criteria prepared by the concerned regulatory agency
3.   Specified contact time and corresponding flow condition (peak or average)
4.   Minimum chlorine residual at the end of the given contact time, and degree of precision for maintaining the minimum chlorine residual
5.   Chlorine demand of the effluent, including minimum and maximum chlorine dosages to maintain the minimum residual
6.   Maximum and minimum quantities of chlorine feed (kg/d)

7. Chlorine supply—gaseous, liquid, or hypochlorites, and container size
8. Equipment manufacturers and equipment selection guide
9. Type of mixing device and shape of contact basin
10. Requirements against solids settling in the contact basin
11. Information on existing facility if plant is being expanded
12. Head loss constraints through the unit
13. Existing site plan with contours and location of the chlorination system

14-6 DESIGN EXAMPLE

14-6-1 Design Criteria Used

The following design criteria shall be used for the design of the chlorination facility in the Design Example.

1. Provide two identical mixing and contact chambers for independent operation, each capable of handling peak wet weather flow and being removed from service for cleaning.
2. Peak design flow $= 1.321$ m^3/s
   Average design flow $= 0.440$ m^3/s
   Minimum initial flow $= 0.152$ m^3/s.
3. Contact period shall be at least 20 min at peak design flow. The contact chamber shall have longitudinal baffles with flow around the ends.
4. Chlorine residual shall be greater than 0.5 mg/ℓ under all hydraulic conditions.
5. Maximum chlorine dosage shall not be less than 8 mg/ℓ to give the required amount of chlorine residual.
6. The chlorination equipment shall be capable of delivering 1.5 times the maximum dosage under all hydraulic conditions.
7. The chlorination equipment shall have all automatic controls.
8. Provision shall be made for draining and cleaning the contact chambers.
9. Provision shall be made to maintain uniform velocity through the contact chamber preferably above 2 m/min at all flows to reduce settling of solids in the chamber. This will be achieved by using a proportional weir. Arrangements shall also be made to flush out the settled solids without draining the contact chamber.

14-6-2 Design Calculations

A. Dimensions of Contact Chamber

1. Compute the volume at peak design flow

   Peak design flow $= 1.321$ m^3/s
   Contact period $= 20$ min
   Volume $= 1.321$ m^3/s $\times$ 20 min $\times$ 60 s/min
   $\qquad\quad = 1585.2$ m^3

2. Select chamber configuration and dimensions
   Provide two identical contact chambers. Each chamber shall have three-pass-

around-the-end baffled arrangement with the following dimensions: The arrangement is shown in Figure 14-4(a) and (b)

Total length of the pass-around-the-end baffles	= 100 m
Width	= 2.5 m
Depth at peak design flow	= 3.2 m
Provide a freeboard	= 0.6 m
Total volume of two contact chambers	= 100 m × 2.5 m × 3.2 m × 2
	= 1600 m^3
Assume the length of the tank, m	= L
100 m	= (L − 2.5 m) + 2.5 m
	+ (L − 2.5 m) + 2.5 m
	+ (L − 1.25 m)
	= 3L + 5.0 m − 6.25 m
3L	= 100 m + 6.25 m − 5.0 m
L	= 33.75 m

3. Compute the contact time at peak design flow

Contact time at peak design flow when two chambers are in operation $= \dfrac{1600 \text{ m}^3}{1.321 \text{ m}^3/\text{s} \times 60 \text{ s/min}} = 20.2 \text{ min}$

B. Influent Structure

1. Select the arrangement of the influent structure

The influent structure consists of a 2-m-wide rectangular channel that brings the flow from the Parshall flume into the chlorine contact chamber. The channel splits the flow equally on either side. Submerged openings 1.5 m wide and 1 m high will discharge the flow into the contact chambers. Motorized sluice gates are provided to close the openings in case one chamber is taken out of service [Figure 14-4(c)].

2. Calculate the head loss across the influent structure at peak design flow when one chamber is out of service

The differential elevation between the water surface in the influent channel and the water surface in the contact chamber is calculated from Eq. (11-2): Using

$Q = 1.321 \text{ m}^3/\text{s}$
$A = 1.5 \text{ m} \times 1 \text{ m} = 1.5 \text{ m}^2$
$C_d = 0.6$

$$\Delta z = \left[\frac{1.321 \text{ m}^3/\text{s}}{0.6 \times \sqrt{2} \times 9.81 \text{ m/s}^2 \, (1 \times 1.5) \text{ m}^2} \right]^2 = 0.11 \text{ m}$$

Therefore, the differential elevation between the water surface in the influent channel at peak design flow and in the contact chamber when only one chamber is in operation is 0.11 m.

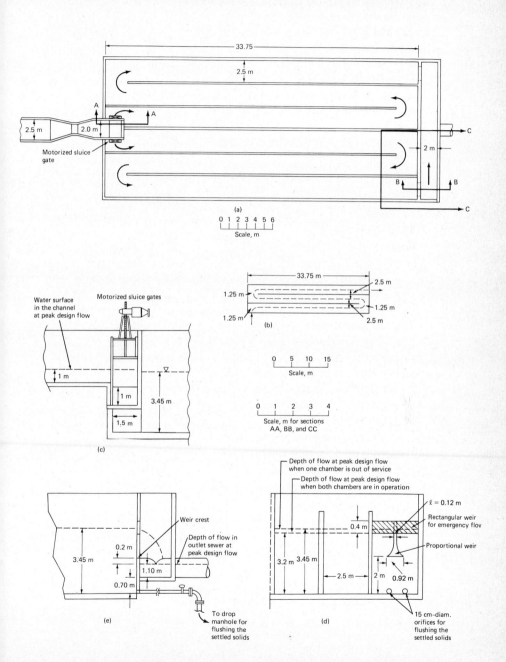

Figure 14-4 Details of Chlorine Contact Chamber. (a) Plan. (b) Flow path through chlorine contact chamber. (c) Section AA. (d) Section CC. (e) Section BB.

C. Effluent Structure

1. Select the arrangement of the effluent structure

   The effluent structure consists of a proportional weir, solids-flushing orifices, effluent channel, and outlet sewer. Proportional weirs can maintain uniform velocity at variable-flow conditions if the weir crest is located close to the bottom of the chamber. However, greater head loss will occur at the effluent structure if the weir crest is lowered. In this design the weir crest is kept 2 m above the bottom of the chamber. Solids settled at lower velocities will be flushed out by the orifices provided at the base of the proportional weir.

2. Develop the design of the proportional weir

   The flow through the proportional weir is expressed by Eq. (14-12):[9]*

   $$Q = 1.57C \sqrt{2g} \, LH^{3/2} \qquad\qquad (14-12)$$

   where

   Q = flow through the proportional weir (also through the contact chamber), m³/s
   H = head over weir, m
   C = coefficient of discharge = 0.6
   L = length of the weir opening at a height H above the weir crest, m
   g = acceleration due to gravity = 9.81 m/s²

   In order that a nearly uniform velocity be maintained through the chlorine contact chamber under variable-flow conditions, the depth of flow in the chamber must be proportional to the flow through the chamber or the head H over the weir. This is achieved by keeping the factor $LH^{1/2}$ in Eq. (14-12) constant. The design calculations of the proportional weir as well as velocity and detention time in the chlorine contact chamber are given in Table 14-5. Details of the proportional weir are given in Figure 14-4(d). The ends of the weir are cut off by a vertical line of about 2 cm. The area of the weir under the curved walls thus cut off is small, and is neglected. The length of the weir crest is 0.92 m as calculated in Table 14-5.

3. Provide a rectangular portion 2.5 m × 0.4 m over the proportional weir for emergency condition when one chlorine contact chamber is out of service

   Assume the depth of flow in the contact chamber at peak design flow when one unit is out of service   = 3.45 m

   The total head over the proportional weir   = 3.45 m − 2.00 m

   = 1.45 m

---

*Other design equations of proportional weir and Sutro weir may be found in Refs. 3, 10, and 11.

TABLE 14-5 Design Calculations of Proportional Weir, Velocity and Detention Time in the Chlorine Contact Chamber

Total Flow (m³/s)	Flow through Each Chamber (m³/s)	Flow Conditions[a]	Depth over Weir Crest (m)	Length of the Weir L (m)	Water Depth in the Chamber (m)	Velocity in the Chamber (m/min)	Contact Time (min)
1.321	0.661	Peak design flow	1.20[b]	0.12[c]	3.20	5.0[d]	20.2[e]
0.916	0.458	Max design dry weather flow	0.84[f]	0.15	2.84	3.9	25.8
0.440	0.220	Design average flow	0.40	0.21	2.40	2.2	45.5
—	—	At small flow	0.02	0.92[g]			

[a] See Table 6-9 for different flow conditions.

[b] The crest of the weir is kept 2 m above the bottom of the chamber. Therefore, the head over the weir at peak design flow is $(3.2 \text{ m} - 2.0 \text{ m}) = 1.2$ m.

[c] L is calculated from Eq. (14-12), $0.661 \text{ m}^3/\text{s} = 1.57 \times 0.6 \times \sqrt{2 \times 9.81 \text{ m/s}^2} \times L \ (1.2 \text{ m})^{1.5}$ or $L = 0.12$ m.

[d] Horizontal velocity $= \dfrac{0.661 \text{ m}^3/\text{s} \times 60 \text{ s/min}}{2.5 \text{ m} \times 3.2 \text{ m}} = 5.0$ m/min.

[e] Contact time $= \dfrac{\text{volume}}{\text{flow}} = \dfrac{100 \text{ m} \times 2.5 \text{ m} \times 3.2 \text{ m}}{0.661 \text{ m}^3/\text{s}} \times \dfrac{1}{60 \text{ s/min}} = 20.2$ min.

[f] $0.458 \text{ m}^3/\text{s} = 1.57 \times 0.6 \times (\sqrt{2 \times 9.81 \text{ m/s}^2})(0.12 \text{ m} \sqrt{1.2 \text{ m}}) \times H$; or $H = 0.84$ m.

[g] The crest length is obtained from equation $L_1 H_1^{1/2} = L_2 H_2^{1/2} = 0.12 \text{ m} \times \sqrt{1.2 \text{ m}} = 0.13 \text{ m}^{1.5}$.

At $H_1 = 0.02$ m, $L_1 = \dfrac{0.13 \text{ m}^{1.5}}{\sqrt{0.02 \text{ m}}} = 0.92$ m.

The head over the rectangular portion at the proportional weir
$$= 1.45 \text{ m} - 1.2 \text{ m}$$
$$= 0.25 \text{ m}$$

The top width L of the proportional weir at 1.45 m head
$$= \frac{\sqrt{1.2 \text{ m}} \times 0.12 \text{ m}}{\sqrt{1.45 \text{ m}}}$$
$$= 0.11 \text{ m}$$

Total flow through the proportional weir under emergency conditions
$$= 1.57 \times 0.6 \times \sqrt{2 \times 9.81} \text{ m/s}^2$$
$$\times 0.11 \text{ m} \times (1.45 \text{ m})^{1.5}$$
$$= 0.802 \text{ m}^3/\text{s}$$

The flow through the rectangular portion at the top of the proportional weir is calculated from Eq. (11-3):

Using $C = 0.624$, $H = 0.25$ m, and $L^* = 2.5$ m $- 0.11$ m $= 2.39$ m,

$$L' = 2.39 \text{ m} - 0.2 \times 0.25 \text{ m} = 2.34 \text{ m}$$

$$Q = \frac{2}{3} \times 0.624 \times 2.34 \text{ m} \times \sqrt{2 \times 9.81} \text{ m/s}^2 \times (0.25 \text{ m})^3$$

Total flow through the contact chamber at peak design flow when one unit is out of service
$$= 0.802 \text{ m}^3/\text{s} + 0.539 \text{ m}^3/\text{s}$$
$$= 1.341 \text{ m}^3/\text{s}$$

The total flow through one contact chamber at 1.45 m head over the weir is slightly greater than the peak design flow. Additional discharge capability through the remaining rectangular portion (2.5 m × 0.15 m) is available for emergency situations.

4. Design the orifices for solids flushing
    The settled solids will be flushed through the orifices provided at the floor level below the proportional weir. The flow through each orifice will be controlled by gate valves provided into the pipe connecting the orifice and discharging into the drop manhole. Provide two openings each connected with a 15-cm-diameter cast iron pipe. When the gate valves are opened, the pipes will discharge into the drop manhole at a high velocity to scour the settled solids in the chlorine contact chamber. The Hazen–Williams equation [Eq. (9-7)] is used to calculate the flushing velocity and discharge through the orifice at a total drop of 4 m.

    L = equivalent length of cast iron pipe 10 m (see Figure 14-4)

    $$V = 0.355 \times 120 \times (0.15\text{m})^{0.63} \left(\frac{4 \text{ m}}{10 \text{ m}}\right)^{0.54} = 7.9 \text{ m/s}$$

    $$Q = VA = 7.9 \text{ m/s} \frac{\pi}{4} (0.15 \text{ m})^2 = 0.14 \text{ m}^3/\text{s}$$

---

*The length of the rectangular weir is 2.5 m, but 0.11-m central portion of the weir has been included into the proportional weir. Therefore $L = 2.5$ m $- 0.11$ m $= 2.39$ m.

To flush the bottom of the chlorine contact chamber these pipes shall be opened 2–3 times a week for a period of 5–10 min. The solids will be discharged into the drop manhole and carried to the wet well under gravity flow.

5. Design the effluent box and outlet sewer
   The effluent channel is common to both the contact chambers. It is 2 m wide and has an outlet sewer located in the middle of the channel. The diameter and slope of the outlet sewer are selected such that a depth of 1.10 m is maintained in the effluent channel at peak design flow. Also, a 0.2-m free fall between the crest of the proportional weir and the water surface in the effluent channel at peak design flow is provided [Figure 14-4(e)]. The design of the outlet sewer is given in Chapter 21.

D. Chlorinator Design

1. Compute average quantity of chlorine usage

   Assume average chlorine dosage $= 5$ mg/ℓ (g/m^3)

   Average design flow $= 0.44$ m^3/s

   Average chlorine usage $= 0.44$ m^3/s $\times$ 5 g/m^3 $\times$ 86,400 s/d $\times$ (1000 g/kg)$^{-1}$

   $= 190$ kg/d

2. Compute maximum chlorine requirement

   Maximum chlorine dosage $= 8$ mg/ℓ (g/m^3)

   Maximum flow $= 1.321$ m^3/s

   Maximum chlorine usage $= 1.321$ m^3/s $\times$ 8 g/m^3 $\times$ 86,400 s/d $\times$ (1000 g/kg)$^{-1}$

   $= 913.1$ kg/d

3. Compute the number of chlorinators required

   Provide chlorinators of 450 kg/d capacity

   $$\text{Number of chlorinators required} = \frac{913.1 \text{ kg/d}}{450 \text{ kg/d}} = 2.03$$

   Provide three chlorinators, one to be used as a standby unit

4. Select chlorine-mixing system
   The chlorine solution will be discharged through diffusers at the diverging section of the Parshall flume. The chlorine will be dispersed by natural mixing. Additional mixing will be achieved at the influent structure of the chlorine contact chamber.

E. Chlorine Storage Facility

1. Select gaseous chlorine withdrawal rate
   The gaseous chlorine withdrawal rate per container at room temperature for the selected equipment is 180 kg/d

2. Compute the number of chlorine containers attached to the headers

$$\text{The number of chlorine containers to be attached to the headers} = \frac{\text{maximum chlorine requirement}}{\text{chlorine withdrawal rate per container}}$$

$$= \frac{913.1 \text{ kg/d}}{180 \text{ kg/d container}} = 5 \text{ containers}$$

3. Compute storage requirements for 2-week supply

Container size used = 907 kg (1-ton)

Maximum chlorine requirement in 2 weeks

= 913.1 kg/d × 2 weeks × 7 d/week

= 12,783 kg

$$\text{Number of containers required in 2 weeks} = \frac{\text{chlorine quantity required in 2 weeks}}{\text{quantity available per container}}$$

$$= \frac{12,783 \text{ kg/week}}{907 \text{ kg/container}}$$

= 14 containers each 907-kg (1-ton) size

Provide chlorine storage facility for 14 containers. The chlorine storage facility is shown in Figure 14-5. Advanced automatic switchover system is illustrated in Figure 14-6.

F. Instrumentation and Control. The instrumentation is an integral part of the control system. Components of the instrumentation include (1) flow measurement and (2) automatic chlorine residual control. These systems are discussed below.

(1) Design flow measurement system

The flow data are necessary to control the feed rate of chlorine solution. A standard Parshall flume design is used in this example. Design details are shown in Figure 14-7.[12,13]

a. Select the dimensions of the channel upstream of the Parshall flume

Channel section	= rectangular
Width of the channel	= 2.0 m
Depth of flow at peak design flow, y_1	= 0.8 m

The slope of the channel is obtained from the Manning Equation [Eq. (7-1)].

Cross-sectional area = 2.0 m × 0.8 m = 1.60 m^2

$$R = \frac{\text{area}}{\text{perimeter}} = \frac{1.60 \text{ m}^2}{\text{width} + 2 \text{ depth}}$$

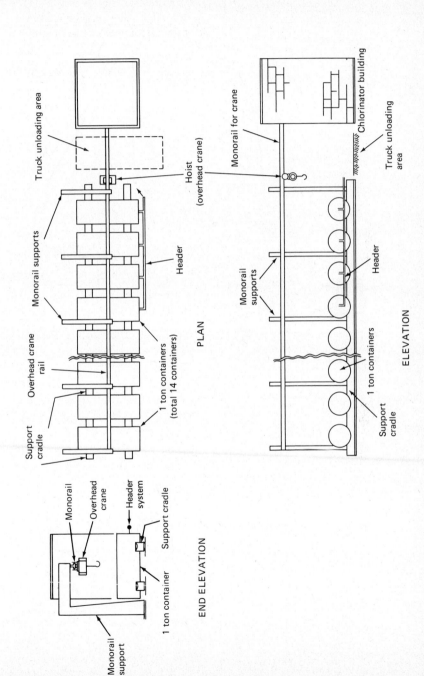

Figure 14-5 Chlorine Storage Facility

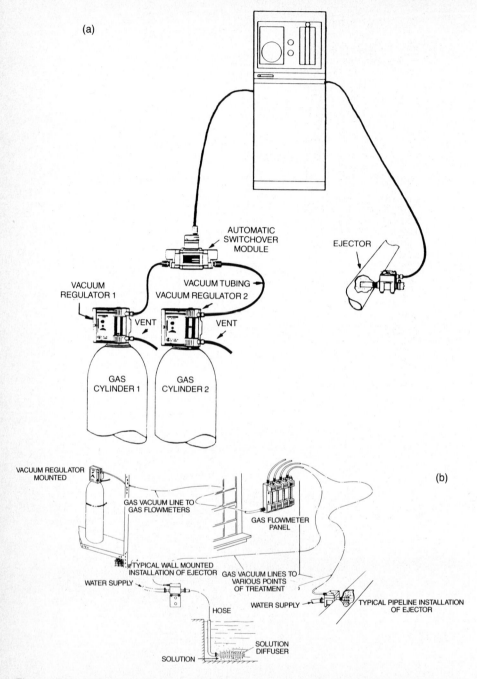

Figure 14-6   *Automatic Compound Loop Chlorination System and Accessories. (a) Advance automatic switchover system using chlorine gas cylinder. (b) Typical chlorine metering
and injection system. (c) Manifold for chlorination system. (d) Automatic switchover chlorination system (Courtesy Capital Controls Company).*

(c)

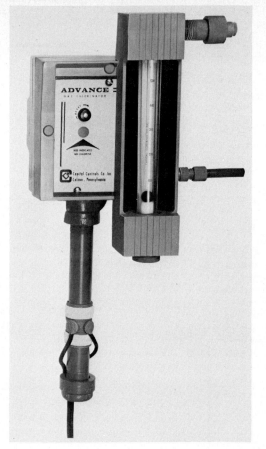

(d)

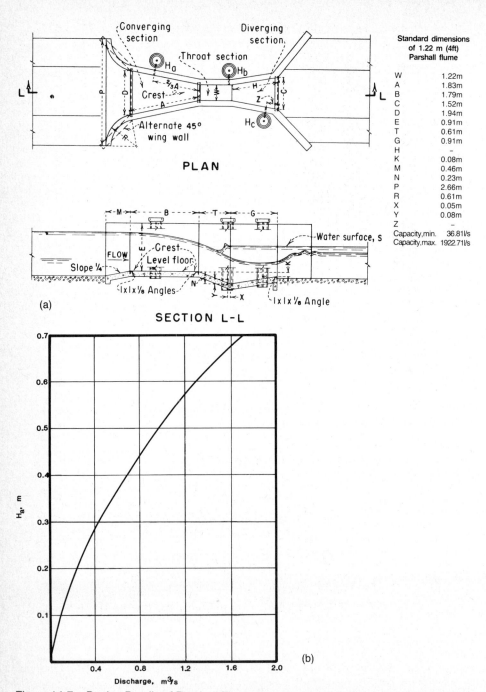

Standard dimensions
of 1.22 m (4ft)
Parshall flume

W	1.22m
A	1.83m
B	1.79m
C	1.52m
D	1.94m
E	0.91m
T	0.61m
G	0.91m
H	–
K	0.08m
M	0.46m
N	0.23m
P	2.66m
R	0.61m
X	0.05m
Y	0.08m
Z	–
Capacity,min.	36.81l/s
Capacity,max.	1922.71l/s

Figure 14-7 *Design Details of Parshall Flume Used for Flow Measurement in the Design Example. (a) Standard dimensions of Parshall flume. [From Refs. 12 and 13.] (b) Calibration curve of 1.22 m (4 ft) Parshall flume. (c) Head loss through Parshall flume. Second-ft = ft³/s. [From Refs. 12 and 13.] (d) Water surface profile through Parshall flume at peak design flow.*

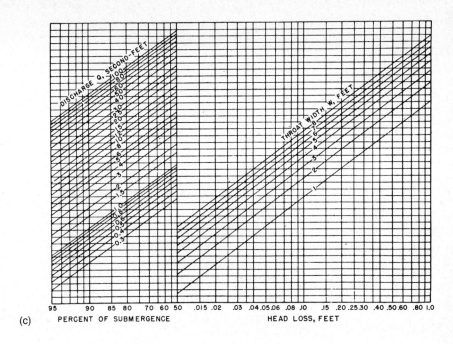

(c)   PERCENT OF SUBMERGENCE   HEAD LOSS, FEET

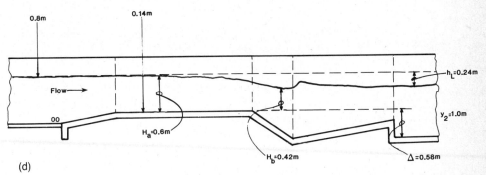

(d)

$$= \frac{1.60 \text{ m}^2}{2.0 \text{ m} + 2 \times 0.8 \text{ m}} = 0.44 \text{ m}$$

$$1.321 \text{ m}^3/\text{s} = 1.60 \text{ m}^2 \times \frac{1}{0.013} \times (0.44 \text{ m})^{2/3} S^{1/2}$$

$$S = 0.000344$$

b.   Select the dimensions of the rectangular channel downstream of the Parshall flume

Width   = 2 m
Depth of flow at peak design flow y_2 = 1 m

The slope of the channel is obtained from Eq. (7-1)

$$\text{Cross-sectional area} = 2.0 \text{ m} \times 1 \text{ m} = 2.0 \text{ m}^2$$

$$R = \frac{2.0 \text{ m}^2}{2 \text{ m} + 2 \times 1 \text{ m}} = 0.5 \text{ m}$$

$$1.321 \text{ m}^3/\text{s} = 2.0 \text{ m}^2 \times \frac{1}{0.013} \times (0.5 \text{ m})^{2/3} \times S^{1/2}$$

$$S = 0.000186$$

c. Select the dimensions of the Parshall flume

Throat width = 1.22 m (4 ft)
Submergence at peak design flow = 70 percent

The dimensions of various components of the Parshall flume are given in Figure 14-7(a)

d. Select the discharge equation for the Parshall flume
At submergence below 70 percent, the flow through a 1.22-m Parshall flume is essentially the same as for free flow conditions.[12] Free flow discharge for a Parshall flume is given by Eq. (14-13):

$$Q = 4WH_a^{1.522W^{0.026}} \tag{14-13}$$

where

Q = free flow, cfs
W = throat width, ft
H_a = depth of water at upstream gauging point, ft (Figure 14-7)

e. Compute H_a and H_b (depth of water at upstream and downstream gauging points).

H_a at peak design flow of 1.321 m^3/s (46.7 cfs) is calculated from Eq. (14-13)

$$46.7 \text{ cfs} = 4 \times 4 \text{ ft} \times H_a^{1.522(4 \text{ ft})^{0.026}}$$

Solving this equation,

$$H_a = 1.97 \text{ ft (0.6 m)}$$
$$H_b \text{ at 70 percent submergence} = 0.7 \times 1.97 \text{ ft} = 1.38 \text{ ft (0.42 m)}$$

f. Compute head loss through Parshall flume at peak design flow
The head loss is computed from Figure 14-7(c). At peak design flow of 46.7 cfs (1.321 m^3/s) and 70 percent submergence, the head loss is 0.77 ft (0.24 m).

g. Compute the downstream channel bottom from the flume crest Δ
The water surface in the flume at the H_b gauge is essentially level with the surface in the downstream channel. Therefore, $\Delta = y_2 - H_b = 1.00 \text{ m} - 0.42 \text{ m} = 0.58 \text{ m}$

    h.   Prepare water surface profile through the Parshall flume
         The water surface profile is shown in Figure 14-7(d).
    i.   Prepare the calibration curve for the Parshall flume
         The calibration curve for the Parshall flume is prepared from Eq. (14-13).
         At lower flows the water depth in the contact chamber will be reduced and
         therefore the submergence at the downstream side of the Parshall flume
         will increase. The calibration curve is shown in Figure 14-7(b).

  2.   Automatic chlorine residual control
       Many types of automatic chlorine residual control systems are available in the
       market. Select an automatic compound loop control chlorination system applic-
       able to the situations where both the chlorine demand and wastewater flow rate
       vary. The system contains a feedback element which senses flow readings and
       chlorine residuals, and feeds this information back into the system for compari-
       son with desired levels of chlorine residual. The control system consists of (1)
       chlorine residual analyzer, (2) chlorinator with automatic chlorine gas valve,
       (3) control system to receive both flow and chlorine residual signals, and (4)
       automatic chlorine feed system to maintain a preset chlorine residual. The
       system is shown in Figure 14-8.

G. Head Losses and Hydraulic Profile.   The head loss calculations through various
treatment units were provided earlier. Following is the summary of head losses that are
encountered through each unit at peak design flow when only one chlorine contact
chamber is in service. The hydraulic profile for such a condition is illustrated in Figure
14-9:

Head loss through Parshall flume	= 0.24
Head loss through influent channel due to friction, turbulence, entrance, etc. (assume)	= 0.03 m
Head loss through influent structure	= 0.11 m
Head loss through chlorine contact chamber	= small
Head loss through proportional weir	= 1.65 m
Head loss through effluent channel	= small
Total	= 2.03 m

    A total head loss of 2.03 m at a chlorination facility is disproportionately large;
however, such head losses are not uncommon with proportional weirs. In situations where
adequate head is not available, a rectangular weir may be provided. Settling of solids in
contact basin will be a problem with rectangular weirs. Therefore, consideration must be
given to mechanical equipment for removal of solids, or aeration may be provided to keep
the solids in suspension.

H. Design Details.   The design details of the chlorination facility are provided in Figures
14-4–14-9.

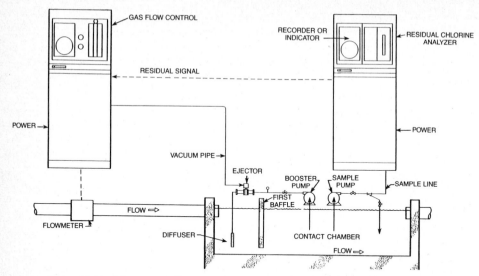

Figure 14-8 *Typical Automatic Compound Loop Chlorination Assembly (Courtesy Capital Controls Company.)*

14-7 OPERATION AND MAINTENANCE, AND TROUBLESHOOTING AT CHLORINATION FACILITY

Routine operation and maintenance of a chlorination facility includes the following:

1. Check the chlorine solution, service pump, and pressure tank each day for proper chlorine feed. Low chlorine gas pressure at the chlorinator or no chlorine feed indicates insufficient number of cylinders connected to the system, stoppage or flow restriction, chlorine gas cylinders nearly empty, plugged or damaged pressure reducing valve, chlorine cylinder hotter than chlorine control apparatus, insufficient evaporator capacity, heating element malfunction, water pump or ejector clogged, leak in vacuum relief valve, etc.

2. Monitor the chlorine residual daily. Inability to maintain adequate chlorine feed rate or wide variation in chlorine residual indicate malfunction or deterioration of water supply pump, chlorine proportion meter capacity inadequate, malfunctioning auto control electrodes, chlorine analyzer-recorder not working, loop time needs adjustment, high solids accumulation in the contact chamber, etc.

3. Inspect chlorine solution diffusers. Inadequate mixing indicate plugged diffusers.

4. Perform coliform count periodically and compare with the standards for disinfection. Inadequate kill indicates insufficient chlorine supply, short circuiting, solids buildup in the contact chamber, etc.

5. Drain chlorine contact chamber annually and inspect underwater portion of concrete structure. Patch defective concrete area as necessary. Remove buildup of sediments from the chamber floor.

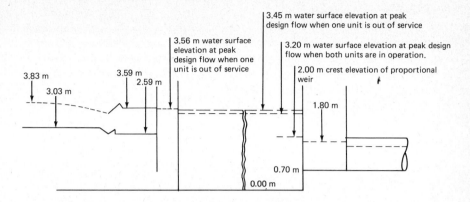

Figure 14-9   *Hydraulic Profile through the Chlorination Facility.*

14-8 SPECIFICATIONS

Specifications on a chlorination facility are briefly presented in this section. The purpose of these specifications is to describe many components that could not be fully covered in the design. The design engineer should use these specifications only as a guide. Detailed specifications for each unit or component should be prepared in consultation with the equipment manufacturers.

14-8-1 General

The manufacturers shall furnish and install chlorination equipment comprising chlorinators, chlorine residual analyzer, indicators and recorder, and auxiliary items as covered herein.

14-8-2 Chlorinator

Provide three chlorinators for compound loop control. Each chlorinator shall have a maximum capacity of 45 kg/d with gas fed under vacuum from the remote vacuum regulator automatic switchover system. The chlorinators shall be housed in a freestanding Fiberglass plastic housing. Chlorine ejector shall be mounted near contact chamber and shall be equipped with check valve to close tight on loss of vacuum. Selector switch shall be provided in chlorinators to permit using any unit in the compound loop mode. The automatic dosage controller shall adjust dosage required at 20–100 percent of maximum feed rate.

14-8-3 Residual Chlorine Analyzer

The chlorine residual analyzer shall have electrodes to measure total residual chlorine in a range of 0–10 mg/ℓ. The electrodes shall be continuously mechanically cleaned. The analyzer shall be automatically temperature-compensated over a range of 4–33°C such that any temperature change in the sample will not affect cell output. The analyzer shall have a gravity flow control system for maintaining a constant sample flow and solution feed through the cell. Reagent bottles shall have a 7-d capacity, and a 1-year supply of all reagents shall be furnished. Chlorine residual shall be indicated on at least 17-cm circular dial from 0–10 mg/ℓ, and signals shall be transmitted to a remote recorder located in the administration building.

14-8-4 Automatic Switchover System

The automatic switchover system shall consist of two vacuum regulators equipped with chlorine gas flow meter and chlorine pressure gauge, and one automatic switchover module. The chlorine flow meter shall show which chlorine cylinder is in use. Vacuum shall be controlled by a regulator which shall close tight on loss of vacuum. Pressure buildup shall be prevented by emergency relief valves located in each vacuum regulator. The automatic switchover module shall be operated that will switch from the empty cylinder to full cylinder. No manual reset shall be required. The automatic switchover module shall be factory set and shall not require field adjustment.

14-8-5 Chlorine Gas Vacuum Gauges and Loss of Vacuum Alarm Switches

Chlorine gas vacuum gauges and loss of vacuum alarm switch shall be integrally mounted in the control chlorinators. The alarm signal shall be transmitted to the remote panel.

14-8-6 Leak Detectors

The chlorine gas leak detector shall have low chlorine level warning light; high chlorine level alarm light; gasketed door with observation window and magnetic door catch; and remotely located solid state sensor. All instrumentation shall be contained in a high-impact-resistant plastic case to protect the solid state electronics. The leak detector shall be of a failsafe design and shall manually reset once the leak is corrected.

14-8-7 Gas Mask and Repair Kit

Breathing apparatus complete with cylinder (with one 30-min spare cylinder) be furnished in a waterproof cabinet. The unit shall be approved for chlorine service by the U.S. Bureau of Mines. Leakage repair kit for 1-ton chlorine containers approved by the Chlorine Institute shall be provided.

14-8-8 Gas Manifold, Valves, and Fittings

Four chlorine gas manifolds shall be supplied. Each manifold shall contain header valves with connectors, isolating valves, and heaters. Necessary piping, tees, valves, and fittings shall be furnished by the contractor.

14-8-9 Diffusers

The diffusers for chlorine solution shall be installed below the Parshall flume. The diffusers shall be placed horizontally in the channel. Each diffuser shall be 2.0 m long and 5 cm outside diameter, supported in channel by end brackets and connected to a vertical riser that shall be connected to the chlorine solution line. The diffusers shall have uniform perforations to dispense the chlorine solution. Total dispensing capacity of the diffusers shall be not less than 950 kg/d chlorine in solution.

14-8-10 Recorder

The recorder shall be located in the administration building. The recorder shall have a strip chart contained in a high-impact case with scratch-proof window, synchronous drive chart motor, high and low adjustable residual alarm contact, and contact switches.

14-9 PROBLEMS AND DISCUSSION TOPICS

14-1. Calculate the dimensions of a chlorine contact basin that has four-pass-around-the-end baffled arrangement. The contact time at 0.5 m^3/s flow is 20 min. The clear width of the basin is 2.0 m, and depth is 3 m.

14-2. Design a proportional weir for velocity control in the bar-screen chamber designed in Chap. 8. The screen chamber is 1.74 m wide, and the depth of flow downstream of the bars is 1.25 m when peak design flow of 1.321 m^3/s is passing through the chamber. The crest of the weir is at the bottom of the chamber. Use Eq. (14-12) and the design procedure given in Table 14-5. The proportional weir is located at the edge of the wet well so that there is a free fall into the wet well.

14-3. A Parshall flume is designed for measuring the flow at a wastewater treatment facility. The expected peak design and minimum initial flows are 1.2 m^3/s and 0.3 m^3/s, respectively. The throat is 1.22 m. Both the influent and effluent channels to the Parshall flume are 3 m wide and have a slope of 0.000054. The Parshall flume has maximum submergence of 70 percent. Determine the head loss at peak and minimum flows. Draw the hydraulic profiles through the flume.

14-4. Calculate the number of chlorinators, number of chlorine containers attached to the heads, and the number of containers required for 3-week chlorine supply. Use the following data:

Max flow   = 2.00 m^3/s
Maximum chlorine feed rate = 9 mg/ℓ

The chlorinators are 450 kg/d capacity.

Gaseous chlorine withdrawal rate per container = 180 kg/d

Use 1-ton container size

14-5. Hydraulic profile at peak design flow was developed in the Design Example. Draw the hydraulic profile through the same chlorination facility at average design flow when both chlorine contact chambers are in operation. The average design flow is 0.44 m^3/s.

14-6. The residual chlorine and chlorination data is given below. Plot the chlorination curve. Obtain the break point chlorination dosage. What will be the initial chlorine demand, and total chlorine demand (kg/d) to give a free chlorine residual of 1.2 mg/ℓ? The flow is 1800 ℓ/s.

Chlorine Dosage (mg/ℓ)	Chlorine Residual (mg/ℓ)
1	0
2	0.8
3	1.4
4	1.0
5	1.1
6	2.0
7	3.0

14-7. The reduction of organism in a chlorination process is expressed by Eq. (14-14)

$$\ln (N_t/N_o) = -kt \tag{14–14}$$

where

N_t = number of coliform organisms at time t
N_o = initial number of coliform organisms
k = disinfection rate constant per min
t = contact time, min

Using the mid-point chlorine dosage of 3 mg/ℓ, and 130 coliform organisms per 100 mℓ remaining in the secondary effluent after 30 min contact time, calculate k. $N_o = 10^6$ coliform per 100 mℓ. Also calculate the number of coliform organisms remaining after 20 min of contact time.

14-8. Calculate the volume of a chlorine contact basin, and the quantity of chlorine needed in kg/d. The average design flow is 0.2m³/s, contact time is 18 min., total chlorine demand is 17 mg/ℓ, and chlorine residual maintained is 1.5 mg/ℓ.

14-9. Define the following terms: free chlorine residual, combined chlorine residual, total chlorine residual, pasteurization and disinfection.

14-10. The laboratory data for a chlorination study is given below. Using Eq. (14-14) calculate the contact period to reduce the coliform count from 10^5 organism/100 mℓ in the influent to 50 organism/100 mℓ in the effluent. The chlorine dosage is the same as used in the laboratory study.

Contact Time, at Chlorine Dosage of 10 mg/ℓ, min	Number of Coliform Organisms Remaining per 100 mℓ
0	10^5
5	5000
10	250
15	12

14-11. Write the advantages and disadvantages of ozonation over chlorination.

14-12. What is the significance of breakpoint chlorination in effluent disinfection? Compare the chlorine dosage requirements to reach the breakpoint chlorination for two effluent samples: one has high ammonia nitrogen, and the other is well nitrified.

REFERENCES

1. Qasim, S. R., "Treatment of Domestic Sewage by Using Solar Distillation and Plant Culture," *Journal Environmental Science and Health,* Vol. 13, No. 8, 1978.
2. Metcalf and Eddy, Inc. *Wastewater Engineering: Treatment, Disposal, Reuse,* McGraw-Hill Book Co., New York, 1979.
3. Joint Committee of the Water Pollution Control Federation and the American Society of Civil Engineers, *Wastewater Treatment Plant Design,* MOP/8, Water Pollution Control Federation, Washington, D.C., 1977.
4. U.S. Environmental Protection Agency, *Innovative and Alternative Technology Assessment Manual,* Municipal Environmental Research Laboratory, Office of Research and Development, USEPA, CD-53, 430/9-78-009, February 1980.
5. Reynolds, T. D., *Unit Operations and Processes in Environmental Engineering,* Brooks/Cole Engineering Division, Monterey, California, 1982.
6. Sundstrom, D. W., and H. E. Klei, *Wastewater Treatment,* Prentice-Hall, Englewood Cliffs, N.J., 1979.
7. Chlorine Institute, Inc., *Chlorine Manual,* New York, 1969.
8. Thalhamer, Michael G., "A Site-Specific Design of Chlorination Facilities," *Journal of the Environmental Engineering Division, Proceedings of the American Society of Civil Engineers,* Vol. 107, No. EE3, June 1981.
9. Babbitt, H. E., and E. R. Baumann, *Sewerage and Sewage Treatment,* Wiley & Sons, New York, 1958.
10. Rex Chain Belt Co., *Weir of Special Design for Grit Channel Velocity Control,* Binder No. 315, Vol. 1, September 1963.
11. Rao, N. S. K., and D. Chandrasekaran, "Outlet Weirs for Trapezoidal Grit Chambers," *Journal Water Pollution Control Federation,* Vol. 44, No. 3, March 1972, p. 459.
12. U.S. Department of the Interior Bureau of Reclamation, *Water Measurement Manual,* U.S. Government Printing Office, Washington, D.C., 1967.
13. Parshall, R. L., *Measuring Water in Irrigation Channels with Parshall Flumes and Small Weirs,* U.S. Soil Conservation Service, Circular 843, May 1950.

15

Effluent Disposal

15-1 INTRODUCTION

The proper disposal of treatment plant effluent is an essential part of planning and designing wastewater treatment facilities. In most cases the ultimate effluent disposal or reuse requirements dictate the selection of the wastewater treatment flow scheme. In Chapter 6 several methods of ultimate disposal of secondary effluents were compared. These methods included (1) natural evaporation, (2) groundwater recharge, (3) irrigational uses, (4) recreational lakes, (5) aquaculture, (6) municipal uses, (7) industrial uses, and (8) discharge into natural waters. In this chapter, each of the above methods of effluent disposal is briefly discussed. Also effluent disposal into natural waters and the design procedure for river outfalls are discussed in considerable detail.

15-2 METHODS OF EFFLUENT DISPOSAL AND REUSE

15-2-1 Natural Evaporation

The process involves large impoundments with no discharge. The amount of evaporation from water surface depends on temperature, wind velocity, and humidity. There are substantial variations in the average daily evaporation rate from month to month and year to year. However, depending on the climatic conditions, large impoundment may be necessary if precipitation exceeds evaporation over several months. Therefore, considerations must be given to net evaporation, storage requirements, and possible percolation and groundwater pollution.

    Natural evaporation has been used for industrial wastewater in arid climates, particularly for disposal of brines. This method is particularly beneficial where recovery of residues is desirable. Impervious liners are necessary in many instances.

15-2-2 Groundwater Recharge

In many areas excessive pumping of groundwater for municipal, industrial, and agricultural uses has resulted in lowering of the groundwater table. Adverse effects, such as subsidance, saltwater intrusion, and, in general, shortage of water, has indicated a need for artificial recharging of the aquifer. Methods of groundwater recharge include rapid infiltration by effluent application or impoundment, intermittent percolation, and direct injection. In all cases risks for groundwater pollution exists. Furthermore, direct injection implies high costs of treating effluent for injection and high costs of injection facilities. Also, information regarding underground environment and geological risk levels of injection must be evaluated. Oil companies have used effluent for recharging oil-bearing strata to increase its yield.[1]

15-2-3 Irrigational Uses

The use of municipal effluents for irrigation is an acceptable practice in many parts of the United States. Irrigation has been practiced primarily as a substitute for scarce natural waters or sparse rainfall in arid areas. Health regulations govern the types of crops that are irrigated with effluents. In most states food chain crops (crops consumed by humans, and those animals whose products are consumed by humans), may not be irrigated by effluent. However, field crops such as cotton, sugar beets, and crops for seed production are grown with wastewater effluent.[1] Availability of suitable farm land and pumping costs for delivering effluent to cultivated acreage where demand may exist are perhaps the main factors for selection of this method. Land application of wastewater is discussed in Chapter 24.

Wastewater effluent has been used for watering parks, golf courses, and highway medians. These possibilities should be considered as part of overall effluent disposal alternatives.

15-2-4 Recreational Lakes

Wastewater effluent has been used successfully for development of artificial lakes used for fish and wildlife propagation, and fishing and boating. An interesting example of such projects is at Santee, California.[2] The effluent from a secondary treatment facility is stored in a lagoon for approximately 30 days. The effluent from the lagoon is chlorinated and then percolated through an area of sand and gravel, through which it travels for approximately 0.5 km, and is then collected in an interceptor trench. It is discharged into a series of lakes used for swimming, boating, and fishing. Indian Creek Reservoir (Lake Tahoe Project) is another example of such a project.[3,4]

15-2-5 Aquaculture

Aquaculture, or the production of aquatic organisms (both flora and fauna), has been practiced for centuries primarily for production of food, fiber, and fertilizer. Water hyacinth, duckweed, seaweed, alligator weed, midge larvae, *Daphnia,* and a host of other organisms have been cultured. Water hyacinths are large, fast-growing floating aquatic plants that thrive on raw as well as on partially treated wastewaters.[5] Wastewater treatment by water hyacinths for BOD, solids, nutrients, and heavy-metals removal has been considered. Potential other applications of water hyacinth culture include upgrading of lagoons, renovation of small lakes and reservoirs, pretreatment of surface waters for domestic water supply, storm water treatment, and recycling of fish culture water. Water hyacinths harvested from these systems have been investigated as a fertilizer/soil conditioner after composting, animal feed, and a source of methane when anaerobically digested. Heavy metals generally accumulate in water hyacinths and limit their suitability as a fertilizer or feed material.

Lagoons are used for aquaculture, although artificial and natural wetlands are also being considered. However, the uncontrolled spread of water hyacinths is itself a great concern because the flora can clog waterways and ruin water bodies.

15-2-6 Municipal Uses

Technology is now available to treat wastewater to the extent that it will meet federal drinking water quality standards. However, direct reuse of wastewater for human consumption depends on economics, governmental regulations, and public acceptance.

Many natural bodies of water that are used for municipal water supply are also used for effluent disposal. In this way, the natural purification processes that are continually at work in natural waters result in further purification of wastewater effluents. Thus, water treatment plants would employ only the essential treatment technologies to produce potable water supply.[6]

Direct reuse of treated wastewater is practicable only on an emergency basis. At the same time, disposal of high-quality effluent into water supply sources will continue to grow. This will be done to supplement the natural water resources by reusing the effluent many times before it finally flows to the sea.

15-2-7 Industrial Uses

The number of industrial uses of treatment plant effluent are increasing. Effluent has been successfully used as cooling water and boiler feed water. Since about 50 percent of all industrial water is used for cooling or feeding a boiler, an enormous potential exists for industrial applications.[6] Deciding factors for effluent reuse by the industry include (1) availability of natural water, (2) quality and quantity of effluent, and cost of processing, (3) pumping or transport cost of effluent, and (4) industrial process water that does not involve public health considerations.

15-2-8 Discharge into Natural Waters

Discharge of effluents into natural waters is the most common disposal practice. The self-purification or assimilative capacity of the natural waters is thus utilized to provide the remaining treatment. The 1972 and 1977 Clean Water Acts established certain effluent limitation requirements—and deadlines for reaching them in order to "restore and maintain the chemical, physical and biological integrity of the nation's waters."[7,8] All states of the United States were required by the government to adopt water quality standards.

As part of the overall water quality management and setting of the effluent standards for conservative and nonconservative pollutants, various types of water quality models are widely employed.* These models are used to provide predictive answers to such questions as:

---

*Conservative pollutants are those that do not decay. Examples of such pollutants are dissolved salts and heavy metals. The nonconservative pollutants are those that exhibit time-dependent decay, such as BOD, ammonia nitrogen, and radioactive wastes.

- What will be the net effect of a given level of treatment on the water quality of receiving body of water?
- How will a particular watershed management practice affect the water quality?
- What are the long- and short-term effects of a certain course of action?
- Is a certain course of action worth the costs of implementation?

Water quality models utilize constituents such as total dissolved solids, total suspended solids, coliform organisms, nutrients, oxidation of carbonaceous and nitrogenous compounds, and dissolved oxygen. Many excellent references are available on this subject.[1,9-11] Relationships between contaminants and their probable impacts are summarized in Table 15-1.

Effluent disposal in natural waters is a most common practice. Therefore, the remainder of this chapter is devoted to the design aspects of the outfalls in the receiving waters. Requirements and design factors for river, lake, estuarine, and ocean outfalls are briefly discussed. The Design procedure for a river outfall is presented in great detail in the Design Example.

Requirements of Outfall. The outfall structures are designed to properly disperse the effluent into the receiving waters. The design of outfall is related to the characteristics of the receiving waters whether they are rivers, lakes, estuaries, or coastal waters of seas and oceans. In all cases consideration must be given to proper dispersion, avoid localized pollution problem, utilization of total available assimilative capacity, and avoidance of shore contamination. Requirements of river, lake, estuarine, and ocean outfalls are given below. In most cases the discharge is made below the water surface through one or multiple outlets to reduce foam formation and achieve good dispersion.

River Outfall. Shoreline releases of effluents into the rivers are poorly dispersed and hug the banks for long distances with little mixing. Installation of diffusers across the width of the river is preferred.* Sewage effluent is generally warmer than most receiving waters and therefore it rises while being swept along by dominant currents. Formation of telltale signs such as sleek or detergent froth should be avoided.

Lake Outfall. In lakes attempt should be made to achieve the best possible dispersion to avoid shore contaminations. Shallow lakes and reservoirs are subject to significant mixing due to wind-induced currents, and effluents are dispersed well. In stratified lakes there is a possibility that the effluent may deplete the dissolved oxygen in the lower stratum.[13] Warmer effluents may also destroy the thermoclines.†

Estuarine Outfall. The effluents discharged into the tidal rivers and estuaries may oscillate due to freshwater flow, and tide- or wind-produced seiches. The mixing is due to dispersion. If stratification occurs due to salinity or temperature, the effluent plume will rise to the surface.

---

*In navigable rivers construction of pipelines across the bottom is subject to approval by U.S. Army Corps of Engineers.

†Thermocline is a zone of significant temperature change and is extremely resistant to mixing.

TABLE 15-1   Relationship between Contaminants and Their Probable Impacts on Water Quality

	Aesthetics	DO Depletion	Sediment Deposits	Excessive Aquatic Growth	Public Health Threats	Impaired Recreational Value	Ecological Damage	Reduced Commercial Value
Floatables and visual contaminants	X					X	X	
Bacteria, virus					X	X		
Degradable organics		X					X	
Suspended solids	X		X			X		
Nutrients	X			X				
Dissolved solids	X							
Toxic materials					X		X	X

Source: Ref. 9.

Marine or Ocean Outfall. The marine outfall consists of long pipes to transport the waste to long distances from the shore and then release it. The function of a diffuser is to mix the low-density waste with high-density seawater at the bottom. The plume will rise under prevailing currents until the density of the mixture is equal to that of seawater below the surface. At this point the plume will spread horizontally and disperse with little or no visual effects. Some excellent references on marine outfalls are available in the literature.[1,14–16]

Design Factors for Outfall. Depending on the characteristics of the receiving waters, many factors are considered for proper mixing and dispersion of effluent. These factors include (1) flow velocity, (2) stratification due to salinity or temperature, (3) depth (shallow or deep), (4) shape (wide or narrow), (5) reversal of current (tidal or wind-produced seiches), (6) wind circulation, and (7) temperature and salinity of effluent. For design and construction of outfalls, three principal factors are considered:

1. Material of the outfall pipe
2. Means of anchoring the outfall in the body of water
3. The design of diffusers

Each of these factors is briefly discussed below.

Material. The construction material for outfall may be steel, concrete, plastic, or any material that is compatible with the waste stream, receiving water, site characteristics, and economics of fabrication and installation.

Installation and Anchoring. Installation and anchoring of outfall pipes and diffusers is especially important if strong currents prevail and additional problems such as errosion and scour, and transport of bedload, particularly such objects as tree trunks, may be expected. Outfall pipes are secured by rock ballast, cement or sand bags, or chain and anchors. Often pipes may be partly embedded into concrete apron.[14,17]

Diffusers. Diffuser must disperse the effluent efficiently and evenly to achieve good mixing. Knowledge of density and temperature gradients as well as currents and wind conditions that influence the dispersion in receiving waters is necessary. Procedure for determining initial dilution and dispersion for ocean outfalls is given in Ref. 14.

   The diffusers may have a total area approximately half the cross-sectional area of the pipe. The diameter of the port is governed by considerations of clogging, hydraulic performance, and spacing. Typical diffusers have port diameters 10–30 cm, spaced over a desired length. Two ports are provided in the bulkhead at the end of the discharge pipe. One port is on the top and the other is at the bottom. These serve not only to discharge waste, but also provide outlets for settleable and floatable materials. Other ports are normally alternated on the opposite side of the pipe at center line.[17] The minimum center-to-center spacing of the port is 10 times the port diameter.[14]

Design Procedure. The hydraulic analysis of multiport diffusers is basically a problem in manifolds and is somewhat complex. A discussion of this subject may be found elsewhere.[18,19] A simplified procedure is given in Ref. 14. It is a process of hydraulic iteration using a trial-and-error solution. The procedure is illustrated in the Design Example.

The diffusers are generally designed for maximum flow. The performance of the diffuser at lower flows must also be determined. In case low velocities are encountered under initial flow conditions, it is customary to plug the inboard ports during initial flow periods and open them later as the flow increases.

15-3 INFORMATION CHECKLIST FOR DESIGN OF OUTFALL STRUCTURE

The following information is necessary to properly design an outfall system:

1. Type of receiving water (river, lake, estuary, or ocean) to be used for disposal of effluent
2. Uses of receiving water (water supply, propagation of aquatic life, recreation, etc.)
3. Treatment plant design criteria and effluent disposal requirements for the receiving waters prepared by the concerned regulatory agencies
4. Peak, average, and minimum flows for the design and initial years
5. Characteristics of receiving water (topography, salinity and temperature profiles, water currents, wind circulation, and water elevations at low flow and high flood conditions)
6. Available head, that is, the differential elevation between the water surface elevation in the outfall channel or pipe, and the high flood level in the receiving water
7. Length of outfall channel or pipe (distance from the last treatment unit to the receiving waters)
8. Postaeration requirements to meet the effluent discharge requirements
9. Existing topographic map from the treatment plant site to the receiving water

15-4 DESIGN EXAMPLE

15-4-1 Design Criteria Used

The following design criteria are used for the design of outfall structures:

1. Provide a shallow trapezoidal channel from the chlorination facility to the outfall structures. Postaeration is not required.
2. Provide a collection box to connect the outfall channel and the outfall pipe that contains the diffuser.
3. The outfall pipe shall be designed for peak design flow of 1.321 m^3/s.
4. The diffusers shall be designed for the peak design flow of 1.321 m^3/s.
5. The diffuser pipe shall be provided longitudinally along the centerline of the river confluence. The selected alignment is shown in Figure 15-1(a) and (b).

15-4-2 Design Calculations

A. Trapezoidal Outfall Channel. The outfall channel is a concrete, lined trapezoidal channel that transports the effluent from the chlorination basin to the outfall structure. The channel has a bottom width of 0.5 m and side slope of 3 horizontal to 1 vertical. A shallow depth of effluent is maintained into the channel to effect natural reaeration. Details of the outfall channel are shown in Figure 15-1(c). The slope of the trapezoidal channel and depth of flow calculations are given in Chapter 21.

B. Collection Box. A 2 m × 2 m collection box is provided to collect the flow from the trapezoidal channel and discharge it into the outfall pipe. A smooth transition shall be provided from the trapezoidal section to the square collection box. The depth of the collection box is determined from the maximum liquid level into the box.

The collection box has a sluice gate to be used in case high flood waters warrant protection of the treatment facility against flooding. Emergency pumps will be required to pump the effluent into the receiving water. Design procedures for a pump station are provided in Chapter 9. Details of collection box are shown in Figure 15-1(a) and (b).

C. Outfall Pipe. The outfall pipe is a cast iron circular pipe that delivers the effluent into the diffuser pipe.

1. Compute diameter and select length
    Assume a velocity of 2 m/s through the outfall pipe at peak design flow of 1.321 m³/s.

    $$\text{Area of the outfall pipe} = \frac{1.321 \text{ m}^3/\text{s}}{2 \text{ m/s}} = 0.661 \text{ m}^2$$

    $$\text{Diameter} = \sqrt{0.661 \text{ m}^2 \times 4/\pi} = 0.92 \text{ m}$$

    The outfall pipe is approximately 50 m long and has several bends as shown in Figure 15-1. It will discharge under pressure at peak design flow.

2. Compute head losses
    The head losses in straight pipe, bends, and velocity head are calculated from Eqs. (9-2), (9-4), and (9-5) general discussion may be found in Chapter 9.

    $$\text{Head loss due to entrance } K = 0.5 \text{ (Appendix B)} = \frac{0.5 \times (2 \text{ m/s})^2}{2 \times 9.81 \text{ m/s}^2}$$

    $$= 0.1 \text{ m}$$

    Head loss due to bends, 2 bends 22.5° each, $K = 0.11$ (Appendix B)

    $$= 2 \times \frac{0.11 \times (2 \text{ m/s})^2}{2 \times 9.81 \text{ m/s}^2} = 0.05 \text{ m}$$

    Head loss due to friction ($f = 0.024$)

    $$= \frac{0.024 \times 50 \text{ m} \times (2 \text{ m/s})^2}{2 \times 9.81 \text{ m/s}^2 \times 0.92 \text{ m}}$$

    $$= 0.27 \text{ m}$$

    Total head loss $= 0.10 \text{ m} + 0.05 \text{ m} + 0.27 \text{ m} = 0.42 \text{ m}$

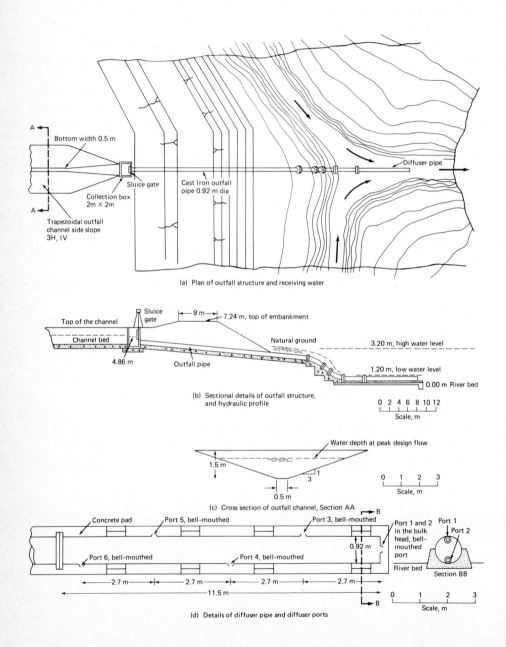

(a) Plan of outfall structure and receiving water

(b) Sectional details of outfall structure,
and hydraulic profile

(c) Cross section of outfall channel, Section AA

(d) Details of diffuser pipe and diffuser ports

Figure 15-1   Details of Outfall Structure.

D. Diffuser Pipe

1. Select a design procedure for diffuser pipe design

   The design of a diffuser pipe is basically a trial-and-error procedure. Once a tentative configuration is selected, it is designed by the procedure described below.[14,17] The procedure consists of selecting an operating head at the end of the diffuser and then calculating the flow through the outermost ports. Using the flow the head loss in the pipe between the outermost port and the next upstream port is calculated. The head loss plus the velocity head, the density differential head, and the slope of the diffuser will give the operating head at the next inboard port. The flow at this port is then calculated and the procedure is continued over the entire length of the diffuser. The sum of discharges through all ports should equal the design flow through the outfall sewer. If they are not equal, the operating head is changed and the calculations are repeated. The process can be tedious if done manually.

2. Select design equations

   Equations (15-1)-(15-4) are used for the design and analysis of the diffusers.[14]

$$q_1 = C_{D_1} \, a_1 \, \sqrt{2gE_1} \tag{15-1}$$

$$E_1 = \frac{\Delta P}{\gamma} = ELV_1 - ELV_0 \tag{15-2}$$

$$E_2 = E_1 + h_{f_{1-2}} + \frac{\Delta S_g}{S_g} \times \Delta Z_{1-2} \tag{15-3}$$

$$h_{f_{1-2}} = f \frac{L_{1-2} \times V_{1-2}^2}{D_{1-2} \times 2g} \tag{15-4}$$

where

q_1 = discharge through port 1 (Figure 15-2). Port 1 is the outermost port

C_{D_1} = coefficient of discharge for port 1. The value of C_{D_1} is a function of $(V_{1-2}^2/2g)/E_1$. This relationship is graphically shown in Figure 15-2.[14]

a_1 = area of port 1

E_1 = assumed operating head for port 1

ELV_1 = free water surface elevation in the diffuser pipe upstream of the port 1

ELV_0 = free water surface elevation in the receiving water

E_2 = total head in the diffuser pipe at port 2. Port 2 is upstream of port 1

$h_{f_{1-2}}$ = head loss due to friction in the diffuser pipe between ports 1 and 2

L_{1-2} = length of pipe between ports 1 and 2

D_{1-2} = diameter of the diffuser pipe between ports 1 and 2

V_{1-2} = velocity in the diffuser pipe between ports 1 and 2 $= \dfrac{q_1}{\frac{\pi}{4} D_{1-2}^2}$

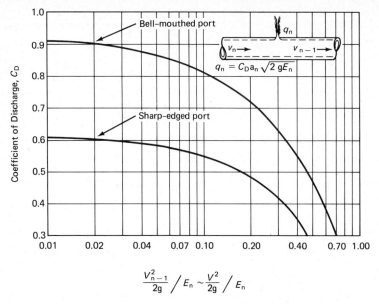

$$\frac{v_{n-1}^2}{2g} \Big/ E_n \sim \frac{v^2}{2g} \Big/ E_n$$

Figure 15-2 Coefficient of Discharge for Small Ports on the Side of a Pipe. [From Ref. 14. Used with permission of American Society of Civil Engineers.]

$\dfrac{\Delta P}{\gamma}$ = difference in pressure head between inside and outside of diffuser port 1

f = Darcy friction factor

ΔS_g = difference in specific gravity between receiving water and effluent

S_g = specific gravity of the receiving water

3. Select a tentative diffuser configuration

Provide diffuser pipe equal in diameter to the outfall pipe (0.92 m)

Provide diffuser port area approximately half the area of the diffuser pipe $= \dfrac{\pi}{4} (0.92 \text{ m})^2 \times \dfrac{1}{2} = 0.33 \text{ m}^2$

Provide each port 27 cm in diameter

Area of each port $= \dfrac{\pi}{4} (0.27 \text{ m})^2 = 0.057 \text{ m}^2$

Number of ports $= \dfrac{0.33 \text{ m}^2}{0.057 \text{ m}^2} = 6$

Provide two ports in the bulkhead at the end and four ports on the sides of the pipe. Total length of the diffuser pipe using center-to-center spacing 10 times the diameter of the port = 4(10 × 0.27 m) = 10.8 m (35.4 ft). Provide 11.5-m-long diffuser pipe. The details and diffuser configuration are shown in Figure 15-1. All ports have a divergent mouthpiece.

4. Compute head losses
    The head loss calculations for the diffuser pipe involve an iterative process. The step-by-step calculations for the final trial at peak design flow are summarized in Table 15-2. In freshwater outfalls (rivers and lakes), the effect of density gradient or the factor $[(\Delta S_g/S_g)\Delta Z]$ in Eq. (15-3) is ignored. Also, the total discharge through all ports should equal the design flow. In this case total flow through six ports is 1.37 m³/s against the peak design flow of 1.321 m³/s. Perhaps another trial will be necessary to achieve a closer agreement. The total head loss in the diffuser pipe is 1.24 m. This head loss includes the operational head of 1.2 m at the end of the diffuser pipe.

E. Head Losses and Hydraulic Profile.   The head loss calculations for various components of outfall structure are given earlier. Following is the summary of head losses that

TABLE 15-2   Design Calculations for Diffuser Pipe Used to Disperse Effluent into the River

Factor		1[a]	2[a]	3	4	5	6
				Port Numbers			
1.	Distance from the end, m	0	0	2.700	5.400	8.100	10.800
2.	Port diameter d, m	0.270	0.270	0.270	0.270	0.270	0.270
3.	Port area a, m²	0.057	0.057	0.057	0.057	0.057	0.057
4.	Total head E, m	1.200[b]	1.200	1.202	1.206	1.214	1.225
5.	Pipe diam D, m	0.920	0.920	0.920	0.920	0.920	0.920
6.	Area of pipe A, m²	0.660	0.660	0.660	0.660	0.660	0.660
7.	Port velocity v, m/s	4.850[c]	4.850	4.856	4.864	4.880	4.902
8.	Discharge through pipe, m³/s	—	0.486[d]	0.724[e]	0.949	1.163	1.370
9.	Velocity in pipe V, m/s	—	0.736	1.100	1.438	1.762	2.076
10.	Velocity head $V^2/2g$, m	—	0.028	0.062	0.105	0.158	0.220
11.	$V^2/2g/E$	—	0.023	0.052	0.087	0.130	0.179
12.	C_D	0.880	0.880[f]	0.860	0.810	0.770	0.740
13.	Port discharge q, m³/s	0.243	0.243	0.238	0.225	0.214	0.207
14.	Friction factor	—	0.024	0.024	0.024	0.024	0.024
15.	Pipe length to next port L, m	—	2.700	2.700	2.700	2.700	2.700
16.	Head loss in pipe h_f, m	—	0.002[g]	0.004	0.008	0.011	0.016
17.	Density head $[(\Delta S_g/S_g)\Delta Z]$, m	0	0	0	0	0	0
18.	Total head E, m	—	1.202[h]	1.206	1.214	1.225	1.240

[a]Ports 1 and 2 are provided in the bulkhead at the end of the discharge pipe.
[b]Assumed operating head at the end of the diffuser.
[c]$v = \sqrt{2gE} = \sqrt{2 \times 9.81 \text{ m/s}^2 \times 1.2 \text{ m}} = 4.85$ m/s.
[d]Total flow through the ports downstream of the pipe section = flow from ports 1 and 2.
[e]Total flow in the pipe = $(0.243 + 0.243 + 0.238)$m³/s = 0.724 m³/s. The flow through port 3 (0.238 m³/s) is obtained from Eq. (15-3) by using a trial-and error-solution.
[f]Value is obtained from Figure 15-2, $V^2/2g/E = 0.023$, and the curve for Bell-mouthed port.
[g]$h_f = 0.02 \times 2.7$ m $\times 0.028$ m/0.92 m = 0.002 m.
[h]Total head = 1.2 + 0.002 = 1.202 m.

are encountered at peak design flow when the differential operating head between the water surface at high flood level and inside of diffuser pipe = 1.2 m.

1. Total head loss in diffuser pipe = 1.24 m
2. Total head loss in outfall pipe = 0.42 m

Total   1.66 m

The hydraulic profile of the outfall structure under peak design flow condition is shown in Figure 15-1. This profile is prepared with respect to the channel bed.

F. Design Details.   The design details of the outfall structure are shown in Figure 15-1.

15-5 OPERATION AND MAINTENANCE, AND TROUBLESHOOTING AT OUTFALL STRUCTURE

The operation and maintenance of the outfall structure include the following:

1. Check the trapezoidal effluent channel periodically for algae growth on the side slopes. Large masses of algae growth should be removed.
2. Check the collection box for obstructions or excessive slime growth on a regular basis. Any rise of water surface in the outfall channel is an indication of flow obstruction in the outfall structure.
3. The diffuser pipe is designed for the peak design flow. Under initial flow conditions low velocities will be encountered. Therefore, a sufficient number of inboard ports shall be plugged during initial flow periods and opened later as the flow is increased with time.
4. Ports 1 and 2 are provided in the bulkhead at the top and bottom. These ports will facilitate self-cleaning of the diffuser pipe from the floating and settled solids.

15-6 SPECIFICATIONS

The effluent structure includes effluent channel (rectangular), outfall channel (trapezoidal), collection box with sluice gate, CI outfall pipe, and steel diffuser pipe. The contractor shall be responsible for furnishing, construction, installation, and testing of all concrete channels, cast and ductile iron pipe and fittings, manholes, and specials as shown on the plans and provided for in these specifications. The plans show sizes and general arrangements of channel, pipes, and appurtenances. Responsibility for furnishing exact lengths of all channels, pipes for proper makeup, and transitions rests with the contractor. All concrete, cast iron, and steel pipe fittings shall conform to the standards (see specifications in Chapters 7 and 12). The diffuser and outfall pipes shall be properly anchored deep in the channel bed and in the embankment to protect them from errosion, scour and uplift. The concrete apron shall be secured by in place poured concrete of sufficient depth below the channel bed.

15-7 PROBLEMS AND DISCUSSION TOPICS

15-1. An outfall pipe 15 cm in diameter is discharging into a lake. The pipe is at the bottom of the lake that has an elevation of 102.00 m. The water surface elevation is 105.50 m. The length of the pipe is 280 m, and it has four 45° bends ($K=0.2$). Calculate the water surface elevation in the upstream unit. $Q = 27$ ℓ/s, $f=0.024$, $K_{entrance} = 0.5$, and $K_{exit} = 1.0$.

15-2. Develop at average design flow of 0.44 m³/s the hydraulic profile through the outfall system provided in the Design Example.

15-3. During the initial years when the flow is small many inboard ports are tightly plugged. Calculate the head loss through the outfall structure in the Design Example at initial peak flow of 966 ℓ/s. Assume ports 6 and 5 are plugged.

15-4. Calculate the head loss through the outfall system given in the Design Example. Assume sharp-crested ports. Peak design flow is 1.321 m³/s, and number of ports are six as given in Figure 15-1.

15-5. Describe different methods of effluent disposal. Prepare a checklist to indicate advantages and disadvantages of each method, and conditions where each method may or may not be suitable for effluent disposal.

15-6. What are various factors that an engineer must take into consideration for designing an outfall structure?

15-7. Develop a design procedure for designing a marine outfall structure where specific gravity term cannot be ignored. List your computational procedure.

15-8. Visit the wastewater treatment facility in your community. Describe the method of effluent disposal and its impact upon the environmental quality. How could the impacts be minimized?

REFERENCES

1. Metcalf and Eddy, Inc. *Wastewater Engineering: Treatment, Disposal, Reuse,* McGraw-Hill Book Co., New York, 1979.
2. Merrill, J. C., et al., *Santee Recreational Project, Santee, California,* Final Report FWPCA Report WP-20-7, U.S. Department of the Interior, Federal Water Pollution Control Administration, Cincinnati, 1967.
3. U.S. Environmental Protection Agency, *Wastewater Purification at Lake Tahoe,* Technology Transfer, Washington, D.C.
4. U.S. Environmental Protection Agency, *Indian Creek Reservoir: A New Fishing and Recreational Lake from Reclaimed Wastewater,* Technology Transfer, Washington, D.C.
5. U.S. Environmental Protection Agency, *Innovative and Alternative Technology Assessment Manual,* CD-53, Office of Research and Development (MERL), Cincinnati, Ohio, February 1980.
6. Clark, J. W., W. Viessman, and M. J. Hammer, *Water Supply and Pollution Control,* IEP-A Dunn-Donnelley Publisher, New York, 1977.
7. *Federal Water Pollution Control Act Amendments of 1972* (PL 92-500), 92nd Congress, October 18, 1972.
8. *Clean Water Act of 1977* (PL 95-217), 95th Congress, December 1977.
9. Hydroscience, Inc., *Water Quality Management Planning Methodology for Hydrographic Modification Activities,* Texas Water Quality Board, Austin, 1976.
10. Hydroscience, Inc., *Simplified Mathematical Modeling of Water Quality,* Environmental Protection Agency, Washington, D.C., 1971.

11. Nemerow, N L., *Scientific Stream Pollution Analysis,* McGraw-Hill Book Co., New York, 1974.
12. Fair, G. M., J. C. Geyer, and D. A. Okun, *Water and Wastewater Engineering,* Vol. I, Wiley and Sons, New York, 1966.
13. Hill, K. V., "Waste Disposal Outfalls Into Fresh Waters," *Water and Sewage Works,* Reference Number, Vol. 113, no. 11, p. 238, November 1966.
14. Rawn, A. M., F. R. Bowerman, and N. H. Brooks, "Diffusers for Disposal of Sewage in Sea Water," *Journal of the Sanitary Engineering Division, Proceedings of the American Society of Civil Engineers,* Vol. 86, No. SA2, March 1960.
15. Burchell, M. E., A. Tchobanaglous, and A. J. Burdoui, "A Practical Approach to Submarine Outfall Calculations," *Public Works,* May 1967.
16. Falk, L. L., "Factors Affecting Outfall Design," *Water and Sewage Works,* Reference Number, Vol. 113, no. 11, p. 233, November 1966.
17. Rambow, C. A., "Submarine Disposal of Industrial Wastes," *Proceedings of the 24th Industrial Waste Conference,* Purdue University, Part 2, May 6, 7, and 8, 1969.
18. McNown, J. S., "Mechanics of Manifold Flow," *Trans. ASCE,* Vol. 119 (1954), p. 1111.
19. Benefield, L. D., J. F. Judkins, and A. D. Parr, *Treatment Plant Hydraulics for Environmental Engineers,* Prentice-Hall, Inc., Englewood Cliffs, N.J., 1984.

16

Sources of Sludge and Thickener Design

16-1 INTRODUCTION

The principal sources of sludge at municipal wastewater treatment plants are the primary sedimentation basin and the secondary clarifiers. Additional sludge may also come from chemical precipitation, nitrification-denitrification facilities, screening and grinder, and filtration devices if the plant has these processes. Many times the sludge produced in these processes is recycled through the primary or secondary treatment systems so that the sludge is removed as either primary or secondary sludge. In some cases, secondary sludge is returned to the primary settling tank, ultimately giving a single sludge stream consisting of combined sludges.

Sludge contains large volumes of water. The small fraction of solids in the sludge is highly offensive. Thus the problems involved with handling and disposal of sludge are complex. Common sludge management processes include thickening, stabilization, dewatering, and disposal. Chapters 16–19 are devoted exclusively to various sludge-processing systems. The main purpose of this chapter is to (1) describe the basic characteristics of sludge produced in the primary and secondary treatment processes, (2) present a general overview of sludge-processing systems and environmental control measures, (3) describe various sludge-thickening methods, and (4) present the step-by-step design procedure of a gravity thickener in the Design Example.

16-2 CHARACTERISTICS OF MUNICIPAL SLUDGE

Sludge consists of organic and inorganic solids. Primary sludge contains solids present in the raw wastewater, while secondary sludge contains chemical or biological solids produced during the treatment processes. The specific gravity of inorganic solids is about 2–2.5 and that of organic fraction is 1.2–1.3. The characteristics of wastewater residues and their production rates from various treatment processes are summarized in Table 4-3. The typical chemical compositions of raw and digested sludges are given in Table 16-1.

TABLE 16-1   *Typical Chemical Composition of Raw and Digested Sludges*

Constituents	Primary Sludge	Waste Activated Sludge	Digested Sludge
pH	5.0–6.5	6.5–7.5	6.5–7.5
Total dry solids, percent	3–8	0.5–1.0	5.0–10.0
Total volatile solids, percent dry wt.	60–90	60–80	30–60
Specific gravity of individual solids particles	1.3–1.5	1.2–1.4	1.3–1.6
Bulk specific gravity	1.02–1.03	1.0–1.005	1.03–1.04
BOD_5/TVS	0.5–1.1	—	—
COD/TVS	1.2–1.6	2.0–3.0	—
Alkalinity, mg/ℓ as $CaCO_3$	500–1500	200–500	2500–3500
Cellulose, percent dry wt.	8–15	5–10	8–15
Hemicellulose, percent dry wt.	2–4	—	—
Lignin, percent dry wt.	3–7	—	—
Grease and fat, percent dry wt.	6–35	5–12	5–20
Protein, percent dry wt.	20–30	32–41	15–20
Nitrogen N, percent dry wt.	1.5–4.0	2.5–7.0	1.6–6.0
Phosphorus P, percent dry wt.	0.8–2.8	2.0–7.0	1.4–4.0
Potash, percent dry wt.	0.1–1.0	0.2–0.5	0.1–3.0
Heating value, kJ/kg	15,000–24,000	12,000–16,000	6000–14,000
Cadmium, mg/kg	16	—	76
Chromium, mg/kg	110	—	160
Copper, mg/kg	200	—	340
Lead, mg/kg	500	—	—
Nickel, mg/kg	46	—	63
Zinc, mg/kg	620	—	930
Hexachlorobenzene, mg/kg	0.4	0.8	
Lindane, mg/kg	0.6	1.0	
Chlordane, mg/kg	2.6	4.4	

Source: Adapted in part from Refs. 1–3.
1 kJ/kg = 0.43 Btu/lb.

16-3   SLUDGE-PROCESSING SYSTEMS AND ENVIRONMENTAL CONTROL

Various sludge-processing systems include (1) thickening, (2) stabilizaton or digestion, (3) dewatering, and (4) disposal. These systems were briefly discussed in Chapter 4 (see Tables 4-6–4-8 and Figure 4-3). More detailed discussion on thickener design is given in this chapter. Other processes are covered in Chapters 17–19.

Large volumes of liquid wastes are produced when sludge is processed. These liquid wastes contain high concentrations of suspended solids and BOD_5, and therefore are returned to the primary or secondary treatment facility. A material mass balance analysis for the liquid and sludge treatment processes is performed at average design flow. The material mass balance analysis is discussed in Chapter 13. The quantities of sludge thus calculated from the material mass balance analysis is generally increased to accommodate the temporary increases in the sludge production due to sustained hydraulic or organic loadings. Many regulatory agencies require sludge-processing systems capable of handling an average of 30 consecutive days of sustained maximum flow or organic loading conditions (see Chapter 3). The quantities of sludge under such sustained loadings is normally 5–20 percent of the material mass balance analysis performed at average design flow. Therefore, proper allowances for sustained flow conditions should be made for sludge-handling and disposal facilities.[1,4]

Odor control devices must also be considered for design of sludge management systems. These devices generally fall into three categories: containment, destruction, and masking. Containment of odors is accomplished by covering or enclosing process equipment. Destruction of odors is achieved by chemical oxidation or scrubbing. Masking agents are used as a temporary measure or as a last resort to camouflage widespread odors.

16-4 SLUDGE THICKENING

Sludge contains large volumes of water. Thickening of sludge is used to concentrate solids and reduce the volume. Thickened sludge requires less tank capacity and chemical dosage for stabilization and smaller piping and pumping equipment for transport. Common methods of sludge thickening used at medium to large plants are (1) gravity thickening, (2) dissolved air flotation, and (3) centrifugation. Each of these methods of thickening are discussed below.

16-4-1 Gravity Thickening

Process Description.   Gravity thickening is accomplished in circular sedimentation basins similar to those used for primary and secondary clarification of liquid wastes. Solids coming to the thickener separate into three distinct zones as shown in Figure 16-1. The top layer is a relatively clear liquid. The next layer is the sedimentation zone, which usually contains a stream of denser sludge moving from the influent end toward the thickening zone. In the thickening zone the individual particles of the sludge agglomerate. A sludge blanket is maintained in this zone where the mass of sludge is compressed by material continuously added to the top. Water is squeezed out of interstitial spaces and flows upward to the channels. Deep trusses or vertical pickets are provided to gently stir the sludge blanket and move the gases and liquid toward the surface.

The supernatant from the sludge thickener passes over an effluent weir and is returned to the plant. The thickened sludge is withdrawn from the bottom. For good operation a sufficient sludge blanket is maintained. The sludge volume ratio (SVR) is the volume of sludge blanket held in the thickener divided by the volume of the thickened sludge removed per day. Accepted value of SVR is 0.5–2 d. In general, a higher SVR provides thicker sludge. However, excessive retention times may lead to gasification and buoying of the solids.

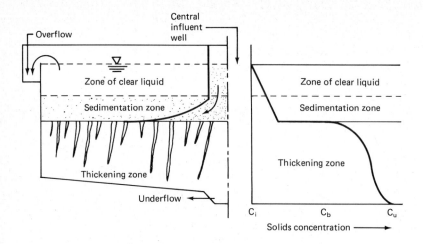

C_i — Influent solids concentration
C_b — Lowest concentration at which sludge thickening begins
C_u — Underflow concentration

Figure 16-1   Typical Concentration Profile of Municipal Wastewater Sludge in a Continuous-flow Gravity Thickener. [From Ref. 1.]

Gravity thickening is used to concentrate solids in sludges from the primary clarifier, trickling filter, and activated sludge. Combined and chemical sludges are also thickened into the gravity thickeners. The degree of thickening may vary from 2 to 5 times the concentration of solids in the incoming sludge. Maximum solids concentration achieved in gravity thickening is normally less than 10 percent. The chemical and waste activated sludges are generally difficult to thicken under gravity.

Design Criteria.   The design criteria for gravity thickeners include the following: (1) minimum surface area based on hydraulic and solids loading, (2) thickener depth, and (3) floor slope. Some of the design criteria are summarized in Table 16-2. Generally, the sludge thickeners are designed with side water depth of 3 m (10 ft), a detention period of 24 h, and hydraulic loading of 10–30 m^3/m^2·d (250–740 gpd/ft^2). To achieve such hydraulic loading, secondary effluent is often blended with the sludge fed into the thickener. The sludge-blending tank may utilize mechanical mixing or air mixing.

Equipment.   Gravity thickeners are generally circular concrete tanks with bottoms sloping toward the center. The equipment includes (1) rotating bottom scraper arm, (2) vertical pickets, (3) rotating scum-collecting mechanism with scum baffle plates, and (4) overflow weir. Other configurations of the tank include circular steel tank as well as rectangular concrete and steel tanks. Circular tanks are generally cheaper because of simplicity of construction, equipment installation, and operation and maintenance.

TABLE 16-2 Design Criteria for Gravity Thickeners

Type of Sludge	Influent Solids Concentration (Percent)	Thickened Solids Concentration (Percent)	Hydraulic Loading $(m^3/m^2 \cdot d)$	Solids Loading $(kg/m^2 \cdot d)$	Solids Capture (Percent)	Overflow, TSS (mg/ℓ)
Primary	1.0–7.0	5.0–10.0	24–33	90–144	85–98	300–1000
Trickling filter	1.0–4.0	2.0–6.0	2.0–6.0	35–50	80–92	200–1000
Waste activated sludge	0.2–1.5	2.0–4.0	2.0–4.0	10–35	60–85	200–1000
Combined primary and waste activated sludge	0.5–2.0	4.0–6.0	4.0–10.0	25–80	85–92	300–800

Source: Adapted in part from Refs. 1–3.
$m^3/m^2 \cdot d \times 24.57 = gal/ft^2 \cdot d$.
$kg/m^2 \cdot d \times 0.2048 = lb/ft^2 \cdot d$.

16-4-2 Dissolved Air Flotation (DAF)

Process Description. Air flotation is primarily used to thicken the solids in chemical and waste activated sludge. Separation of solids is achieved by introducing fine air bubbles into the liquid. The bubbles attach to the particulate matter which then rise to the surface. In a dissolved air flotation system, the air is dissolved in the incoming sludge under a pressure of several atmospheres. The pressurized flow is discharged into a flotation tank that operates at one atmosphere. Fine air bubbles rise that cause flotation of solids. The principal advantage of flotation over gravity thickening is the ability to remove more rapidly and completely those particles that settle slowly under gravity. The amount of thickening achieved is 2–8 times the incoming solids. Maximum concentration of solids in the float may reach 4–5 percent.

Two variations of the dissolved air flotation process include (1) pressurizing total or only a small portion of the incoming sludge and (2) pressurizing the recycled flow from the flotation thickener. The latter method is preferred because it eliminates the need for high-pressure sludge pumps, which are generally associated with maintenance problems. Chemicals such as alum and iron salts and organic polymers are often added to aid the flotation process.

Design Parameters. The important design and operation parameters for dissolved air flotation system include (1) air/solids ratio, (2) solids loading rate, (3) hydraulic loading rate, and (4) polymer dosage. These parameters are established by laboratory tests using dissolved air flotation apparatus.[1] The air/solids ratio is expressed by Eq. (16-1).

$$\frac{A}{S} = \frac{1.3 \, s_a \, (fP - 1)q}{S_a Q} \tag{16–1}$$

where

$\dfrac{A}{S}$ = air/solids ratio

s_a = solubility of air at the required temperature, ml/ℓ*

S_a = solids in incoming sludges, mg/ℓ

f = fraction of air dissolved at pressure P, usually 0.5–0.8

P = pressure in atmospheres = $\dfrac{p + 101.35}{101.35}$ (SI units)

$\qquad\qquad\qquad\qquad\qquad\quad = \dfrac{p + 14.7}{14.7}$ (U.S. Customary units)

p = gage pressure, kPa (lb/in.^{2} gage)

q = recycled flow, or a portion of incoming flow pressurized, m^3/d (mgd)

Q = sludge flow to the thickener, m^3/d (mgd)

---

*The values of s_a at temperatures 0, 10, 20, and 30°C are 29.2, 22.8, 18.7, and 15.7 ml/ℓ respectively.

TABLE 16-3   *Typical Ranges of Design Parameters for Dissolved Air Flotation*

Type of Sludge	Air/Solids Ratio	Solids Loading Rate $(kg/m^2 \cdot d)$	Hydraulic Loading Rate $m^3/m^2 \cdot d$	Polymer Added (mg/kg)	Solids Captured (Percent)	TSS in Side Stream (mg/ℓ)
Primary	0.04–0.07	90–200	90–250	1000–4000	80–95	100–600
Waste activated sludge	0.03–0.05	50–90	60–180	1000–3000	80–95	100–600
Trickling filter	0.02–0.05	50–120	90–250	1000–3000	90–98	100–600
Primary + waste activated sludge	0.02–0.05	60–150	90–250	1000–4000	90–95	100–600

Source: Adapted in part from Refs. 1–3.
$m^3/m^2 \cdot d \times 24.57 = gal/ft^2 \cdot d$.
$kg/m^2 \cdot d \times 0.2048 = lb/ft^2 \cdot d$.

When the entire flow is pressurized (and recycled flow is zero), $q = Q$. The factor 1.3 is the weight of air in mg/ml. The use of Eq. (16-1) is illustrated in many textbooks.[4-6] Typical ranges for various design parameters for dissolved air flotation thickener are summarized in Table 16-3.

Equipment.   The common equipment for dissolved air flotation thickener includes the following: (1) sludge feed pump with air compressors, (2) flotation tank with skimmer, (3) chemical storage and feed system, and (4) thickened sludge pump. A schematic flow scheme for a dissolved air flotation system pressurizing a portion of return flow is shown in Figure 16-2.

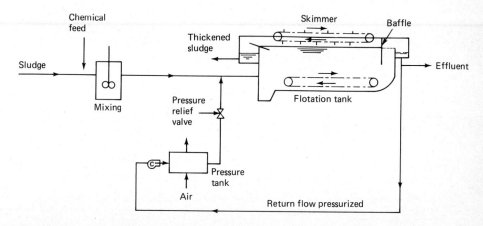

Figure 16-2   *Schematic Flow Scheme of Dissolved Air Flotation System Pressurizing a Small Portion of the Effluent.*

16-4-3 Centrifugation

Process Description.   Centrifugation is a process by which solids are thickened or dewatered from the sludge under the influence of a centrifugal field many times the force of gravity.

There are three basic types of centrifuges available for sludge thickening: (1) basket, (2) disc nozzle, and (3) solid bowl (or scroll-type decanter). The basket centrifuge operates on a batch basis. The disc-nozzle centrifuges are the continuous type but require extensive and careful prescreening and grit removal from the sludge. The solid bowl centrifuges offer continuous operation and have received widespread application in sludge thickening.

Centrifugal thickening of sludge requires high power and high maintenance costs. Use should be limited to plants where space is limited, skilled operation is available, and sludge is difficult to thicken by other means.

Design Parameters.   The important design parameters for solid bowl decanter type centrifuges are summarized in Table 16-4.

Equipment.   The typical equipment for centrifuge thickening are: centrifuge, sludge feed pump, centrate pump, and thickened sludge pump.

16-5   EQUIPMENT MANUFACTURERS

The equipment manufacturers of gravity thickeners, air flotation systems, and centrifuges are listed in Appendix D. Each equipment component must be evaluated for compatibility, operational flexibility, maintenance requirements, and design criteria. Equipment selection considerations and responsibilities of the design engineers are covered in Sec. 2-10.

16-6   INFORMATION CHECKLIST FOR THICKENER DESIGN

Before designing a thickener, the design engineer must develop design data and make many important decisions. Adherence to a carefully developed data for thickener design

TABLE 16-4   Design Parameters for Solid Bowl, Decanter-Type Centrifuges

Parameter	Range of Values
Bowl diameter	36–152 cm (14–60 in.)
Capacity	38–600 ℓ/min (10–160 gal/min)
Gravitational force	1400–2300 times gravity
Feed solids, waste activated sludge	0.3–2.0 percent
Thickened solids	5–8 percent
Solids recovery	85–95 percent
Polymer usage	0–3 g/kg dry solids (0–6 lb/ton)

Source: Adapted in part from Refs. 1 and 2.

will minimize project delays and loss of engineers time in redesigning the units and equipment. A checklist of such design activities is presented below.

1. Conduct material mass balance analysis at average daily design flow, and establish characteristics of primary, secondary, or any other type of sludges reaching the thickener.
2. If primary, secondary, or other treatment facilities have been designed, use the actual quantities and solids concentrations of the sludges reaching the thickeners.
3. Increase the quantities of sludges produced under average daily flow conditions by a certain factor to accommodate the sustained peak flow conditions reaching the plant. The increase in sludge production is normally assumed 5–20 percent.
4. Select the thickening processes for different types of sludges. Dissolved air flotation thickening is generally suited for waste activated sludge, chemical sludges, or solids that settle slowly. On the other hand, gravity thickening is simple, economical to operate, and works well with primary and combined sludges. Centrifugal thickening is effective but requires high power and maintenance costs. At medium-size secondary treatment facilities, the gravity thickening of combined sludges is normally most cost-effective.
5. Develop design parameters such as solids and hydraulic loading rates, air/solids ratio, tank geometry, etc. Laboratory studies may be needed to develop the design parameters.
6. Obtain the design criteria from the concerned regulatory agency.
7. Select equipment manufacturers and equipment selection guides.

16-7 DESIGN EXAMPLE

16-7-1 Design Criteria Used

The following design assumptions and criteria are used for the design of thickeners:

1. Provide two gravity thickeners for thickening of combined primary and waste activated sludge.
2. Design the combined sludge thickener for the following loadings:
   a. Solids loading not to exceed 46.9 kg/m^2·d (9.6 lb/ft^2·d).
   b. Hydraulic loading not less than 9.0 m^3/m^2·d (221 gal/ft^2·d).
3. The design flow and combined sludge solids are summarized in Table 16-5. It may be noted that these quantities are 10–20 percent higher than the values obtained from the material mass balance analysis (Figure 13-12) performed at average design daily flow. These increases are necessary to allow for the maximum sustained loading conditions that may occur at the plant. Procedure and justification for these allowances are also indicated in Table 16-5.
4. The influent structure shall consist of an influent well that will receive the primary and secondary sludges and plant effluent for blending to maintain proper solids consistency and hydraulic loading. The blended sludge shall be pumped into the central well of each thickener.
5. The effluent structure for thickener overflow shall consist of V notches, effluent

TABLE 16-5   *Characteristics of Primary, Secondary, and Combined Sludges for Design of Gravity Thickener*

Types of Sludge	Dry Solids (kg/d)	Solids (Percent)	Flow (m³/d)
A. Average design condition developed from material mass balance analysis given in Chap. 13 (Fig. 13-12)			
Primary sludge	6227	4.5	134
Waste activated sludge	1911	0.375	510
Combined sludge	8138	1.24[a]	644
B. Peak design condition with allowance for peak sustained loadings			
Primary sludge	6725[b]	4.5	145[c]
Waste activated sludge	2332[d]	0.375	622
Combined sludge	9057[e]	1.16[a]	767[f]

[a]Specific gravity of combined sludge = 1.02.
[b]Dry solids have been increased by 8 percent of the quantity of primary sludge produced at average daily design flow.
[c]Volume of primary sludge $= \dfrac{6725 \text{ kg/d} \times 1000 \text{ g/kg}}{0.045 \text{ g/g} \times 1.03 \times 1 \text{ g/cm}^3 \times 10^6 \text{ cm}^3/\text{m}^3} = 145 \text{ m}^3/\text{d}.$
[d]The quantity of waste activated sludge was calculated in Chapter 13. This quantity was obtained after increasing the influent flow and BOD_5 to the aeration basin. Therefore, no additional allowance for waste activated sludge under sustained maximum loading conditions were necessary.
[e]The quantity of solids in combined sludge is 11.3 percent higher than that obtained by material mass balance analysis.
[f]The volume of combined sludge is 19.1 percent higher than that at average design flow.

launder, and a common sump. The design of effluent structure shall be similar to that of the final clarifier described in Chapter 13. The thickener overflow shall be returned to the aeration basin.

6.   The thickened sludge from both thickeners shall be pumped into the anaerobic digesters.

16-7-2 Design Calculations

A. Thickener Area and Diameter

1.   Compute the surface area of the thickener based on solids loading
     The surface area is computed using the solids loading and hydraulic loading

$$\text{Total surface area for thickening at solids loading of 46.9 kg/m}^2 \cdot \text{d} = \frac{9057 \text{ kg/d}}{46.9 \text{ kg/m}^2 \cdot \text{d}} = 193.1 \text{ m}^2$$

2. Compute hydraulic loading and volume of dilution water

$$\text{Hydraulic loading} = \frac{\text{sludge volume/d}}{\text{surface area}}$$

$$= \frac{767 \text{ m}^3/\text{d}}{193.1 \text{ m}^2} = 3.97 \text{ m}^3/\text{m}^2 \cdot \text{d}$$

This is significantly lower than the normal range of the hydraulic loading rate for gravity thickeners for combined sludges. A decrease in surface area will exceed the solids loading outside the acceptable limit. Therefore, dilution water must be blended with the incoming sludge. Quantity of dilution water added to achieve a hydraulic loading greater than $9.0 \text{ m}^3/\text{m}^2\cdot\text{d}$ is calculated as follows. Start with an assumption that hydraulic loading $= 9.8 \text{ m}^3/\text{m}^2\cdot\text{d}$

Total flow to the thick-
ener $= 9.8 \text{ m}^3/\text{m}^2 \cdot \text{d} \times 193.1 \text{ m}^2 = 1892 \text{ m}^3/\text{d}$

Combined sludge flow $= 767 \text{ m}^3/\text{d}$

Dilution water needed $= 1892 \text{ m}^2/\text{d} - 767 \text{ m}^3/\text{d} \qquad = 1125 \text{ m}^3/\text{d}$

Total solids concentra-
tion in the blended sludge[*] $= \dfrac{9057 \text{ kg/d} \times 1000 \text{ g/kg}}{1.01 \times 1 \text{ g/cm}^3 \times 10^6 \text{ cm}^3/\text{m}^3 \times 1892 \text{ m}^3/\text{d}}$

$$= 0.0047$$

$$= 0.0047 \times 100 = 0.47 \text{ percent}$$

3. Select geometry of the gravity thickener

Provide two circular thickeners

$$\text{Area of each thickener} \quad = \frac{193.1 \text{ m}^2}{2} = 96.6 \text{ m}^2$$

$$\text{Diameter of each thickener} = \sqrt{\frac{4}{\pi} \times 96.6 \text{ m}^2} = 11.09 \text{ m} \ (36.4 \text{ ft})$$

Provide two thickeners each of 11.6 m (38 ft) diameter[†]

$$\text{Surface area of each thick-}\atop\text{ener} \quad = \frac{\pi}{4} (11.6)^2 = 105.7 \text{ m}^2$$

4. Check solids and hydraulic loading at peak condition when both thickeners are operating

Solids loading when both
units are operating $= \dfrac{9057 \text{ kg/d}}{2 \times 105.7 \text{ m}^2} = 42.8 \text{ kg/m}^2 \cdot \text{d}$

Hydraulic loading when both
units are operating $= \dfrac{1892 \text{ m}^3/\text{d}}{2 \times 105.7 \text{ m}^2} = 9.0 \text{ m}^3/\text{m}^2 \cdot \text{d}$

[*]Specific gravity of blended sludge is assumed to be 1.01.
[†]Manufacturers usually supply equipment in 2 ft increments of diameter.

5. Check solids and hydraulic loadings at average flow when one thickener is out of service

With only one unit in service and at average daily design flows, the solids and hydraulic loadings are calculated as follows:

$$\text{Solids loading*} = \frac{8138 \text{ kg/d}}{105.7 \text{ m}^2} = 77.0 \text{ kg/m}^2 \cdot \text{d}$$

The proportionate blended flow to the thickener at average design flow
$$= \frac{1892 \text{ m}^3/\text{d}}{9057 \text{ kg/d}} \times 8138 \text{ kg/d}$$

$$= 1700 \text{ m}^3/\text{d}$$

$$\text{Hydraulic loading}^\dagger = \frac{1700 \text{ m}^3/\text{d}}{105.7 \text{ m}^2} = 16.1 \text{ m}^3/\text{m}^2 \cdot \text{d}$$

B. Thickener Depth. The total sidewater depth of the gravity thickener comprises three separate items: clear liquid zone, the settling zone, and the thickening zone. Provide a freeboard of 0.6 m (2 ft). Generally, in gravity thickeners the clear liquid zone of 1.0 m (3 ft), and settling zone of 1.5 m (5 ft) are considered sufficient.

The thickening zone is generally sized to allow at least 1 d detention time for the sludge. An estimate of average sludge concentration must be made.

1. Determine the solids concentration in the thickener at the upper part of the thickening zone

The solids concentration in the combined and blended sludge = 0.47 percent. Assume that the blended sludge reaches its original solids concentration of 1.16 percent at the upper part of the thickening zone (see Table 16-5).

2. Determine the solids concentration at the bottom of the thickened zone

The desired concentration of thickened sludge is at least 6 percent. The concentration of solids at the bottom of thickened zone is therefore 6 percent.

3. Compute average sludge concentration in the thickening zone

$$\text{Percent average solids concentration} = \frac{1.16 + 6.00}{2} = 3.58 \text{ percent}$$

4. Compute the side water depth of the thickening zone for an average retention time of 1 d

Assume the depth of thickening zone
$$= h \text{ (m)}$$

Volume of sludge blanket per thickener
$$= \frac{\pi}{4} (11.6 \text{ m})^2 h \text{ (m)}$$
$$= 105.7 \, h \text{ (m}^3)$$

---

*The solids at average daily design flow is obtained from Table 16-5.
†The volume of dilution water is reduced from 1125 m³/d to 1056 m³/d at average daily design flow (1056 m³/d = (1700 − 644) m³/d, where 1700 m³/d is total flow to the thickener and 644 m³/d is the average volume of combined sludge from Table 16-5.

Amount of solids
in the thickening zone $= \dfrac{(105.7\ h)\ \text{m}^3 \times 0.0358\ \text{g/g}}{\times\ 1.03 \times 1\ \text{g/cm}^3 \times 10^6\ \text{cm}^3/\text{m}^3} \times \dfrac{1}{1000\ \text{g/kg}}$
at 3.58 percent solids
$= 3898\ h\ (\text{kg})$

Quantity of solids held
in the thickening zone $= \dfrac{9057}{2}\ \text{kg/d} = 4529\ \text{kg/d}$
per thickener

At one day of solids retention period,

$$\frac{(3898\ h)\ \text{kg}}{4529\ \text{kg/d}} = 1\ \text{d}$$

$$h = \frac{4529\ \text{kg}}{3898\ \text{kg}} = 1.2\ \text{m}\ (3.8\ \text{ft})$$

Provide 1.4 m depth of thickening zone. This will give 17 percent additional allowance against unusual conditions that may occur due to sustained loadings or equipment breakdown.

Total depth of the thickener is as follows:

Freeboard	0.6 m (2.0 ft)
Clear liquid zone	1.0 m (3.0 ft)
Settling zone	1.5 m (4.9 ft)
Thickening zone	1.4 m (4.5 ft)
Total	4.5 m (14.5 ft)

5. Compute the depth of thickener at the central well
   The bottom slope for most thickeners with central well sludge withdrawal and with sludge scrapers is 17 cm/m (2 in./ft).

   Therefore total drop
   to the central well $= \dfrac{17\ \text{cm/m}}{100\ \text{cm/m}} \times \dfrac{11.6\ \text{m}}{2}$ (radius) $= 1.0\ \text{m}$

   The depth of the thickener at the central well $= 4.5\ \text{m} + 1\ \text{m} = 5.5\ \text{m}\ (18.1\ \text{ft})$
   Note that the bottom cone adds volume to the thickening zone. No allowance is made for this in calculating the side water depth. The added volume will give extra storage for the thickened sludge in the thickener. The design details of gravity thickener are given in Figure 16-3.

C. Blending Tank. The primary and waste activated sludge must be blended thoroughly to achieve a consistent feed to the thickener. The blending unit also provides a convenient place to meter dilution water, and add pH adjusters, thickening aids, flocculants, and other chemicals. The size of the blending tank depends on the pumping schedule of primary and waste activated sludge, and dilution water.

1. Compute the dimensions of the blending tank
   Provide sludge storage and blending period of 2 h under peak design sludge loading

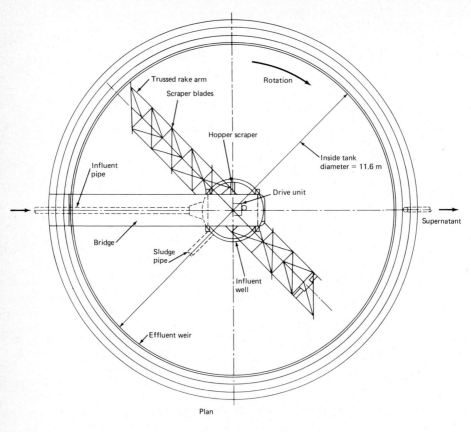

Plan

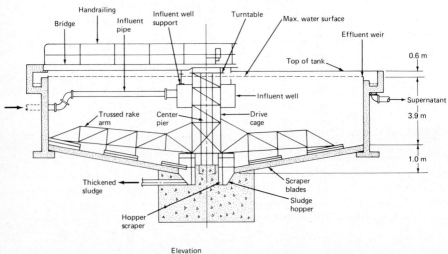

Elevation

Figure 16-3   Design Details of Gravity Thickener.

Quantity of primary and waste activated sludge and dilution water	$= 1892 \text{ m}^3/\text{d}$
Volume of blending tank	$= 1892 \text{ m}^3/\text{d} \times \dfrac{2 \text{ h}}{24 \text{ h/d}}$
	$= 158 \text{ m}^3$
Provide liquid depth	$= 3 \text{ m}$
Provide freeboard	$= 0.6 \text{ m}$
Area of the blending tank	$= \dfrac{158 \text{ m}^3}{3 \text{ m}} = 52.7 \text{ m}^2$
Diameter of the blending tank	$= \sqrt{52.7 \text{ m}^2 \times \dfrac{4}{\pi}} = 8.2 \text{ m}$

Select 8.2-m (27-ft) diameter and 3-m (10-ft) deep blending tank. Design details of the blending tank are shown in Figure 16-4.

2. Select sludge-mixing and blending system
Mixing keeps solids in suspension, and blends primary and waste activated sludge. Mixing may be accomplished by (1) coarse air diffusers, (2) liquid recirculation, or (3) mechanical paddles. Diffuse air mixing has the advantage

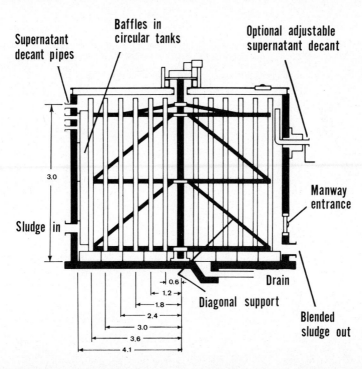

Figure 16-4 Details of Sludge-blending Tank, and Blender Mixer. (Courtesy EIMCO Process Equipment Company.) All dimensions are in meters.

of freshening the sludge that will minimize odors in the thickener dome and may improve thickening. However, due to odor problems the blending tank should be covered and the captured air should be scrubbed.

The design of mixing equipment by gas and sludge recirculation is given in Chapter 17. The same design procedure can also be used for the blending tank. In this design, however, a mechanical paddle mixer is provided to illustrate the procedure of a different design.

3. Select the design equations

Equations (16-2) and (16-3) are commonly used equations for design of flocculation paddles:[4-6]

$$P = G^2 \mu V \tag{16-2}$$

$$A = \frac{2P}{C_D \rho v^3} \tag{16-3}$$

where

P = power requirement for mixing, W (ft · lb/s)

G = mean velocity gradient, s^{-1}. Typical values for G vary from 30 to 85 per s

μ = dynamic viscosity, N s/m^2 (lb s/ft^2)

V = volume of the tank, m^3 (ft^3)

A = area of paddle, m^2 (ft^2)

v = relative velocity of paddles in fluid, m/s (ft/s), usually 0.7–0.8 of paddle tip speed, v_p

C_D = Coefficient of drag of flocculator paddles moving perpendicular to the fluid. For rectangular paddles, the C_D is about 1.8

ρ = Mass fluid density, kg/m^3

4. Compute power requirement

Volume of blending tank V = 158 m^3

Assume G = 60/s

Assume μ = 2 times the viscosity of water at 20°C

$\qquad$ = 2 × 1.002 × 10^{-3} N s/m^2

$\qquad$ = 2.004 × 10^{-3} N s/m^2

$\qquad P$ = (60/s)^2 × 2.004 × 10^{-3} N s/m^2 × 158 m^3

$\qquad$ = 1140 N m/s = 1140 W

$\qquad$ = 1.14 kW (1.53 HP)

Motor power at 75 percent efficiency = 1.14 kW/0.75 = 1.5 kW

5. Compute area and dimensions of the paddles

Provide 12 vertical flat paddles (6 on each side of the central shaft as shown in Fig. 16-4)

The distances to the middle of the paddles from the center of the column are 3.6, 3.0, 2.4, 1.8, 1.2, and 0.6 m

Assume rotational speed of the paddle shaft, n = 0.06 revolutions/s

Average paddle speed v_p = 2π × distance from center × n

$$P = \tfrac{1}{2} C_D \, \rho A v^3$$

If a = area of each vertical paddle, and v is its relative velocity $P = \tfrac{1}{2} C_D \rho \Sigma a v^3$ using C_D = 1.8

$$\rho = 1.01^* \times 1 \text{ g/cm}^3 \times 10^6 \text{ cm}^3/\text{m}^3 \ (1000 \text{ g/kg})^{-1} = 1010 \text{ kg/m}^3$$
$$v = 0.75 \, v_p$$
$$P = \tfrac{1}{2} C_D \rho \, [2a(0.75\pi \times n \times 3.6 \times 2)^3 + 2a \, (0.75\pi \times n \times 3.0 \times 2)^3$$
$$+ \ 2a(0.75\pi \times n \times 2.4 \times 2)^3 + 2a(0.75\pi \times n \times 1.8 \times 2)^3$$
$$+ \ 2a(0.75\pi \times n \times 1.2 \times 2)^3 + 2a(0.75\pi \times n \times 0.6 \times 2)^3]$$
$$P = \tfrac{1}{2} \times 1.8\rho \times 2a(0.75\pi \times 0.06 \times 2)^3 \, (3.6^3 + 3.0^3 + 2.4^3 + 1.8^3$$
$$+ \ 1.2^3 + 0.6^3)$$
$$= \tfrac{1}{2} \times 1.8 \times 1010 \times 2a(0.023)(95.256)$$
$$1140 = 3983a$$
$$a = 0.29 \text{ m}^2$$

The height of the paddle = 2.5 m

Width of the paddle   = 0.29 m²/2.5 m = 0.12 m (4.7 in.)

Provide 12 redwood vertical paddles, each 2.5 m × 12 cm flat. The area of the diagonal support is ignored in this design.

6. Select pumping arrangement

   The primary and waste activated sludges shall be pumped from their respective areas to the sludge-blending tank. Design and operational details of the primary sludge and waste activated sludge pumps are given in Chapters 12 and 13, respectively.

   The dilution water shall be obtained from the chlorine contact basin.[†] Two constant-speed centrifugal pumps shall be provided. Each pump shall be synchronized with the operation of the primary sludge pump. Thus, the volume of the dilution water reaching the blending tank shall be proportional to the flow of primary sludge.

   The blended sludge shall be pumped from the blending tank to the thickeners. Two identical constant-speed centrifugal pumps (one for each thickener) shall be provided to transfer blended sludge to the thickener. An additional pump shall be provided as a standby unit. The control system shall include high-level start and low-level stop, and an extra high-level alarm (see Chapter 9 for more details). Metering for all pumps shall be provided in order to control the

---

*Specific gravity of the blended sludge.

†Many designers have successfully utilized raw wastewater, primary or secondary effluents, or side streams from the sludge-processing areas as dilution water.

blending process and feed the thickeners. Figure 16-5 is a schematic layout of the blending and thickening area.

D. Influent Structure.   The influent structure to the thickener consists of a central well similar to that of the final clarifier. Design details are given in Figure 16-3.

E. Thickened Sludge Withdrawal
    1.   Compute quantity of thickened sludge.

Quantity of sludge reaching both thickeners	= 9057 kg/d
Solids capture efficiency	= 85 percent
Quantity of solids withdrawn from both thickeners	= 9057 kg/d × 0.85
	= 7699 kg/d

    2.   Compute thickened sludge-pumping rate

$$\text{Volume of thickened sludge withdrawal rate from each thickener at 6 percent solids} = \frac{3849 \text{ kg/d} \times 1000 \text{ g/kg}}{0.06 \text{ g/g} \times 1.03 \times 1 \text{ g/cm}^3 \times 10^6 \text{ cm}^3/\text{m}^3}$$

$$= 62.3 \text{ m}^3/\text{d}$$

    3.   Select thickened sludge pumps
        Provide one plunger-type pump for each thickener to transfer thickened sludge to the digester. Each pump shall have variable time clock control to pump for 5 min at 96-min intervals. The constant pumping rate shall be approximately 0.83 m³/min (219 gpm). Each sludge pump shall have cross-connections such that if one pump fails, the other pump shall serve both thickeners. Each pump will have low-level override to stop pumping operation if water surface in the thickener drops below a certain predetermined level.

    4.   Check the sludge volume ratio (SVR)
        The sludge volume ratio (SVR) is the volume of the sludge blanket held into the thickener divided by the volume of the thickened sludge removed per day.

$$\text{Volume of sludge blanket held in each thickener (thickening zone)} = \frac{\pi}{4} (11.6 \text{ m})^2 \times 1.4 \text{ m}$$

$$= 148 \text{ m}^3$$

$$\text{Volume of thickened sludge withdrawn per thickener} = 62.3 \text{ m}^3\text{d}$$

$$\text{SVR} = 148 \text{ m}^3/62.3 \text{ m}^3/\text{d} = 2.4 \text{ d}$$

This is slightly higher than the value generally used for normal operation of the thickener. Increase sludge withdrawal rate or decrease the depth of thickening zone if necessary.

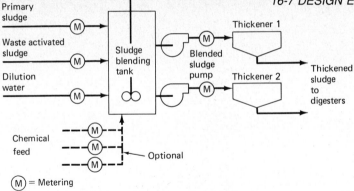

(a) Sludge metering system

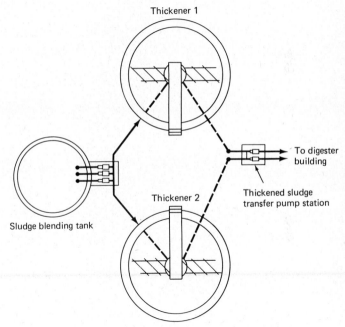

(b) Layout of sludge blending tank and gravity thickeners

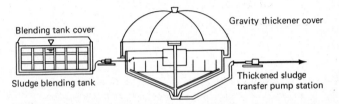

(c) Sectional view of sludge blending tank and gravity thickeners

Figure 16-5   Schematic of Metering and Layout of Sludge-Blending and Thickening Facilities.

F. Effluent Structure. The effluent structure consists of V-notch weirs around the periphery, effluent launder, and outlet pipe. The entire arrangement is similar to that of the final clarifiers. The design procedure for V notch, effluent launder, outlet pipe, and hydraulic profile may be found in Chapters 12 and 13.

The effluent will be discharged under gravity into a sump. In this sump, the supernatants from digesters and dewatering facilities will also be collected. The combined returned flow will be pumped to the aeration basin. The dimensions of the sump are arbitrary. Provide 6 m × 6 m × 4 m (deep) sump. The bottom of the sump shall be sloped to provide a hopper at the pump suction.

G. Quality of the Supernatant from the Thickener Overflow

1. Compute the volume of thickener overflow

$$\text{Average volume of thickener overflow}^* = 1892 \text{ m}^3/\text{d} - 2 \times 62.3 \text{ m}^3/\text{d}$$
$$= 1767 \text{ m}^3/\text{d}$$

2. Compute the concentration of solids in the thickener overflow

$$\text{Amount of solids lost in the thickener overflow}^\dagger = 9057 \text{ kg/d} \times 0.15$$
$$= 1359 \text{ kg/d}$$

$$\text{TSS concentration}^\ddagger = \frac{1359 \text{ kg/d} \times 1000 \text{ g/kg} \times 1000 \text{ mg/g}}{1767 \text{ m}^3/\text{d} \times 1000 \text{ } \ell/\text{m}^3}$$
$$= 769 \text{ mg/}\ell$$

3. Compute the concentration of BOD_5 in the thickener overflow

$$BOD_5 = 769 \text{ mg/}\ell \times 0.541 \text{ [ratio of } BOD_5 \text{ and TSS (see Sec. 13-7-3, step A7]}$$
$$= 416 \text{ mg/}\ell$$

H. Design Details. The design details of gravity thickener are given in Figures 16-3–16-5.

16-8 OPERATION AND MAINTENANCE, AND TROUBLESHOOTING AT SLUDGE-THICKENING FACILITY

Gravity sludge thickening can create serious odors if the units are not carefully designed, operated, and maintained. Poorly thickened sludge may hydraulically overload the sludge

---

*The actual volume of thickener overflow (1767 m³/d) is considerably higher than that developed in the material mass balance analysis. Higher overflow is due to dilution water.

†Total amount of solids lost in the thickener overflow is 11 percent larger than that estimated in the material mass balance analysis at average design flow [(1359 kg/d − 1221 kg/d) × 100/1221 kg/d = 11%]. This increase is due to higher values of flow and BOD_5 assumed in Sec. 13-7-4, step A1.

‡Due to dilution water, TSS in the thickener overflow is considerably smaller than that obtained from material mass balance analysis.

digester, and high solids in the thickener overflow will increase the loading to the plant. Therefore, proper operation and maintenance of the facility is essential. Important operation and maintenance considerations of a sludge-thickening facility are briefly described below.

16-8-1 Common Operational Problems and Troubleshooting Guide

1. Septic odors or rising sludge in thickener is generally due to low or infrequent thickened sludge pumping rate, low thickener overflow rate, or too high depth of sludge blanket. The problem can be overcome by pumping thickened sludge more frequently, increasing the dilution for overflow rate, chlorinating influent, or adding air to the blending tank.

2. Too thin thickened sludge may be due to high overflow rate, high underflow rate, or short circuiting through the tank. This situation is overcome by reduction in influent sludge-pumping rate, reduction in dilution water, reduction in pumping of thickened sludge, and maintenance of high sludge blanket. Short circuiting in gravity thickeners can be detected if uneven discharge of solids occurs over the effluent weir. Weir leveling and change in baffle arrangements may be necessary.

3. Torque overload of sludge-collecting equipment may be due to accumulation of dense sludge or heavy foreign object jamming the scraper. The problem may be solved by agitation of sludge blanket in front of the collector arms with rod or water jets. Foreign objects must be removed by grappling device or by draining the basin.

4. Plugging of sludge lines and pump may be due to too thick sludge. The lines should be flushed and all valves should be fully opened.

5. Hard-to-remove sludge may be due to too much grit. Problem can be reduced by removing grit efficiently.

6. Excessive growth on weirs may be due to accumulation of solids and the resultant growth. The weirs and all surfaces should be frequently and thoroughly cleaned by a water jet.

16-8-2 Routine Operation and Maintenance

1. Clean all vertical walls and channels by squeegee daily, and hose down and clean sludge spills without delay.

2. Check the sludge level daily. The sludge level should be kept well below the top of the thickener. Sludge wasting should be controlled to maintain proper sludge blanket.

3. Check daily electrical motor for overall operation, bearing temperature, overload detector, and unusual noises.

4. Check oil level in gear reducers weakly and add as needed. Change oil quarterly and lubricate worn gears weekly.

5. Drain the thickener annually and inspect underwater portion of concrete structure and mechanisms. Inspect the mechanical equipment for wear and corrosion, adjust mechanism, set proper clearance for flights at tank walls. Patch defective concrete. Metal surfaces should be inspected for corrosion, cleaning, and painting.

16-9 SPECIFICATIONS

The specifications of gravity sludge thickeners are briefly covered here to describe the general features of the equipment. Detailed specifications of similar equipment may be found in Chapters 12 and 13.

16-9-1 General

The gravity sludge thickeners shall consist of complete assembly, including blending tank, feed pumps, collector mechanism with flights, above water center drive mechanisms, influent central well, drive cage, access bridge, bridge support, center pier support, and overload alarm system. Following is the summary of thickener components:

Blending tank
    Number of units   1
    Dimensions   8.2-m diam, 3-m side water depth, 0.6-m freeboard
    Mixing arrangement Mechanical paddle mixer
Gravity thickener
    Number of units   2
    Dimensions   11.6-m diam, 3.9-m side water depth, 0.6-m freeboard, bottom floor slope 17 cm/m
    Center column   0.38 m diam
    Feed well   2-m diam, 1.5 m deep

16-9-2 Materials and Fabrication

All structural steel, iron castings, and concrete shall conform to the current ASTM standards.

16-9-3 Sludge Blending

The mechanical paddle assembly shall consist of a fabricated structural steel frame, and redwood paddle blades attached to the frame as shown on the drawings. The frame shall be designed to resist static and dynamic stresses under all operating conditions. The drive unit shall consist of a primary variable-speed motor reducer, a secondary worn gear reduction unit, and a motor reducer transfer device.

16-9-4 Gravity Thickener

All equipment specified herein is intended for use with combined primary and waste activated sludge. The gravity thickener mechanisms shall be designed to handle thickened sludge up to a maximum of 10 percent solids concentration and be capable of continuously plowing the thickened sludge and moving the settled sludge to the center channel for removal. Each thickener shall be of the center feed and peripheral overflow type, with a central driving mechanism which shall support and rotate two attached rake arms. Rake collector blades attached to the arms shall be arranged to move the settled sludge on the tank bottom to a concentric sludge channel surrounding the center column. The scrapers attached to the arms shall provide 100 percent coverage of the tank floor.

All gravity-thickening equipment shall be designed so that there will be no chains, sprockets, bearings, or operating mechanisms below the liquid surface. The drive assembly shall comprise an electric motor connected to a primary gear reducer, drive and driver sprockets with drive chain, an intermediate-worn gear reducer, pinion gear, turntable base and main spur gear, and complete automatic overload-actuating system.

16-10 PROBLEMS AND DISCUSSION TOPICS

16-1. Calculate air to solids ratio, and dimensions of the flotation tank. Use the following data:

Combined sludge	$= 760$ m³/d
Solids concentration	$= 1.2$ percent
Operating temperature	$= 30°C$
The solids loading rate not to exceed	$= 75$ kg/m²·d
Hydraulic loading rate not to exceed	$= 80$ m³/m²·d
Length to width ratio	$= 5:1$
Depth	$= 2.5$ m
Operating pressure	$= 19$ atm

16-2. Calculate the dimensions of a gravity thickener to thicken combined sludge. The sludge volume is 500 m³/d and has solids concentration of 1.0 percent, and sp. gr. of 1.008. The solids loading must not exceed 50 kg/m²·d, and hydraulic loading must not be less than 6 m³/m²·d. The underflow concentration of thickened sludge is 6 percent (sp.gr $= 1.03$). Calculate the volume of thickened sludge, volume of dilution water, volume of supernatant, and TSS concentration in the supernatant. Assume solids capture efficiency of 85 percent.

16-3. Design the effluent structure as described in Sec. 16-7-2, step F. Use the design procedure as given for the secondary clarifier in Chap. 13. Effluent launder is 0.3 m wide and 90° V notches are arranged on one side of the launder around the periphery.

16-4. Determine the power requirement, and the paddle area required to achieve $G = 60$/s in a tank that has volume $= 4000$ m³, water temperature $= 20°C$, and $C_D = 1.8$. The paddle-tip velocity is 0.6 m/s.

16-5. Determine the motor power required for a sludge blending tank using redwood paddles. There are 12 paddles, 6 on each side of the central shaft. Each paddle is 0.10 m wide × 2.75 m high. The distances to the middle of each paddle from the center of the shaft are 0.5, 1.0, 1.5, 2.0, 2.5 and 3.0 m. Assume rotational speed of the shaft $= 0.06$ rpm, $C_D = 1.8$, motor efficiency $= 75$ percent, and specific gravity of blended sludge $= 1.01$.

16-6. Discuss advantages and disadvantages of gravity thickening over dissolved air flotation.

16-7. Under what conditions centrifugation may be selected for sludge thickening. Describe various types of centrifuges that may be used for sludge thickening.

16-8. Calculate the volume of dilution water needed to achieve a hydraulic loading of 9.5 m³/m²d in a gravity thickener. The quantity of dry solids in waste sludge is 8000 kg/d and the sludge volume is 900 m³/d. Assume solids loading in the thickener is 40 kg/m²d.

REFERENCES

1. U.S. Environmental Protection Agency, *Process Design Manual, Sludge Treatment and Disposal,* USEPA, Technology Transfer, EPA-625/1-79-011, September 1979.

2. U.S. Environmental Protection Agency, *Process Design Manual for Sludge Treatment and Disposal,* USEPA, Technology Transfer, EPA-625/1-74-006, October 1974.
3. U.S. Environmental Protection Agency, *Process Design Manual Municipal Sludge Landfills,* Environmental Research Information Center, Technology Transfer, Office of Solid Waste, EPA-625/1-78-010, SW705, October 1978.
4. Metcalf and Eddy, Inc., *Wastewater Engineering: Treatment, Disposal, and Reuse,* McGraw-Hill Book Co., New York, 1979.
5. Reynolds, T. D., *Unit Operations and Processes in Environmental Engineering,* Brooks/Cole Engineering Division, Monterey, Cal., 1982.
6. Schroeder, E. D., *Water and Wastewater Treatment,* McGraw-Hill Book Co., New York, 1977.
7. Culp, G. L., and N. F. Heim, *Performance Evaluation and Troubleshooting at Municipal Wastewater Treatment Facilities,* U.S. Environmental Protection Agency, MO-16 EPA-430/9-78-001, Washington, D.C., January 1978.

17

Sludge Stabilization

17-1 INTRODUCTION

The principal purposes of sludge stabilization are to reduce pathogens, eliminate offensive odors, and control the potential for putrefication of organic matter. Sludge stabilization can be accomplished by biological, chemical, or physical means. Selection of any method depends largely on the ultimate sludge disposal method. As an example, if the sludge is dewatered and incinerated, frequently no stabilization procedure is employed. On the other hand, if the sludge is applied on land, stabilization is necessary to control odors and pathogens.

Various methods of sludge stabilization are (1) anaerobic or aerobic digestion (biological), (2) chemical oxidation or lime stabilization (chemical), and (3) thermal conditioning (physical). In recent years, because of its inherent energy efficiency and normally low chemical requirements, anaerobic digestion process is the most widely selected stabilization process at medium- and large-size municipal plants.

In this chapter each of the above methods of sludge stabilization is briefly discussed. Because of the increasing use of anaerobic and aerobic sludge digestion, these processes are discussed in greater detail. Also, step-by-step design procedure, design details, and operation and maintenance for an anaerobic digester is presented in the Design Example.

17-2 ANAEROBIC DIGESTION

17-2-1 Process Description

Anaerobic digestion utilizes airtight tanks in which anaerobic microorganisms stabilize the organic matter producing methane and carbon dioxide. The digested sludge is stable, inoffensive, low in pathogen count, and suitable for soil conditioning. Major difficulties with anaerobic digestion are high capital cost, vulnerability to operational upsets, and tendency to produce poor supernatant quality.

Anaerobic digestion involves a complex biochemical process in which several groups of facultative and anaerobic organisms simultaneously assimilate and break down organic matter. The process may be divided into two phases: acid and methane.

451

In acid phase, facultative acid forming organisms convert the complex organic matter to organic acids (acetic, propionic, butyric, and other acids). In this phase little change occurs in the total amount of organic material in the system, although some lowering of pH results. The methane phase involves conversion of volatile organic acids to methane and carbon dioxide.

The anaerobic process is essentially controlled by the methane-forming bacteria. Methane formers are very sensitive to pH, substrate composition, and temperature. If the pH drops below 6.0, methane formation essentially ceases, and more acid accumulates, thus bringing the digestion process to a standstill. Thus, pH and acid measurements constitute important operational parameters. The methane bacteria are highly active in the mesophilic (27–43°C) and thermophilic (45–65°C) ranges. Anaerobic digesters are most commonly operated in the mesophilic range (35–40°C). Recent thinking, however, is to operate the digester in the thermophilic range, for which the main advantages are increased efficiency and improved dewatering.[1–3]

17-2-2 Types of Anaerobic Digesters

The anaerobic sludge digesters are of two types: standard rate and high rate. In the standard-rate digestion process the digester contents are usually unheated and unmixed. The digestion period may vary from 30 to 60 d. In a high-rate digestion process, the digester contents are heated and completely mixed. The required detention period is 10–20 d.

Often a combination of standard- and high-rate digestion is achieved in two-stage digestion. The second stage digester mainly separates the digested solids from the supernatant liquor: although additional digestion and gas recovery may also be achieved. Schematic flows of standard- and high-rate as well as single- and two-stage digestion processes are given in Figure 17-1.

17-2-3 Process Design and Control

The most important factors controlling the design and operation of anaerobic digestion are digester capacity, digester heating and temperature control, mixing, gas production and utilization, digester cover, supernatant quality, and sludge characteristics. Each of these factors is discussed below.

Digester Capacity. The digester capacity is generally based on (1) digestion period, mean cell residence time, or solids retention time, (2) volumetric loading, (3) population basis, and (4) observed volume reduction. Each of these bases are discussed below.

Digestion Period, Mean Cell Residence Time, or Solids Retention Time. Most of the standard-rate digesters are designed for a digestion period of 30–60 d. A high-rate anaerobic digester is heated and is a completely mixed biological reactor (complete-mixed reactors are defined in Chapter 13). These digesters are generally designed for a digestion period of 10–20 d as solids are adequately stabilized in this period. Many kinetic equations have been proposed that describe the biochemical reactions.[3–5] These equations and design calculations are shown in the Design Example.

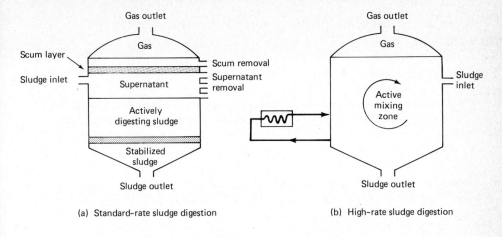

(a) Standard-rate sludge digestion

(b) High-rate sludge digestion

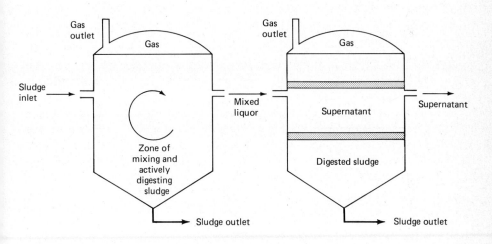

(c) Two-stage sludge digestion

Figure 17-1   Typical Anaerobic Sludge Digesters.

Volumetric Loading.   The digester capacity is also estimated using the volumetric loading. The volumetric loading is generally expressed as kg total volatile solids added per d per m³ of the digester capacity. Typical organic loadings for design of standard- and high-rate digesters are summarized in Table 17-1.

Population Basis.   The digester capacity is estimated on population basis using 120 g of solids per capita per d. The design values on population basis are given in Table 17-1.

Observed Volume Reduction.   During digestion the volume of solids is generally re-duced and a certain amount of supernatant may be returned to the plant. Thus the volume

TABLE 17-1   *Typical Design Criteria for Standard-Rate and High-Rate Digesters*

Parameter	Standard Rate	High Rate
Solids retention time (SRT), days	30–60	10–20
Sludge loading, kg VS/m^3·d	0.64–1.60	2.40–6.41
lb VS/ft^3·d	0.04–0.10	0.15–0.40
Volume criteria		
Primary sludge,		
m^3/capita	0.03–0.04	0.02–0.03
ft^3/capita	2–3	1–2
Primary sludge + waste activated sludge,		
m^3/capita	0.06–0.08	0.02–0.04
ft^3/capita	4–6	1–3
Primary sludge + trickling filter,		
m^3/capita	0.06–0.14	0.02–0.04
ft^3/capita	4–5	1–3
Sludge feed solids concentration, percent dry wt.		
Primary + waste activated sludge	2–4	4–6
Digested solids underflow concentration,		
percent dry wt.	4–6	4–6

Source: Refs. 1 and 2.

of sludge remaining in the digester decreases exponentially. The required volume of the digester is calculated from Eq. (17-1):[3]

$$V = [Q_{in} - \tfrac{2}{3}(Q_{in} - Q_{out})] D_T \qquad (17\text{–}1)$$

where

V = volume of digester, m^3 (ft^3)
Q_{in} = sludge feed rate, m^3/d (ft^3/d)
Q_{out} = sludge withdrawal rate, m^3/d (ft^3/d)
D_T = digestion period, d

In many designs, the digester capacity should be checked to make sure that the minimum digestion period during periods of high flow does not fall below the critical value of 10 d. At less than 10 d of digestion period the production of methane bacteria cannot keep up with the production of acids; thus the digestion process begins to slow down. The digester capacity, therefore, must be checked when the following conditions occur simultaneously:

1. *Peak hydraulic loading.* Peak condition should be estimated by combining poor

thickener operation with maximum expected 7-d sustained hydraulic or organic loading.

2. *Maximum grit and scum accumulation.* Considerable amounts of grit and scum may accumulate before a digester is cleaned. This reduces the active volume of the tank.
3. *Liquid level.* Approximately 0.5–1 m variability in liquid level must be provided in the design. This is necessary to allow for differences in the rate of feeding and withdrawal, scum accumulation, and to provide reasonable operational flexibility.

Digester Heating and Temperature Control. The rates of biological growth and solids stabilization increase and decrease, respectively, with temperature within certain limits. For mesophilic and thermophilic digestion, the optimum temperatures are around 35 and 54°C, respectively. It is therefore important to maintain proper temperature by heating the incoming sludge and heating the digester content. The total amount of input heat should balance the heat losses from the digester. The heat loss sources from the digester are digester walls, floor, roof, piping, etc. Proper heat loss calculations must be made to design the heating system. Common digester heating methods are (1) internal heat exchanger coils, (2) steam injection, and (3) external heat exchangers. Each of these methods of heating is discussed below.

Internal Heat Exchanger. Digester heating in early days was done by heat exchanger coils placed inside the digester. Serious problems developed when the coils became encrusted, reducing the heat transfer. To minimize caking of sludge on the coils, water recirculating through the coils is kept between 45 and 55°C.

Steam Injection Heating. Steam is pumped into the digester for heating. The benefit of this system is that no heat exchanger is needed. The problems, however, are dilution of the sludge and 100 percent boiler makeup water.

External Heat Exchangers. Three types of external heat exchangers are commonly used for sludge heating: water bath, jacket pipe, and spiral exchanger. In the water bath exchanger the boiler tubes and sludge piping are located in a common, water-filled container. In a jacketed pipe exchanger, hot water is pumped countercurrent to pipe surrounding the sludge pipe. The spiral exchangers also utilize countercurrent flow design; however, the sludge and water passageways are cast in a spiral. The heat transfer coefficients for design of external heat exchangers range between 3000 and 5640 kJ/h·m²·°C (150–275 Btu/h·ft²·°F).

The hot water or steam used to heat digesters is most commonly generated in a boiler fueled by sludge gas. Up to 80 percent heat value of the gas can be recovered in a boiler. Provision for burning an alternate fuel source (natural gas) must be made. Often sludge gas is used in engines to generate electricity and the waste heat from these engines is sufficient to meet digester heating requirements. Figure 17-2 shows the external heating and mixing arrangement in a digester building.

Digester Mixing. Anaerobic digesters must be mixed properly to provide optimum performance. Mixing has the following beneficial effects:

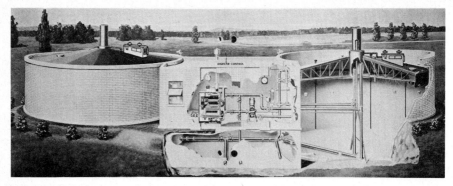

Figure 17-2   Control Room of an External Sludge Heating Unit Using Two-Stage Digester System. (Courtesy Envirex Inc., a Rexnord Company.)

1. Maintains intimate contact between feed sludge and active biomass
2. Creates physical, chemical, and biological uniformity throughout the digester
3. Rapidly disperses metabolic end products produced, and any toxic chemicals entering the digester
4. Prevents formation of surface sum

A certain amount of natural mixing occurs in an anaerobic digester caused by both the rise of sludge gas bubbles and the thermal convection currents created by the addition of heated sludge. However, natural mixing is not sufficient, and therefore additional mixing is needed. Methods used for mixing include external pumped circulation, internal mechanical mixing, and internal gas mixing. Each of these methods is discussed below.

External Pumped Circulation   A large volume of sludge is pumped out and then returned to the digester. Besides circulation, external heating of digester contents is also possible. The sludge is generally withdrawn from the mid-depth of the digester and then pumped back through two nozzles located at the base of the digester on opposite sides, or pumped at the top to break the scum. This method of mixing has a high energy demand.

Internal Mechanical Mixing   Mechanical mixers are generally installed into a shaft tube to promote vertical mixing. Mechanical mixing has not been very successful as large amounts of raggy materials in sludge result in fouling of the propellers and subsequent failure of the mechanisms.

Internal Gas Mixing.   This is an effective method of digester mixing. Many types of gas-mixing arrangements have been successfully designed.[1,2,6] These include:

- Injection of sludge gas bubbles at the bottom of a central-draft-tube to create piston-pumping action and provide surface agitation
- Injection of sludge gas sequentially through a series of lances suspended from the digester cover to as great a depth as possible
- Release of gas from a ring of spargers mounted on the floor of the digester

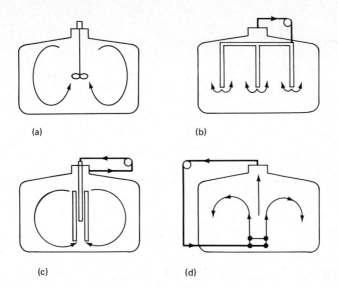

(a)   (b)

(c)   (d)

Figure 17-3   Different Types of Mixing Arrangements Used in an Anaerobic Digester. (a) Mechanical mixing. (b) Gas mixing from suspended pipes. (c) Gas mixing with central draft tube. (d) Gas mixing from rings of spargers mounted on floor.

Some of the digester mixing arrangements are shown in Figure 17-3. The design procedure for a gas-mixing system is covered in the Design Example.

Gas Production and Utilization.   There is a great interest in the utilization of sludge gas as an energy source. The digester gas contains approximately 60–70 percent methane, 25–30 percent carbon dioxide, and small amounts of hydrogen, nitrogen, hydrogen sulfide, and other gases. The gas has a heating value of 21,000–25,000 kJ/m³ (500–650 Btu/ft³)* and density of 86 percent of air.† The digester gas has been successfully used to heat the digesters, buildings, and drive engines.

The methane generation rate may be estimated from the kinetic equations developed for the anaerobic digester. Equations (17-2) and (17-3) express these kinetic relationships:[1-3]

$$P_x = \frac{YQES_o \, (10^3 \text{ g/kg})^{-1}}{1 + k_d \, \theta_c} \tag{17–2}$$

$$V = 0.35 \text{ m}^3/\text{kg} \{[EQS_o \, (10^3 \text{ g/kg})^{-1}] - 1.42(P_x)\} \tag{17–3}$$

where

P_x = net mass of cell produced, kg/d
Y = yield coefficient, g/g. For municipal sludge it ranges 0.04 to 0.1 mgVSS/ mgBOD utilized

---

*For comparison, the heating values of methane and natural gases are 35,800 and 37,300 kJ/m³ respectively.
†Air weighs approximately 1.162 kg/m³ (0.07251 lb/ft³).

E = efficiency of waste utilization (0.6–0.9)
Q = flow rate of sludge, m^3/d
S_o = ultimate BOD_L of the influent sludge g/m^3 (mg/ℓ)
k_d = endogenous coefficient, d^{-1}. For municipal sludge it is 0.02–0.04 d^{-1}
θ_c = mean cell residence time, d. This is also equal to the digestion period
V = volume of methane gas produced, m^3/d
0.35 = theoretical conversion factor for the amount of methane produced from the conversion of 1 kg of BOD_L
1.42 = conversion factor for cellular material to BOD_L

Other rules of thumb for estimating digester gas volume are:

1. 0.5–0.75 m^3/kg (8–12 ft^3/lb) of volatile solids loading
2. 0.75–1.12 m^3/kg (12–18 ft^3/lb) of volatile solids reduced
3. 0.03–0.04 m^3 per person per day (1.1–1.4 ft^3 per person per d)

The gas collection system include fixed or floating covers in the digesters, gas pipings and pressure relief valves, adequate flame traps, gas compressors, gas meters, and gas storage tank. It should be noted that the digester gas makes an explosive mixture with air. Therefore, necessary safety precautions must be utilized to prevent explosion.

Digester Cover. Anaerobic digester covers are provided to keep out oxygen, contain odors, maintain operating temperature, and collect digester gas. The digester covers can be either fixed or floating.

Fixed cover digesters are less expensive and are designed to maintain a constant surface level in the tank. Often rapid withdrawals of digested sludge can draw air into the tank, producing an explosive mixture of sludge gas and oxygen.[3] The explosive range of sludge gas in air is 5–20 percent by volume. In addition, if the liquid level in the digester is increased, it can damage the cover structurally. Fixed cover designs are shown in Figure 17-4.

Floating covers are more expensive but allow independent addition and withdrawal of the sludge, reduce gas hazards, and can be designed to control formation of a scum mat. There are two common types of floating cover designs: (1) Pontoon or Wiggins type, and (2) the Downes type. Both types of floating covers float directly on liquid and commonly have a maximum vertical travel of 2–3 m (6–8 ft). These covers differ primarily in the method used to maintain buoyancy, which, in turn, determines the degree of submergence. These types of floating covers are shown in Figure 17-4. General discussion on each type may be found in Refs. 1 and 2.

A variation of the floating cover is the floating gas holder. Basically the gas holder cover is a floating cover with an extended 3-m (10-ft) skirt to allow storage of gas during periods when gas production exceeds the demand. Gas-holding covers are less stable because they are supported on gas cushion, and they also expose a large side area to lateral wind loads. A gas holder cover is shown in Figure 17-4.

Typical appurtenances for a digester cover include sampling ports; manholes for access, ventilation, and debris removal during cleaning; a liquid overflow system; and a

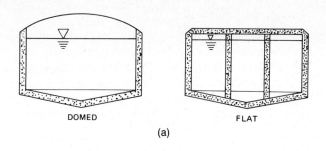

DOMED   FLAT

(a)

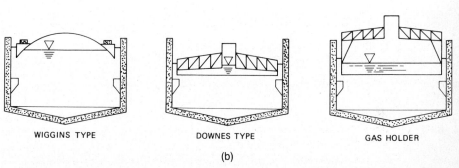

WIGGINS TYPE   DOWNES TYPE   GAS HOLDER

(b)

Figure 17-4   Types of Digester Covers. (a) Fixed covers. (b) Floating covers. [From Ref. 1.]

vacuum pressure relief system equipped with a flame trap. The gas pressure under a digester cover is typically $0–3.7$ kN/m^2 (0–15 in. of water).

Supernatant Quality.   The digester supernatant quality is dependent on whether the digester has one or two stages, whether it is mixed, and how well the solids separate from the liquor. The supernatant is generally returned to the plant for treatment. It may often add significant loading to the plant. Characteristics of supernatant from the anaerobic digester are given in Table 17-2.

TABLE 17-2   Characteristics of Supernatant from Anaerobic Digester Treating Thickened Primary and Waste Activated Sludge

Parameter	Concentration (mg/ℓ)
TS	3000–15,000
BOD$_5$	1000–10,000
COD	3000–30,000
Ammonia nitrogen as N	400–1000
Total phosphorus as P	300–1000

Source: Refs. 1 and 2.

Sludge Characteristics and Other Factors.   pH, volatile acids, nutrients, and toxic materials are important for process control. Some of these factors are discussed in digester operation (Sec. 17-8).

17-3 AEROBIC DIGESTION

Aerobic sludge digestion is commonly used at small plants to stabilize the organic matter in the sludge. The process involves aeration of sludge for an extended period in open tanks. The process is similar to an activated sludge and involves the direct oxidation of biodegradable matter and oxidation of microbial cellular material (endogenous respiration). Stabilization is not complete until there has been an extended period of primarily endogenous respiration (10–20 days). The process has the following advantages: (1) it is simple to operate, (2) it involves low capital cost, (3) the digested sludge is odorless, biologically stable, and has excellent dewatering properties, and (4) the supernatant is low in BOD_5. The digested sludge is normally dewatered on sand drying beds. The disadvantage of aerobic digestion is high operating cost.

The important process parameters are (1) air or oxygen requirement, (2) aeration period, (3) sludge age, (4) temperature, (5) biodegradable volatile solids, (6) processing requirements of the digested sludge, and (7) supernatant quality. Current practice is to provide approximately 15 days of detention time to achieve 40–50 percent reduction in volatile solids. Oxygen requirements, exclusive of nitrification, can vary from 3 to 30 mg per h per g of volatile solids under aeration.[1] When nitrification occurs, both pH and alkalinity are reduced. A schematic flow diagram of an aerobic digestion system is given in Figure 17-5. Basic design parameters of aerobic digesters are summarized in Table 17-3. Characteristics of aerobic digester supernatant are given in Table 17-4.

17-4 OTHER SLUDGE STABILIZATION PROCESSES

Other sludge stabilization processes utilize chemical and physical means. Commonly used chemical and physical stabilization processes are chemical oxidation, lime stabilization, and thermal conditioning. Each of these processes is briefly discussed below.

TABLE 17-3   Aerobic Digestion Design Parameters

Parameter	Value	Remark
Solids retention time (SRT), d	10–15[a] 15–20[b]	Depending on temperature, type of sludge, mixing, etc., the sludge age may range from 10 to 40 days
Volume allowance		Depending on temperature, type of sludge, mixing, etc.
m^3/capita	0.085–0.113	
ft^3/capita	3–4	

TABLE 17-3   Aerobic Digestion Design Parameters

Parameter	Value	Remark
VS loading		Depending on temperature, type of sludge, mixing, etc.
kg/m^3·d	0.384–1.600	
lb/ft^3·d	0.024–0.10	
Air requirements		Enough to keep the solids in suspension and to maintain a DO of 1–2 mg/ℓ
Diffused system[a]		
m^3/m^3·min	0.020–0.035	
ft^3/1000 ft^3·min	20–35[a]	
Diffused system[b]		
m^3/m^3·min	>0.06	
ft^3/1000 ft^3·min	>60	
Mechanical system		
kW/m^3	0.0263–0.0329	
HP/1000 ft^3	1.0–1.25	
Minimum DO, mg/ℓ	1.0–2.0	
Temperature, °C	>15	If sludge temperatures are lower than 15°C, additional detention time should be provided because the digestion will occur at the lower biological reaction rates
VSS reduction, percent	35–50	
Tank design		Aerobic digestion tanks are open and generally require no special heat transfer equipment or insulation. For small treatment systems 0.044 m^3/s (1.0 mgd) or less, the tank design should be flexible enough so that the digester tank can also act as a sludge-thickening unit. If thickening is to be utilized in the digestion tank, sock-type diffusers should be used to minimize clogging
Power requirements		
kW per 10,000 population equivalent	6–7.5	
BHP per 10,000 population equivalent	8–10	

[a]Waste activated sludge only.
[b]Primary plus waste activated sludge, or primary sludge only.
Source: Adapted in part from Refs. 1 and 2.

TABLE 17-4   Characteristics of Supernatant from an Aerobic Digester[a]

Parameter	Range	Typical
pH	5.9–7.7	7.0
BOD$_5$	9–1700	500
Soluble BOD$_5$	4–183	51
COD	288–8140	2600
TSS	46–11,500	3400
Kjeldahl N	10–400	170
Total P	19–241	98
Soluble P	2.5–64.0	26

[a]All units in mg/ℓ except pH.

17-4-1 Chemical Oxidation

Chlorine oxidation involves treatment of sludge with high dosages of chlorine gas or chlorine compounds. The sludge is deodorized and microbiological activity slowed down. The sludge is then dewatered. The chlorine oxidation units usually have a prefabricated modular design and are generally applicable to small plants only. Other chemicals, such as hydrogen peroxide and ozone, have also been used.[1,3]

17-4-2 Lime Stabilization

Lime is added to raw sludge to raise the pH to 12 or higher.[1,3] At high pH, the sludge does not putrefy, create odors, or pose a health hazard. A high lime dosage and about a 3-h contact time is required to get a significant level of pathogen kill. Because the organics are

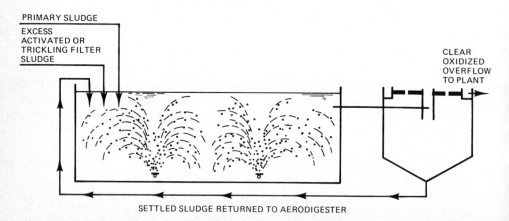

Figure 17-5   Schematic of Aerobic Sludge Digestion. [From Ref. 2.]

not destroyed by this process, the sludge must be disposed of before the organic matter starts to putrefy again.

17-4-3 Heat Treatment or Thermal Conditioning

Heat treatment is both a conditioning and stabilization process. It involves heating the sludge (140°–200°C) for a short time under pressure. Heat coagulates the solids, breaks down the gel structure, and reduces the water affinity of the sludge solids. The resulting sludge is sterilized, practically deodorized, and dewatered readily on vacuum filters or filter presses without the need of chemicals.[1,2] This process is most applicable to biological sludges that are normally difficult to condition or stabilize by chemicals. Because of high capital costs the process is generally limited to large plants.

Sludge stabilization is also achieved by wet oxidation. In this process the sludge solids are incinerated in liquid phase. This process is discussed in Chapter 19.

17-5 EQUIPMENT MANUFACTURERS OF SLUDGE STABILIZATION SYSTEMS

The equipment manufacturers of various sludge stabilization systems are given in Appendix D. These systems include anaerobic and aerobic sludge digestion, chemical oxidation and lime stabilization, and thermal conditioning and incineration. Proper selection of the equipment is necessary to assemble the required components that will provide the operational flexibility and maintenance requirements. Basic considerations for manufacturer selection are given in Sec. 2-10.

17-6 INFORMATION CHECKLIST FOR DESIGN OF SLUDGE STABILIZATION FACILITY

It is important that the design engineers make the necessary design decisions prior to starting the design calculations for sludge stabilization facility. The following checklist may be helpful in developing the necessary predesign data.

1. Select the ultimate sludge disposal method as the degree of sludge stabilization will depend on the requirements of the disposal practice.
2. Develop the characteristics of thickened sludge that will reach the sludge stabilization facility. This includes flow and solids contents under average design flow conditions and under maximum sustained loading conditions including poor thickener operation. Material mass balance analysis similar to that given in Chapter 13 may be necessary.
3. Select the sludge stabilization method that is compatible with the influent sludge characteristics, dewatering, and ultimate disposal method.
4. Develop design parameters for the selected sludge stabilization facility. The design parameters may include organic loading, hydraulic loading, chemical dosage, reaction period, etc. Laboratory tests may be necessary to develop some of the design parameters.

5. Obtain the design criteria from the concerned regulatory agency.
6. Obtain necessary manufacturers' catalogs and equipment selection guides.

17-7 DESIGN EXAMPLE

17-7-1 Design Criteria Used

The following design standards and criteria are used for the sludge stabilization facility:

1. Select anaerobic sludge digestion for stabilization of organic solids.
2. Provide two complete-mixed, high-rate anaerobic heated digesters with digestion temperature of 35°C (95°F).
3. The design flow to the sludge digester shall be equal to thickened sludge under the daily design flow condition. The digester shall have the flexibility to adequately handle the thinnest sludge at the highest solids production rate and the thickest sludge at the lowest solids production rate. The characteristics of sludge reaching the digester are summarized in Table 17-5.
4. Total volatile solids loading to the digester shall not exceed 3.0 kg/m³·d under extreme high loading condition.
5. The solids retention time at extreme high-flow condition shall not be less than 10 days.
6. The digester mixing shall be achieved by internal gas mixing.

TABLE 17-5 Characteristics of Sludge Under Average and Extreme Conditions Reaching the Anaerobic Digester

Factors	Average Flow (Figure 13-12)	Extreme Low Flow	Extreme High Flow
Sludge production, kg/d	6917	5879[a]	7699[b]
Concentration of solids dry wt. basis, per-cent	6	8	3.5
Specific gravity	1.03	1.04	1.02
Average daily flow rate, m³/d	112	71	216[c]
Pumping rate into each digester during the pumping cycle	0.83[d]	0.83	0.83
Influent temperature, °C	21	30	10
Volatile solids fraction before digestion	0.75	0.80	0.70

[a] Extreme low solids to the digester is 85 percent of the average solids loading.
[b] Extreme high solids to the digester = quantity of thickened sludge withdrawn under sustained loading = 9057 kg/d (Table 16-5) × 0.85 (solids capture) = 7699 kg/d.
[c] $\dfrac{7699 \text{ kg/d} \times 1000 \text{ g/kg}}{0.035 \text{ g/g} \times 1.02 \times 1 \text{ g/cm}^3 \times 1000 \text{ cm}^3/\ell \times 1000 \ \ell/m^3} = 216 \text{ m}^3/\text{d.}$
[d] The pumping rate of 0.83 m³/min gives a velocity of 0.786 m/s in the 15 cm diameter pipe (see Sec. 16-7-2, step E2).

7. The digester heating shall be achieved by recirculation of sludge through external heat exchanger. The sludge recirculation system shall also be designed to provide digester mixing.
8. Provide floating digester cover for gas collection. Separate gas storage sphere shall be provided to store excess gas.
9. The heat loss from the digester cover, side walls, and floor shall be calculated using the standard heat transfer coefficients for the digester construction material.
10. Provide gas-fired hot water boiler for external heat exchanger. The fuel-burning equipment shall include the necessary accessories for burning sludge gas. At the time of insufficient gas supply the burner shall change over automatically to the natural gas.
11. Explosion prevention devices shall be provided to minimize the possibility of an explosive mixture being developed inside the floating covers. Proper flame traps shall be provided to assure protection against the passage of flame into the digester, gas storage sphere, and supply lines.
12. The digester design shall include supernatant withdrawal system, sight glass, sampler, manhole, etc.
13. Arrangement shall be provided to break the scum that may form on the sludge surface.

17-7-2 Design Calculations

A. Digester Capacity and Dimensions.   The digester capacity may be calculated using (1) digestion period, (2) mean cell residence time equations, (3) volumetric loading, (4) population basis, and (5) observed volume reduction. The digester capacity using each of the above methods is computed below.

1. Compute digester capacity at average flow condition using 15 days digestion period

$$\text{Average flow to the digester} = 112 \text{ m}^3/\text{d}$$
$$\text{Digester volume} = \text{flow} \times \text{digestion period}$$
$$= 112 \text{ m}^3/\text{d} \times 15 \text{ d} = 1680 \text{ m}^3$$

2. Compute digester capacity using volatile solids-loading factor

$$\begin{array}{l}\text{Assume VS loading at} \\ \text{average flow condition}\end{array} = 2.2 \text{ kg/m}^3 \cdot \text{d}$$

$$\begin{array}{l}\text{Total volatile solids} \\ \text{reaching the digester}\end{array} = 6917 \text{ kg/d} \times 0.75 = 5188 \text{ kg/d}$$

$$\text{Digester capacity} = \frac{5188 \text{ kg/d}}{2.2 \text{ kg/m}^3 \cdot \text{d}} = 2358 \text{ m}^3$$

3. Compute digester capacity using volume per capita allowance

Assume 0.030 m^3 digester capacity per capita

$$\text{Population served} = 80,000 \text{ (Chapter 6)}$$
$$\text{Digester capacity} = 80,000 \times 0.030 = 2400 \text{ m}^3$$

4.   Compute digester capacity using volume reduction method

Volume of digested sludge (Figure 13-12)   = 81 m³/d
Volume of raw sludge to the digester (Figure 13-12) = 112 m³/d

Digester capacity is calculated from Eq. (17-1)

$$V = (112 \text{ m}^3/\text{d} - \tfrac{2}{3}(112 \text{ m}^3/\text{d} - 81 \text{ m}^3/\text{d}))\ 15\ \text{d} = 1370\ \text{m}^3$$

5.   Select digester capacity
The digester capacity calculated from the volume reductions method is signifi-cantly smaller than the capacity obtained from other methods. It should be mentioned that in completely mixed, high-rate digesters the supernatant with-drawal is generally achieved by stopping all mixing devices and letting the solids settle in the digester for 1–2 h. The supernatant and digested sludge are then withdrawn from the digester. Frequently, this cannot be done and the digested sludge is withdrawn under completely mixed conditions. It is therefore desirable to provide excess capacity for operational flexibility. Select active digester capacity of 2350 m³.

B. Digester Dimensions and Geometry

1.   Correct for volume displaced by grit and scum accumulations, and floating cover level

Provide 1-m depth for grit accumulation in the bottom cone
Provide 0.6-m depth for scum blanket
Provide 0.6-m minimum space between floating cover and maximum digester level

Total displaced height   = 1 m + 0.6 m + 0.6 m = 2.2 m

If the side water depth without bottom cone is 7.6 m (25 ft), additional volume will be available in the cone

Assume active digester volume   = 2500 m³

Provide two digesters

Volume of each digester   = 1250 m³

$$\text{Area of each digester} = \frac{1250 \text{ m}^3}{7.6 \text{ m}} = 164.5 \text{ m}^2$$

$$\text{Diameter of each digester} = \sqrt{\frac{4}{\pi} \times 164.5 \text{ m}^2} = 14.5 \text{ m (47.5 ft)}$$

If floating covers come in 1.5-m (5-ft)-diameter increments, provide digesters with 13.7-m (45-ft) diameter

$$\text{Revised side water depth} = \frac{1250 \text{ m}^3}{\dfrac{\pi}{4} \times (13.7 \text{ m})^2} = 8.5 \text{ m (27.9 ft)}$$

Provide two digesters each 13.7 m (45 ft) diameter and 8.5 m (28 ft) side water depth. The digester details are given in Figure 17-6.

2. Check the active volume of the digesters, including volume of cone
The floor of the digester is sloped at 1 vertical to 3 horizontal
The bottom cone depth of 2.3 m adds additional volume
Therefore,

Active digester
volume*
$= \text{(volume of active cylindrical portion)}$
$+ \text{(total volume of the cone)}$
$- \text{(allowance for grit accumulation)}$

$$= \frac{\pi}{4} (13.7 \text{ m})^2 \times 7.3 \text{ m} + \frac{1}{3}\left(\frac{\pi}{4}\right)(13.7 \text{ m})^2$$

$$\times 2.3 \text{ m} - \frac{1}{3}\left(\frac{\pi}{4}\right)(6.0 \text{ m})^2 \times 1.0 \text{ m}$$

$$= 1076.1 \text{ m}^3 + 113.0 \text{ m}^3 - 9.4 \text{ m}^3 = 1179.7 \text{ m}^3$$

Active volume
of two digesters
$= 2 \times 1179.7 \text{ m}^3 = 2359.4 \text{ m}^3$

Total volume
of two digesters
$= 2\left[\left(\frac{\pi}{4}\right)(13.7 \text{ m})^2 \times 8.5 \text{ m} + 113.0 \text{ m}^3\right]$

$$= 2732.0 \text{ m}^3$$

Active volume
ratio including
cone
$= \dfrac{2359.4 \text{ m}^3}{2732.0 \text{ m}^3} = 0.86$

C. Actual Solids Retention Time and Solids Loading

1. Compute actual digestion period at average, extremely low, and extremely high flows

$$\text{Digestion period at average flow} = \frac{2359.4 \text{ m}^3}{112 \text{ m}^3/\text{d}} = 21.1 \text{ d}$$

---

*Active depth of cylindrical portion = side water depth (8.5 m) − scum blanket (0.6 m) − space below floating cover (0.6 m) = 7.3 m.
Grit accumulation allowance up to 1 m in the bottom cone will reduce the effective volume of the bottom cone.

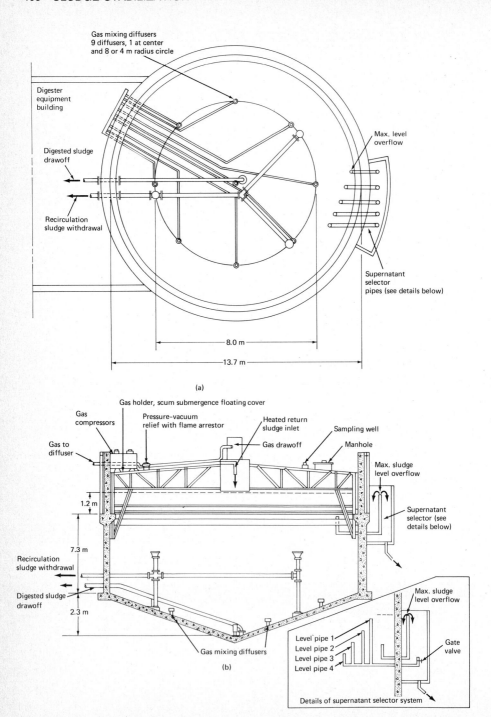

Figure 17-6   Design Details of Anaerobic Sludge Digester. (a) Plan. (b) Section.

$$\text{Digestion period at extreme high flow} = \frac{2359.4 \text{ m}^3}{216 \text{ m}^3/\text{d}} = 10.9 \text{ d}$$

$$\text{Digestion period at extreme low flow} = \frac{2359.4 \text{ m}^3}{71 \text{ m}^3/\text{d}} = 33.2 \text{ d}$$

2. Compute actual solids loadings at average, extreme low, and extreme high conditions

Solids loading at average load-ing condition
$$= \frac{6917 \text{ kg/d} \times 0.75 \text{ VS}}{2359.4 \text{ m}^3}$$

$$= 2.2 \text{ kg VS/m}^3 \cdot \text{d}$$

Solids loading at extreme minimum loading condition
$$= \frac{5879 \text{ kg/d} \times 0.8 \text{ VS}}{2359.4 \text{ m}^3}$$

$$= 2.0 \text{ kg VS/m}^3 \cdot \text{d}$$

Solids loading at extreme high loading condition
$$= \frac{7699 \text{ kg/d} \times 0.7 \text{ VS}}{2359.4 \text{ m}^3}$$

$$= 2.3 \text{ kg VS/m}^3 \cdot \text{d}$$

D. Gas Production.   The gas production rate is estimated from several relationships. Calculations are given below.

1. Calculate gas production from Eqs. (17-2) and (17-3)

Average quantity of sludge reaching the digester
$$= 6917 \text{ kg/d}$$

Average volume of sludge reaching the digester
$$= 112 \text{ m}^3/\text{d}$$

Concentration of solids
$$= \frac{6917 \text{ kg/d} \times 1000 \text{ g/kg}}{112 \text{ m}^3/\text{d}}$$

$$= 61{,}759 \text{ g/m}^3$$

Assume 65 percent solids are biodegradable and 1 g of biodegradable solids = 1.42 g BOD_L, $Y = 0.05$, $k_d = 0.03/\text{d}$, and $E = 0.8$

BOD_L in sludge
$$= 61{,}759 \text{ g/m}^3 \times 0.65 \times 1.42 \text{ g/g} = 57{,}004 \text{ g/m}^3$$

$$P_x = \frac{YQES_o \, (10^3 \text{ g/kg})^{-1}}{1 + k_d \theta_c}$$

$$P_x = \frac{0.05 \times 112 \text{ m}^3/\text{d} \times 0.8 \times 57{,}004 \text{ g/m}^3 \times (10^3 \text{ g/kg})^{-1}}{1 + 0.03/\text{d} \times 21.1 \text{ d}}$$

$$= 156 \text{ kg/d}$$

The volume of methane gas is calculated from Eq. (17-3).

$$V = 0.35 \text{ m}^3/\text{kg} \{(0.8 \times 112 \text{ m}^3/\text{d} \times 57{,}004 \text{ g/m}^3 \ (1000 \text{ g/kg})^{-1}$$
$$- 1.42 \text{ g/g} \times 156 \text{ kg/d}\}$$

$$= 1710 \text{ m}^3/\text{d}$$

If methane is 66 percent in the digester gas,

$$\text{Digester gas production} = 1710 \text{ m}^3/\text{d} \times \frac{1}{0.66} = 2591 \text{ m}^3/\text{d}$$

2. Estimate gas production from other rules of thumb
    a. Based on volatile solids loading using VS = 0.75 of total solids and gas production rate of 0.50 m^3/kg VS

    Gas produced   = 6917 kg/d × 0.75 × 0.50 m^3/kg
                       = 2594 m^3/d

    b. Based on VS reduction
        Assume average VS reduction of 52 percent and gas production of 0.94 m^3/kg VS reduced

    Total VS reduced = 6917 kg/d × 0.75 × 0.52 = 2698 kg/d
    Gas produced   = 2698 kg/d × 0.94 m^3/kg = 2536 m^3/d

    c. Based on per capita

    Total population
    served   = 80,000

    Use gas production rate of 0.032 m^3/capita

    Gas produced   = 80,000 persons × 0.032 m^3/person · d
                       = 2560 m^3/d

Based on the above analysis assume a conservative gas production rate of 2550 m^3/d at standard conditions (0°C and 1 atm)

E. Digested Sludge Production
1. Compute the quantity of solids in digested sludge

    TVS   = 6917 kg/d × 0.75 = 5188 kg/d
    TVS destroyed   = 5188 kg/d × 0.52 = 2698 kg/d
    TS remaining after digestion = Nonvolatile solids + VS remaining
                                  = (6917 − 5188) kg/d + 0.48
                                    × 5188 kg/d
                                  = 1729 kg/d + 2490 kg/d = 4219 kg/d

2. Compute total mass reaching the digester

    Total solids reaching the digester = 6917 kg/d
    Total solids in thickened sludge   = 6 percent by weight
    Total mass reaching digester   = (6917 kg/d)/(0.06 kg/kg)
                                       = 115,283 kg/d

3. Compute total mass leaving the digester

$$\text{Mass leaving the digester} = \text{mass reaching the digester} - \text{mass lost in gas}$$

$$\text{Gas produced} = 2550 \ \text{m}^3/\text{d}$$

Assume the density of digester gas is 86 percent of that of air, and air weighs 1.162 kg/m³ (0.07251 lb/ft³)

$$\text{Total quantity of digester gas produced} = 2550 \ \text{m}^3/\text{d} \times 1.162 \ \text{kg/m}^3 \times 0.86$$

$$= 2548 \ \text{kg/d}$$

$$\text{Total mass leaving the digester} = 115{,}283 \ \text{kg/d} - 2548 \ \text{kg/d}$$

$$= 112{,}735 \ \text{kg/d}$$

4. Compute supernatant flow from the digester

$$\text{Assume digester supernatant solids} = S \ \text{kg/d}$$

$$\text{Assume total solids in the digester supernatant} = 4000 \ \text{mg}/\ell$$

$$\text{Specific gravity of digester supernatant} = 1.0$$

$$\text{Total solids in the digested sludge} = 5 \text{ percent by weight}$$

$$\frac{S}{0.004} + \frac{4219 \ \text{kg/d} - S}{0.05} = 112{,}735 \ \text{kg/d}$$

$$S = 123 \ \text{kg/d}$$

$$\text{Supernatant flow rate} = \frac{123 \ \text{kg/d} \times 1000 \ \text{g/kg}}{0.004 \ \text{g/cm}^3 \times 10^6 \ \text{cm}^3/\text{m}^3}$$

$$= 31 \ \text{m}^3/\text{d}$$

$$\text{Concentration of solids in the supernatant} = \frac{123 \ \text{kg/d} \times 1000 \ \text{g/kg} \times 1000 \ \text{mg/g}}{31 \ \text{m}^3/\text{d} \times 1000 \ \ell/\text{m}^3}$$

$$\approx 4000 \ \text{mg}/\ell$$

5. Select a supernatant selector system

The purpose of the supernatant selector is to withdraw liquid from the top. Ideally, it should serve the following purposes: (1) allow direct visual inspection of the sludge, (2) allow removal of clear liquid from the top, (3) permit operation by one person, (4) be extremely reliable, with a minimum of moving parts, (5) minimize the danger of allowing air to enter the digester, (6) be easy to service in case of blockage by grease, scum, or sludge.

Many types of supernatant selector and control systems are available from equipment manufacturers. One simple type is used in this design and is shown in Figure 17-6. The clogged supernatant selector pipe is cleaned by pumping hot water into the digester through the pipe.

6. Compute the quantity of digested sludge

$$\text{Quantity of digested sludge} = \frac{\text{TS remaining after digestion}}{-\text{ TS lost in supernatant}}$$

$$= 4219 \text{ kg/d} - 123 \text{ kg/d}$$
$$= 4096 \text{ kg/d}$$

$$\text{Volume of digested sludge}^* = \frac{4096 \text{ kg/d} \times 1000 \text{ g/kg}}{0.05 \text{ g/g} \times 1.02 \times 1 \text{ g/cm}^3 \times 10^6 \text{ cm}^3/\text{m}^3}$$

$$= 80 \text{ m}^3/\text{d}$$

F. Influent Sludge Line to the Digester.   The sludge from the thickeners is pumped into the digester control building. The sludge pipe is 15 cm (6 in.) in diameter and pump operation is intermittent with a constant pumping rate of 0.83 m³/min from each thickener. The pumping operation is controlled by a time clock that can be manually varied to achieve the thickest possible sludge. The pumping cycle is adjusted such that both pumps do not operate simultaneously. The thickened sludge from each digester is divided equally in the digesters. Details of the influent sludge pipe are shown in Figure 17-7.

G. Digester Heating Requirements
1. Compute heating required for raw sludge
Heating required for sludge is calculated from Eq. (17-4)

$$H_R = \text{flow} \times C_p(T_2 - T_1) \tag{17-4}$$

where

H_R = heat required, J/d
C_p = specific heat of sludge (same as for water = 4200 J/kg · °C or 1.0 Btu/lb · °F)
T_2 = digestion temperature, °C
T_1 = temperature of the thickened sludge, °C

The critical heat requirement for raw sludge is reached when sludge flow is maximum and influent temperature is lowest:

$$\text{Heat required} = \frac{7699 \text{ kg/d} \times 4200 \text{ J/kg°C } (35 - 10)°\text{C}}{0.035 \text{ kg/kg}}$$

$$= 2.31 \times 10^{10} \text{ J/d}$$

---

*Volume of digested sludge is also calculated from the volume of thickened sludge reaching and the volume of supernatant leaving the digester. The volume of digested sludge plus the volume of water vapor lost with the gas = 112 m³/d − 31 m³/d = 81 m³/d.

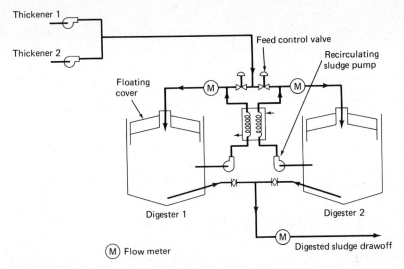

Figure 17-7   *Schematic Flow Scheme of Sludge Heating and Recirculation System.*

2. Compute heat loss from the digester
    Heat loss is calculated from Eq. (17-5)

$$H_L = UA(T_2 - T_1) \qquad (17\text{--}5)$$

where

H_L = heat loss, J/h (Btu/h)
U^* = overall coefficient of heat transfer, J/s · m² · °C (Btu/h · ft² · °F)
A = area through which heat loss occurs, m² (ft²)
T_2 = digester operating temperature, °C (°F)
T_1 = outside air temperature, °C (°F)[†]

Heat losses from the digester occur from the roof, bottom, and side walls
a. Compute area of roof

$$\text{Roof area} = \pi D \left(\frac{\text{slant length}}{2} \right)$$

$$\text{Slant height} = \sqrt{\left(\frac{D}{2}\right)^2 + (\text{vertical rise of cover})^2}$$

$$= \sqrt{\left(\frac{13.7 \text{ m}}{2}\right)^2 + (0.46 \text{ m})^2} = 6.87 \text{ m}$$

$$\text{Roof area} = (\pi \times 13.7 \text{ m} \times 6.87 \text{ m})/2 = 147.9 \text{ m}^2$$

*J/s · m² · °C × 0.1763 = Btu/h · ft² · °F.
†Critical average air and ground temperatures are 0 and 5°C, respectively.

b. Compute area of side walls
Area of side wall above ground level = $\pi D \times$ (exposed height)
Assume 50% side wall is exposed

$$\text{Side wall area above ground} = \pi \times 13.7 \text{ m} \times \frac{8.5}{2} \text{ m} = 182.9 \text{m}^2$$

$$\text{Area of side wall below ground} = 182.9 \text{ m}^2$$

c. Compute bottom area
Digester bottom is generally sloped at 1 vertical to 3 horizontal
Therefore,

$$\text{total drop of the bottom slope at the center} = \frac{D}{2 \times 3} = \frac{13.7 \text{ m}}{2 \times 3} = 2.3 \text{ m}$$

$$\text{Bottom area} = \pi \times 13.7 \text{ m} \times \tfrac{1}{2} \times \sqrt{\left(\frac{13.7 \text{ m}}{2}\right)^2 + (2.3 \text{ m})^2}$$

$$= 155.5 \text{ m}^2$$

d. Select overall coefficients of heat transfer for different areas
Digester floating covers and roofing consist of 6.5-mm ($\frac{1}{4}$-in.) plate steel, 76-mm (3-in.) rigid foam insulation,* inside air space, and buildup roofing 1236 kg/m^2 (70 lb/ft^2) U† = 0.90 J/s·m^2·°C
Exposed digester side 300-mm (12-in.) concrete, 76-mm (3-in.) urethane foam insulation, 100-mm (4-in.) brick siding, U = 0.68 J/s·m^2·°C
Buried digester side 300-mm (12-in.) concrete surrounded by moist soil, U = 0.80 J/s·m^2·°C
Digester bottom surrounded by moist soil, U = 0.62 J/s·m^2·°C

e. Compute heat loss from the digester

Heat loss from the cover and roofing

$$= 147.9 \text{ m}^2 \times 0.9 \text{ J/s} \cdot \text{m}^2 \cdot \text{°C} (35 - 0)\text{°C} \times 86,400 \text{ s/d}$$
$$= 4.03 \times 10^8 \text{ J/d}$$

Heat loss from exposed wall

$$= 182.9 \text{ m}^2 \times 0.68 \text{ J/s·m}^2\cdot\text{°C} \times (35 - 0)\text{°C} \times 86,400 \text{ s/d}$$
$$= 3.76 \times 10^8 \text{ J/d}$$

Heat loss from buried wall

$$= 182.9 \text{ m}^2 \times 0.8 \text{ J/s·m}^2\cdot\text{°C} \times (35 - 0)\text{°C} \times 86,400 \text{ s/d}$$
$$= 4.43 \times 10^8 \text{ J/d}$$

---

*Common insulating materials are glass wool, insulation board, urethane foam, lightweight insulating concrete, and dead air space, etc.
†J/s · m^2 °C × 0.1763 = Btu/h · ft^2 · °F.

Heat loss from bottom

$$= 155.5 \text{ m}^2 \times 0.62 \text{ J/s·m}^2\text{·°C} \times (35 - 5)°\text{C} \times 86,400 \text{ s/d}$$
$$= 2.50 \times 10^8 \text{ J/d}$$

Total heat loss from each digester $= 14.72 \times 10^8$ J/d
Total heat loss from both digesters, including 23 percent minor losses, and 50 percent emergency condition $= 14.72 \times 10^8$ J/d $\times 2 \times 1.73 = 50.93 \times 10^8$ J/d

f. Compute the heating requirements for the digester

Heating requirements for
raw sludge under critical
condition $= 2.31 \times 10^{10}$J/d
Heat loss from the digester $= 50.93 \times 10^8$ J/d
Total heating requirement $= 2.82 \times 10^{10}$ J/d
$= 1.175 \times 10^9$ J/h
$= 1.175 \times 10^6$ kJ/h

H. Selection of Heating Units and Energy Balance

1. Select heating units for external heat exchanger
   Provide two heating units each rated at 1.25×10^6 kJ/h (1.19 million BTU/h) with natural gas. The digester gas has approximately 65 percent of the heating value of the natural gas.* Therefore, each unit will be derated at 0.813×10^6 kJ/h (0.77 million Btu/h). Total heat provided by two units $= 2 \times 0.813 \times 10^6$ $= 1.626 \times 10^6$ kJ/h.

$$\frac{\text{Percent extra}}{\text{capacity available}} = \frac{(1.626 \times 10^6 - 1.175 \times 10^6) \times 100}{1.175 \times 10^6}$$

$$= 38 \text{ percent}$$

The heating requirements are determined at the critical conditions of sludge flow and external temperatures. Therefore, these values are used to size the equipment. The actual average heat requirements would be substantially less.

2. Compute digester gas requirements
   At 75 percent efficiency of heating units the digester gas requirements are calculated as follows:

$$\text{Digester gas needed} = \frac{1.626 \times 10^6 \text{ kJ/h}}{0.75 \times 24,300 \text{ kJ/m}^3}$$

$$= 89.22 \text{ m}^3\text{/h}$$
$$= 89.22 \text{ m}^3\text{/h} \times 24 \text{ h/d} = 2141 \text{ m}^3\text{/d}$$

$$\frac{\text{Total quantity of di-}}{\text{gester gas produced}} = 2550 \text{ m}^3\text{/d}$$

*Heating value of natural gas $= 37,300$ kJ/m^3. Heating value of digester gas $= 24,300$ kJ/m^3.

This gives approximately 20 percent excess gas under the most critical condition when the digester heating demand is greatest. Under average and minimum digester-heating requirements, a considerably larger quantity of excess gas will be available. Excess gas will also be used to produce heated water for other plant uses.

3. Design makeup heat exchangers for external sludge heating

The heating of the digesters and thickened sludge is achieved by recirculating digester contents through external heat exchangers

a. Compute average temperature rise of the sludge through the external exchangers

Provide 23-cm (9-in.) diameter sludge recirculation pipe, and a constant flow recirculation pump for each digester. A common external jacketed type heat exchanger will be used to heat the recirculated sludge. If velocity of 1 m/s is maintained in the pipe,

$$\text{the sludge pumping rate from each digester} = \frac{\pi}{4} \times (0.23 \text{ m})^2 \times 1 \text{ m/s} \times (86,400) \text{ s/d}$$

$$= 3590 \text{ m}^3/\text{d}$$

$$= 3590 \text{ m}^3/\text{d} \times 1.02 \times 1 \text{ g/cm}^3 \times 10^6 \text{ cm}^3/\text{m}^3 \times (1000 \text{ g/kg})^{-1}$$

$$= 3.662 \times 10^6 \text{ kg/d}$$

Average sludge temperature entering the external heat exchanger = 35°C
Assume average sludge temperature increase after passing through the heat exchanger = ΔT°C
Assume specific heat of sludge is 4200 J/kg°C (same as for water)

$$\text{Total heat supplied to the sludge} = 4200 \frac{\text{J}}{\text{kg°C}} \times \Delta T \text{°C} \times 3.662 \times 10^6 \text{ kg/d}$$

$$= (1.538 \times 10^{10}) \times \Delta T \text{ J/d}$$

$$\text{Total heat loss from each digester} = 1.41 \times 10^{10} \text{ J/d}$$

If the efficiency of the heat exchangers is 80 percent,

$$(1.538 \times 10^{10}) \times \Delta T \text{ J/d} \times 0.8 = 1.41 \times 10^{10} \text{ J/d}$$

where

$$\Delta T = \frac{1.41 \times 10^{10} \text{ J/d}}{1.538 \times 10^{10} \text{ J/d} \times 0.8} = 1.15 \text{°C}$$

Average temperature of the sludge entering heat exchanger = 35°C
Average temperature of the sludge leaving the heat exchanger = 36.15°C
Sludge recirculation of 3590 m³/d (660 gpm) in each digester will also provide digester mixing.

b. Compute hot water recirculation rate through the external heat exchanger Provide one jacketed pipe heat exchanger for both digesters. The hot water is pumped countercurrent to the sludge flow through the jacket surrounding the sludge pipes. Assume that the water enters the jacket pipe at 95°C and leaves at 60°C.

Drop in heating water temperature $= 95°C - 60°C = 35°C$

Total heating required for each digester $= 1.41 \times 10^{10}$ J/d

If 25 percent additional heating is provided to account for heat losses,

Total heat required per digester $= 1.41 \times 10^{10} \times 1.25 = 1.76 \times 10^{10}$ J/d

Total heat required for both digesters $= 3.52 \times 10^{10}$ J/d

Using specific heat of water $= 4200$ J/kg°C

Total heat supplied by water $= 4200$ J/kg°C $\times 35°C = 147,000$ J/kg

Hot water recirculation rate through the common heat exchanger $= \dfrac{3.52 \times 10^{10} \text{ J/d}}{147,000 \text{ J/kg}}$

$= 2.40 \times 10^5$ kg/d

Volume of water recirculated $= \dfrac{2.4 \times 10^5 \text{ kg/d} \times 1000 \text{ g/kg}}{1 \text{ g/cm}^3 \times 10^6 \text{ cm}^3/\text{m}^3}$

$= 240$ m³/d (44 gpm)

c. Compute the length of sludge pipe in the heat exchanger jacket

Average temperature of the sludge in the heat exchanger $= \dfrac{35°C + 36.15°C}{2} = 35.58°C$

Average temperature of the heating water in the heat exchanger $= \dfrac{95°C + 60°C}{2} = 77.5°C$

Assume heat transfer coefficient of external water jacketed heat exchanger $= 4000$ kJ/h·m²·°C (196 Btu/h·ft²·°F)

Total heat radiated from the heating water $= \dfrac{(77.5 - 35.58)°C \times 4000 \text{ kJ/h·m}^2 \cdot°C}{\times 24 \text{ h/d} = 4.02 \times 10^6 \text{ kJ/d·m}^2}$

$$\text{Total area of the sludge pipe for each heat exchanger} = \frac{1.76 \times 10^{10} \text{ J/d}}{4.02 \times 10^6 \text{ kJ/d} \cdot \text{m}^2 \times 1000 \text{ J/kJ}}$$

$$= 4.38 \text{ m}^2$$

$$\text{Length of 23-cm (9 in) diameter jacketed pipe} = \frac{4.38 \text{ m}^2}{\pi(0.23 \text{ m})} = 6.0 \text{ m}$$

Provide 6-m long, 23-cm diameter heat exchanger sludge pipe per digester into a common hot water jacket.

The flow scheme of makeup heat exchanger is shown in Figure 17-7.

I. Gas Storage and Compressor Requirements

1.   Compute the diameter of the gas storage sphere
     Provide a total of 3-d gas storage to serve the digester heating requirements and other plant uses.

$$\begin{aligned}
\text{Total gas stored} \quad &= 3 \text{ d} \times 2550 \text{ m}^3/\text{d} \\
&= 7650 \text{ m}^3 \text{ (standard condition; 0°C and 1 atm)} \\
\text{Storage pressure} \quad &= 5.1 \text{ atm (assume)} \\
\text{Storage temperature} &= 50°C \text{ summer} \\
\text{Storage volume } V_2 \quad &= \frac{P_1 V_1 T_2}{P_2 T_1}
\end{aligned}$$

where P_1, V_1, and T_1 are pressure, volume, and absolute temperature of the digester gas produced; and P_2, V_2, and T_2 are the storage pressure, volume, and absolute temperature of the digester gas stored.

$$V_2 = \frac{1 \text{ atm} \times 7650 \text{ m}^3 \, (273 + 50)°K}{5.1 \text{ atm} \, (273 + 0)°K}$$

$$= 1774.7 \text{ m}^3$$

Provide a high-volume gas storage sphere

$$\text{Volume of sphere} = \frac{\pi}{6} (\text{diam})^3$$

$$\text{Diam of sphere} = \left[\frac{1774.7 \text{ m}^3 \times 6}{\pi} \right]^{1/3}$$

$$= 15.0 \text{ m (49 ft)}$$

Provide 15.0-m (49-ft) diameter sphere for gas storage.

2.   Compute size of high-pressure gas compressors
     Compressors are used to compress the digester gas into the gas storage sphere. Compressor power requirement is calculated from Eq. (13-26).
     Total weight of digester gas produced under standard conditions = 2548 kg/d (see Sec. 17-7-2, step E3)

Assuming weight of gas compressed is 200 percent of production rate, then

$$w = 2 \times 2548 \text{ kg/d} \times \frac{1}{24 \text{ h/d} \times 3600 \text{ s/h}} = 0.0590 \text{ kg/s}$$

$R = 8.314$ kJ/kmole °K

$e =$ compressor efficiency of 75%

$T_o =$ inlet temperature $= (273 + 35)$°K

$P_o = 1.03$ atm (gas pressure inside the floating cover is normally less than 0.4 m of water)

$P = 5.10$ atm

$$P_w = \frac{0.0590 \text{ kg/d} \times 8.314 \text{ kJ/kmole °K} \times (273 + 35) \text{ °K}}{8.41 \times 0.75 \text{ kg/kmole}}$$

$$\times \left[\left(\frac{5.1}{1.03} \right)^{0.283} - 1 \right]$$

$$= 13.7 \text{ kW (18.3 HP)}$$

Provide two constant-speed compressors each driven by 7.5-kW (10-HP) electric motor.

J. Digester Gas Mixing

1. Compute power requirements for gas mixing

    Power requirements for gas mixing of the digester is obtained from Eq. (16-2). The volume of each digester $V = 1179.7$ m³.

$$\mu \qquad = 2 \text{ times the viscosity of water at 35°C}$$
$$= 2 \times 0.73 \times 10^{-3} \text{ N s/m}^2 \ (2 \times 1.51 \times 10^{-5} \text{ lb·s/ft}^2)$$
$$= 1.46 \times 10^{-3} \text{ N s/m}^2 \ (3.02 \times 10^{-5} \text{ lb·s/ft}^2)$$

Velocity gradient for sludges above 5 percent solids is over 75/s. Use $G = 85$/s for this problem.

$$P \qquad = (85/\text{s})^2 \times 1.46 \times 10^{-3} \text{ N·s/m}^2 \times 1179.7 \text{ m}^3$$
$$= 12,444 \text{ N m/s (9178 ft lb/s)}$$
$$= 12.4 \text{ kW (16.6 HP)}$$

Total power required for two digesters $= 2 \times 12.4 \text{ kW} = 24.8 \text{ kW (33.2 HP)}$

Provide three compressors each driven by 15-kW (20-HP) motor.

Total power provided for mixing $= 45 \text{ kW (60 HP)}$

Two compressors will deliver the required power, while the third compressor will be a stand-by unit, serving both digesters.

2. Compute gas flow

    The digester gas flow rate for mixing is calculated from Eq. (13-26).

$$P_w = 30 \text{ kW}$$
$$R = 8.314 \text{ kJ/kmole } °\text{K}$$
$$T_o = (273 + 35) = 308°\text{K}$$

P = 2.4 atm. This is sufficient to overcome the static sludge head in the digester and head losses in the piping.

P_o = 1.03 atm (gas pressure inside the floating digester cover)

e = compressor efficiency of 75%

w = mass of gas compressed per digester, kg/s

$$w = \frac{15.0 \text{ kW} \times 8.41 \text{ kg/kmole} \times 0.75}{8.314 \text{ kJ/kmole } °\text{K} \times 308 \text{ }°\text{K}\left[\left(\dfrac{2.4}{1.03)}\right)^{0.283} - 1\right]}$$

$$= 0.14 \text{ kg/s}$$

$$\frac{\text{Gas flow* per}}{\text{digester}} = \frac{0.14 \text{ kg/s}}{1.162 \text{ kg/m}^3 \times 0.86} = 0.14 \text{ m}^3\text{/s}$$

3. Select digester-mixing arrangement

   The digester mixing will be achieved by flow recirculation, raw sludge, and internal gas mixing. The purpose of digester mixing is to transfer the required quantity of heat and aid in mixing of digester contents so that the digestion efficiency is maintained.

   The sludge recirculation of 3590 m³/d (658 gpm) in each digester was calculated in step H3. The sludge will be withdrawn from the mid-depth and discharged above the scum blanket level to assist in scum mixing (see Figure 17-6).

A multipoint gas-mixing system is provided for effective use of the gas. The gas is withdrawn from the top and recirculated by means of nine ports for gas injection. One gas injection diffuser is located in the center, and eight are equally spaced around a 4-m radius circle. Details are shown in Figure 17-6. Details of diffusers are covered in the specifications (Sec. 17-9-2).

17-8 OPERATION AND MAINTENANCE, AND TROUBLESHOOTING AT ANAEROBIC SLUDGE DIGESTION FACILITY

The operation and control of anaerobic digesters is difficult because it depends not only on the results of the laboratory tests, but also on operators' judgement and skills, treatment plant loadings, industrial wastes, and weather conditions. The routine operations are further complicated by the need for repairs, shutdown, cleaning, and startup. EPA has published an operations manual for anaerobic sludge digestion which provides very

---

*The density of digester gas is 86 percent of air. Weight of air is assumed 1.162 kg/m³ (0.07251 lb/ft³) [see step E3].

detailed information on process and troubleshooting.[7] Following is the summary informa-tion on digester startup, troubleshooting, and routine operation and maintenance.[7,8]

17-8-1 Digester Startup

If sufficient digested sludge seed is available from a nearby treatment plant, the startup operation is simplified. The following steps are necessary:

1. Haul approximately 500 m^3 seed and transfer it into one digester.*
2. Fill the digester with raw wastewater.
3. Start heating and mixing, and bring to operating temperature.
4. Begin feeding raw sludge at a uniform rate approximately 25 percent of daily feed per digester. Increase the loading gradually.
5. Maintain the following records:
   (a) Quantity of TVS fed daily
   (b) TVS, volatile acids and alkalinity ratio (VA/ALK), and pH
   (c) Temperature, gas production, and CO_2 content in gas
6. At low feeding rate it is possible to bring normal operation without adding chemicals for pH control. If VA/ALK ratio rises to 0.8 or more, and pH is below 6.5, addition of chemicals such as lime or soda ash may be considered.
7. Fairly stable conditions should be reached in 30–40 days if loading is kept below 1.0 kg VS/m^3·d.

17-8-2 Common Operational Problems and Troubleshooting Guide

1. A rise in VA/ALK ratio (greater than 0.3), increase in CO_2 content, decrease in pH, rancid or H_2S odors are indications of hydraulic or organic overloading, excessive withdrawal of digested sludge, or incoming toxic materials. The problem may be over-come by decreasing organic loading, adding seed from another digester, decreasing sludge withdrawal rate, increasing mixing rate and mixing time, exerting proper tempera-ture control, and instituting an industrial pretreatment program.

2. Poor supernatant quality may be due to excessive mixing, insufficient settling time before sludge withdrawal, too low supernatant drawoff point, and insufficient sludge withdrawal rate. The problem may be solved by reducing mixing, allowing longer time for settling, using higher supernatant withdrawal ports, and increasing the digested sludge withdrawal.

3. Foam in supernatant may be due to scum blanket breaking up, excessive gas recirculation, and organic overload. Stop withdrawal supernatant, throttle compressor output, and reduce feeding rate.

4. Thin digested sludge may be due to short circuiting, excessive mixing, or too high sludge-pumping rate. Stop mixing several hours before sludge withdrawal and withdraw the sludge by short pumping cycle.

---

*Amount of seed required is approximately 4–5 times the anticipated volatile solids in the raw daily sludge.
Raw TVS per digester = 112 m^3/d (Table 17-5)
Seed quantity = 4.5 × 112 m^3 = 500 m^3

5. Falling sludge temperature may be due to plugged sludge recirculation lines, inadequate mixing, hydraulic overload, lower water feed rate in heat exchangers and boiler burner not firing. The problem can be overcome by back flushing the sludge recirculation lines by heated and digested sludge, and checking and correcting the boiler and heat exchangers.

6. Too high sludge temperature may be due to faulty controller, boiler and hot water temperature too high, and high hot water recirculation rate. Check and take proper action.

7. Insufficient mixing may be due to gas mixer feed lines plugged and gas flow too small. Clean gas lines and valves. Increase capacity of compressor.*

8. Low gas pressure in digester may be due to gas leaking from pressure relief valve (PRV), digester cover, gaslines, and hoses; too high gas-withdrawal rate; and high supernatant withdrawal. Check and repair leaks, and control gas and supernatant withdrawal rates.

9. High gas pressure in the digester may be due to insufficient gas withdrawal and PRV stuck shut due to freezing or defect. Increase gas withdrawal and correct PRV.

10. High scum blanket generally is due to plugged supernatant overflow. Lower the liquid level in the digester using bottom drawoff pipes, then rod the supernatant line.

11. Too thick scum is due to lack of mixing, and high grease contents. Break the scum manually, increase the mixing, increase sludge recirculation to discharge liquid above scum, or use chemicals to soften the blanket.

12. Tilting floating cover may be due to uneven distribution of load, thick scum accumulation around the edges, rollers or guide broken or rollers out of adjustment. The problem can be overcome by distributing the ballast or weight around, breaking the scum, and readjusting of the rollers. It may be necessary to lower the level of liquid in the digester.

13. Binding cover (even when rollers and guides are free) may be due to damaged internal guides or guy wires; lower the liquid level. If cover does not go all the way down, use crane to secure the cover in one position, then drain the digester and repair the guides.

17-8-3 Routine Operation and Maintenance

Routine digester operation must utilize the laboratory results to protect the digester from upset. The key operational goals are to (1) minimize excess water, (2) control organic loading, (3) control temperature, (4) control mixing, (5) reduce accumulation of scum, and (6) withdraw supernatant that is low in solids.

Monitoring Program for Process Control. The following important tests must be performed daily for control of the digestion process:

1. Volatile solids (VS) and total alkalinity (TA)
2. Gas production rate and composition (CH_4 and CO_2)
3. pH

---

*Gas recirculation rate of 5–10 $m^3/\cdot min \cdot 1000 \ m^3$ of digester capacity is sufficient. In some cases recirculation rate of 20 $m^3/min \cdot 1000 \ m^3$ may be necessary.

4. Volatile solids reduction
5. Digester temperature
6. Feed sludge volume and VS
7. Supernatant volume, and TSS and BOD
8. Digested sludge volume and VS
9. Visual gas test (a yellow flame with blue at the base is normal: too much blue and inability to stay lit indicates too much CO_2; orange flame with smoke indicates H_2S)
10. Sniff test; simply smelling the gas, supernatant, and digested sludge may give indication of septic, sour, putrid, well digested, or presence of chemicals such as oils, solvents, sulfides, etc.

Results of these tests should be fully utilized in operation and control and in troubleshooting.

Routine Operation and Maintenance Checklist.   The following checklist should be used for routine operation and maintenance of high-rate anaerobic digesters:

Feed Sludge

1. Record daily volume pumped for a 24-h period.
2. Perform daily total solids test, and make sure that too much water is not being fed.
3. Check daily pump operation for packing gland leaks, proper adjustment of cooling water, unusual noises, undue bearing heat, and suction and discharge pressures.
4. Monitor feed pump time clock operation for on-and-off and running time cycle. Also check the sludge consistency with these time cycles.

Recirculated Sludge

1. Record daily temperature and flow of recirculated sludge.
2. Collect 2–3 times per week samples of recirculated sludge and determine pH, alkalinity, TS, TVS, etc.
3. Check daily boiler temperatures, burner flame, and exhaust fan for proper operation.
4. Check daily temperature and flow of recirculating hot water.
5. Check daily and record heat exchanger inlet and outlet temperatures.
6. Check weekly for leaks in sludge lines.
7. Check daily pump operations—packing gland leaks, proper adjustment of cooling water, unusual noises, undue bearing temperatures, and suction and discharge pressures.

Digesters

1. Check daily gas manometers for proper digester gas pressure.
2. Drain daily the condensate traps.
3. Drain daily sediment traps.
4. Check daily gas burner for proper flame.
5. Record daily floating cover position, check cover guides, and check for gas leaks.

6. Record daily digester and natural gas meter readings.
7. Check daily and record fuel oil.
8. Check daily gas-mixing equipment for flow of gas to all feed points.
9. Check daily pressure relief and vacuum breaker valves. Verify operation with manometer and check for leaking gas.
10. Check daily supernatant tubes for proper operation, collect sample, and hose down supernatant box.
11. Check daily the level and condition of water seal on digester cover.
12. Check daily the flow meters for correct flow, leaks, and vibration.
13. Check daily the feed sludge density meter for correct density, leaks, and other items specified by the manufacturer.
14. Check daily the scum blanket through sight glass.
15. Check daily the gas storage tank for gas leaks and odors. Record readings on pressure gauges and drain condensate traps.

17-9 SPECIFICATIONS

Anaerobic digesters utilize complex equipment and controls. Brief specifications of various components are given below to describe many components that could not be fully covered in the Design Example. The design engineer should work closely with the equipment manufacturers to develop the detailed specifications for the selected components.

17-9-1 Digester Cover

General.   Furnish two floating covers for 13.7-m (45-ft) diameter digesters. The cover shall be arranged to float directly on the liquid contents of the tank, thereby providing positive submergence of all materials that tend to float. The gas holder covers shall be provided with spiral guides to engage rollers set in the concrete wall to assure that the cover is level at all times. The cover shall be designed to receive a welded steel-plate roof and shall be made up of fabricated assemblies which shall be piecemarked for erection. The top cover shall be designed to accommodate a housing containing gas-mixing compressors and all appurtenances.

Material and Fabrication

1. All structural steel shall comply with the ASTM specifications for structural steel.
2. The digester cover shall be designed for the following loadings: dead load = 98 kg/m^2 (35 lb/ft^2), and live snow and vacuum load = 140 kg/m^2 (50 lb/ft^2).
3. The materials to form the completed floating cover shall be shipped in subassemblies, and the gas dome and appurtenances shall be assembled at the site.
4. All shop welding shall be shielded arc welding and shall conform to the latest standards of the American Welding Society.
5. The installed weight of the floating cover shall be sufficient to provide a static gas pressure of at least 0.2 m (8 in.) of water.

6. After erection the gas holder shall be tested for gas tightness and leaks as specified by the designer.

Appurtenances.   In addition to the main assemblies, the following accessories shall be provided in each digester:

*   A pressure equalization unit.
*   A 76-cm (30-in.) entrance hatch.
*   A 1.22-m (48-in.) manhole with bolted and gasketed cover.
*   Six 20-cm (8-in.) sampling wells with gas-tight, quick-opening cover.
*   A 60-cm (24-in.) gas pipe housing and combination pressure and vacuum relief assembly of the unit-weighted diaphragm-operated type. The weights provided shall allow adjustment of the relief pressure in increments of 0.6 cm.

17-9-2 Mixing

General.   Furnish and install a complete gas-mixing system utilizing an external compressor furnishing gas at a controlled rate through feed lines to diffusers mounted on the digester floor.

Gas Diffusers.   The gas-mixing equipment shall comprise nine 2.5-cm (1-in.) stainless steel gas lines through the digester wall as shown in Figure 17-6; nine gas diffusers mounted on the floor with PVC connecting pipes and gas manifold; gas filter, flame trap shutoff valves; and gas metering for each line.

Compressor.   Provide three parallel combination, identical gas compressors, each driven by a 15-kW motor. Each compressor shall deliver a minimum of 9 m^3/min (317 cfm) gas against a pressure of 2.4 atm with suction conditions of 1.03 atm. Each compressor shall have explosion-proof motor with belt drive, guard, pressure gauge, pressure relief valve, and shutoff valve for intake and discharge side of the compressor. Furnish each compressor with (1) a combination flame trap and thermal shutoff valve with fusible elements, sediment trap assembly, and gas filter on inlet side, and (2) a motorized valve on the discharge line to provide a positive shutoff when the compressor is without power supply. All gas compressors, valves, controls, internal piping, and wiring shall be contained in fabricated steel housings.

17-9-3 Heater and Heat Exchanger

General.   Furnish and install boiler and heat exchanger units for heating the recirculated sludge. The boiler and heat exchangers must be an integral unit, designed for a working pressure of 207 kPa* (30 psi) and constructed in accordance with the ASME Low Pressure Heating Boiler Code.

*1 kPa = kN/m^2 = 0.145 psi.

Boiler. Provide two boilers each having a rating of 0.813×10^6 kJ/h (0.77 million Btu/h) with a digester gas consumption of 45 m^3/h. The boiler shall have refractory lined access at the rear for easy servicing of fire tubes and fire box. The boiler shall contain inspection ports, combination pressure gauge, low water cutoff, safety devices, sensing elements, blowers and hot water recirculating pumps, and all other appurtenances and controls for complete installation and operation as required. Each unit shall be factory-assembled, factory-wired, and factory-tested.

The fuel-burning equipment shall include the necessary accessories for the burning of either digester gas or natural gas, or a mixture of the two gases with automatic switchover. Pressure regulators, pressure check valves, and ignition transformer shall be provided.

Recirculating Sludge Pipe. The recirculating sludge pipe from each digester shall be of standard weight steel, 23 cm (9 in.) in diameter. The heating length of the pipe shall be 6 m. The constant speed sludge pump shall have a rated capacity of 2.5 m^3/min (660 gpm) to recirculate the sludge through the heat exchanger.

Heat Exchanger. Provide a common jacketed pipe heat exchanger to heat the recirculating sludge from each digester. The heat exchangers shall be designed for a working pressure of 207 kPa (30 psi). The sludge-heating capacity of the heat exchanger shall be 3.52×10^7 kJ/d. Indicating thermometers shall be provided at the inlet and outlet of the sludge tubes, and hot water line. Automatic controls shall be provided to vary the flow of hot water through the water jacket in order to maintain uniform heating temperature of the recirculating sludge.

17-10 PROBLEMS AND DISCUSSION TOPICS

17-1. An aerobic digester was designed to stabilize combined sludge from a secondary treatment facility serving a population of 10,000. The combined sludge is not thickened, and reaches the digester at a solids concentration of 1.5 percent. TS reaching the digester is 1100 kg/d. Calculate the dimensions of the digester, volume allowance m^3/capita, VS loading kg/m$^3 \cdot$d, diffused air requirement, and TSS in the effluent. The digested sludge has 5 percent solids. Provide digestion period of 17 d.

17-2. An anaerobic sludge digester has average liquid depth of 7 m and a diameter of 10 m. It receives 1600 kg/d thickened sludge at 6 percent solids. The specific gravity of the thickened sludge is 1.03. Calculate (a) volatile solids loading factor, (b) digestion period, (c) capacity per capita if the population served is 19,000. Assume VS in sludge solids is 79 percent.

17-3. An anaerobic sludge digester receives 2000 kg/d solids at 76 percent VS. Calculate total gas production per day using Eqs. (17-2) and (17-3). Compare gas production using other rules of thumb. The population served is 35,000. 1 gm of biodegradable solids = 1.42 g BOD$_L$, Y = 0.06, k_d = 0.04/d, and E = 0.79. Biodegradable solids = 70 percent. Incoming sludge is 5.5 percent solids and sp. gr. = 1.025. Average digestion period is 18 d.

17-4. Pilot plant experiments using complete-mix anaerobic digester without recycle were conducted on an industrial sludge. The data are given below.

Test Run	BOD$_L$ of the Sludge Solids, dS/dt (kg/d)	Net Mass of Active Solids in the Digester X, kg	Sludge Solids Withdrawn from Digester dX/dt (kg/d)
1	450	200	39
2	220	55	21

If the digestion period (θ_c) is 10 d, and waste-utilization efficiency E is 0.8, calculate the percent stabilization of BOD$_L$. Use Eqs. (17-2) and (17-3), and

$$\frac{dX}{dt} = Y\frac{dS}{dt} - k_d X$$

Also calculate the percent BOD$_L$ stabilized in a digester receiving 4500 kg/d BOD$_L$.

17-5. Compute the power requirements for gas mixing in a digester. The volume of the digester is 2000 m^3 and velocity gradient for proper mixing is 88/s, and viscosity of sludge at operating temperature is 1.53×10^{-3} N s/m^2

17-6. An anaerobic digester receives 2000 kg total dry solids per day. The sludge contains 5.8 percent solids at a specific gravity of 1.03. The digested sludge contains 4.2 percent solids and has specific gravity of 1.02. Calculate the BOD$_5$ and TS of the digester supernatant. Use the following data:

TVS in raw sludge	= 75 percent
TVS stabilized	= 56 percent
Digested sludge withdrawal rate	= 24 m^3/d
BOD$_L$ of digested sludge	= 0.65 × TVS in digested sludge
BOD$_5$/BOD$_L$	= 0.68

17-7. Thickened sludge is heated from 15 to 36°C before reaching an anaerobic digester. Average quantity of sludge is 100 m^3/d, specific heat of sludge is 4200 J/kg°C (same for water), and heat exchange capacity of jacketed heat exchanger is 3060 kJ/h·m^2·°C. Calculate the following:
  (a) Length of sludge pipe in the jacket
  (b) Volume of hot water recirculated through the heat exchanger if inlet and outlet temperatures are 85 and 40°C. Assume heat transfer efficiency of 80 percent.
  (c) Digester gas needed in the boiler (m^3/d). Assume 75 percent boiler efficiency; and heating value of the digester gas is 24,500 kJ/m^3.

17-8. What is the purpose of sludge mixing in an anaerobic digester? Describe various methods of sludge mixing and give advantages and disadvantages of each method.

17-9. Discuss different factors that affect the performance of an anaerobic digester. List various tests that must be performed daily for control of the digestion process. What is the significance of each test in relation to the normal digester operation?

17-10. Describe various types of digester covers and advantages and disadvantages of each type.

17-11. Describe various process parameters, and advantages and disadvantages of an aerobic digester over anaerobic digester.

17-12. An aerobic digester is designed to receive 4000 kg solids at 7 percent by weight. The digester roof, bottom and side wall areas are 450, 470 and 550 m^3 having overall coefficients of heat transfers of 0.9, 0.62 and 0.68 J/s·m^2·°C, respectively. Assume the incoming sludge, digester operating, and average outside temperatures are 25, 38 and 21°C respectively. Minor heat losses are 25 percent of total heat loss. Calculate the total heat requirement of the digester. The specific gravity of the incoming sludge is 1.03.

REFERENCES

1. U.S. Environmental Protection Agency, *Process Design Manual for Sludge Treatment and Disposal,* Technology Transfer, EPA-625/1-74-006, October 1974.
2. U.S. Environmental Protection Agency, *Process Design Manual for Sludge Treatment and Disposal,* Technology Transfer, EPA-625/1-79-011, September 1979.
3. Metcalf and Eddy, Inc., *Wastewater Engineering: Treatment, Disposal, and Reuse,* McGraw-Hill Book Co., New York, 1979.
4. Benefield, L. D., and C. W. Randall, *Biological Process Design for Wastewater Treatment,* Prentice-Hall, Englewood Cliffs, N.J., 1980.
5. Reynolds, T. D., *Unit Operations and Processes in Environmental Engineering,* Brooks/Cole Engineering Division, Monterey, Cal., 1982.
6. Clark, J. W., W. Viessman, and M. J. Hammer, *Water Supply and Pollution Control,* 3rd Edition, IEP-A Dun-Donnelley, New York, 1977.
7. Zickefoose, C., and R. B. Joe Hayes, *Operations Manual: Anaerobic Sludge Digestion,* U.S. Environmental Protection Agency, Washington, D.C., EPA 430/9-76-001, February 1976.
8. Culp, G. L., and N. F. Heim, *Performance Evaluation and Troubleshooting at Municipal Wastewater Treatment Facilities,* U.S. Environmental Protection Agency, MO-16 EPA-430/9-78-001, Washington, D.C., January 1978.

18

Sludge Conditioning and Dewatering

18-1 INTRODUCTION

Sludge dewatering is necessary to remove moisture so that the sludge cake can be transported by truck and can be composted or disposed of by landfilling or incineration. The solid particles in municipal sludge are extremely fine, are hydrated, and carry electrostatic charges. These properties of sludge solids make dewatering quite difficult. Sludge conditioning is necessary to destabilize the suspension so that proper sludge-dewatering devices can be effectively used.

Sludge-dewatering systems range from very simple devices to extremely complex mechanical processes. Simple devices involve natural evaporation, and percolation from sludge lagoons or drying beds. Complex mechanical systems utilize sludge conditioning followed by centrifugation, vacuum filtration, filter presses, and belt filters. The selection of any device depends on the quantity and type of sludge and the method of ultimate disposal. At smaller plants sludge-drying beds or lagoons are frequently selected. Vacuum filters are used extensively at large plants. Filter presses, while widely used in Europe, have only been introduced in the United States in recent years. However, due to their considerably lower capital costs and operational reliability, they are currently very popular. They produce cake with 20–30 percent solids.

In this chapter various methods of sludge conditioning and dewatering are briefly presented. Step-by-step design procedure, operation and maintenance, and equipment specifications for a sludge-dewatering facility using chemical conditioning and filter presses are covered in the Design Example.

18-2 SLUDGE CONDITIONING

Conditioning involves chemical and/or physical treatment of the sludge to enhance water removal. In addition, some conditioning processes also disinfect sludge, control odors, alter the nature of solids, provide limited solids destruction, and improve solids recovery. Several chemical and physical conditioning methods are discussed below.

18-2-1 Chemical Conditioning

Chemical conditioning is associated principally with mechanical sludge-dewatering systems such as vacuum filter, filter press, belt filter, and centrifugation. Chemical conditioning is achieved by inorganic or organic chemicals.

Inorganic Chemicals.   Ferric chloride and lime are the commonly used inorganic chemicals, although ferrous sulfate and alum have also been used. Ferric chloride that hydrolyzes in water is added first, forming positively charged, soluble iron complexes. The iron complexes then neutralize the negatively charged sludge solids, thus causing them to aggregate. Ferric chloride also reacts with the bicarbonate alkalinity in the sludge to form hydroxides that causes flocculation. Hydrated lime is usually used in conjunction with ferric iron salts. Lime provides pH control, odor reduction, and disinfection. $CaCO_3$ produced in the reaction of lime and bicarbonate provides a granular structure which increases sludge porosity and reduces sludge compressibility.

Both power plant fly ash and sludge incinerator ash have also been used successfully in conditioning of sludge. Ash improves dewatering of sludge due to partial solubilization of the metallic constituents in ash, its sorptive capabilities, and its irregular particle size. When ash is used for sludge stabilization, the quantity of sludge cake is increased considerably.

Organic Chemicals.   Organic polymers (polyelectrolytes) are widely used in wastewater treatment and sludge conditioning. These are long-chain, water soluble, specialty chemicals. These materials differ greatly in chemical composition and functional effectiveness.[1,2]

Organic polyelectrolytes dissolve in water to form solutions of varying viscosity. The electrolytes in solution act by adhering to the sludge particle surfaces, thus causing (1) desorption of bound surface water, (2) charge neutralization, and (3) agglomerization by bridging between the particles.

Proper dosage and mixing of chemicals are necessary to achieve sludge conditioning. The chemical dosage is determined by laboratory studies. Filter leaf tests or Büchner funnel tests are commonly used to determine proper chemical dosage, filter yield, and suitability of various filtering media.[3–5] Chemical dosage requirements for different dewatering processes and types of sludges are discussed in Sec. 18-3. In general, the dosage of iron salts is 2–6 percent of the dry solids in the sludge. Lime dosage usually varies from 7 to 15 percent of dry solids. The dosage of polyelectrolytes may vary from 0 to 0.5 percent of dry solids in the sludge. The amount of fly ash for sludge stabilization is 25–50 percent of the sludge solids.

Intimate mixing of sludge with coagulating chemicals is necessary to properly condition the sludge. In order to minimize floc shearing, mixing should provide just enough energy to disperse the conditioner chemicals throughout the sludge. Design criteria and design procedure for the mixing equipment are presented in the Design Example.

18-2-2 Physical Conditioning

Important physical sludge-conditioning methods include elutriation and thermal conditioning. Other, less commonly used methods are freezing, ultrasonic vibration, solvent extraction, and irradiation. These methods are briefly discussed below.

Elutriation. Elutriation is washing of sludge in order to remove certain soluble organic or inorganic compounds that would consume large amounts of chemicals for conditioning of sludge. Wastewater effluent is normally used for elutriation. The elutriation of sludge generates large volumes of liquid that contain high concentration of suspended solids. This liquid when returned to the plant increases the loadings.

Generally, the volume of washwater is 2–6 times the volume of sludge. Elutriation tanks are designed to act as gravity thickeners with solids loading of 39–50 kg/m^2·d (8–10 lb/ft^2·d). The cost of additional tanks and equipment for washing and solids separation may not justify the savings in the cost of chemicals. Additional information on sludge elutriation can be obtained in Refs. 1–5.

Thermal Conditioning. The thermal conditioning process involves heating the sludge to a temperature of 140–240°C (248–464°F) in a reaction vessel under pressures of 1720–2760 kN/m^2 (250–400 psig) for periods of 15–40 min. One modification of the process involves the addition of a small amount of air. Heat coagulates the solids, breaks down the gel structure, and reduces the affinity for water. The resulting sludge is sterilized, practically deodorized, and dewatered readily on vacuum filters or filter presses without the need of chemicals. This process is most applicable to biological sludges that are normally difficult to condition or stabilize by chemicals. Because of high capital costs, this process is generally limited to large plants. Process side streams include gases and liquids. The gases are malodorous. Methods of odor control include combustion, adsorption, scrubbing, and masking. Liquid side streams contain high BOD and when returned to the plant may significantly increase the loading to the aeration basin.[2] The characteristics of liquor from thermal conditioning system are summarized in Table 18-1.

Other Conditioning Processes. Freezing and thawing improves sludge filterability. In northern regions natural freezing has been used for sludge conditioning.[1] Sludge irradiation also improves filterability. Discussion on irradiation of wastewater may be found in Chapter 14. Electric current when passed through sludge using graphite anodes and iron cathodes can also condition the sludge.[1] Ultrasonic vibration has also improved filterability of sludge. All these processes are in experimental stages.[1–3]

18-3 SLUDGE DEWATERING

A number of sludge-dewatering techniques are currently used. The selection of any sludge-dewatering system depends on (1) characteristics of sludge to be dewatered, (2) space available, and (3) moisture content requirements of the sludge cake for ultimate disposal. When land is available and the sludge quantity is small, natural dewatering systems are most attractive. These include drying beds and drying lagoons. The mechanical dewatering systems are generally selected where land is not available. Common

TABLE 18-1   Characteristics of Returned Liquor from
Thermal Conditioning of Sludge

Parameters	Concentration
BOD_5, mg/ℓ	5000–15,000
COD, mg/ℓ	10,000–30,000
TSS, mg/ℓ	300–20,000
Total nitrogen, mg/ℓ	600–1000
Total phosphorus, mg/ℓ	150–200
pH	5.0–6.5

Source: Refs. 1 and 2

mechanical sludge-dewatering systems include centrifuge, vacuum filter, filter press, and horizontal belt filter. Each method is discussed below.

18-3-1 Drying Beds

Sludge drying beds are the oldest method of sludge dewatering. These are still used extensively in small- to medium-size plants to dewater digested sludge. Typical sand beds consist of a layer of coarse sand 15–25 cm in depth and supported on a graded gravel bed that incorporates selected tiles or perforated pipe underdrains. Paved drying beds are also used. Each section of the bed (8 m × 30 m) contains water-tight walls, underdrain system, and vehicle tracks for removal of sludge cake. Sludge is placed on the bed in 20- to 30-cm (8- to 12-in.) layers and allowed to dry. The underdrained liquid is returned to the plant. The drying period is 10–15 days and moisture content of the cake is 60–70 percent. Poorly digested sludge may cause odor problems. Depending on the climatic condition and odor control requirements, the drying bed may be open or covered. Typical design criteria for sludge drying beds receiving digested primary and secondary sludges are given below:

1. Sludge drying bed area
    0.14–0.28 m^2 per capita (1.5–3.0 ft^2/capita) for uncovered beds
    0.10–0.20 m^2 per capita (1.0–2.0 ft^2/capita) for covered beds
2. Sludge loading rate
    100–300 kg dry solids per m^2 per year (20–61 $lb/ft^2 \cdot yr$) for uncovered beds
    150–400 kg dry solids per m^2 per year (30–82 $lb/ft^2 \cdot yr$) for covered beds

The sludge cake from drying beds contains 20–40 percent solids; almost 90–100 percent solids capture occurs. The design details of the sludge drying beds are given in Figure 18-1. An excellent discussion of sludge drying beds may be found in Refs. 1 and 2.

18-3-2 Drying Lagoons

Sludge lagoons are an economical method for sludge dewatering where sufficient land is available. They are similar to drying beds because the sludge is periodically removed and the lagoon refilled. Sludge must be stabilized to reduce odor problems.

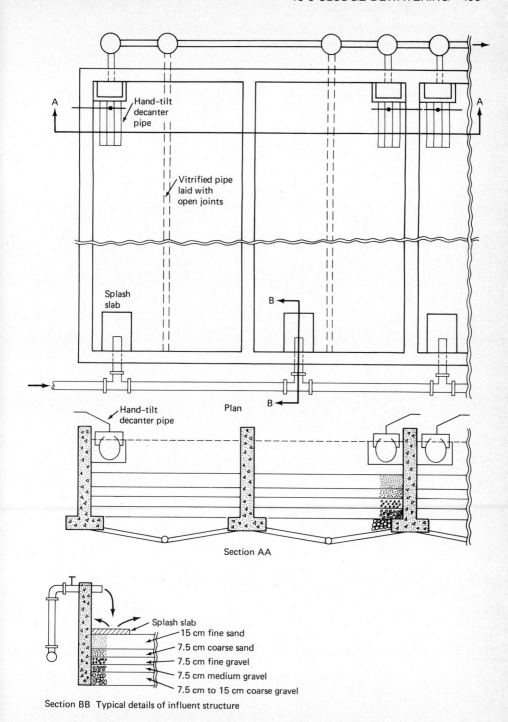

Plan

Section AA

Section BB Typical details of influent structure

Figure 18-1   Details of Sludge Drying Beds.

Sludge-drying lagoons consist of shallow earthen basins. Earthen dykes (0.7–1.4 m high) enclose the sludge lagoon. Sludge 0.7–1.4 m (2–4 ft) in depth is applied. The supernatant is decanted from the surface and returned to the plant. The sludge liquid is allowed to evaporate. Sludge drying time depends on the climatic conditions and the depth of sludge application. Generally 3–6 months is required to reach 20–40 percent solids in the sludge cake. Solids capture in drying lagoons is 90–100 percent. Sludge cake is removed by mechanical equipment.

The suggested solids-loading rates for drying lagoons are 37 kg/m^3·yr (2.3 lb/ft^3·yr) of lagoon capacity. Some designers provide lagoon capacity of 0.3–0.4 m^2/capita (3–4 ft^2/capita) for primary and secondary sludge. The proper design of sludge drying lagoons requires a consideration of the following factors: climate, subsoil permeability, sludge characteristics, lagoon depth, and area management practices.

18-3-3 Centrifugal Dewatering

The centrifuge uses centrifugal force to speed up the sedimentation rate of sludge solids. Sludge dewatering can be achieved by solid-bowl and basket centrifuges. In a typical unit, the conditioned sludge is pumped into a horizontal or cylindrical "bowl" rotating at 1600–2000 rpm. The solids are spun to the outside of the bowl where they are scraped out by a screw conveyor. The liquid or "centrate" is returned to the wastewater treatment plant for treatment. The centrifuging process is comparable to the vacuum filtration in cost and performance. Centrifuges are compact, entirely enclosed (which may reduce odors), require small space, and can handle sludges that might otherwise plug filter cloth. The disadvantages include complexity of maintenance, abrasion problems, and centrate high in suspended solids.

The sludge cake from centrifuge contains 20–35 percent solids, and solids capture of 85–90 percent is achieved. The polymer dosage for sludge conditioning prior to centrifuge is 0.1–0.7 percent of dry solids in the feed. A centrifuge dewatering system is shown in Figure 18-2. Excellent discussion of centrifuge dewatering may be found in Refs. 1 and 2.

18-3-4 Vacuum Filter

Rotary vacuum filters are widely used for dewatering of both raw and digested sludges. Vacuum filters consist of a cylindrical drum covered with cloth of natural or synthetic fabric. The drum remains partly submerged in a vat of sludge and rotates slowly. Internal vacuum that is maintained inside the drum draws the sludge to the filter medium and water is withdrawn from the sludge. The cake-drying zone represents from 40 to 60 percent of the drum surface and terminates at the cake discharge zone where the cake is removed. In a drum-type rotary vacuum filter, the sludge cake is scraped off. Compressed air may be blown through the media to release the cake prior to scraping. In belt-type rotary vacuum filters the covering or media belt leaves the drum and sludge cake is released by use of two stainless steel coils arranged around the drum. A typical vacuum filter arrangement is shown in Figure 18-3.

A variation of conventional rotary drum filter is the top feed drum filter. In this case the sludge is fed to the vacuum filter through a hopper located above the filter.

The important design factors for rotary vacuum filters include characteristics of

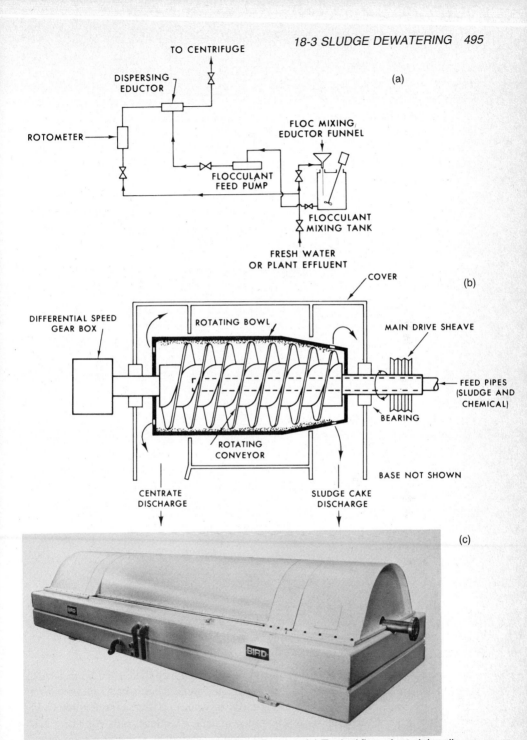

Figure 18-2   Centrifugal Sludge Dewatering System. (a) Typical flocculant piping diagram
of centrifuge used for sludge dewatering. [From Ref. 2.] (b) Schematic of typical solid-bowl
centrifuge. [From Ref. 2.] (c) Continuous flow solid-bowl centrifuge for sludge dewatering.
(Courtesy Bird Machine Company.)

(a)

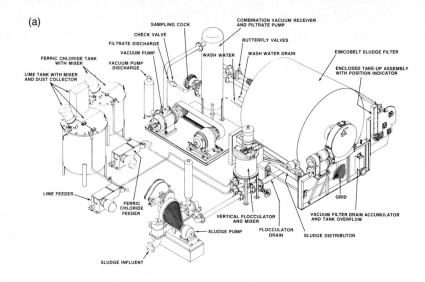

(b)

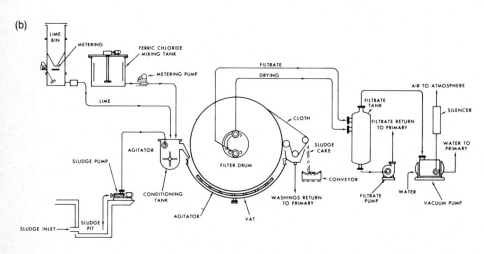

Figure 18-3   Installation Details of Rotary Vacuum Filter Assembly. (a) Typical rotary vacuum filter Installation. (Courtesy EIMCO Process Equipment Company.) (b) Schematic flow diagram of rotary vacuum filter system. [From Ref. 2.]

conditioned sludge, cake formation time, viscosity, vacuum applied, specific resistance of the sludge cake, type of filter medium, and filter yield. Many equations have been developed to express the filtration rate, specific resistance of sludge, and filter yield. Basic theory of vacuum filtration may be found in some excellent publications.[3–5] Two test procedures used for determining the filterability of sludges are the Büchner funnel method and the filter leaf techniques. The Büchner funnel method enables determination of the relative effects of various chemical conditioners, and the calculation of the specific resistance of the sludge, but it is seldom used for the calculation of required filter area. The filter leaf test is used to determine the required filter area, evaluate filter medium, and it permits an accurate prediction of the operation of a full-scale filter. Detailed procedures of Büchner funnel and filter leaf tests may be found in Refs. 2–5.

Rotary vacuum filters range in size from 5 to 60 m^2 (50–600 ft^2) and are normally supplied by equipment manufacturers. The system includes vacuum pump, filtrate receiver and pump, filtering medium, and sludge-conditioning apparatus. The vacuum pump requirements are normally 0.5 m^3 of air per min per m^2 of filter area at 69 kN/m^2 (1.5–2.0 $ft^3/min \cdot ft^2$ at 20 in. of mercury vacuum).[1] The typical dewatering performance data for rotary vacuum filters using cloth media is summarized in Table 18-2. Other filter media include cotton, wool, nylon, Dacron, etc. An evaluation of various filtering media was

TABLE 18-2 *Dewatering Performance Data for Rotary Vacuum Filters Using Cloth Media*

Type of Sludge	Feed Solids (Percent)	Chemical Dosage (Percent)[a]		Filter Yield $(kg/m^2 \cdot h)$[b]	Cake Solids (Percent)
		FeCl₃	CaO		
Raw sludge					
Primary	4–9	2–4	8–10	17–40	27–35
Primary and activated sludge	3–7	2–4	9–12	12–30	18–25
Primary and trickling filter	4–8	2–4	9–12	15–35	23–30
Anaerobically digested					
Primary	4–8	3–5	10–13	15–35	25–32
Primary and activated sludge	3–7	4–6	15–20	10–25	18–25
Primary and trickling filter	5–10	4–6	13–18	17–40	20–27
Aerobically digested					
Primary and activated sludge	3–6	3–7	8–12	8–20	16–23
Elutriated sludge					
Anaerobically digested					
Primary and activated sludge	4–8	3–6	0–7	15–18	18–25
Thermally conditioned					
Primary and activated sludge	6–15	0	0	20–40	35–45

[a]Chemical dosage percent of dry solids, 1 percent = 10g/kg (20 lb/ton).
[b]1 $kg/m^2 \cdot h \times 0.205 = lb/ft^2 \cdot h$.
Source: Adopted in part from Ref. 1.

done at the Chicago Sanitary District.[2] The results indicated that Dacron, a polyester, was most suitable for their use. Other treatment plants have found polypropylene media to be satisfactory. Polyethylene media tend to stretch when wet and require constant operator vigilance of belt tension. Minneapolis-St. Paul has reported a life of 12,400 h for a Saran medium. Monofilament fabrics are most resistant to blinding and have been used exclusively in recent installations of drum or belt filter.[2]

18-3-5 Plate and Frame Filter Press

Plate and frame presses are also called filter presses or recessed plate pressure filters. Typical installation of plate and frame presses is shown in Figure 18-4. These consist of round or rectangular recessed plates which when pressed together form hollow chambers. On the face of each individual plate is mounted a filter cloth. In a fixed-volume filter press, the sludge is pumped under high pressure 350–1575 kN/m^2 (50–225 psi) into the chamber. The water passes through the cloth while the solids are retained and form a cake on the surface of the cloth. The sludge filling continues until the press is effectively full of cake. The entire filling operation takes 20–30 min. The pressure at this point is generally the designed maximum and is maintained for a 1- to 4-h period. During this time more filtrate is removed, and the desired cake solids level is reached (filter presses can attain up to 40 percent solids). The filter is then mechanically opened, and the dewatered cake drops from the chamber onto a conveyor belt for removal. Cake breakers are usually required to break up the rigid cake into conveyable form. Figure 18-5 shows the details of a fixed volume recessed plate pressure filter.

Figure 18-4   *Typical Installation of Plate and Frame Filter Press Assembly. (Courtesy Envirex Inc., a Rexnord Company.)*

A variation of the fixed-volume filter press (discussed above) is the variable-volume recessed plate pressure filter. A diaphragm is placed behind the filter cloth that provides air or water pressure to squeeze the sludge. Generally 10–20 min is required to fill the press with the conditioned sludge. When the end point is reached, the sludge feed pump is automatically turned off. Water or air under high pressure is then pumped into the space between the diaphragm and the plate, thus squeezing the already formed and partially dewatered cake to the desired solids content. At the end of the cycle, the water is returned to a reservoir, plates are automatically opened, and sludge cake is discharged. The operational details of a variable-volume recessed plate filter assembly are shown in Figure 18-5. The typical dewatering performance of filter presses is summarized in Table 18-3.

18-3-6 Belt Filter Press

Belt filter presses employ single or double moving belts to dewater sludge continuously. Belt filter presses are currently very popular in the United States. The main advantage of belt filter presses is drier cake, low power requirement, and continuous operation. The main disadvantages are short media life and a filtration rate sensitive to incoming sludge.

The belt filtration process involves three basic operational stages: (1) chemical conditioning, (2) gravity draining of excess water, and (3) compaction of predewatered sludge. The system components of belt filter presses are shown in Figure 18-6. The conditioned sludge is discharged over the moving belt. Typically 1 or 2 min is required for drainage of the excess water. The sludge is then subject to increased pressure due to either the compression of the sludge between the carrying belt and cover belt or the application of a vacuum on the carrying belt. The sludge cake is squeezed between the two belts as it passes between various rollers. The compaction pressure can be widely varied by using variable belt and roller arrangement as shown in Figure 18-7.[6] A continuous belt press filter using endless belt around a system of rollers for sludge dewatering are shown in Figure 18-8. The design and operating data for belt filter press are summarized in Table 18-4.

18-3-7 Other Mechanical Dewatering Systems

The concept of filtration has been utilized to develop many other types of dewatering devices. These devices include:

1. Moving screen concentrators
2. Capillary dewatering system
3. Rotating gravity concentrator
4. Twin-roll press
5. Screw press
6. Gravity bag filter

Each of these systems are briefly described below. Additional details may be found in Refs. 1–3.

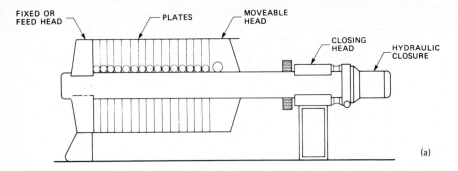

Figure 18-5   Details of Recessed Plate Pressure Filteration System. (a) Details of a fixed-volume recessed pressure plate filter assembly. [From Refs. 1.] (b) Cross section of a fixed-volume recessed plate filter assembly. [From Ref. 1.] (c) Operation of a variable volume recessed plate filter assembly. (Courtesy Envirex Inc., a Rexnord Company.)

TABLE 18-3   Typical Dewatering Performance of Filter Presses

| Type of Sludge | Feed Solids (Percent) | Chemical Dosage (Percent Dry Solids) | | Filter Yield (kg/m²·h) | Cake Solids (Percent) |
		FeCl₃	CaO		
Primary and secondary	4	5	15	5	40
Anaerobically digested, primary and secondary	4	6	16	5	40
Thermally conditioned, primary and secondary	14	0	0	12	60

Source: Ref. 1.

TABLE 18-4   Design and Operation Data of Belt Filter Press

Condition	Data
Solids in feed sludge	3–10 percent dry wt.
Solids in cake	20–40 percent dry wt.
Polymer for conditioning	0.2–0.5 percent of dry solids
Total suspended solids in filtrate	100–1000 mg/ℓ
Solids capture	90–95 percent
Filter yield	20–40 kg/m²·h
Sludge-dewatering rate with 1-m-wide belt filter press	3000–6000 kg in 8 h

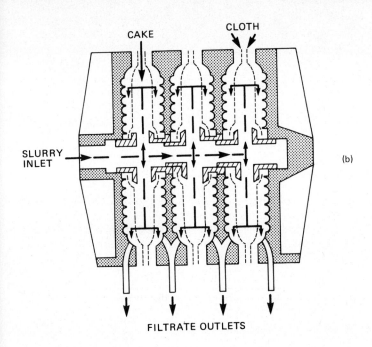

(b)

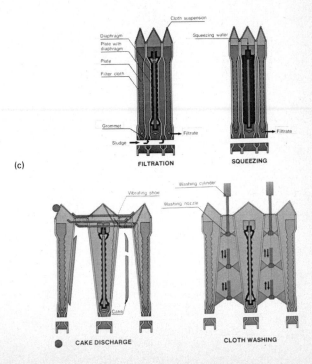

(c)

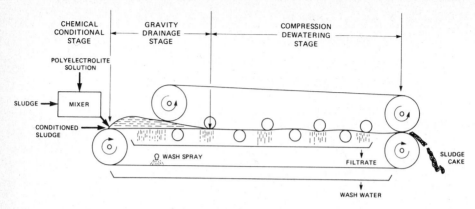

Figure 18-6   Belt Filter Press Used for Sludge Dewatering, Showing Three Basic Stages of Operation. [From Ref. 1.]

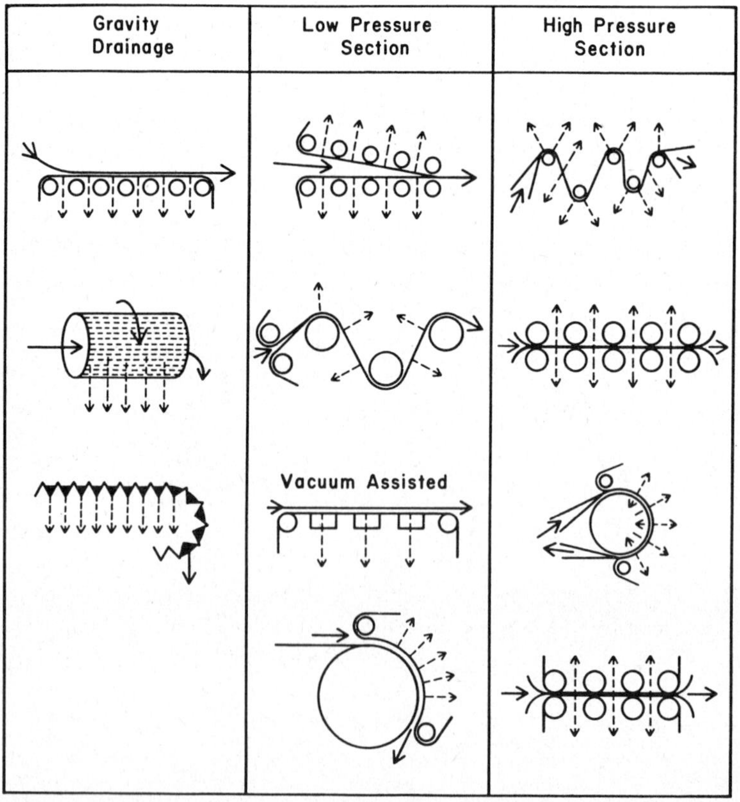

Gravity Drainage	Low Pressure Section	High Pressure Section

Figure 18-7   Alternative Designs for Obtaining Water Releases with Belt Filter Presses. [From Ref. 6. Used with permission from National Council of the Paper Industry for Air and Stream Improvement, Inc.]

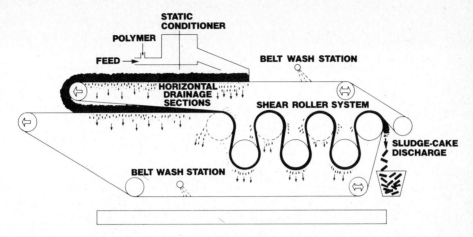

Figure 18-8   Operational Schematic of Belt Filter Press. (Courtesy Ashbrook-Simon-Hartley.)

Moving Screen Concentrator.   The device uses a variable-speed moving screen. The thickened and conditioned sludge goes through gravity dewatering, is then passed onto the second screen where solids are passed through rollers with each successive set applying higher pressure. The sludge cake has solids content of 20–30 percent. Solids capture is approximately 90 percent.

Capillary Dewatering.   This device is similar to a horizontal belt filter. The conditioned sludge is applied over a portion of the screen where free water is released and solids are concentrated. The sludge cake is then passed through final compression zone where additional water is squeezed out. The screen is washed and the cycle restarted. The sludge cake contains 16–20 percent solids, and 85–90 percent solids captured is achieved. The solids-loading rate is 26 kg/m^2·h.^{2}

Rotating Gravity Concentrator.   A rotating gravity concentrator (also called dual-cell gravity filter) consists of two independent cells. The cells are formed by a fine-mesh nylon filter cloth which travels continuously over front and rear guide wheels. Dewatering occurs in the first cell and cake formation takes place in the second cell. The partially dewatered solids are then carried over the drive roll separator, where the cake of relatively low moisture content is produced. The dewatering is entirely by gravity and without the application of either pressure or vacuum. If more complete dewatering is needed, a multiroll press is used. The sludge cake contains 10–16 percent solids without multiroll press. The system components are shown in Figure 18-9.

Twin-Roll Press.   The twin-roll press consists of a pair of perforated rolls, one roll fixed and the other movable so that the nip (or space) between the roll can be varied. The sludge is pumped into a vat under a slight pressure. The sludge is moved into the nip where it is dewatered at a pressure of 36–72 kg/lineal cm (200–400 lb/lineal in.) of the roll.[1] The sludge cake contains 80–90 percent solids.

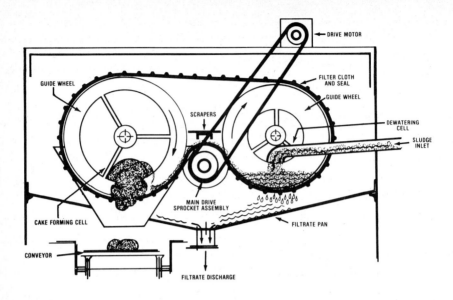

Figure 18-9 *Details of Dual-Cell Gravity Filter Assembly. (Courtesy The Permutit Company, Inc.)*

Screw Press. The screw press employs a screen surrounded by a perforated steel cylinder.[1] The sludge is pumped inside the screen. The screw moves the progressively dewatered sludge against a containment. Polymer dosage is 0.2–0.5 percent of dry solids. The sludge cake solids are 18–25 percent. Almost 90 percent solids capture is achieved.

Gravity Bag Filter. The conditioned sludge is pumped into the suspended porous bags. The weight of the sludge forces the water out of the bag from sides and the bottom. Sludge is retained for a maximum of 24 h depending on the desired dryness and is then released through the bottom opening. The sludge solids are generally 15–20 percent.

18-4 EQUIPMENT MANUFACTURERS OF SLUDGE-CONDITIONING AND DEWATERING SYSTEMS

A list of sludge and dewatering equipment suppliers is given in Appendix D. This list includes the suppliers of chemical feeders, mixers, and controls; and various types of sludge-dewatering equipment. Important considerations for equipment selection are presented in Sec. 2-10.

18-5 INFORMATION CHECKLIST FOR DESIGN OF SLUDGE-CONDITIONING AND DEWATERING FACILITIES

The design of a sludge-conditioning and dewatering facility should be started after the following basic information has been developed and necessary decisions have been made.

1. Develop sludge characteristics. This includes average quantity of sludge dewatered per day, solids concentration, and whether the sludge is raw or stabilized. Material mass balance analysis at average daily design flow must be conducted to establish this information.
2. Select an ultimate sludge disposal method, and specify the moisture content of the sludge cake.
3. Evaluate site conditions such as space, access roads, noise, and other environmental limitations.
4. Establish operational characteristics of the equipment, including energy requirements, specialized maintenance requirements, performance reliability, and simplicity of operation.
5. Establish nature of operation, that is, whether continuous or intermittent.
6. Determine type of chemicals and the dosages needed for proper dewatering. Laboratory bench scale or pilot testing may be necessary to establish chemical dosages, mixing, and reaction period.
7. Develop design data for the selected equipment for dewatering. This includes operational period, solids-loading rate, or dewatering rate.
8. Obtain the design criteria from the concerned regulatory agency.
9. Obtain manufacturers' catalogs and equipment selection guides.

18-6 DESIGN EXAMPLE

18-6-1 Design Criteria Used

The following design assumptions and criteria are used for the design of a sludge-dewatering facility:

1. The combined primary and secondary sludge is anaerobically digested. The characteristics of anaerobically digested sludge under average daily design flow conditions is given below:

Total solids	= 4100 kg/d*
Total volatile solids	= 2490 kg/d
Flow	= 80 m³/d
Concentration of solids	= 5.0 percent
pH	= 6.5–7.5
Specific gravity	= 1.02

2. The minimum dry solids allowed in the sludge cake shall be 25 percent.
3. Select variable-volume, recessed plate filter press assembly for dewatering.
4. Based on the pilot testing with a variable-volume, recessed plate filter assembly, the following design data were developed:

Chemical Conditioning

Optimum lime dosage (CaO)	= 5 percent of dry solids
Ferric chloride (FeCl₃)	= 0
Organic polymer (cationic)	= 2 percent of dry solids

Dewatering Equipment

Operating time	= 8 h/d, 5 d/week
Cycle time	
Feed	= 20 min at 345 kN/m² (50 psig)
Compression	= 15 min at 1550 kN/m² (225 psig)
Extraction	= 25 min
Total	= 60 min (1 h per cycle) or 8 cycles per d
Filtration rate	= 10 kg/m²·h (2 lb/ft²·h)
Cake solids (minimum)	= 25 percent

Mechanical Requirement

Water needed per cycle	
Wash	= 3420 ℓ
Cake extraction	= 380 ℓ
Diaphragm losses	= 190 ℓ
Total	= 3990 ℓ
Air blowdown per cycle	= 28 m³ at 210 kN/m² (1000 scf at 30 psig)

*Actual quantity of digested solids calculated in Sec. 17-7-2, step E6 = 4096 kg/d.

5. Filtration system shall be housed inside a building. The filter building design shall include filtrate removal, sludge cake removal, chemical conditioning tanks, chemical feed equipment, chemical storage, and pressure filter assembly. The digested sludge-holding well shall be kept outside the filter building.
6. The number of pressure filter units shall be selected such that the normal filtration process continues when the largest single unit is out of service. When all units are in operation, the maximum filtration capacity shall be 125 percent of the average daily design capacity.

18-6-2 Design Calculations

A. Pressure Filter Area
1. Compute total solids processed per hour of filter operation

$$\text{Total solids dewatered} = \text{sludge} + \text{lime} + \text{polymer}$$

$$\text{Sludge solids} = \frac{4100 \text{ kg/d} \times 7 \text{ d/week}}{5 \text{ d/week (operation)}}$$

$$= 5740 \text{ kg/d } (718 \text{ kg/h})$$

Lime (CaO) $= 5740 \text{ kg/d} \times 0.05 = 287 \text{ kg/d}$ (36 kg/h)
Polymer $= 5740 \text{ kg/d} \times 0.02 = 115 \text{ kg/d}$ (14 kg/h)
Total solids $= 6142 \text{ kg/d}$

$$\text{Total solids processed per hour} = \frac{6142 \text{ kg/d}}{8 \text{ h/d}} = 768 \text{ kg/h}$$

2. Compute filter area

$$\text{Filter area} = \frac{768 \text{ kg/h}}{10 \text{ kg/m}^2 \cdot \text{h}} = 76.8 \text{ m}^2$$

B. Selection of Efficient Filter Unit.
1. Determine the pressure filter sizes available
   From the manufacturer's catalogs determine the sizes of various filter units. Tabulate the filter area available with and without the single largest unit. Such information for the Design Example is provided in Table 18-5.
2. Select proper filter unit
   The most efficient and manageable filter unit assembly is the one which has relatively fewer operating units and provides nearly 100 percent operating capacity when one unit is out of service, and almost 25 percent extra capacity when all units are in operation. Select unit assembly D. It has a total of 5 units including one standby. This assembly will provide 99 percent capacity of average daily requirement when one unit is not operating. With the standby unit in operation, 24 percent extra dewatering capacity is achieved. A typical layout of the filter units is shown in Figure 18-10.

TABLE 18-5   Selection of Pressure Filter Units

Unit Identification	Filter Size (mm)	No. of Chambers	Area of Each Unit (m²)	Min. Units Required	Actual Area (m²)	Total Units with One Standby	Total Area with One Standby Unit (m²)	Area without Standby Unit (Percent)	Area with Standby Unit (Percent)
A	600	14	7	11	77	12	84	100.3[a]	109.4[b]
B	600	20	10	8	80	9	90	104.2	117.2
C	800	14	13	6	78	7	91	101.6	118.5
D	800	20	19	4	76	5	95	99.0	123.7
E	800	26	24	4	96	5	120	125.0	156.3
F	1000	20	30	3	90	4	120	117.0	156.3
G	1000	26	39	2	78	3	117	101.6	152.3
H	1250	26	48	2	96	3	144	125.0	187.5

[a] $\dfrac{77 \text{ m}^2}{76.8 \text{ m}^2} \times 100 = 100.3$ percent.

[b] $\dfrac{84 \text{ m}^2}{76.8 \text{ m}^2} \times 100 = 109.4$ percent.

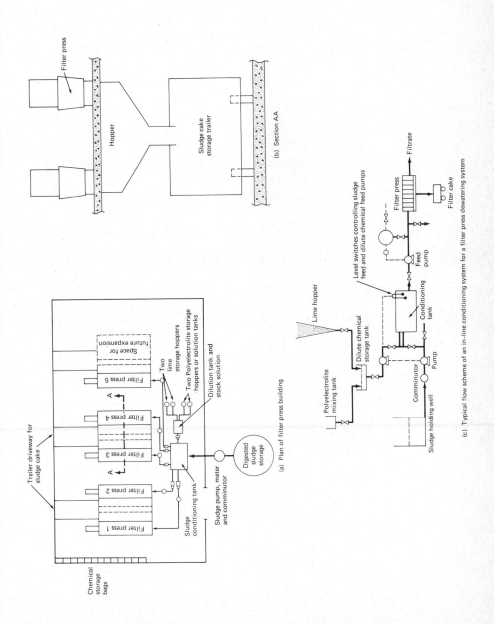

Figure 18-10 *Details of Filter Press for Sludge Dewatering. (a) Plan of filter press building. (b) Section AA. (c) Schematic flow diagram of in-line conditioning.*

C. Sludge Storage Conditioning and Feed System

1.   Develop the system components

The flow scheme of the sludge-conditioning and dewatering system design for this example is shown in Figure 18-10. The system has the following major components:

a.   Digested sludge-holding well outside the filter building.

b.   Lime storage and lime mixing. Store hydrated lime in bags with a capacity to hold a 30-d lime requirements under average conditions. Two volumetric feeders* are provided for continuous daily delivery of measured amounts of dry lime to the dilution tank where a 10 percent slurry of $Ca(OH)_2$ by weight is produced. The milk of lime is metered to the conditioning tank.

c.   Polyelectrolyte storage and mixing tank. Store a 30-d supply of the polyelectrolyte. Provide a mixing tank where 5 percent solution will be prepared for feeding into the conditioning tank.

d.   Sludge-feeding arrangement. Prior to conditioning, all sludge is passed through an in-line grinder. Grinding improves sludge mixing and flow characteristics, and protects downstream pumping and dewatering equipment.

e.   Conditioning system. In the past, most sludge-conditioning units were designed in such a way that the chemicals were added in batches to the sludge contained in an agitated tank. The sludge was then pumped from the tank into the pressure filter as required. Prolonged agitation of sludge resulted in deteriorating dewaterability. For this reason an in-line conditioning system is designed in this example. The in-line conditioning tank is small in which chemicals are added and mixed. The conditioned sludge is pumped into the filters. This reduces prolonged agitation and deterioration in filterability.

f.   Feed pump system. The conditioned sludge is pumped in each filter assembly under a pressure of 345 kN/m^2 (50 psig).[†] The feed cycle is 30 min long. After the completion of the filter feed, the compression cycle at 1550 kN/m^2 (225 psig) will be resumed by pumping water into the diaphragms.

g.   Cake breaker. Design of suitable cake breakers is necessary for rapid removal of the cake. Various types of cake breakers using breaker wire, bars, or cables beneath the filter are of common practice. Manufacturers provide details for their system.

---

*There are two types of automatic dry feeders. Volumetric feeders deliver a constant preset volume of chemical in a unit time. Gravimetric feeders discharge a constant weight of chemical. Gravimetric feeders are more costly and require more maintenance.

[†]It may be noted that in this example variable-volume recessed plate filter presses are provided. In fixed-volume recessed plate filter assembly, high pressure is applied by the feed pump. Rapid filling at high pressure may not permit uniform cake formation. As a result the filtrate may have high solids, or filter cloth clogging may result. Therefore, in new designs a pressure vessel is used to give high filling rate at lower pressures. More discussion may be found in Ref. 1.

h. Cloth washing and cleaning. The filter cloth must be washed by high-pressure water jet to remove the solids. Equipment suppliers provide information on the frequency of washing as well as quantity and pressure of water needed for filter cloth cleaning.

2. Compute the size of digested sludge storage well
    Provide a sludge storage period of 2.5 d.

$$\text{Volume of sludge storage well} = 80 \text{ m}^3/\text{d} \times 2.5 \text{ d} = 200 \text{ m}^3$$

Provide 7.5 m diameter × 4.5 m deep digested sludge storage well

3. Compute the size of lime storage facility
    Provide 30-d storage at average condition.

Lime needed per 8-h day   = 287 kg (CaO)
Total lime needed per week = 5 × 287 = 1435 kg
Total lime needed per month = 1435 kg/week* × 4.33 week/month
                            = 6214 kg

Since hydrated lime of 90 percent purity is used,

$$\text{30-d storage quantity} = \frac{6214 \text{ kg}}{0.9} \times \frac{74 \text{ (mol. wt. of Ca(OH)}_2)}{56 \text{ (mol. wt. of CaO)}}$$

$$= 9124 \text{ kg}$$

$$\text{Number of bags of hydrated lime needed per month} = \frac{9124 \text{ kg}}{23 \text{ kg/bag}}$$

$$= 397 \text{ bags.}$$

Provide storage for 400 bags of lime. The storage facility must be covered to prevent rain from wetting the bags. The capacity of the lime hopper shall be large enough to contain one day's supply of lime.

$$\text{Number of bags needed per day} = \frac{287 \text{ kg/d}}{23 \text{ kg/bag}} = 13 \text{ bags.}$$

Bulk density of hydrated lime = 481 kg/m^3 (30 lb/ft^3). Provide two hoppers to hold seven bags of hydrated lime each.

$$\text{Volume of each hopper} = \frac{7 \text{ bags} \times 23 \text{ kg/bag}}{481 \text{ kg/m}^3} = 0.34 \text{ m}^3.$$

Each hopper shall feed lime into a mixing tank and 10 percent slurry (milk of lime) shall be metered into the conditioning tank.

*Average 4.33 weeks per month.

4.   Compute the size of polyelectrolyte storage facility

Provide 30-d storage at average condition.

$$\begin{array}{ll}\text{Polyelectrolyte needed} \\ \text{per 8-h day}\end{array} = 115 \text{ kg}$$

$$\begin{array}{ll}\text{Total quantity needed} \\ \text{per month}\end{array} = 115 \text{ kg/d} \times 5 \text{ d/week} \times 4.33 \text{ week/month}$$

$$= 2490 \text{ kg}$$

Store the polyelectrolyte in bags and unload daily into the hopper for feeding to the conditioning tank. Provide two hoppers for feeding polyelectrolyte into the mixing tank and then into the sludge-conditioning tank. Bulk density of polyelectrolyte $= 321 \text{ kg/m}^3$ (20 lb/ft^3).

$$\text{Volume of each hopper } = \frac{115 \text{ kg/d}}{2 \times 321 \text{ kg/m}^3} = 0.2 \text{ m}^3.$$

5.   Compute the size of sludge pump to the conditioning tank
The conditioning tank is connected to five pressure filters. Assuming all pressure filters are filled at one time,

$$\begin{array}{ll}\text{Total sludge pumped per} \\ \text{operating day}\end{array} = \frac{80 \text{ m}^3/\text{d} \times 7 \text{ d/week}}{5 \text{ d/week (operation)}}$$

$$= 112.0 \text{ m}^3/\text{d}$$

$$\begin{array}{ll}\text{Pumping rate per cycle} \\ \text{(filling time is 30 min)}\end{array} = \frac{112.0 \text{ m}^3/\text{d}}{8 \text{ cycles/d}} \times \frac{1}{0.33 \text{ h/cycle}}$$

$$= 42.4 \text{ m}^3/\text{h}$$

Provide pumping capacity of 45 m^3/h (198 gpm).

6.   Compute the size of conditioning tank
Provide in-line conditioning tank with 10 min detention time.

$$\text{Volume of conditioning tank } = 45 \text{ m}^3/\text{h} \times \frac{10 \text{ min}}{60 \text{ min/h}}$$

$$= 7.5 \text{ m}^3$$

Provide level switches to control the operation of sludge and dilute chemical feed pumps.

7.   Compute the size of sludge feed pump to the pressure filter
Each pressure filter has one feed pump. Total quantity of conditioned sludge plus chemical solutions to the pressure filter is calculated as follows:

$$\text{Sludge quantity} \qquad = 112.0 \text{ m}^3/\text{d}$$

Water in lime at 10 percent solution $= \dfrac{287 \text{ kg/d}}{0.1 \times 1000 \text{ kg/m}^3} = 2.9 \text{ m}^3/\text{d}$

Water in polyelectrolyte at 5 percent solution $= \dfrac{115 \text{ kg/d}}{0.05 \times 1000 \text{ kg/m}^3}$

$= 2.3 \text{ m}^3/\text{d}$

Total quantity $= 112.0 \text{ m}^3/\text{d} + 2.9 \text{ m}^3/\text{d} + 2.3 \text{ m}^3/\text{d}$

$= 117.2 \text{ m}^3/\text{d}$

Pumping rate of each pump when four filter units are in operation $= \dfrac{117.2 \text{ m}^3/\text{d}}{4 \text{ pumps} \times 8 \text{ cycles/d} \times 0.33 \text{ h/cycle}}$

$= 11 \text{ m}^3/\text{pump} \cdot \text{h} \ (48 \text{ gpm})$

Provide reciprocating high-pressure pump to transfer conditioned sludge at a pressure of 345 kN/m² (50 psig).

D. Sludge Cake

1. Compute sludge cake solids
   The filter cake contains sludge solids, lime, and polymers. It is assumed that 75 percent of the conditioning chemicals are incorporated into the sludge cake. The dewatering facility provides 95 percent solids capture.

Sludge solids in cake	$= 0.95 \times 5740 \text{ kg/d}$	$= 5453 \text{ kg/d}$
Lime	$= 0.75 \times 287 \text{ kg/d}$	$= 215 \text{ kg/d}$
Polymers	$= 0.75 \times 115 \text{ kg/d}$	$= \underline{86 \text{ kg/d}}$
Total solids in sludge cake		$5754 \text{ kg/d} \ (719 \text{ kg/h})$

2. Compute the volume of sludge cake
   The sludge cake has 25 percent solids and specific gravity is 1.06.

   Volume of sludge cake on an operating day $= \dfrac{5754 \text{ kg/d} \times 1000 \text{ g/kg}}{0.25 \times 1.06 \times 1 \text{ g/cm}^3 \times 10^6 \text{ cm}^3/\text{m}^3}$

   $= 21.7 \text{ m}^3/\text{d}^*$

   Volume of cake on an average basis $= 21.7 \text{ m}^3/\text{d} \times \dfrac{5 \text{ d/week}}{7 \text{ d/week}} = 15.5 \text{ m}^3/\text{d}^*$

   The sludge cake shall be discharged into a hopper for direct dumping into the transport trucks. The sludge cake transport system is shown in Figure 18-10.

---

*Based on 5 d/week and 8 h/d operation. The quantity of sludge cake obtained in Figure 13-12 is based 7 d/week operation.

E. Filtrate Quality.   The filtrate in the sludge-dewatering area shall be piped into a common sump containing the thickener overflow and digester supernatant. The combined flow is then returned to the aeration facility. The filtrate return system is shown in plant layout (Chapter 20).

    1.   Compute the volume of filtrate

$$\text{Filtrate volume} = 112.0 \text{ m}^3/\text{d} - 21.7 \text{ m}^3/\text{d} = 90.3 \text{ m}^3/\text{d}$$

$$\begin{array}{l}\text{Water in lime at}\\ \text{10 percent solution}\end{array} = 2.9 \text{ m}^3/\text{d}$$

$$\begin{array}{l}\text{Water in poly-}\\ \text{electrolyte at 5}\\ \text{percent solution}\end{array} = 2.3 \text{ m}^3/\text{d}$$

$$\begin{array}{l}\text{Other return flow}\\ \text{3990 } \ell \text{ per cycle}\\ \text{per filter assembly}\end{array} = 3990 \text{ }\ell/\text{cycle} \cdot \text{unit} \times 8 \text{ cycle/d} \times 4 \text{ units}$$

$$\times \frac{1}{1000 \text{ }\ell/\text{m}^3} = 127.7 \text{ m}^3/\text{d}$$

$$\text{Total return flow} = 90.3 \text{ m}^3/\text{d} + 2.9 \text{ m}^3/\text{d} + 2.3 \text{ m}^3/\text{d} + 127.7 \text{ m}^3/\text{d}$$

$$= 223.2 \text{ m}^3/\text{d}$$

    2.   Compute total solids in the return flow

$$\begin{array}{l}\text{Total solids in}\\ \text{conditioned sludge}\end{array} = 6142 \text{ kg/d}$$

$$\begin{array}{l}\text{Total solids in}\\ \text{returned flow}\end{array} = (6142 - 5754) \text{ kg/d} = 388 \text{ kg/d}$$

    3.   Compute total solids concentration in return flow

$$\begin{array}{l}\text{Total solids in return}\\ \text{flow}\end{array} = 388 \text{ kg/d}$$

$$\text{Total return flow} = 223.2 \text{ m}^3/\text{d}$$

$$\begin{array}{l}\text{Concentration of total}\\ \text{solids}\end{array} = \frac{388 \text{ kg/d}}{223.2 \text{ m}^3/\text{d}} \times 10^3 \text{ g/kg} = 1738 \text{ g/m}^3 \text{ (mg/}\ell\text{)}$$

    4.   Compute the BOD_5 in the return flow

Total volatile solids (TVS) in digested sludge (see Sec. 17-7-2, step E1)

$$= 2490 \text{ kg/d} \times \frac{7 \text{ d/w}}{5 \text{ d/w operation}} = 3486 \text{ kg/d}$$

Assume that
8 percent TVS is lost in the filtrate
54 percent TVS is biodegradable

$$\begin{array}{ll}
\text{1 gram of biode-} & = 1.42 \text{ g ultimate BOD}\\
\text{gradable solids} &
\end{array}$$

$$\text{BOD}_5 \quad = 0.68 \times \text{ultimate BOD}$$

$$\begin{array}{ll}
\text{Total BOD}_5 \text{ in} & = 3486 \text{ kg/d} \times 0.08 \times 0.54 \times 1.42 \times 0.68\\
\text{the filtrate} & = 145 \text{ kg/d}
\end{array}$$

$$\begin{array}{ll}
\text{Concentration} & = \dfrac{145 \text{ kg/d}}{223.2 \text{ m}^3/\text{d}} \times 1000 \text{ g/kg} = 650 \text{ g/m}^3 \text{ (mg/}\ell)\\
\text{of BOD}_5 &
\end{array}$$

F. Design Details. The design details of the filter press assembly, chemical feed systems, and conditioning units are given in Figure 18-10.

18-7 OPERATION AND MAINTENANCE, AND TROUBLESHOOTING AT FILTER PRESS FACILITY

Important operation and maintenance considerations of filter press assembly are briefly described below. Additional details may be found in Ref. 7.

18-7-1 Common Operational Problems and Troubleshooting Guide

1. Difficulty in sealing the filter plate is due to poor alignment or inadequate shimming. The problem is overcome by realigning the plates and adjusting the shimming of stay bosses.
2. Difficulty in discharging the cake is due to inadequate precoat and improper conditioning of sludge. Increase precoat, feed at 172–276 kPa (25–40 psig), and change conditioner chemical or dosage based on filter leaf tests.
3. Excessive filter cycle time may be due to improper conditioning or feed solids too low. Change the chemical dosage and improve solids concentration in digested sludge.
4. Sticking filter cake to solids conveying equipment may be due to insufficient inorganic chemicals used in conditioning. Increase conditioning dosage, particularly the inorganic conditioners.
5. Gradual increase in precoat pressure is due to improper sludge conditioning and improper precoat feed, or filter media plugged by calcium buildup. This problem is overcome by acid washing the media, changing the chemical dosage and decreasing the precoat.
6. Frequent media blinding may be due to inadequate precoat or too high initial feed rate. Increase precoat time, and develop initial cake slowly.
7. Excessive moisture in cake may be due to improper conditioning or too short filtration cycle. Change chemicals or dosages, and lengthen filter cycle.
8. Sludge blowing out of press may be due to obstruction between plates. Shut down feed pump, hit press closure drive, restart feed pump, and clean feed eyes of plates at end of the cycle.
9. Leaks around lower face of plates may be due to excessive wet cake soiling the

media on lower faces. Check moisture content of the feed sludge, increase precoat, and develop initial cake slowly.

10. If improper filter media is specified, the cake will stick causing media cleaning difficulties. Change the media and specify relatively coarse media.
11. Cake transport conveyor will not operate properly if the slope exceeds 15°.

18-7-2 Routine Operation and Maintenance

A good preventive maintenance program will reduce breakdowns, and make the operation clean and pleasant.

1. Check the following components for wear, corrosion, and proper adjustment: drive and gear reducers; drive chains and sprockets; closing mechanism; bearing brackets; electrical contacts in starters and relays; and suction lines and sumps.
2. Hose down all spilled sludge immediately.
3. Wash the cloth daily as specified by the equipment supplier. If high-pressure jet is specified, use plastic cover draped over the filter to confine spray. When acid wasting is provided, a recirculating system may be needed for scrubbing and achieving the desired acid effects.
4. All rubber surfaces of the plates should be draped only with soft plastic to avoid damage.

18-8 SPECIFICATIONS

The specifications of filter press equipment are briefly presented to describe the general features of the equipment. Detailed specifications should be developed in consultation with the equipment suppliers.

18-8-1 General

The manufacturer shall furnish and install five fully automatic sequencing and cake-discharging high-pressure diaphragm-type recessed chamber filter presses complete with sludge feed pumps, sludge-conditioning tanks, and sludge reaction mix tanks.

18-8-2 Filter Press Equipment

Each filter press shall be of the fully automatic sequencing diaphragm type employing a two-stage pressure filtration cycle, a fully automatic positive cake discharge mechanism, fully automatic wash system that allows high-pressure washing of both sides of the cloth media, and an integrated washwater drain system. Each press frame shall be fabricated of carbon steel. All integral process pipings and manifolding shall be supported off the press frame. The typical process feed connections shall include sludge feed, diaphragm pressurization water, press wash water, washwater discharge, core blow feed, compressed air blow, etc. A typical automatic operation sequence is as follows:

1. The hydraulic ram shall be pressurized closing all chambers of the filter press.
2. Feed sludge shall be introduced at the specified pressure.
3. After completion of the sludge feed cycle, high-pressure water shall be pumped into the diaphragms at minimum specified pressure causing expansion of the diaphragms to dewater the sludge to terminal dryness.
4. After the sludge has been pressed to proper dryness, the feed pipe shall be purged with compressed air and water shall be removed from behind the diaphragms to return the diaphragms to their original position.
5. The hydraulic cylinder shall release all pressure and completely open all filter chambers and discharge the cake.
6. The filter cloth shall be washed on both sides and then returned to its original position.
7. The press will be ready to start the next cycle.

18-8-3 Filter Press Frame, Plates, and Diaphragm

The filter press assembly shall be designed to provide a completely integrated structure sufficient to support the entire weight of the filter plate assemblies and withstand the operating press pressures. The frame shall be provided with side rails to support the plates, and leveling screws to level the side rails. The filter plates shall be located within the pressure frame assembly alternating with the diaphragm plate assemblies. The plate shall provide necessary feed opening, drainage surface, and shall seal against operating pressure.

Each diaphragm plate assembly shall have connections for high-pressure diaphragm water, air blow, and filtrate drainage. The diaphragm shall ensure a long service life and be easy to replace.

18-8-4 Filter Media Cloth and Automatic Cloth Washing

The filter cloth media provided shall be suitable for the sludge type specified. Each filter chamber shall have its own individual filter media assembly formed from a single-filter cloth folded at the top, with precut holes for sludge-feeding pipe, and aligned with diaphragm plate assemblies. The filter media cloth drive assembly shall have cloth-lowering drive mechanism. Cake shall be discharged automatically into the hopper. Each filter press shall be equipped with an integral automatic cloth wash system designed to wash both sides of the filter cloth. The washwater system shall also contain drains, and washwater drainage troughs.

18-8-5 Filter Press Hydraulic System

A hydraulic cylinder shall open and close the filter press frame, and shall be designed to develop the operating clamping pressure. The hydraulic system shall consist of an oil reservoir, hydraulic pumps, motors, oil filtration system, control switches, pressure gauges, and contact switches necessary for monitoring.

18-9 PROBLEMS AND DISCUSSION TOPICS

18-1. Assume that 2000 kg of digested sludge reaches drying beds for dewatering each day. The total solids in the sludge are 5 percent and specific gravity is 1.025. The sludge drying beds are uncovered and solids loading is 110 kg/m^{2}year. Calculate (a) the number of beds if each paved bed is 4 m $\times$ 20 m, (b) how many beds will be filled each day if average sludge application rate is 25 cm, and (c) what is the sludge drying bed area in m^2 per capita if digested sludge production is 0.05 kg per capita per day.

18-2. A vacuum filter is designed for dewatering combined digested sludge. The influent digested sludge contains 5 percent solids and specific gravity is 1.025. The volume of influent sludge is 200 m^3/d. Laboratory tests show that filter yield is 20 kg/m^2·h and optimum chemical dosage is 9.0 percent lime and 3 percent ferric chloride. The filter cake has 22 percent solids including conditioning chemicals. If 75 percent of the added chemicals are fixed into the cake solids, and solids capture efficiency is 90 percent, calculate (a) the dimensions of the vacuum filter, (b) the chemical dosage kg/d and kg/kg of solids, and (c) the weight of filter cake produced per day (kg/d). Assume filter operation 16 h/d, and the specific gravity of the filter cake is 1.06.

18-3. Calculate the number of variable volume recessed plate filter units to dewater 5000 kg/d digested sludge under average flow condition. The filter operation is 8 h/d and 5 d/week. The lime, ferric chloride, and polymer dosages are 6, 0.5, and 1.3 percent of dry solids, respectively. The filter cycle time is 80 min, and filtration rate is 15 kg/m^2·h. Filter area of each chamber is 0.5 m^2. Also calculate how many chambers in each filter unit shall be provided so that 100 percent filtration can be achieved with one unit out of service, and 124 percent filtration capability is available when all units are in operation.

18-4. Determine the volume and concentration of TS and BOD$_5$ in return flow from a dewatering facility using filter presses. The average volume of digested sludge reaching the dewatering facility is 100 m^3/d. The specific gravity of the sludge is 1.025, total solids are 5 percent, and biodegradable portion of total solids is 35 percent. The filters operate 8 h/d and 5 d/week. The cycle time is 60 min and there are 5 filter units. Lime and polymer dosages are 6 and 3 percent of dry solids. Assume 75 percent added chemicals are fixed into the sludge cake. Biodegradable solids/BOD$_L$ = 1.42, and BOD$_5$/BOD$_L$ = 0.68. Also calculate the quantity and volume of sludge cake if average moisture content in sludge cake is 75 percent by weight, and specific gravity is 1.06. Solids capture efficiency of filter presses is 88 percent. The lime and polymer solutions used in chemical conditioning are 10 and 5 percent respectively. The water used per cycle for filter washing and leakage through the diaphragms is 4 m^3 per cycle per filter.

18-5. Anaerobically digested sludge is conditioned using hydrated lime and polyelectrolyte. Calculate the volume of conditioned sludge (m^3/d), and capacity of the sludge feed pump (m^3/h) used to pump conditioned sludge into the recessed plate filter press. Use the following data:

Average flow of anaerobically digested sludge = 50 m^3/d
    Solids in sludge = 4000 kg/d
    Lime dosage = 5.5 percent of dry solids
    Lime slurry solution = 12 percent
    Polyelectrolyte = 2 percent of dry solids
    Polyelectrolyte solution = 5 percent
    Sludge pump operate 8 cycles/d and 0.5 h/cycle
    Filter press operation 6 d per week and 8 h/d

18-6. Describe various sludge conditioning methods. Give advantages and disadvantages and application of each method.

18-7. Describe various sludge dewatering methods. Which methods would be best for dewatering of sludge from two-stage lime treatment process?

18-8. 1000 kg per day digested sludge reaches the sludge drying earthern cells for dewatering. The total solid in the sludge is 7 percent, and specific gravity of the sludge is 1.025. Solids loading rate for the cells is 37 kg/m^3-yr. Calculate (1) the number of cells if each cell is 0.02 hectare in area and 1 m deep (2) in how many days each cell will be filled (3) the cell area in m^2 per capita if digested sludge production rate is 0.05 kg per capita per day.

REFERENCES

1. U.S. Environmental Protection Agency, *Process Design Manual, Sludge Treatment and Disposal,* Technology Transfer, EPA-625/1-79-011, September 1979.

2. U.S. Environmental Protection Agency, *Process Design Manual for Sludge Treatment and Disposal,* Technology Transfer, EPA-625/1-74-006, October 1974.

3. Metcalf and Eddy, Inc., *Wastewater Engineering: Treatment, Disposal, and Reuse,* McGraw-Hill Book Co., New York, 1979.

4. Reynolds, T. D., *Unit Operations and Processes in Environmental Engineering,* Brooks/Cole Engineering Division, Monterey, Cal., 1982.

5. Schroeder, E. D., *Water and Wastewater Treatment,* McGraw-Hill Book Co., New York, 1977.

6. National Council of the Paper Industry for Air and Stream Improvement, Inc., *A Review of the Operational Experience with Belt Filter Presses for Sludge Dewatering in The North American Pulp and Paper Industry,* Technical Bulletin 315, October 1978.

7. Culp G. L., and N. F. Heim, "*Performance Evaluation and Troubleshooting at Municipal Wastewater Treatment Facilities,* U.S.Environmental Protection Agency, MO-16 EPA-430/9-78-001, Washington, D.C. January 1978.

Sludge Disposal

19-1 INTRODUCTION

Ensuring the safe disposal of municipal sludge and other residues, such as screenings, grits, and skimmings, is considered an integral part of good planning, design, and management of municipal wastewater treatment facilities. Acceptable sludge disposal practices include (1) conversion processes (incineration, wet oxidation, pyrolysis, composting, etc.) and (2) land disposal (land application and landfilling).

Experience has shown that conversion processes are costly because they are either energy- or labor-intensive. Composting produces soil conditioners that can be used by farmers, homeowners, and at city parks and highway medians. Composting is also labor-intensive and can only be successful if there is a market for the soil conditioner. Land spreading of municipal sludge requires a large amount of land. Also, there is a concern that some sludges may cause accumulation of toxic chemicals in land or in crops grown on land. This leaves landfilling as an attractive alternative for disposal of municipal sludges and other residues.

In this chapter a general overview of various methods of sludge disposal is presented. Since landfilling of municipal sludges is most widely used in the United States, this topic is covered in greater detail. Step-by-step procedure for planning design and operation of a sludge disposal landfill is given in the Design Example.

19-2 OVERVIEW OF SLUDGE DISPOSAL PRACTICES

Both conversion and land disposal methods require sludge processing. Thickening, stabilization, conditioning, and dewatering are the common methods of sludge processing. These processes essentially prepare the sludge for safe ultimate disposal. Discussion of thickening, stabilization, conditioning, and dewatering of sludge may be found in Chapters 4 and 16–18.

The sludge conversion processes and land disposal methods are discussed below.

19-2-1 Sludge Conversion Processes

Sludge conversion processes are generally thermal techniques and are intended to reduce the solids required for final disposal or to recover a resource. Composting is thermal or biological conversion of organic matter into usable soil conditioner. The sludge conver-

TABLE 19-1 Summary of Conversion Processes for Disposal of Municipal Sludges

Conversion Process	Recommended Pretreatment	Codisposal of Other Residue[a]	Additional Processing Requirements
Incineration	Thickening and de-watering	Yes	Ash landfilling
Wet oxidation	Thickening	No	Separation of ash, treatment of returned liquid, landfilling of ash
Pyrolysis	Thickening	No	Utilization of by-products (gas, liquid, carbon, etc), disposal of residues
Recalcining	Thickening and de-watering	No	Recovery of lime, landfilling of ash
Composting by heat drying	Thickening and de-watering	No	Utilization or sale of compost
Composting by microbial action	Thickening, digestion, and dewatering	No	Utilization or sale of compost

[a]Other residues include screenings, grits, and skimmings.
Source: Adapted in part from Ref. 1.

sion and resource recovery processes are summarized in Table 19-1.[1] A general description is given below.

Incineration. Incineration involves drying of sludge cake followed by complete combustion of organic matter. A minimum temperature of 700°C is needed to deodorize the stack emissions. Excess air 50–100 percent over the stoichiometric air requirement is necessary. Natural gas or fuel oil is provided as auxiliary fuel for ignition and to maintain the proper temperature. Often sludge is incinerated with municipal solid waste and other residues. Two major incineration systems are the multiple-hearth furnace and the fluidized bed reactor. The multiple-hearth furnace contains several hearths arranged in a vertical stack. Sludge cake enters the top and proceeds downward through the furnace from hearth to hearth. The fluidized bed incinerator utilizes a hot sand reservoir in which hot air is blown from below to expand and fluidize the bed. Air pollution control equipment is needed in both cases to clean the emission. A discussion of sludge incineration may be found in Refs. 2 and 3.

Wet Oxidation. Wet air oxidation (Zimmerman process) is also used to incinerate the sludge. Dewatering of sludge is not required. The sludge and sufficient air is pumped into a reactor where high temperature (200–300°C) and high pressure (5–20 kN/m^2) are maintained. The organic matter is oxidized in liquid phase. The liquid and solid residues are separated by settling or filtration. The liquid is high in BOD, nitrogen, and phosphorus, which must be returned to the plant. The ash is landfilled. A general discussion of wet oxidation may be found in Refs. 1 and 4.

Pyrolysis. Pyrolysis or destructive distillation is heating of organic matter in an oxygen-free atmosphere. At high temperatures of 300–700°C and the absence of oxygen, the organic matter undergoes cracking, producing gases, oil, and tar, and charcoal. All three components are combustible and can be used as energy source. Currently, pyrolysis of municipal solid wastes on a large scale is in practice. Sludge pyrolysis or copyrolysis with municipal solid wastes will probably be used in the future.[1,5]

Recalcining. The recalcining and reuse of lime sludges in wastewater treatment plants is new. A high rate of lime addition in one or two stages produces large quantities of sludge. Recalcining involves heating of dewatered lime sludge to about 1000°C in a multiple-hearth furnace. Moisture and CO_2 are driven off. Often CO_2 is used in the recarbonation process. At the bottom outlet of the furnace large pieces of recalcined lime are obtained which are ground and taken to the storage area. Lime recovery at the wastewater treatment plant may not be feasible owing to the presence of undesirable impurities, particularly nitrogen and phosphorus. A portion of lime sludge must be wasted to avoid buildup of impurities. The cost of recalcining at a wastewater treatment plant must be weighed against the cost of new lime, savings in cost from reclaimed lime, and reduction in sludge quantities for disposal.[6–8]

Composting by Heat Drying. The purpose of heat drying is to remove the moisture from the wet sludge and partially combust the organics. Sludge drying occurs at temperatures of approximately 370°C. At this temperature part of the volatile matter is removed and the sludge is stable enough for use as compost. The gases evolved in the drying process are reheated to about 750–800°C to eliminate odors. The most common types of heat-drying systems for sludge are (1) flash dryers, (2) spray dryers, (3) rotary dryers, (4) multiple-hearth dryers, and (5) the Carver-Greenfield process.[2] The sludge after drying is processed into soil conditioner. Drying permits grinding, weight reduction, and prevention of continued microbial decomposition.[4]

Composting by Microbial Action. Composting is a process in which organic matters undergo biological decomposition to produce a stable end product that is acceptable as a soil conditioner. There are two methods of composting: open windrow and mechanical process. Windrow composting uses long, narrow piles 1.2 m high and 2.5 m wide. These piles are turned every few days. Moisture is maintained at 55–70 percent. It takes 4–6 weeks to produce stable compost. Mechanical systems of composting produce stable compost in 5–10 d. There has been a limited market for compost in this country. A discussion of composting may be found in Refs. 2, 9, and 10.

19-2-2 Land Disposal of Municipal Sludge

Land disposal for municipal sludge includes (1) land application of liquid sludge or cake, and (2) disposal by landfilling. Direct land disposal of digested liquid sludge to cropland has the advantage of eliminating the dewatering cost; but the disadvantage is the large volume of sludge that must be transported and applied over land. Small cities often haul the liquid sludge in trucks while large cities utilize pumping. Dewatered sludge may be disposed of by spreading over farmlands and plowing after it has dried. Wet sludge may be incorporated into the soil directly by injection. The factors used for selection and design of land application system and types of crops grown are discussed in Chapter 24. A normal sludge-loading rate is 10,000–25,000 kg solids per ha per year.[2,9,10]

Raw or digested sludge is buried if a suitable site is available within an economical hauling distance. Sludges are often buried at municipal sanitary landfills along with other solid wastes. Site selection and design considerations require that the sites have distance from populated areas, leachate protection, runoff and erosion control, and protection against gas movement. A monitoring program and future land use planning is essential.

As mentioned earlier, landfilling of municipal sludge and other residues from wastewater treatment facilities is the most widely used method in the United States. It is expected that this practice will continue to be a major disposal option in this country. Therefore, the rest of this chapter is devoted to detailed discussion on planning, design, and operation of landfills for municipal sludge.

19-3 PLANNING, DESIGN, AND OPERATION OF MUNICIPAL SLUDGE LANDFILLS

Planning, design, and operation of municipal sludge landfills requires detailed investigation. Besides soil erosion, dust, vectors, and noise and odor problems, leachate and gases continue to be produced for years to come. There is the danger of ground and surface water contamination as well as migration of explosive gases to the nearby structures if landfills are not properly designed and operated. Therefore, considerations should be given to the following factors for proper planning, design, and operation of sludge landfills:

1. Sludge characteristics
2. Site selection
3. Regulations and permits, and public participation
4. Methods of sludge landfilling
5. Design considerations
6. Operation and maintenance
7. Monitoring of completed landfills

Each of these factors is discussed below.

19-3-1 Sludge Characteristics

Not all sludges are suitable for landfilling due to either odor or operational problems. Moisture contents and chemical composition of sludge solids are important in the design and operation of sludge landfills. Sludges having solids contents of 15 percent will not support cover material and therefore dry soil as bulking agent must be used. Soil-bulking operations are generally not cost-effective if solids content in sludge is less than 15 percent. Severe operational problems could occur at the landfills if sludge-dewatering facilities are unable to produce sludge cake at desired solids content.

Sludges from chemical precipitation facilities create different types of problems. These sludges are viscous, sticky, and slippery (particularly if polymers are used). Handling of these sludges is difficult even when bulking agents are used. Undigested sludges produce large amounts of methane gas that poses serious explosion problems to the nearby structures.

Heavy metals in wastewater generally come from industrial discharges. A large portion of the heavy metals is removed by conventional treatment processes and is concentrated in the sludge. Typical concentrations of heavy metals in municipal sludge is summarized in Table 16-1. Heavy metals from sludges do not leach out as long as the pH remains above 10. For this reason and others, lime is added to the sludge prior to dewatering (Chapter 18). As anaerobic decomposition progresses in the fill, organic acids are produced that lower the pH, and release of heavy metals may begin.

Nitrogen release from landfills is also of particular interest. Nitrogen in the form of ammonia, nitrite, and nitrate are relatively mobile and cause serious risk of groundwater pollution.

In determining the suitability of sludges for landfilling, considerations must be given to (1) the source and characteristics of wastewater, (2) wastewater treatment processes, (3) sludge-handling and disposal methods, (4) concentration of heavy metals, (5) organic contents, and (6) nitrogen content. Each of these factors should be evaluated carefully in light of landfill site, surface and groundwater hydrology, and geology of the area. Proper safeguard techniques must be utilized in the design and operation of the fill to ensure minimal environmental consequences.

19-3-2 Site Selection

The technical considerations involved in site selection of sludge landfills include factors that span many disciplines. These disciplines include engineering; land use; social and political science; and economics. These disciplines are briefly discussed below.

Engineering

Useful Life.   The selected site should last for several years to justify the cost of permanent structures. The land needed for access roads, buffer land, and stockpiles should be considered within the gross total site area. The calculation procedure for estimating the fill life is included in the Design Example.

Topography.   Flat sites (slope less than 1 percent) tend to pond. Steeper sites (slope greater than 20 percent) could erode, thus creating operational difficulties.

Surface Water.   Surface runoff resulting from landfill areas must comply with NPDES permit issued for the facility. Proper control methods must be considered in the design.

Soil and Geology. Soil serves as cover and bulking material, attenuates potential contaminants, and controls runoff and leachates. Therefore the chemical, physical, and hydrological properties of soils must be studied. Important properties are pH, cation-exchange capacity (CEC), texture, structure, distance from the groundwater table, and permeability. A detailed discussion of soil characteristics for sludge landfilling may be found in Ref. 1. Basic information is summarized in Table 19-2. Fine-textured soils are considered desirable for landfilling as they have lower permeability and higher attenuation capacity against heavy metals and other pollutants. Sites operating in clay and clay loams, for instance, have operated successfully with as little as 1–2 m (3–6 ft) of soil separating sludge deposits from the highest groundwater table. In impermeable soils, a leachate collection system must be installed.

Highly organic (peaty) soils and formations with faults, major fractures, and joint sets are not recommended for sludge landfilling. Impervious linears with leachate collection and treatment system must be used in such situations.

Groundwater. The information concerning depth of groundwater (high and low water tables), the hydraulic gradient, the quality of groundwater, current and projected water use, and location of primary recharge zones should be obtained as part of the site selection process.

Site Access. The existing and future access roads should be accessible in all weather conditions. The probable impacts of traffic on the neighborhoods should be fully evaluated.

Ecology. Sludge landfills should not be located in environmentally sensitive areas such as wetlands, flood plains, permafrost, critical habitats of endangered species, and recharge zones of sole source aquifers. Vegetation in the site should also be considered. Vegetation can serve as a buffer and reduce dust, noise, odor, and visibility.

TABLE 19-2 Properties of Soils and Their Suitability for Sludge Landfilling

Soil Type	Grain Size (mm)	Permeability (cm/s)	CEC (meq/100 g)	Suitability for Landfilling
Clay	0.002 and less	10^{-8}–10^{-6}	Over 20	Excellent. Impermeable liner is not needed
Silt loam	0.002–0.05	10^{-6}–10^{-3}	12–20	Fair. Impermeable liner may be needed
Sandy soils	0.05–0.25	10^{-3}–10^{-1}	1–10	Poor. Impermeable liner needed for protection of groundwater

Source: Adapted in part from Ref. 1.

Land Use.   The current and future land use patterns of the site should be fully studied. The state, regional, county, and local planning agencies and zoning authorities should be consulted for site evaluation and selection.

Social and Political Science.   Waste disposal sites have serious social and political implications. The effect of disposal sites on the neighborhood and the extent of public resistance or acceptance should be fully evaluated. These issues will certainly come up during public participation programs discussed later in this chapter.

The archaeological or historical significance of the land involved in the potential site should be ascertained. These issues are generally evaluated in the environmental impact assessment report (see Chapters 2 and 6).

Economics.   Early in the site selection process the relative costs for different sites should be developed for comparison purposes. These costs involve land acquisition, permanent structure, utilities, and equipment purchase, and operation. A detailed discussion of economics may be found in Ref. 1.

19-3-3 Regulations and Permits, and Public Participation

Many regulatory agencies require a permit for construction and operation of sludge landfills. The design of the fill must follow the requirements of the permit that are generally available from the federal, state, or local agencies. The federal requirements relevant to sludge landfills are contained in the proposed criteria for the classification of solid waste disposal facilities.[11,12] The criteria address the following topic areas:

1.   Environmentally sensitive areas
2.   Surface water
3.   Groundwater
4.   Odors
5.   Application on land used for the production of food-chain crops
6.   Safety

The objective of a public participation program in the establishment of sludge landfills is to give the public a participatory role throughout planning, design, and operation. The public and the concerned citizens must therefore be kept informed as to the status of site selection, development of design criteria, and operation phases of the project, through news releases, mailing lists, and public hearings. A public participation program for the Design Example has been presented in Chapter 6. Additional details may be found in Refs. 1 and 13.

19-3-4 Methods of Sludge Landfilling

The purpose of this section is to identify, describe, and develop design criteria for different methods of sludge landfilling. Common methods are (1) trenching (narrow and wide) and (2) area fill (mound, layer, and dike containment). Each of these methods is briefly discussed below. This discussion is adapted in part from Refs. 1 and 14.

Trenching. Stabilized or unstabilized sludge is placed in a trench and covered with soil. Trench operations are more specifically categorized as narrow and wide. Narrow trenches have widths less than 3 m (10 ft); wide trenches have widths greater than 3 m (10 ft). The width of the trench is determined by the solids content of the receiving sludge, its capability of supporting cover, need for using the filled area for soil stockpiles, and operating equipment and haul vehicles. Design considerations should include provisions to control leachate and gas migration, dust, vectors, and/or aesthetics. Leachate control measures include the maintenance of 0.6 to 1.5 m (2–5 ft) of soil thickness between trench bottom and highest groundwater level or bedrock (0.6 m for clay to 1.5 m for sand), or membrane liners and leachate collection and treatment system. Installation of gas control facilities may be necessary if buildings are nearby.

Narrow Trench. Sludge is placed in a single application and then a single layer of cover soil is applied atop. Trenches are usually excavated by equipment based on solid ground adjacent to the trench. Backhoes, excavators, and trenching machines are particularly useful. The excavated material is usually immediately applied as cover over an adjacent sludge-filled trench. Sludge is placed in trenches either directly from haul vehicles through a chute extension, or by pumping. The main advantage of a narrow trench (0.6–1 m) is its ability to handle sludge with a relatively low solids content (15–20 percent). Instead of sinking into the sludge, the cover soil bridges over the trench and receives support from undisturbed soil along each side of the trench. One to three meters width is more appropriate for sludge with solids content of 20–28 percent, which is high enough to support cover soil. The basic design factors and trench operation are presented in Table 19-3 and Figure 19-1.

Wide Trench. Wide trenches are usually excavated by equipment operating inside the trench. Track loaders, draglines, scrapers, and track dozers are suitable equipment. The excavated material is stockpiled on solid ground adjacent to the trench for subsequent application as cover material. If the sludge is incapable of supporting equipment, cover is applied by equipment based on undisturbed ground adjacent to the trench. A front end loader is suitable for trenches up to 3 m (10 ft); a dragline is suitable for trench widths up to 15 m (50 ft). If sludge can support equipment, a track dozer applies cover from within the trench. Sludge is placed in trenches by one of the following methods: haul vehicles directly entering the trench or haul vehicles dumping from the top of the trench. Dikes can be used to confine sludge to a specific area in a continuous trench. The basic design factors are given in Table 19-3. Wide-trench operation is shown in Figure 19-1.

Area Fill. Area fill is a sludge disposal operation in which sludge is placed above the original ground surface and subsequently covered with soil. To achieve stability and soil bearing capacity, sludge is mixed with a bulking agent, usually dry soil. The soil absorbs excess moisture from the sludge and increases its workability. The large quantities of soil needed may require importing from elsewhere. Provisions must be made to keep the dry soil stockpiled. Installation of a liner is generally required for protection of groundwater, and provisions made for surface drainage control, gas migration, dust, vectors, and/or aesthetics. Area fills are more specifically categorized as follows:

Figure 19-1 *Operational Details of Sludge Disposal by Landfilling Using Trench Method.*
(a) Narrow trench landfilling operation. (b) Wide trench landfilling operation with interior
dikes. [From Ref. 1.]

TABLE 19-3 Design Criteria for Sludge Landfills

Design Factors or Conditions	Trench		Area Fill		
	Narrow	Wide	Mound	Layer	Dike
Sludge solids, percent	15–28	20–28	≥ 20	≥ 15	20–28
Bulking ratio,[a] vol. of soil / vol. of sludge	—	—	0.5–1.00	0.25–1.00	0.25–0.50
Cover thickness[b]					
Intermediate, m	Not used	Not used	1	0.1–0.3	0.3–0.6
Final, m	0.6–1.2	1–1.5	1–1.5	0.6–1.2	1–1.2
Imported soil needed	No	No	Yes	Yes	Yes
Trench width, m[c]	0.6–3	>3	—	—	—
Sludge characteristics	Stabilized or unstabilized	Stabilized or unstabilized	Stabilized	Stabilized or unstabilized	Stabilized
Hydrology	Deep groundwater and bedrock	Deep groundwater and bedrock	Shallow groundwater and bedrock	Shallow groundwater and bedrock	Shallow groundwater and bedrock
Ground slope, percent	< 20	< 10	Suitable for steep terrain as long as level area is prepared for mounding	Suitable for medium slope but level ground preferred	Suitable for steep terrain as long as a level area is prepared inside dikes
Sludge application (in actual fill area, m³/ha(yd³/acre)	2200–10,000 (1200–5600)	6000–27,400 (3200–14,500)	5500–27,000 (3000–14,000)	3500–17,000 (2000–9000)	9000–28,000 (4800–15,000)
Equipment	Backhoe with loader, excavator, trenching machine	Track loader, dragline, scraper, track dozer	Track loader, backhoe with loader, track dozer	Track dozer, grader, track loader	Dragline, track dozer, scraper

[a]More bulking agent is needed at lower sludge solids.
[b]Higher depths of intermediate and final covers are used at lower sludge solids.
[c]Narrower trench is used at lower sludge solids.
Source: Adapted in part from Ref. 1.

Area Fill Mound. Sludge is mixed with a bulking agent, usually soil, and the mixture is hauled to the filling area, where it is stacked in mounds approximately 2 m (6 ft) high. Cover material is then applied in a 1-m (3-ft) thickness. The cover thickness may be increased to 1.5 m (5 ft) if additional mounds are applied atop the first lift. The appropriate sludge/soil bulking ratio and soil cover thickness depends on the solids content of the sludge as received, the need for mound stability, and the bearing capacity as dictated by the number of lifts and equipment weight. Lightweight equipment with swamp pad tracks is appropriate for area fill mound operation; heavier wheel equipment is appropriate in transporting bulking material to and from stockpiles. Construction of earthen containments is useful to minimize mound slumping, and for sloping sites. Design factors are summarized in Table 19-3 and Figure 19-2.

Area Fill Layer. Sludge is mixed with soil on- or off-site and spread evenly in consecutive layers 15 cm to 1 m (0.5–3 ft) thick. Interim cover between layers may be applied to 15- to 30-cm (0.5–1 ft) thick applications. Layering may continue to an indefinite height before final cover is applied. Lightweight equipment with swamp pad tracks is appropriate for area-fill layer operations; heavier wheel equipment is appropriate for hauling soil. Slopes should be relatively flat to prevent sludge from flowing downhill. However, if sludge solids content is high or sufficient bulking soil is used, the effect can be prevented and layering may be performed on mildly sloping terrain. Design factors are given in Table 19-3 and Figure 19-2.

Diked Containment. Dikes are constructed on level ground around all four sides of a containment area. Alternatively, the containment area may be placed at the toe of the hill so that the steep slope can be utilized as containment on one or two sides. Dikes would then be constructed around the remaining sides. Access is provided to the top of the dikes so that haul vehicles can dump sludge directly into the containment. A 0.3- to 1-m (1–3 ft) intermediate cover may be applied at certain points during the filling; a 1- to 2-m (3–5 ft) final cover should be applied when filling is discontinued. Cover material is applied either by a dragline based on solid ground, atop the dikes, or by track dozers directly on top of the sludge. Usually, operations are conducted without the addition of soil bulking agents, but occasionally soil bulking is added. Typical dimensions are 15–30 m (50–100 ft) wide, 30–60 m (100–200 ft) long, and 3–10 m (10–30 ft) deep. Design parameters are presented in Table 19-3 and Figure 19-2.

19-3-5 Design Considerations

The sludge landfills should be designed to ensure adequate protection of the environment and cost-effective utilization of site, storage volume, equipment, and soil. The important environmental factors that should be considered in the design include (1) protection of groundwater and leachate control, (2) gas control, (3) storm water management, and (4) other design factors (access road, odor, dust, and vectors control, and aesthetics). Each of these impacts and methods to minimize them are discussed below.

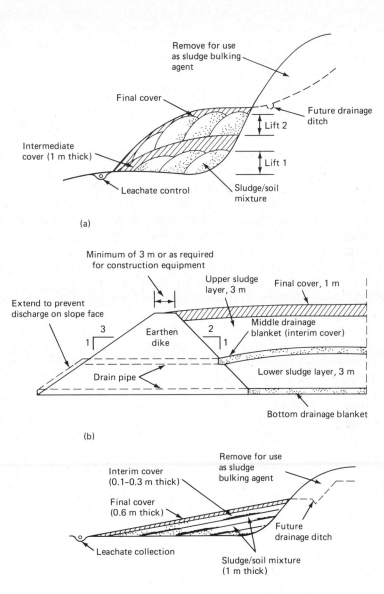

Figure 19-2   Design Details of Sludge Landfill by Area Fill Method. (a) Cross-section of typical area fill mound operation. (b) Cross-section of typical dike containment operation. (c) Cross-section of typical area fill layer operation. [From Ref. 1.]

Protection of Groundwater and Leachate Control. Leachate is produced from the filled areas due to excess moisture in the sludge and the rainwater entering the fill. Leachate may enter the water system essentially through two pathways:

1. Percolation of the leachate laterally or vertically through the soil into the ground-water, and
2. Runoff of leachate outcroppings into the surface water

Leachates contain high concentrations of organic and inorganic pollutants that may cause serious damage to the water quality.[1,15–17] Basic design safeguards are as follows:

1. Slope the surface of the landfill so that the rainwater drains from the filled area. The surface profile of a completed landfill is shown in Figure 19-3.
2. Divert storm water around the landfill.
3. Provide sufficient depth of suitable soil between the groundwater table and the bottom of the fill. Contaminants in the leachate can be attenuated when passing through the soils by physical-chemical, mechanical, and/or biological processes.[1] The cation-exchange capacity of different soils is given in Table 19-2.
4. Provide impervious soil-additive mixture or membrane liner to confine and drain to a sump, then pump, and treat the leachate. The liner for sludge landfills may be of asphalt or portland cement composition, soil sealant, liquid rubber, or synthetic polymeric membrane.[1] Methods of treatment include pumping to existing sewers, evaporation, recycle through the landfill, and on-site treatment.[18,19] The impervious liner and underdrain system for leachate collection are shown in Figure 19-3.

Gas Control. Gas is produced by the decomposition of organic matter in sludge. Generally, methane gas is 50–70 percent and carbon dioxide is 30–45 percent. The remaining portion include nitrogen, oxygen, water vapors, ammonia, and hydrogen sulfide. The rate of gas generation depends on the characteristics of sludge, physical and chemical conditions in the fill, and microbial activity. Undigested sludge produces 0.5–1.0 m^3/kg (8–16 ft^3/lb) of dry solids. Digested sludge produces much less gas. The gas is generated over an extended period.

Methane gas in air at 5–15 percent concentration constitutes an explosive mixture. Therefore, its migration toward existing structures should be prevented. Gas control measures include (1) impermeable methods and (2) permeable methods. Both methods are shown in Figure 19-4. In recent years, there has been some effort to install gas recovery systems from large sanitary landfills.[1,19]

Storm Water Management. All upland drainage should be collected and directed around the landfill. Open earthen ditches, concrete- or stone-lined channels, and corrugated metal pipe are used for drains. Various types of surface drains are shown in Figure 19-5.

The surface of the sludge landfill should be well graded to promote runoff and inhibit ponding. Slopes greater than 2 percent and less than 5 percent are recommended. Grades higher than 5 percent may encourage erosion. Siltation ponds are also used to settle solids.

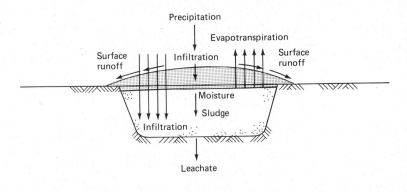

(a)

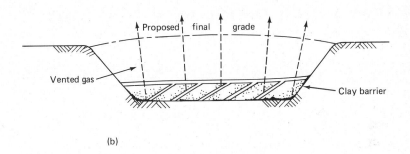

(b)

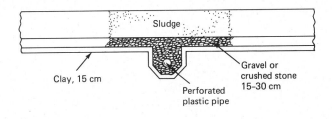

(c)

Figure 19-3   *Moisture Balance and Leachate Control in Sludge Landfills. (a) Surface profile of completed landfill and water balance. (b) Impermeable layer (clay or plastic liner) used as barrier to control migration of leachate. (c) Under drain for collection of leachate into a sump. [From Ref. 1.]*

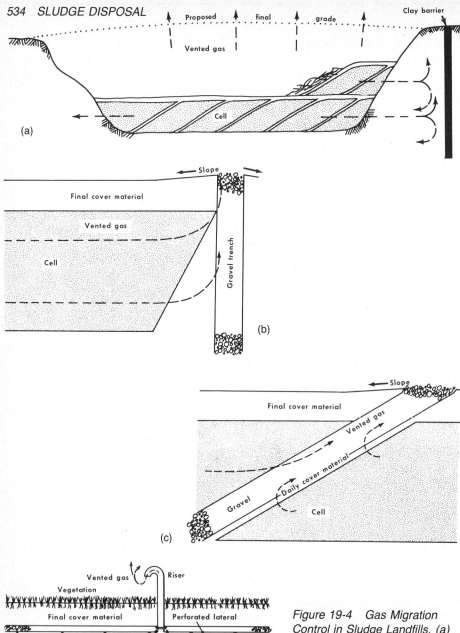

Figure 19-4 Gas Migration Control in Sludge Landfills. (a) Impermeable method of gas migration control. (b) Permeable method using gravel filled trench along the boundary. (c) Permeable method using gravel filled trench inside the landfill. (d) Permeable pipes installed in sludge landfill for venting the gases. [From Ref. 17.]

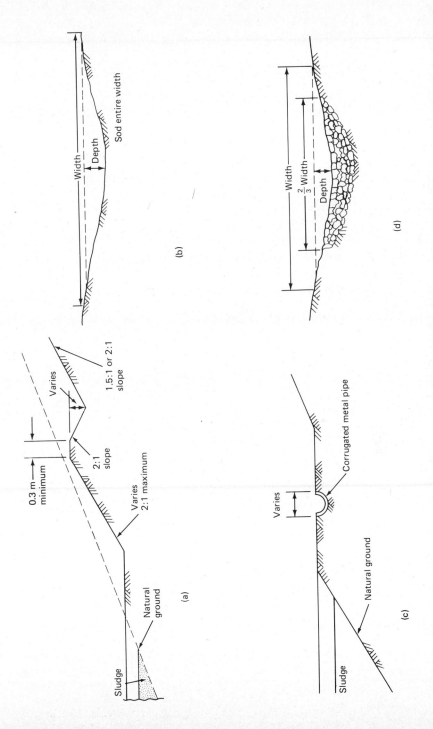

Figure 19-5   Drainage Channels for Control of Surface Runoff. (a) Earthen drainage channel. (b) Earthen drainage channel (sodded). (c) Corrugated metal pipe drainage channel. (d) Stone drainage channel. [From Ref. 1.]

Other Design Factors.   Other important factors for efficient design and operation of sludge landfills include the following:

1. Provide all-weather access roads 6–7 m wide. To minimize dust, access roads should be gravel paved.
2. Consider on-site availability of soil, and its use for bulking and covering. To minimize dust, all covered areas should be vegetated soon after completion. As an alternative, water should be applied to dusty roads.
3. Utilize available fill volume to the fullest extent.
4. Provide toilet facilities for employees.
5. Provide utilities, including water, sewer, electrical, and telephone.
6. Provide fence.
7. Provide proper lighting.
8. Provide wash rack for equipment.
9. Do not leave uncovered sludge for extended time periods. Covering with well compacted soil will reduce odors, and control flies.

19-3-6 Operation and Maintenance

A sludge-landfilling operation is considered as an ongoing construction project that should implement the design plan. An effective operation requires a choice of equipment compatible with sludge characteristics, site conditions, landfilling method, operation plan, and contingency plans.

A sludge-landfilling operation is divided into two parts: (1) specific operational procedures and (2) general operational procedures. The specific operational procedures depend on the landfilling methods and include site preparation, sludge unloading, and sludge handling and covering. A typical equipment selection guide is summarized in Table 19-4.

The general operational procedures include (1) environmental control practices (spillage, erosion, dust, mud, vectors, odors, noise, aesthetics, health, and safety), (2) inclemental weather practices, (3) hours of operation, and (4) special wastes such as grit, screenings, ash, etc. A detailed discussion of operational procedures for different types of fills may be found in Ref. 1.

19-3-7 Monitoring of Completed Landfills

The purpose of monitoring a completed landfill is to detect adverse environmental effects that may develop in the future. This includes monitoring (1) groundwater, (2) surface water, and (3) gas movement. Monitoring ground- and surface water involves sample collection to establish baseline data, detection of contamination, and development of mitigation procedures. Common analytical parameters are pH, electrical conductivity or total dissolved solids (TDS), nitrate, chloride, total organic carbon (TOC) or COD, heavy metals (especially lead, cadmium, and mercury), and methylene-blue-active substances.

Gas monitoring is generally needed near the property line of the existing structures. The gas-sampling device generally consists of gas probes that are installed at various depths within a single hole, facilitating measurement of methane gas concentration.

For proper closure of the entire landfill or segment of the fill that has been filled to the capacity, the following operational procedures must be used. These procedures can be conducted concurrent with ongoing site operation:

1. No sludge should be left exposed.
2. Maximum settlement will occur in the first year. Therefore the area should be regraded to ensure proper drainage. All depressions and cracks must be filled. The final cover shall not be less than the specified thickness.
3. Check the sediment and erosion controls, and modify the surface grade for best results.
4. Seed the area with the appropriate mixture of grasses.
5. Outline timetable to ensure that the following features are inspected on a regular basis: settlement, grading and vegetation cover, sediment and erosion control, leachate and gas control, monitoring of ground- and surface water, fencing, and vandalism.

19-4 INFORMATION CHECKLIST FOR DESIGN OF SLUDGE LANDFILLS

Design of sludge landfills consists of completing existing information and generating new information on the sludge and site conditions. Some of this information is collected during the site selection phase and some during the design of the other sludge-handling facilities. However, additional information must also be collected during the landfill design phase. Adherence to a carefully planned sequence of activities to develop a sludge landfill design minimizes project delays and expenditures. A checklist of design activities is presented below. These activities are listed in their general order of performance. However, in many cases separate tasks can and should be performed concurrently or even out of the order given below:[1]

1. Determine sludge volumes and characteristics for the initial and design years.
2. Determine the quantities and characteristics of other residues that must also be landfilled. Some of these residues are screenings, grits, skimmings, and ash.
3. Compile existing and generate new site information such as area, property boundaries, topography and slopes, surface water, utilities, roads, structure, and land use.
4. Compile hydrogeological information and prepare location map. The general information include (a) groundwater hydrology (average depth and seasonal fluctuations, hydraulic grade, direction and rate of flow, water quality and uses); (b) soil data (depth, texture, structure, porosity, permeability, pH, cation-exchange capacity, ease of excavation, and stability).
5. Compile climatological data (precipitation, evaporation, temperature, wind direction, number of freezing days, etc.).
6. Identify federal, state, and local regulations and design standards. Most of these regulations cover the following items: requirements for sludge stabilization; sludge-loading rates; frequency and depth of cover; distance to residences, roads, surface and groundwater; monitoring requirements; building codes; etc.

TABLE 19-4   Typical Equipment Selection Guide for Sludge Landfill

Sludge Cake Processed (m-ton/d)[a]	Trenching Machine	Backhoe with Loader	Excavator	Track Loader	Wheel Loader	Track Dozer	Scraper	Dragline	Total Equipment
Narrow trench									
up to 10		1							1
10–50		1				1[b]			2
50–100			1[b]			1			2
100–250	1	1				1			3
250–500	2	1				2[b]			5
Wide trench									
up to 10				1					1
10–50				1[b]		1[b]			2
50–100				1		1			2
100–250						1	1[b]		2
250–500				1[b]		2[b]	1		4
Area fill, mound									
up to 10				1					1
10–50		1[b]		1					2
50–100		1[b]		1		1[b]	1[b]		4
100–250		1[b]		1	1	1	1		5
250–500		1		1	1	1	1		5

Area fill, layer

up to 10		1		1
10–50		1	1[b]	2
50–100		1	1[b]	2
100–250		2[b]	1[b]	3
250–500	1[b]	2	1	4

Area fill, dike

up to 10		1			1
10–50		1[b]	1[b]		2
50–100		1[b]	1[b]	1	3
100–250		1	1	1	3
250–500		2[b]	1	1	4

[a]Metric ton/day.
[b]May not have 100 percent utilization.
Source: Adapted from Ref. 1.

7. Select sludge-landfilling methods, values of design parameters, and operational features.
8. The complete design of the facility should include excavation plans; leachate and gas control; erosion control; access roads; special working areas, structures, and utility lines; lighting; wash racks; fencing; landscaping; monitoring wells; equipment; operation plans; and cost estimates.

19-5 DESIGN EXAMPLE

19-5-1 Design Criteria Used

The following design criteria are used for the design of a sludge landfill.

1. Site location

   The sludge-landfilling site is a wooded area located on the northeastern side of the proposed site of the wastewater treatment plant. The sludge landfilling area is flat. On the northern boundary is the flood protection levee enclosing the proposed wastewater treatment plant site. A plan view of the entire plant site is presented in Figure 20-1. The proposed sludge landfill site is shown in Figure 19-6.

2. Soil conditions and hydrology

   Four test borings were performed to determine the soil condition and hydrology. The boring locations are shown in Figure 19-6. The boring results are summarized in Table 19-5.

3. Climate

   The climate of the area was briefly described in Sec. 6-4-2. Important climatological factors are listed below:

Average annual precipitation	= 89 cm (35 in.)
Average annual evaporation	= 76 cm (30 in.)
Number of days minimum temperature 0°C (32°F) and below	= 12 d/year

TABLE 19-5 Average Soil Condition and Hydrology

Condition	Description
Boring depth 0−6 m (0−19.5 ft)	Silt loam
6−9 m (19.5–29.5 ft)	Saturated silt loam
> 9 m (29.5 ft)	Fractured crystalline rock
Depth of groundwater	6 m (19.5 ft)
Direction of flow	North
Analysis of silt loam provided the following data	
Texture	Medium
Permeability	2×10^{-4} cm/s (moderately slow)
pH	6.5
Cation-exchange capacity (CEC)	18 meq/100 g

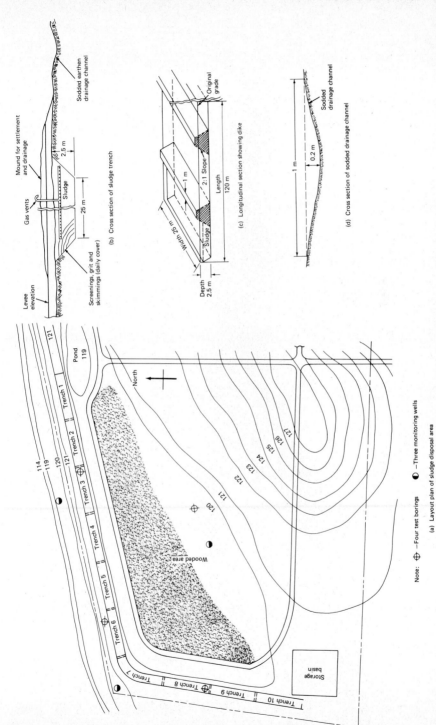

(a) Layout plan of sludge disposal area

(b) Cross section of sludge trench

(c) Longitudinal section showing dike

(d) Cross section of sodded drainage channel

Note: ⊕ —Four test borings ◑ —Three monitoring wells

Figure 19-6 Layout Plan and Design Details of Trenches for Sludge Disposal.

4. Type of wastes landfilled

Disposal of various treatment plant residues will be achieved at the landfill. These residues are (a) screenings, (b) grits, (c) skimming, and (d) anaerobically digested and dewatered sludge. The sludge will be placed in the fill segregated from other residues. This will allow the sludge to be excavated in the future and used as topsoil or soil conditioner.

Part of the digested sludge cake will be used on the city property and park areas. The general characteristics of these residues are summarized in Table 19-6. The estimated qualities of these residues at different time periods are presented in Table 19-7.

5. Landfilling method

Select a wide-trench landfilling method for segregated disposal of sludge from screenings, grits and skimmings. This method is most appropriate as the site is flat, groundwater table is deep, sludge is stabilized, and solids content is approximately 25 percent.

TABLE 19-6 Characteristics of Wastewater Treatment Plant Residues Landfilled

Residues	Orginating Processes	Conditioning Processes Used	Quality Description	Refs.
Screenings	Bar screens	None	Odorous and attracts flies. Moisture contents 80 percent. Unit weight 960 kg/m³ (60 lb/ft³)	Chapters 4 and 8
Grits	Aerated grit chamber	None	Odorous and attract flies. Organic content 3–4 percent. Moisture content 20–30 percent. Unit weight 1600 kg/m³ (100 lb/ft³)	Chapters 4 and 11
Skimmings	Primary sedimentation and final clarifier	None	Highly odorous. Moisture content 40–75 percent. Specific gravity 1–0.95. Quantity of skimmings is 2–13 g/m³ (16.8 to 110 lb/million gallon) of wastewater treated	Chapters 4 and 12
Sludge	Primary sedimentation and final clarifier	Thickening, anaerobic digestion, and dewatering by filter press	Musty odor, not offensive. Sp. gr 1.06, sludge solids 25 percent	Chapters, 12, 13, 16, 17, and 18

TABLE 19-7 *Quantities of Wastewater Treatment Residues for Landfilling*

	Quantities of Residues at the End of Different Periods Since Initial Year				
Residue	0 (Initial Year)[a]	5[a]	10[a]	15 (Design Year)	Ref.
Screenings, m³/d	0.56	0.65	0.72	0.76	Chap. 8
Grit, m³/d	0.84	1.00	1.10	1.14	Chap. 11
Skimmings, m³/d	0.24	0.27	0.30	0.32	Chap. 12
Sludge, m³/d	8.82	10.20	11.40	12.00[b]	Chap. 18
Total	10.46	12.12	13.52	14.22	

[a]Estimated qualities are in proportion to the population (see Table 6-2 for population estimates).
[b]Approximately 23 percent sludge cake is used on city property. Total digested sludge = 15.4 m³/d × 0.77 = 12 m³/d.

6. Groundwater quality shall be monitored for possible pollution resulting from the fill.

19-5-2 Design Calculations

A. Site Development. The site is developed in accordance with the plan shown in Figure 19-5. Features of this plan include the following:

1. Leave maximum width of wooded buffer zone between the landfill and the treatment plant.
2. Provide the sludge trenches parallel to the flood protection levee on the western and northern ends of the site.
3. The excavation depth of the trench is determined initially by the groundwater table. If 2.5 m deep trenches are excavated, a separation depth of the soil from the water table shall be 3.5 m (11.5 ft). At CEC of 18 meq/100 g and permeability classified as "moderately slow," the containment and attenuation of contaminants is seen sufficient to cause no adverse effect on the natural water system.
4. Provide 10 wide trenches. Each trench shall have bottom dimensions 120 m (400 ft) long and 25 m (82 ft) wide. The side slope of the trenches is kept 1 : 1. The trenches shall be constructed parallel to the flood protection levee, with a clear spacing of 6 m between each trench. The trench arrangement is shown in Figure 19-6.
5. In each trench, the screenings, grits, and skimmings shall be placed on the far side (next to the levee). The sludge shall be compacted concurrently in the remaining trench up to the existing ground level. A soil cover of 1 m compacted soil shall be placed over the sludge, thus bringing the finished grade equal to that of the flood protection levee.

6. Provide sodded surface water drain along the side of the trenches. These drains shall carry the surface runoff to a sedimentation pond. The ground surface profile and drainage pattern is shown in Figure 19-6.
7. In accordance with the state regulations, one groundwater-monitoring well shall be located up gradient from the fill area and two monitoring wells shall be located down gradient from the fill area. The monitoring wells are shown in Figure 19-6.
8. The access roads to the sludge landfill shall be paved with asphalt. Site arrangement is shown in Figure 19-6.

B. Life of the Fill

1. Compute total volume available for sludge disposal

    Trench dimensions

    Bottom length = 120 m
    Bottom width   = 25 m
    Average depth = 2.5 m
    Side slope   = 1 : 1
    Total volume
    for each trench = $\frac{1}{3}$ × depth × $[A_1 + A_2 + \sqrt{A_1 \times A_2}]$

    where A_1 and A_2 are bottom and top areas of the trench.

$$\text{Total volume} = \frac{1}{3} \times 2.5 \text{ m } (25 \text{ m} \times 120 \text{ m} + 30 \text{ m} \times 125 \text{ m} + \sqrt{(25 \text{ m} \times 120 \text{ m}) (30 \text{ m} \times 125 \text{ m})}$$

$$= \frac{2.5 \text{ m}}{3} (3000 \text{ m}^2 + 3750 \text{ m}^2 + \sqrt{3000 \times 3750 \text{ m}^4})$$

$$= 8420 \text{ m}^3$$

    If 10 percent volume reduction is assumed due to dikes and intermediate cover and bulking material, total available volume of 10 trenches = $10 \times 7578 \text{ m}^3$ = 75,780 m³.

2. Compute total volume occupied by sludge and residues during the first 15 years since the plant startup

    a. Average volume occupied during the first 5 years (see Table 19-7)

    $$= \frac{10.46 + 12.12}{2} \text{ m}^3/\text{d} \times 365 \text{ d/yr} \times 5 \text{ yr} = 20,604 \text{ m}^3$$

    b. Average volume occupied during the second 5 years

    $$= \frac{12.12 + 13.52}{2} \text{ m}^3/\text{d} \times 365 \text{ d/yr} \times 5 \text{ yr} = 23,397 \text{ m}^3$$

    c. Average volume occupied during the last 5 years

    $$= \frac{13.52 + 14.22}{2} \text{ m}^3/\text{d} \times 365 \text{ d/yr} \times 5 \text{ yr} = 25,313 \text{ m}^3$$

Total volume occupied in 15 years $= 20{,}604 \text{ m}^3 + 23{,}397 \text{ m}^3 + 25{,}313 \text{ m}^3 = 69{,}314 \text{ m}^3$

3. Compute life of the fill

Total volume remaining $= 75{,}780 \text{ m}^3 - 69{,}314 \text{ m}^3 = 6466 \text{ m}^3$

Total life of the fill $= 15 \text{ yr} + \dfrac{6466 \text{ m}^3}{14.22 \text{ m}^3/\text{d} \times 365 \text{ d/yr}}$

$= 15 \text{ yr} + 1.3 \text{ yr} = 16.3 \text{ yr}$

C. Surface Water Drain

A sodded collection ditch shall be provided along the length of the trenches to intercept the uphill drainage. The drain shall be sloped toward the sedimentation pond. The cross-section and alignment of the drainage ditch are shown in Figure 19-6.

D. Equipment Requirements

The equipment need is determined from Table 19-4.

1. Estimate the quantity of residues landfilled per day

Average volume of screenings, grit, skimmings, and sludge during first 5 years $= \dfrac{10.46 + 12.12}{2} \text{ m}^3/\text{d}$

$= 11.29 \text{ m}^3/\text{d}$

Using specific gravity of 1.06, total quantity per day $= 11.29 \text{ m}^3/\text{d} \times 1.06 \times 1 \text{ m-ton/m}^3$

$= 12 \text{ m-ton/d}$

Since the sludge disposal operation will be 5 days per week,

quantity handled per working day $= \dfrac{12 \text{ m-ton/d} \times 7 \text{ d/week}}{5 \text{ working d/week}}$

$= 17 \text{ m-ton/working d}$

2. Determine equipment need from Table 19-4

Provide one track loader and one track dozer with 91-cm-wide buckets. Because of small quantities of sludge and residues that will be handled each day, this equipment will not be fully utilized.

E. Sedimentation Pond and Effluent Structure

Provide a 1.2-ha (2.9-acre) sedimentation pond on the northeastern side of the sludge landfill area. This is the lowest area and therefore the surface runoff from a large portion of the site will drain to the pond. The depth of the pond is 1.2 m below the ground surface. The area surrounding the pond shall be graded to permit natural drainage to the pond. The excavated soil will be utilized in the construction of the flood protection levee.

The storm water structure at the pond consists of an effluent box, effluent baffles, and an outlet pipe to drain the effluent into a sump. From the sump the effluent will be discharged into the river under gravity. A floodgate in the sump and storm water pumps are provided to pump the surface runoff under high flood conditions.

19-6 OPERATION AND MAINTENANCE, AND SPECIFICATIONS FOR SLUDGE LANDFILLS

An asphalt-paved road shall be constructed to provide access to the entire sludge-landfilling area. Runoff, erosion, and sediment control as well as monitoring wells shall be installed. Initially, the area for trench number 1 shall be cleared and grubbed. The trench excavation shall begin from the end nearest to the sedimentation pond and proceed generally toward the farthest end in accordance with the design dimensions. The excavation equipment shall operate inside the trench. Dikes shall be left at 20-m intervals within the trench to form individual cells 20 m × 25 m. These cells shall be used for phase filling and covering. Each cell shall be excavated completely and the soil stockpiled for the filling operation.

19-6-1 Landfilling Operation

Landfilling shall be performed by equipment operating inside the trench. Because the screenings and skimmings are highly odorous, daily covering of such material is considered essential. Therefore, these residues shall be spread on the far side of the trench (toward the levee) and covered daily. The sludge shall be unloaded directly into the cell and spread out evenly throughout the cell area designated for sludge filling. If necessary, the soil-bulking operation shall be conducted inside the trench. After the residue and sludge-filling areas reach the ground level, approximately 1 m final cover shall be applied over the entire cell area. At least 1 month in advance of completion of each cell, a new cell or a new trench area shall be excavated and kept ready for filling.

Operation shall be conducted at the site 8 h/d and 5 d/week to coincide with sludge deliveries and avoid odors generally encountered with storage of sludge and other residues. The equipment shall be operated 7 h/d plus 1 h downtime/d for routine maintenance and cleanup.

19-6-2 Trench Completion

Approximately 1 month after completion of each trench, the surface shall be graded to the smooth ground surface at an elevation of the flood protection levee. The surface shall be sloped inward to drain into the surface water drains. Immediately thereafter, the site shall be hydroseeded if weather conditions permit. Leftover soil shall be utilized in surface grading as well as the construction of levee and other embankments.

19-6-3 Surface Water Drains and Sedimentation Pond

All surface water drains shall be sodded and kept clean from debris. All on-site drainage shall be channeled to the sedimentation pond.

19-6-4 Operation During Inclement Weather

During inclement weather, soil shall be stockpiled and covered with plastic sheets to keep dry and workable. The mud from the haul vehicles and equipment shall be washed in the wash rack area.

19-6-5 Monitoring

Background samples shall be collected from all wells prior to initiating landfilling operation to establish baseline data. Subsequently, samples from each well shall be collected at 3-month intervals and analyzed for the constituents discussed in Sec. 19-3-7.

19-7 PROBLEMS AND DISCUSSION TOPICS

19-1. Discuss the technical considerations that are involved in site selection of a sludge disposal landfill. How they differ from those for site selection of a wastewater treatment plant presented in Chapter 2?

19-2. Compare the trenching and area landfilling methods of sludge disposal. Describe the site conditions for which each method is most suitable.

19-3. Calculate the useful life of a site in which sludge will be landfilled by area fill mound operation. The site is approximately 3 ha, and after removing the cover and bedding material the average depth of disposal site is 3.5 m. The finished grade must be kept in-line with the ground surface above the depression. Approximately 27 m-tons of residues will be landfilled each day. The bedding material and intermediate covers constitute 18 percent volume. The top cover is 1.3 m. Assume that the specific gravity of mixed residue is 1.1. Also determine the equipment needed.

19-4. List different methods of leachate and gas control in a sludge landfill.

19-5. Calculate the average quantity of digested sludge handled over 15 years in m-tons per working day. Use the data given in Table 19-7 to compute the quantity of sludge. Unit weight of sludge is 1.06 m-ton per m^3. Assume the operation of sludge disposal is 5 days per week.

19-6. Calculate the life of a landfill site in which sludge and screenings will be landfilled by dike containment method. The total volume of the sludge and other residues is 3942 m^3/y. The dimensions of the dike containment area are: length = 200 m; width = 60 m; maximum depth = 4.6 m; and ground after scraping of cover material is flat. An intermediate cover of 0.3 m is applied when 1.5 m depth of sludge is reached. Final cover is 1 m. Volume occupied by bulking material is 12 percent. Draw the cross sectional view of the completed fill.

19-7. Study Ref. 1 and discuss advantages and disadvantages of codisposal of sludge and municipal refuse in sanitary landfills.

19-8. What are sludge conversion processes? Compare and list the advantages and disadvantages of each method.

REFERENCES

1. U.S. Environmental Protection Agency, *Process Design Manual: Municipal Sludge Landfills*, Environmental Research Information Center, Technology Transfer, Office of Solid Waste, EPA-625/1-78-010, SW705, October 1978.

2. Metcalf and Eddy, Inc., *Wastewater Engineering: Treatment, Disposal, and Reuse,* Second Edition, McGraw-Hill Book Co., New York, 1979.

3. Corey, Richard C., *Principles and Practices of Incineration,* Wiley-Interscience, New York, 1969.

4. Zimmerman, F. J., "Wet Air Combustion", *Industrial Water and Wastes,* Vol. 6, No. 4, July–August 1961, pp. 102–106.

5. Boucher, F. B., et al., *Pyrolysis of Industrial Wastes for Oil and Activated Carbon Recovery,* U.S. Environmental Protection Agency, Pub. 270–951, May 1977.

6. Clark, J. W., W. Viessman, and J. J. Hammer, *Water Supply and Pollution Control,* IEP-A Dun-Donnelley, New York, 1977.

7. Culp, R. L., G. M. Wesner, and Gordon L. Culp, *Handbook of Advanced Waste Water Treatment,* Van Nostrand Reinhold Co., 1978.

8. Joint Committee of the Water Pollution Control Federation and American Society of Civil Engineers, *Wastewater Treatment Plant Design,* MOP/8 Water Pollution Control Federation, Washington, D.C., 1977.

9. U.S. Environmental Protection Agency, *Environmental Pollution Control Alternatives: Municipal Wastewater,* Technology Transfer, EPA-625/5-76-012, 1976.

10. Council on Environmental Quality and U.S. Environmental Protection Agency, *Municipal Sewage Treatment, A Comparison of Alternatives,* Contract EQC 316, February 1974.

11. U.S. Environmental Protection Agency, *Solid Waste Disposal Facilities,"* Federal Register, Vol. 43, No. 25, February 6, 1978, p. 4942.

12. U.S. Environmental Protection Agency, 40 CFR 260, *Hazardous Waste Management System,* Vol. 45, No. 98, May 19, 1980.

13. U.S. Environmental Protection Agency, *Municipal Wastewater Management Citizen's Guide to Facility Planning,* FRO-6 U.S. EPA, Washington, D.C., February 1979.

14. U.S. Environmental Protection Agency, *Innovative and Alternative Technology Assessment Manual,* EPA-430/9-78-009 CD-53. U.S. EPA, Washington, D.C., February 1980.

15. Qasim, S. R., and J. C. Burchinal, "Leaching of Pollutants from Refuse Beds," *Journal of the Sanitary Engineering Division,* ASCE, Vol. 96, No. SA1, February 1970, pp. 49–58.

16. Qasim, S. R., and J. C. Burchinal, "Leaching from Simulated Landfills," *Journal of Water Pollution Control Federation,* Vol. 42, No. 3, March 1970, pp. 371–379.

17. Brunner, D. R., and D. J. Keller, *Sanitary Landfill Design and Operation,* U.S. Environmental Protection Agency, Report No. SW-65ts, 1972.

18. Palit, T., and S. R. Qasim, "Biological Treatment Kinetics of Landfill Leachate," *Journal of the Environmental Engineering Division,* ASCE, Vol. 103, No. EE2, April 1977, pp. 353–366.

19. Tchobanoglous, G., H. Theisen, and R. Eliassen, *Solid Wastes: Engineering Principles and Management Issues,* McGraw-Hill Book Co., New York, 1977.

Plant Layout

20

20-1 INTRODUCTION

Plant layout is the physical arrangement of designed treatment units on the selected site. Careful consideration must be given to properly locating the treatment units, connecting conduits, roads and parking facilities, administration building, and maintenance shops. The design engineer must integrate the functions of all components. However, it should be remembered that the factors influencing site design are both natural and social. Therefore, efforts should be made to become familiar with the proposed site and its neighborhood. Experience has shown that proper plant layout can (1) enhance the attractiveness of the plant site, (2) fit the operational needs of the processes, (3) suit the maintenance needs of the plant personnel, (4) minimize construction and operational costs, (5) offer flexibility in future process modifications and plant expansion, and (6) maintain the landscaping and plant structures in perfect harmony with the environment.[1-3]

In this chapter the basic considerations for plant layout are discussed. Also, the specific site plan for the plant in the Design Example has been developed.

20-2 FACTORS AFFECTING PLANT LAYOUT AND SITE DEVELOPMENT

A variety of factors should be considered as general design guidelines for plant layout on the selected site. Some of these factors are discussed below. Typical plant layout plans for medium and large wastewater treatment facilities are shown in Figure 20-1.

20-2-1 Unit Construction Considerations

Site Topography and Geology.   Consideration of site topography and geology is important because plant layout should respect such existing site features as character, topography, and shoreline. Site development should take the advantage of the existing site topography to either emphasize or diminish the visual impact of the facility depending on the design goals. The following principles are important to consider:

1. A site on a side-hill slope can facilitate gravity flow that will reduce pumping requirements and locate normal sequence of units without excessive excavation and fill.
2. A side-hill location can be used to advantage as the sludge-handling building can provide direct access to trucks from the ground and upper levels.

(b)

(a)

Figure 20-1   Layout of Two Wastewater Treatment Plants. (a) Ten Mile Creek Regional Wastewater Treatment Plant (Trinity River Authority, Texas). Average design capacity 0.3 m³/s. 1. Primary clarifier. 2. Aeration basin. 3. Final clarifier. 4. Return sludge pumping station and chlorination facility. 5. Primary sludge thickener (gravity). 6. Sludge pumping station. 7. Centrifuge deck. 8. Anaerobic digester. 9. Digester control building. 13–18. Bypass, junction, and division boxes. (b) Southside Wastewater Treatment Plant (City of Dallas, Texas). Average design capacity 1.3 m³/s. 1. Raw wastewater pumping station. 2. Grit basin. 3. Primary clarifier. 4. Aeration basin. 5. Final clarifier. 6. Chlorine contact basin. 7. Chlorine building. 8. Effluent pumping station. 9. Gravity thickener. 10. Anaerobic digester. 11. Aerobic digester. 12. Digester control building. 13. Compressor building. 14. Control building. 15. Backwash surge tank. 16. Coagulation basins. 17. Multi-media filter and clear water well.

3.   When landscaping is utilized it should reflect the character of the surrounding area. Site development should alter existing naturally stabilized site contours and drainage patterns as little as possible. Consideration to limit erosion and siltation must also be given.

4.   The developed site should be compatible with the existing land uses and the comprehensive development plan. In some instances it may be desirable to design the various buildings at a treatment plant to blend the treatment facility with the surrounding developments.

Foundation Considerations.   The results of soil investigation should be utilized in locating the treatment units, buildings, and heavy equipment. Consideration should be given to load bearing capacities, water table and flotation effects, and piling.

Compact and Modular Site Development.   A relatively compact site plan can minimize piping requirements. Modular design and centralization of similar process units, process equipment, and personnel and facilities may reduce total staff size as well as optimize plant supervision and operation features.

Access Roads.   Access roads should be included in plant layout. Access roads should be provided to serve equipment in process areas. Employee parking at the rear and visitors parking in the front of the administration building should be provided. Trucks and service traffic should be separated from visitors and employee traffic early upon site entry. Signs for visitors parking and appropriate building entry should be posted.

Odor and Aerosol Sources.   Processes causing potential odors or aerosol should be located downwind from the public spaces or developments near the site. Such units should be located near each other and should be covered if necessary. Protective barriers, such as heavy plantation of large trees and buffer land, should also be considered.

Noise Sources.   Noise should be controlled to prevent discomfort to plant personnel and neighbors. Equipment such as pumps, ejectors, generators, blowers, and compressors can produce disturbing sound levels and therefore should be isolated. Where necessary, noise sources should be provided with sound barriers, absorptive enclosures, etc. Vehicular noise should also be considered in plant layout.

20-2-2 Buildings

Buildings are needed for plant personnel, process equipment, and visitors. The following considerations should be given to building design and location:

1. Location of equipment at the point of maximum usage will be helpful. However, to avoid cluttered appearance, buildings in groups should be provided with considerations of minimum loss in efficiency.
2. Buildings may be located as barriers to undesired views of the facility.
3. The climate of the area should be considered in building orientation and design to minimize heating, ventilation, lighting, and air-conditioning costs. In cold areas, buildings should not shade trucking and parking areas thus reducing snow and ice clearance problems.
4. Area and space requirements should be based on number of people served, and current and future functions.
5. The administration building should be located near the entrance and should generally be in public view. The administration building should contain offices, laboratory, instrumentation and control room, shower, lavatories, and locker rooms. In addition, the visitors lobby should have educational displays and tour information.
6. Other building requirements include machine shops with tools and storerooms, garages, and equipment buildings, e.g., pump, compressors, chemicals, instrumentation.

20-2-3 Shoreline Planning

The waterfront is a prime recreational zone. Erosion of banks, siltation of waterways, and destruction of valuable ecological niches should be protected from damage. Plant structures should be located as far back from the water edge as possible to permit other types of compatible land uses, particularly recreation and open space. Also, the long axis of the plant structures should be perpendicular to the natural water bodies in order to avoid blocking views and accesses.[4] An example of shoreline planning is illustrated in Figure 20-2.

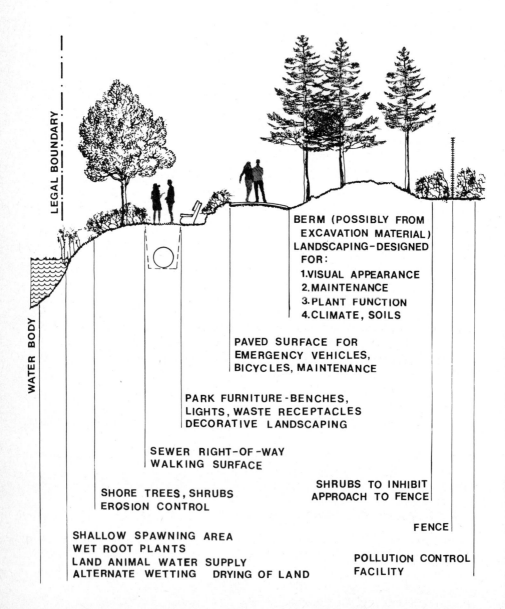

Figure 20-2   Wastewater Treatment Plant Site, Shoreline Planning. [From Ref. 2.]

20-2-4 Flood Plain Avoidance

Plant siting in flood plain should be avoided. If necessary, raising of the facilities and buildings above flood levels, and construction of flood protection levees should be considered. Provision must be made for collection of surface runoff at a central location. A storm water management system with outlet structure, flood gate, storm water pumps, and outfall pipe must be designed for removal of storm water under critical condition when highest flood stage is reached in the receiving water.

20-2-5 Landscaping

Existing site vegetation, trees, and shrubs should be assessed and utilized. Planting should be considered for control of slope erosion, surface runoff, enhanced attractiveness, and to provide sound and odor control barriers. Local soils and climatological and biological conditions should be carefully investigated by a competent landscape architect.

20-2-6 Lighting

Proper lighting at treatment facilities promotes safer operation, efficiency, and security. Considerations should be given to interior, exterior, safety, and security lighting. Illumination to highlight structural and landscape features should also be provided.

20-2-7 Plant Utilities

The utilities at the treatment plant include electrical power, natural gas, water lines, effluent lines, telephone lines, and an intercommunication system. Power generation from wastewater gas is usually not economical except for larger plants. The cost of power generation is generally determined from careful consideration of many economic factors. The wastewater gas is used to heat the buildings and digesters, and to produce hot water and steam for other uses.

   The design of the utility system should conform to the applicable codes and regulations of the municipality concerned and to the operating rules of the concerned utilities. All utility lines should be properly shown on the layout plans, marked on the site (if exposed), and grouped properly to facilitate repairs, modifications, and expansion. Additional discussion of utilities may be found in Refs. 1 and 5.

20-2-8 Occupational Health and Safety

A wastewater treatment facility offers many types of occupational health hazards that must be considered as part of plant design and layout. Important factors for which proper safeguards must be provided include chemicals and chemical handling, biological vectors, toxic gases, fire protection, explosions, burns, electric shocks, rotating machinery parts, material and equipment handling, falls and drowning, and the like. An excellent discussion of the subject can be found in Ref. 1.

20-2-9 Security

All accesses to the treatment plant should be controlled. Fences and other barriers should be provided to enclose the facility. Proper signs should be displayed at all accesses indicating the name and owner of the facility.

20-2-10 Future Expansion

Provisions for future plant expansion must be made. The provisions should include (1) future space requirements, (2) plant expansion with least interruption to plant operation, and (3) process modifications with little interruption to the existing plant operation.

20-3 DESIGN EXAMPLE

The site development plan of the Design Example is shown in Figure 20-3. Various treatment units, pipings, buildings, access roads and parkings, etc., are properly laid utilizing the many principles and factors discussed earlier. Special considerations are given to social and delicate environmental factors.

20-3-1 Description of the Site

The proposed site is classified as a socially and ecologically sensitive suburban site. An inventory of the existing conditions indicates that the site is a 125-ha, undeveloped tract bounded by two small streams that join immediately north of the site. Suburban developments are on the south characterized by industrial park and some single- and multiple-family residences (Figure 6-2). The prevailing winds are northernly, and the site access and influent lines are from the south. The existing aerated lagoon is located on the western side of the site (Figure 20-3).

The principal constraints to the site are waterfronts on three sides somewhat haphazardly bounded by the streams. Most of the waterfronts have severe slopes, erodible soils, flood plane, and drainage pattern encroachment. Site inventory, however, shows no valuable vegetation or special wildlife habitats.

20-3-2 Description of Plant Layout and Site Development

After careful considerations of the site development factors the site plan was prepared and treatment units and buildings were located. The following points may be noted in Figure 20-3:

1. The pump station is located on the western side and the wastewater is pumped to a high location from which flow by gravity is achieved in normal sequence of treatment units.
2. The existing aerated lagoon is utilized for diverting the flow from the collection box in case a power failure occurs (see Sec. 6-10-2).
3. The facility has been located away from the developed area, and existing trees and new landscaping have been used as buffers.

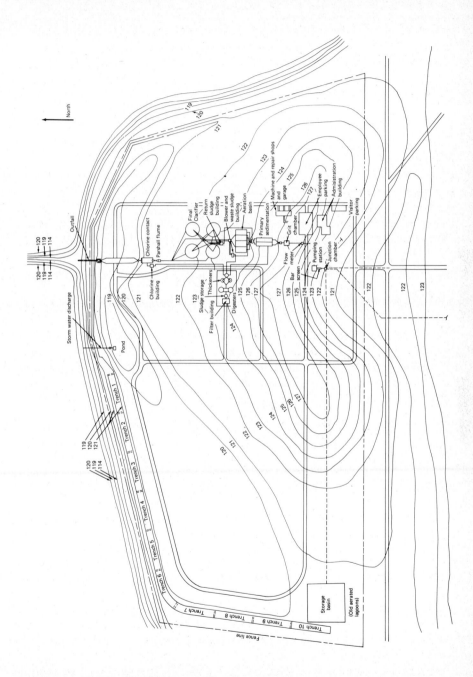

Figure 20-3   Layout Plan and Final Ground Surface Grade of the Treatment Facility Covered in the Design Example.

4. The administration building is located in the front with all treatment units far behind. Visitors parking is provided in the front and employees parking in the back. Each equipment building contains a small work area and necessary tools. A separate machine shop and repairs building is provided for general use.
5. Sludge-processing and disposal areas are kept far on the west and north to give adequate buffer land for odor control.
6. Aeration basins are located with landscaping and vegetation around the facility to provide a buffer against aerosols.
7. Ample space for future process expansion with each treatment unit has been provided. Such expansion would maintain modular design and centralization of similar processing units.
8. The surface is drained away from the main structures, and the entire site is sloped towards the sedimentation pond at the northern edge of the site plan. The pond is located at the lowest location to permit natural drainage.
9. As much as possible, the developed site does not alter existing naturally stabilized site contour and drainage pattern. The landscaping reflects rolling character of the surrounding area. The entire area is sodded to reduce erosion.
10. The stormwater structure at the sedimentation pond utilizes flood control gates and storm water pumps to remove surface runoff under highest flood level on the other side of the levees.
11. The unit's layout is compact, with sufficient length of connecting conduits for future expansion.
12. Paved service roads pass by each facility.
13. The facility has been located such that the use of the river edge as a recreation area can be accomplished by building garden walks and esplanades similar to those shown in Figure 20-2.

With the above design and site development goals, the plant layout has been prepared. Additional considerations were made for design of individual unit operations and processes. Such considerations may be found in Chapters 7–19. The site plan with finished contours is shown in Figure 20-3. Unit details and connecting pipings are shown in Figure 20-4. The plans of the administration and other utility buildings are given in Figure 20-5.

20-4 PROBLEMS AND DISCUSSION TOPICS

20-1. As a design engineer, you have the responsibility to develop the plant layout, site plan, landscaping, and finished contours of a wastewater treatment facility. List the basic rules that you must consider in developing such a plan.

20-2. Study the site plan given in Figure 20-3. Identify those features that do not conform to the basic rules of the site plans that you have listed in Problem 20-1.

20-3. In Figure 20-3, a total area of 80 ha drains toward the storage pond. Assuming rainfall intensity of 0.89 cm/h over the time of concentration of 2 h, and the coefficient of runoff of 0.13, design a suitable storm water outfall system at the storage pond. The outfall system should include an outlet structure with flood gate, outfall pipe, and storm water pumps for emergency condition.

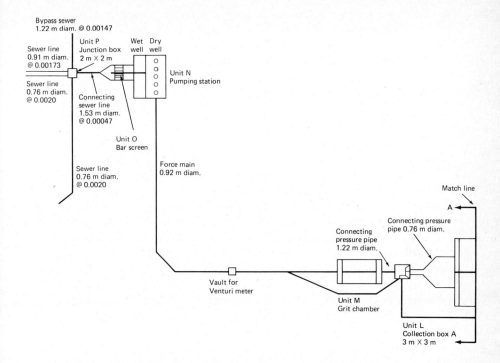

Figure 20-4   Plant Layout and Connecting Yard Pipings.

20-4.   Review the layout plan, finished contours, and landscaping features of the wastewater treatment plant in your community. Compare the important features of this plant layout with that of the Design Example given in Figure 20-3.

20-5.   Design a storm water pumping station for the flow calculated in Problem 20-3. Describe the type of pumps and reasons for their selection. Draw the details of pumping station.

REFERENCES

1. Joint committee of the Water Pollution Control Federation and American Society of Civil Engineers, *Wastewater Treatment Plant Design,* MOP/8, Water Pollution Control Federation, 1977.
2. Leffel, R. E., *Direct Environmental Factors at Municipal Wastewater Treatment Works,* Report prepared for the Environmental Protection Agency, EPA-430/9-76-003, MCD-20, January 1976.
3. Metcalf and Eddy, Inc., *Wastewater Engineering: Treatment, Disposal, and Reuse,* McGraw-Hill Book Co., New York, 1979.
4. *Siting and Design of Municipal Treatment Plants,* Hudson River Valley Commission, New York State Department of Health, Albany, New York, 1970.
5. *Uniform Plumbing Code,* International Association of Plumbing and Mechanical Officials, Los Angeles, Cal., 1973.

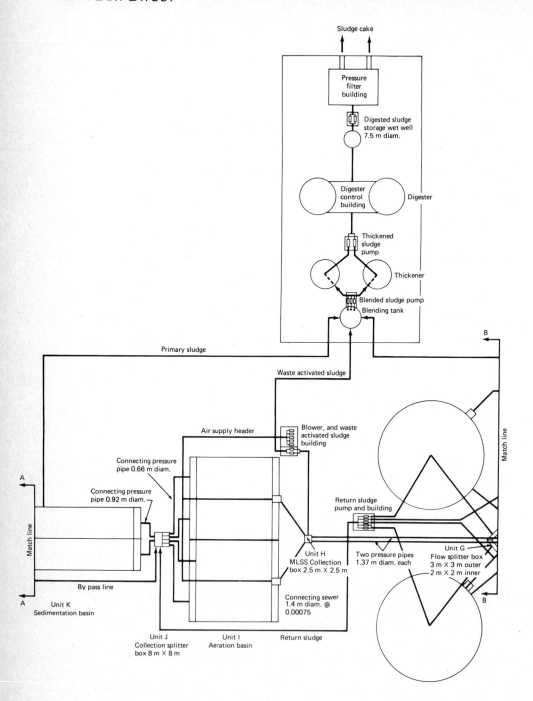

Figure 20-4   (continued)

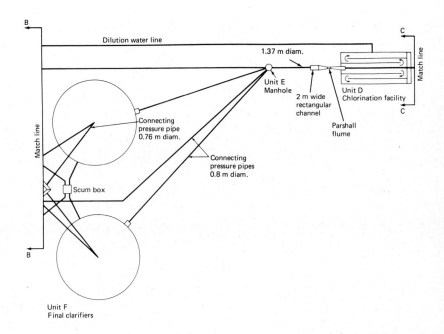

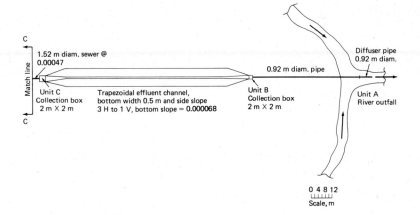

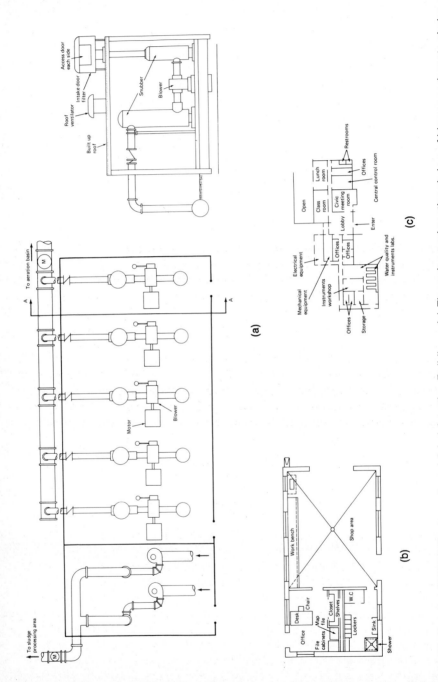

Figure 20-5   Details of Utility, Equipment, and Administration Buildings. (a) Floor plan and sectional view of blower and waste sludge pump building. (b) Maintenance and machine shop. (c) Administration building.

21

Yard Piping and Hydraulic Profile

21-1 INTRODUCTION

Yard piping includes connecting conduits, collection and division boxes, valves and gates, and manholes and appurtenances between various treatment units. After the treatment units and the connecting pipes and appurtenances are marked on the layout plan, the head losses through pipings and treatment units are calculated. Hydraulic profile is the graphical representation of the hydraulic grade line through the plant. The elevations of treatment units and pipings are adjusted to give adequate hydraulic gradient to ensure gravity flow.

Many treatment plants have encountered serious operational problems (even flooding under peak flow) due to inadequate hydraulic gradient through the various units. It is therefore important that the design engineer take proper care in developing the hydraulic profile. In this chapter the basic principles of designing yard piping and preparation of hydraulic profile are presented. Furthermore, the calculation procedure and graphical representation of the hydraulic profile for the Design Example are shown.

21-2 YARD PIPING

The arrangement of channels, pipelines, and appurtenances is important to transmit flows from one treatment unit to the other. Often flows are collected from several treatment units operating in parallel or divided into several units. Hydraulically similar inlet pipings and channels as well as splitting boxes are normally used for flow splitting and solids distribution. There are three basic considerations in preparing the piping layout: (1) convenience of construction and operation, (2) accessibility for maintenance, and (3) ease with which future connections can be made or more lines added.

Small distances between the adjacent treatment units can provide compact plant layout and minimize the costs of connecting pipings. However, it should not be forgotten that the connecting pipings and space between the treatment units are the primary means to expand a plant expeditiously. Many unexperienced designers overlook this fact. Plants with tight pipings and hydraulics have resulted in serious expansion problems; often pumping was needed to route the flows through the new treatment units.

The connecting channels and pipings may be above or below the ground level. However, all pipings must be clearly marked with reference to the treatment units on the layout plans. Provisions for future tie-ins should be made and properly indicated on the plans. Valves or gates should be provided in the connecting piping to isolate or bypass units and modules for routine servicing and maintenance. Clearly marked schematic drawings with treatment units, connecting piping, and valves will assist the designers and operator to understand the flow routing and unit isolation, and in general to understand the operational capability of a well-designed plant.

In many plants underground tunnels (called pipe galleries or operating galleries) are constructed to locate pipings and necessary controls. Although such galleries provide access to pipings and controls and passage between the buildings, such expenditure can be justified only for large plants. In small- and medium-size plants pipes are commonly located above or below the ground. Plans showing yard pipings are very helpful in reaching proper pipes and controls for operation and maintenance needs.

21-3 PLANT HYDRAULICS

Hydraulic profile is the graphical representation of the hydraulic grade line through the treatment plant. If the high-water level in the receiving water is known, this level is used as a control point, and the head loss computations are started backward through the plant. Sometimes the computations are started in the direction of the flow from the interceptor using the water surface as the reference level. In some plants, the hydraulic calculations are started somewhere in the middle using an arbitrary elevation. At the end, the elevations of water surface are adjusted in both directions.

The total available head at a plant is the difference of water surface elevation in the interceptor and the water surface elevations in the receiving water at high flood level. If the total available head is less than the head loss through the plant, flow by gravity cannot be achieved. In such cases pumping is needed to raise the head so that flow by gravity can occur.

There are many basic principles that must be considered when preparing the hydraulic profile through a plant. Some of these principles are listed below:

1. The hydraulic profiles are prepared at peak and average design flows and at minimum initial flow.
2. The hydraulic profile is generally prepared for all main paths of flow through the plant.
3. The total head loss through a treatment plant is the sum of head losses in the treatment units and in the connecting piping and appurtenances.
4. The head losses through treatment unit include the following:

    a. Head losses at the influent structure
    b. Head losses at the effluent structure
    c. Head losses through the unit
    d. Miscellaneous and free-fall surface allowance

The largest head loss through a treatment unit may occur at peak design flow plus recirculation when the largest unit is out of service.

The approximate head losses across different treatment units may be as follows:

Bar screen	0.02–0.3 m
Grit removal	
Aerated grit channels	0.5–1.2 m
Velocity-controlled grit channel	1.0–2.5 m
Primary sedimentation	0.5–1.0 m
Aeration basin	0.3–0.8 m
Trickling filter	
Low rate with dosing tank	3.0–6.0 m
High rate, single stage	2.0–5.0 m
Secondary clarifier	0.5–1.0 m
Chlorination facility	0.2–2.5 m

5.   The total loss through the connecting pipings, channels and appurtenances is the sum of the following:

   a.   Head loss due to entrance
   b.   Head loss due to exit
   c.   Head loss due to contraction and enlargement
   d.   Head loss due to friction
   e.   Head loss due to bends, fittings, gates, valves, and meters
   f.   Head required over weir and other hydraulic controls
   g.   Free-fall surface allowance
   h.   Head allowance for future expansions of the treatment facility

6.   The velocity in connecting pipings and conduits is kept large enough to keep the solids in suspension. A minimum velocity of 0.6 m/s at peak design flow is considered adequate. At minimum initial flow, a velocity of 0.3 m/s is considered necessary to transport the organic solids. Often the ratio of maximum and minimum flows is so large that self-cleaning velocity cannot be maintained under the initial flow conditions. In such cases the flushing actions can only be achieved under higher flows. Frequency of occurrence of flushing actions must be considered in such designs. In many cases aeration is provided to keep solids in suspension. It may even be desirable to provide separate lines from multiple units so that the line could be cleaned when a unit is out of service.

7.   The minor head losses in open channels and conduits are calculated in terms of the velocity head. Detailed discussions may be found in Chapters 7 and 9.

8.   Friction losses in pressure conduits is obtained using Hazen–Williams formula (see Chapter 9).

9.   In channels, the depth of flow varies depending on the flow conditions. Therefore, the depth and grade of open channel is kept in such a way that the water surface at the design flow corresponds to the hydraulic profile (see Chapter 7).

In open channels the flow may be either uniform or nonuniform. Uniform flow occurs in channels that have constant cross-section, flow, and velocity. Manning's equation is generally used to calculate the grade of the water surface. In design of channels it is generally assumed that the flow is uniform at peak design flow.

Nonuniform flow exists in channels when cross section changes or volume of wastewater entering the channel is not constant. Nonuniform flow generally occurs in channels that have free-fall, effluent flumes or launders, or channel junctions with surcharge. Friction formula do not apply for nonuniform flows. Backwater or drawdown analysis is necessary. Computational techniques of nonuniform flows have been given in Chapters 11–15. Readers are referred to some excellent textbooks on hydraulics of open channel for more detailed discussion on the subject.[1–4]

In wastewater treatment plants sufficient allowances are made for the transitions and nonuniform flows by providing invert drops. Head loss through the transitions is generally calculated by using energy equation.

10.   Most of the flow measuring devices used in wastewater treatment plants operate with head loss. Proper head loss calculations should be made for flow-measuring devices (Venturi tube, Parshall flume, orifice plate, weir, etc.) and included in the hydraulic profile.

11.   In preparation of hydraulic profile the vertical scale is intentionally distorted to show the treatment facilities and the elevation of the water surface. Ground surface is also indicated to establish the optimum elevation of the plant structures and the hydraulic controls.

21-4   DESIGN EXAMPLE

The procedure for preparing the hydraulic profile through the treatment plant designed in Chapters 7–15 is presented in this section. The procedure involves determination of head loss as the wastewater flows through various treatment units, the connecting conduits, and appurtenances.

21-4-1 Head Losses Across Treatment Units

The head loss calculations, and hydraulic profiles across various treatment units have been developed in respective chapters. The head loss calculations were made under peak design flow when the largest component was out of service. The hydraulic capacity and the head loss across each treatment unit are summarized in Table 21-1. The head loss across a treatment unit is the difference of water surface elevations in the influent and effluent structures.

21-4-2 Head Losses in Connecting Pipings

The total head loss in the connecting piping is the sum of all head losses encountered in the pipings, channels, collection and division boxes; head over weir; and allowance for free-fall and future expansions. The largest head loss generally occurs at the control points such as weir, flume, drop manhole, etc. The head losses due to friction, bends, entrance, and exits may also be significant. A detailed plant layout with piping schedule showing straight lengths as well as vertical and horizontal bends should be prepared. Then the head loss calculations are started. These calculations are tedious and time consuming. It is

TABLE 21-1   *Hydraulic Capacity and Head Loss through Individual Treatment Units*

Treatment Unit	Total Number of Components	Flow	Hydraulic Capacity	Head Loss Across the Unit, m	Ref.
Bar screen	2	Peak design flow	Flow through one unit (clean rack)	0.03	Figure 8-6
Pump station	5	Peak design flow	Four units	12.67 (operating head at max. station capacity)	Table 9-7
Venturi meter	1	Peak design flow	Flow through one unit	0.45	Sec. 10-7-2, step C
Grit chamber	2	Peak design flow	Flow through one unit	1.07	Figure 11-5
Primary sedimentation[a]	2	Peak design flow	Flow through two units	0.99	Figure 12-15
Aeration basin	4	Peak design flow plus recirculation	Flow through three units	0.65	Figure 13-13
Secondary clarifier	4	Peak design flow plus recirculation	Flow through three units	0.52	Figure 13-15
Chlorination facility	2	Peak design flow	Flow through one unit	2.03	Figure 14-9
Outfall	1	Peak design flow	Flow through one unit	1.66	Figure 15-1

[a]When one unit is out of service, the flow is bypassed to the aeration basin.

time consuming. It is suggested that the design engineer should use a format that is systematic, concise, and easier to perform and check calculations.

Preparation of plant layout has been discussed in Chapter 20. Proposed plant layout and the connecting pipings of the Design Example are illustrated in Figures 20-3 and 20-4. Head loss calculations are started from the high-water elevation in the receiving water and carried backward. The hydraulic profile is shown in Figure 21-1. The head loss calculations are summarized in the following sections. Other examples of head loss calculations through the treatment units and connecting conduits and appurtenances may be found in Ref. 4.

21-4-3 Head Loss Calculations for Connecting Conduits

A. Outfall Conditions (Unit A)
High-water surface elevation in the receiving stream (Fig. 21-1) = 116.98 m
Stream bottom elevation = 113.78 m

B. Water Surface and Invert Elevations in Collection Box (Unit B).   The effluent structure consists of effluent pipe and diffusers

1.   Head loss in the diffuser and outfall pipe (entrance, bends, and friction, Sec 15-4-2, step E) = 1.66 m
2.   Water surface elevation in unit B = 116.98 m + 1.66 m = 118.64 m
3.   Invert elevation if pipe submergence is 0.04 m = 118.64 m – 0.92 m – 0.04 m = 117.68 m

C. Water Surface and Invert Elevations in Collection Box (Unit C).   The connecting conduit is a trapezoidal channel. The channel has bottom width with 0.5 m and side slope of 3 horizontal to 1 vertical. The ends of the channel have smooth transition to connect into a 2-m-wide rectangular opening in the box.

1.   Compute the depth of flow and velocity in the rectangular portion of the channel

Peak design flow	1.321 m³/s
The depth of flow in the rectangular portion of the channel	= 0.96 m
Velocity at the rectangular portion of the channel	$= \dfrac{1.321 \text{ m}^3/\text{s}}{0.96 \text{ m} \times 2 \text{ m}}$
	= 0.69 m/s

2.   Compute the velocity and depth of flow in the trapezoidal channel
Apply energy equation (Eq. 8-4) at trapezoidal and rectangular sections (see Figure 15-1(c)).
For smooth transition of sides and bed slopes, $K_c = 0.2$ and $(Z_1 - Z_2) = 0$

$$y_1 + \frac{(q/a)^2}{2g} = 0.96 \text{ m} + \frac{(0.69 \text{ m/s})^2}{2 \times 9.81 \text{ m/s}^2} + 0.2 \left[\frac{-(q/a)^2}{2g} \right.$$

$$\left. + \frac{(0.69 \text{ m/s})^2}{2 \times 9.81 \text{ m/s}^2} \right]$$

$$q = 1.321 \text{ m}^3/\text{s}, \quad a = \frac{0.5 \text{ m} + (6 \, y_1 + 0.5)\text{m}}{2} \times y_1 \text{ (m)}$$

The solution of the above equation is reached by trial-and-error procedure. Solving the above equation,

$$y_1 = 0.97 \text{ m}$$
$$a = \frac{0.5 \text{ m} + [(2 \times 3 \times 0.97 \text{ m}) + 0.5 \text{ m}]}{2} \times 0.97 \text{ m}$$
$$= 3.31 \text{ m}^2$$

$$\text{Velocity} = \frac{q}{a} = \frac{1.321 \text{ m}^3/\text{s}}{3.31 \text{ m}^2} = 0.40 \text{ m/s}$$

3. Compute the slope of the invert and water surface elevations at the trapezoidal channel near collection box (unit B)

    Invert elevation (same for unit B) = 117.68 m
    Water surface elevation = 117.68 m + 0.97 m = 118.65 m

4. Compute the slope of the trapezoidal channel from Manning equation (Eq. (7-1))
    For trapezoidal channel at depth = 0.97 m,

    $$a = 3.31 \text{ m}^2, \, n = 0.013$$

    $$\text{Wetted parameter} = 0.5 \text{ m} + 2\sqrt{(0.97\text{m})^2 + (3 \times 0.97 \text{ m})^2}$$
    $$= 0.5 \text{ m} + 6.13 \text{ m} = 6.63 \text{ m}$$

    $$r = \frac{a}{\text{wetted perimeter}} = \frac{3.31 \text{ m}^2}{6.63 \text{ m}} = 0.50 \text{ m}$$

    $$s = \left[\frac{0.40 \text{ m/s} \times 0.013}{(0.50)^{2/3}}\right]^2 = 0.000068$$

5. Compute the invert and water surface elevations in the trapezoidal channel near unit C

    Differential elevation of the channel bottom due to slope = channel length × slope
    = 95 m × 0.000068
    = 0.0065m ≈ 0.01 m
    Invert elevation = 117.68 m + 0.01 m = 117.69 m
    Water surface elevation = 117.69 m + 0.97 m = 118.66 m

6. Compute depth of flow and velocity in unit C.
    Use energy equation as in step C2. Raise invert elevation of unit C by 0.01 m and solving for depth of flow and velocity by trial and error

    Depth of flow = 0.96 m
    Velocity = 0.41 m/s

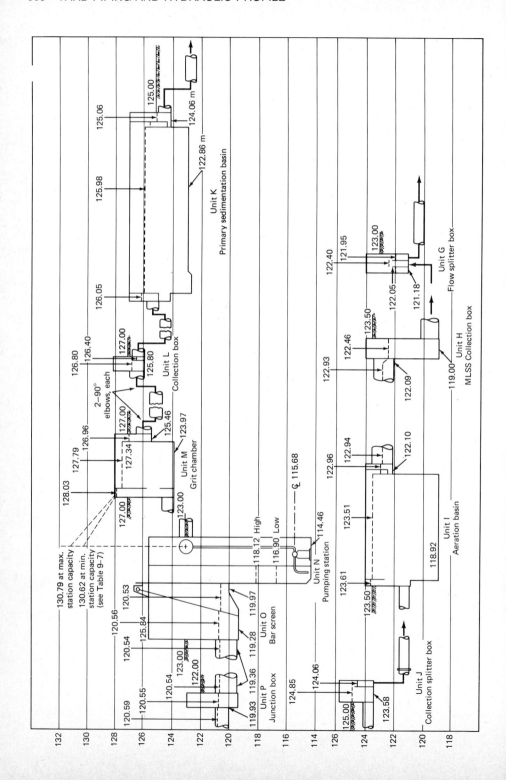

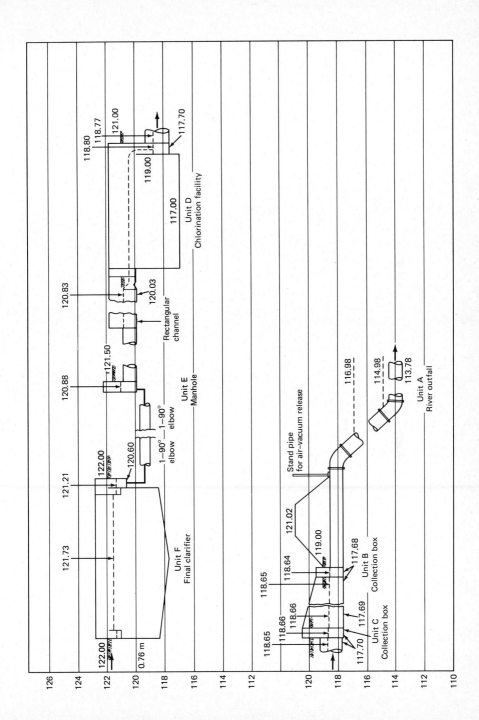

Figure 21-1   Hydraulic Profile through the Treatment Facility in the Design Example.

7. Compute water surface elevation in unit C
    The water surface elevation in

    unit C $= 117.69$ m $+ 0.01$ m $+ 0.96$ m

                  $= 118.66$ m

D. Water Surface and Other Elevations in Chlorination Facility (Unit D).   A 1.52-m
sewer line connects unit C with the effluent channel of the chlorination facility

1. Compute the depth of flow in connecting sewer line at peak design flow
    Diameter and slope of connecting sewer are 1.52 m and 0.00047, respectively.
    The sewer will be flowing partially full. From Manning equation ($n = 0.013$),
    the depth of flow and velocity at peak design flow of 1.321 m³/s are 1.07 m and
    0.97 m/s (see Chapter 7 for calculation procedure).

2. Compute the invert and water surface elevations in the connecting sewer line
    Set the invert of the connecting sewer same as that of the unit C. Using a sudden
    expansing outlet with a coefficient of 0.75, the upstream depth of 0.95 m is
    obtained from the energy equation. This gives subcritical flow in the sewer and
    a M2 profile is indicated. The maximum depth of water cannot exceed the
    uniform depth of 1.07 m. Therefore, an upstream depth of 1.07 m in the sewer
    will give a conservative design.

3. Compute the invert and water surface elevations in the effluent channel of the
    chlorine contact tank

Differential invert elevation of the connecting sewer line	$=$ slope $\times$ length
	$= 0.00047 \times 4$ m $=$ small
Invert elevation of the sewer at the effuent channel	$= 117.70$ m
Water surface elevation in the sewer at the effluent channel	$= 118.77$ m
Entrance loss ($K = 0.15$)	$= \dfrac{0.15 \times (0.97 \text{ m/s})^2}{2 \times 9.81 \text{ m/s}^2} = 0.01$ m
Miscellaneous losses due to change in direction and turbulence	$= 0.02$ m
Invert elevation of the effluent channel	$= 117.70$ m
Water surface elevation in the effluent channel	$= 117.70$ m
	$+ 1.07$ m $+ 0.01$ m $+ 0.02$ m
	$= 118.80$ m

4. Compute various elevations in the chlorination facility
    The hydraulic profile through the chlorination facility was prepared in Chapter
    14. Total head loss through the facility is 2.03 m

$$\begin{array}{l} \text{The water surface elevation in the rectangular} \\ \text{channel upstream of the Parshall flume} \end{array} = 120.83 \text{ m}$$

$$\begin{array}{l} \text{The invert elevation in the rectangular channel} \\ \text{upstream of the Parshall flume} \end{array} = \begin{array}{l} 120.83 \text{ m} - 0.8 \text{ m} \\ \text{(depth of flow)} \end{array}$$

$$= 120.03 \text{ m}$$

E. Water Surface and Invert Elevations in Collection Box (Unit E).   The connecting
conduits between the Parshall flume and unit E are rectangular channel and a pressure
pipe. The rectangular channel is 2 m wide and 4 m long (Figure 20-4). The invert and
water surface elevations in the rectangular channel as calculated above are 120.03 and
120.83 m, respectively.

1. Compute the head loss in the connecting pipe

$$\text{Diameter of the pipe} = 1.37 \text{ m}$$

$$Q = 1.321 \text{ m}^3/\text{s}$$

$$V = \frac{1.321 \text{ m}^3/\text{s}}{(\pi/4) \ (1.37 \text{ m})^2}$$

$$= 0.90 \text{ m/s}$$

Friction loss is calculated from Hazen-Williams equation [Eq. (9-3)].

$$C = 100$$

$$h_f = 6.82 \left(\frac{0.90}{100} \right)^{1.85} \times \frac{22 \text{ m}}{(1.37)^{1.167}}$$

$$= 0.02 \text{ m}$$

$$\text{Entrance loss } (K = 0.5) = \frac{0.50 \times (0.90 \text{ m/s})^2}{2 \times 9.81 \text{ m/s}^2}$$

$$= 0.02 \text{ m}$$

$$\begin{array}{ll} \text{Total exit and miscellaneous losses} & = 0.01 \text{ m} \\ \text{Total losses in the connecting pipe} & = 0.02 \text{ m} + 0.02 \text{ m} + 0.01 \text{ m} \\ & = 0.05 \text{ m} \end{array}$$

2. Compute water surface elevation in unit E

    The water surface elevation in unit E $= 120.83 \text{ m} + 0.05 \text{ m}$

Invert elevation of the manhole (same as sewer invert) $= 120.04$ m

Entrance loss ($K = 0.35$) $= \dfrac{0.35 \times (0.95 \text{ m/s})^2}{2 \times 9.81 \text{ m/s}^2} = 0.02$ m

Water surface elevation in the manhole (unit E) $= 120.86$ m $+ 0.02$ m

$= 120.88$ m

F. Water Surface and Other Elevations in the Final Clarifier (Unit F). There are four pressure pipes connecting the manhole (unit E) with the effluent box of the final clarifiers. The calculations for the longest path are given below.

1. Compute the head loss through the final clarifier

Peak design flow when three clarifiers are in operation $= \dfrac{1.321 \text{ m}^3/\text{s}}{3}$

$= 0.44$ m^3/s

Diameter of the connecting pipe $= 0.80$ m

Velocity $= \dfrac{0.44 \text{ m}^3/\text{s}}{(\pi/4)(0.8 \text{ m})^2} = 0.9$ m/s

Exit loss at unit E ($K = 0.5$) $= \dfrac{0.5 \times (0.9 \text{ m/s})^2}{2 \times 9.81 \text{ m/s}^2} = 0.02$ m

Minor loss (2–90° elbows, see Figure 21-1, $K = 0.3$)

$= \dfrac{2 \times 0.3 \times (0.9 \text{ m/s})^2}{2 \times 9.81 \text{ m/s}^2} = 0.03$ m

Minor loss (2–45° elbows, see plan view Figure 20-4, $K = 0.2$)

$= \dfrac{2 \times 0.2 \times (0.9 \text{ m/s})^2}{2 \times 9.81 \text{ m/s}^2} = 0.02$ m

Entrance loss at unit F ($K = 1.0$) $= \dfrac{1.0 \times (0.9 \text{ m/s})^2}{2 \times 9.81 \text{ m/s}^2} = 0.04$ m

Friction loss from Hazen–Williams equation [Eq. (9-3)] ($C = 100$, $L = 140$ m) $= 0.20$ m

Assume miscellaneous losses $= 0.02$ m

Total loss $= \dfrac{0.02 \text{ m} + 0.03 \text{ m} + 0.02 \text{ m}}{+ 0.04 \text{ m} + 0.20 \text{ m} + 0.02 \text{ m}}$

$= 0.33$ m

2. Compute the water surface and other elevations in the final clarifiers (unit F)

$$\text{Water surface elevation in the effluent box} = 120.88 \text{ m} + 0.33 \text{ m}$$
$$= 121.21 \text{ m}$$

$$\begin{array}{ll}\text{Invert elevation of the effluent box of the} \\ \text{final clarifier}\end{array} = \begin{array}{l}121.21 \text{ m} - 0.61 \text{ m} \\ \text{(see Chapter 13)}\end{array}$$
$$= 120.60 \text{ m}$$

Maximum water surface elevation in the influent central well at peak design flow plus recirculation when one unit is out of service $= \begin{array}{l}121.21 \text{ m} + 0.52 \text{ m} \\ \text{(Table 21-1)}\end{array}$

$$= 121.73 \text{ m}$$

Consult Chapter 13 for other elevations in the final clarifiers.

G. Water Surface and Other Elevations in the Splitter Box (Unit G). The MLSS from the splitter box is piped to the central wells of each final clarifier. The splitter box is a square chamber that has four identical rectangular weirs, one on each side to distribute flow equally to the final clarifiers. Sluice gates are provided over each weir to stop flow and remove any clarifier from service.

1. Compute the head loss in the pressure pipe connecting the splitter box to the central well of the final clarifier

Peak design flow plus recirculation when one clarifier is out of service $= \dfrac{1.321 \text{ m}^3/\text{s} + 0.292 \text{ m}^3/\text{s}}{3}$

$$= 0.538 \text{ m}^3/\text{s}$$

The diameter of the connecting pressure pipe $= 0.76 \text{ m}$

Velocity in the pipe $= \dfrac{0.538 \text{ m}^3/\text{s}}{\pi/4 \ (0.76 \text{ m})^2} = 1.19 \text{ m/s}$

Exit loss at the central well ($K = 1.0$) $= \dfrac{1 \times (1.19 \text{ m/s})^2}{2 \times 9.81 \text{ m/s}^2} = 0.07 \text{ m}$

Minor loss (1–90° elbows, see Figure 21-1, $K = 0.3$) $= \dfrac{1 \times 0.3 \times (1.19 \text{ m/s})^2}{2 \times 9.81 \text{ m/s}^2}$

$$= 0.02$$

Entrance loss at the splitter box ($K = 0.5$) $= \dfrac{0.5 \times (1.19 \text{ m/s})^2}{2 \times 9.81 \text{ m/s}^2}$

$$= 0.04 \text{ m}$$

Friction loss from the Hazen–Williams equation [Eq. (9-3)] ($C = 100$, $L = 30$ m) $= 0.08 \text{ m}$

Assume miscellaneous losses $= 0.01 \text{ m}$

Total head loss in connecting pipe $\quad= 0.07$ m $+ 0.02$ m $+ 0.04$ m
$+ 0.08$ m $+ 0.01$ m

$= 0.22$ m

2. Compute the water surface and invert elevations in the outer trough of the splitter box

Water surface elevation in the outer trough of the splitter box $\quad= 121.73$ m $+ 0.22$ m

$= 121.95$ m

Invert elevation of the outer trough $= 121.95$ m $- 0.77$ m
(depth of water)

$= 121.18$ m

3. Compute the head over the weir and water surface elevations

Length of rectangular weir $= 1.50$ m

Head over weir is calculated from Eq. (11-3) by trial and error [see Sec. 11-9-2, step E2]
Assume effective length of the weir $L' = 1.43$ m

Head over weir $H = \left[\dfrac{0.538 \text{ m}^3/\text{s} \times 3}{2 \times 0.624 \times 1.43 \ \sqrt{2 \times 9.81} \ \text{m/s}^2} \right]^{2/3}$

$= 0.35$ m

Check effective length L'

$L' = 1.50$ m $- 0.2 \times 0.35$ m $= 1.43$ m

Provide 0.10-m allowance for free-fall

Elevation of weir crest $\quad= 121.95$ m $+ 0.10 = 122.05$ m

Elevation of water surface in the central well of the splitter box $\quad= 122.05$ m $+ 0.35$ m $= 122.40$ m

H. Water Surface and Invert Elevations in the MLSS Collection Box (Unit H).   Unit H receives MLSS from four aeration basins. Two pressure pipes are provided to carry the MLSS from unit H to the flow splitter box (unit G). Only one pipe will be used during initial flow conditions. As the flow is increased in the future, both pipes will be used. A portion of the MLSS is wasted from unit H and pumped to the sludge-handling area.

1. Compute head loss in connecting pressure pipe

Peak design flow plus recirculation for each pipe $\quad= \dfrac{1.321 \text{ m}^3/\text{s} + 0.292 \text{ m}^3/\text{s}}{2}$

$= 0.807 \ \text{m}^3/\text{s}$

Pipe diameter	$= 1.37$ m*
Velocity	$= \dfrac{0.807 \text{ m}^3/\text{s}}{(\pi/4) \ (1.37 \text{ m})^2} = 0.55$ m/s
Exit loss at unit G ($K = 1.0$)	$= \dfrac{1.0 \times (0.55 \text{ m/s})^2}{2 \times 9.81 \text{ m/s}^2} = 0.02$ m
Entrance loss at unit H ($K = 0.3$)	$= \dfrac{0.3 \times (0.55 \text{ m/s})^2}{2 \times 9.81 \text{ m/s}^2} = 0.01$ m
Minor loss (1–$90°$ elbow, $K = 0.3$)	$= \dfrac{0.3 \times (0.55 \text{ m/s})^2}{2 \times 9.81 \text{ m/s}^2} = 0.01$ m
Friction loss from the Hazen–Williams equation [Eq. (9-3)] ($C = 100$ and length of pipe $= 72$ m)	$= 0.02$ m
Total head loss in connecting pipe	$= 0.02$ m $+ 0.01$ m $+ 0.01$ m $+ 0.02$ m
	$= 0.06$ m

2.   Compute water surface and invert elevations in unit H

Water surface elevation in unit H $= 122.40$ m $+ 0.06$ m

$= 122.46$ m

Set the invert elevation in unit H at 119.00 m

I. Water Surface and Other Elevations in the Aeration Basin (Unit I).   The connecting conduit between the effluent box of aeration basin, and the MLSS collection box (unit H) is a 1.40-m diameter sewer.

1.   Compute the velocity and depth of flow in the connecting sewer

Peak design flow from two aeration basins plus return flow when only three basins are in operation $= (1.321$ m^3/s $+ 0.292$ m^3/s)/3 $\times 2 = 1.08$ m^3/s

Depth of flow and velocity under partial flow is calculated from Manning equation [see Sec. 13-7-4, step C5]

Assumed slope $= 0.00075$
Depth of flow $= 0.84$ m
Velocity $= 1.14$ m/s

2.   Compute the water surface elevations in the sewer

*It may be desirable to provide 0.91 m diameter pipe to achieve higher velocity in the pipe.

Set the invert elevation in the sewer at 122.09 m

Water surface elevation in the sewer $= 122.09 \text{ m} + 0.84 \text{ m}$

$$= 122.93 \text{ m}$$

There is a free fall of 0.47 m from the sewer into the unit H. This will give a M2 profile and the upstream depth cannot exceed the normal depth in the sewer. Therefore, an upstream depth of 0.84 m in the sewer will provide a conservative design

Differential invert elevation of the sewer line	$= \text{slope} \times \text{length}$
	$= 0.00075 \times 16 \text{ m}$
	$= 0.01 \text{ m}$
Invert elevation of the sewer below effluent box of the aeration basin	$= 122.09 \text{ m} + 0.01 \text{ m}$
	$= 122.10 \text{ m}$
Water surface elevation in sewer below effluent box of the aeration basin	$= 122.10 \text{ m} + 0.84 \text{ m}$
	$= 122.94 \text{ m}$

3.   Compute water surface and other elevations in the aeration basin

Entrance loss at the effluent box of aeration basin ($K = 0.3$)	$= \dfrac{0.3 \times (1.14 \text{ m/s})^2}{2 \times 9.81 \text{ m/s}^2}$
	$= 0.02 \text{ m}$
Water surface elevation in the effluent box	$= 122.94 \text{ m} + 0.02 \text{ m}$
	$= 122.96 \text{ m}$
Invert elevation of the effluent box	$= 122.96 \text{ m} - \text{water depth}$
	$= 122.96 \text{ m} - 0.86 \text{ m}$
	$= 122.10 \text{ m}$
Water surface elevation in the influent channel of the aeration basin (Figure 13-13)	$= 122.96 \text{ m} + 0.65 \text{ m}$
	$= 123.61 \text{ m}$

Other elevations are given in Figures 13-13 and 21-1.

J. Water Surface and Other Elevations in the Collection-Splitter Box (Unit J).   The collection-splitter box (unit J) is located upstream of the aeration basin. This box receives flow from the primary sedimentation basin, return sludge, and sidestreams from the sludge-processing area. The combined flow is then discharged over four identical weirs for division into four pressure pipes that lead the MLSS to four aeration basins. Each weir is adjustable, has stop gate, and is equipped with manual head measurement system over the weirs.

1. Compute the head loss in the pressure pipe connecting the splitter box and aeration basin

    Peak flow in the pipe when one aeration basin is out of service[*] $= \dfrac{1.321 \text{ m}^3/\text{s} + 0.292 \text{ m}^3/\text{s (return flow)}}{3}$

    $= 0.538 \text{ m}^3/\text{s}$

    Pipe diameter $= 0.66 \text{ m (assume)}$

    Velocity $= \dfrac{0.538 \text{ m}^3/\text{s}}{(\pi/4)\,(0.66 \text{ m})^2} = 1.57 \text{ m/s}$

    Exit loss at the aeration basin $(K = 1.0)$ $= \dfrac{1.0 \times (1.57 \text{ m/s})^2}{2 \times 9.81 \text{ m/s}^2} = 0.13 \text{ m}$

    Minor losses (3–90° elbows, one elbow in vertical plane, and two elbows in horizontal plane, $K = 0.3$)

    $= \dfrac{3 \times 0.3 \times (1.57 \text{ m/s})^2}{2 \times 9.81 \text{ m/s}^2} = 0.11 \text{ m}$

    Entrance loss at the splitter box $(K = 0.5)$ $= \dfrac{0.5 \times (1.57 \text{ m/s})^2}{2 \times 9.81 \text{ m/s}^2}$

    $= 0.06 \text{ m}$

    Friction loss $(L = 30 \text{ m}$ longest line, $C = 100)$ $= 0.15 \text{ m}$

    Total head loss $= 0.13 \text{ m} + 0.11 \text{ m} + 0.06 \text{ m} + 0.15 \text{ m}$

    $= 0.45 \text{ m}$

    Water surface elevation $= 123.61 \text{ m} + 0.45 \text{ m} = 124.06 \text{ m}$

    Invert elevation $= 124.06 - 0.48 \text{ m (water depth in the box)}$
    $= 123.58 \text{ m}$

2. Compute the weir crest and water surface elevation in the collection box
    The length of the rectangular weir is 1.50 m. The procedure for calculation of head over weir is same as presented in step G3 above. From Eq. (11-3), head over weir = 0.35 m. Provide an allowance of 0.44 m for free-fall.

    Elevation of weir crest $= 124.06 \text{ m} + 0.44 \text{ m} = 124.50 \text{ m}$

[*]The flow due to returned sidestreams from sludge-processing area is small.

$$\text{Water surface elevation in the collection box} = 124.50 \text{ m} + 0.35 \text{ m}$$
$$= 124.85 \text{ m}$$

K. Water Surface and Other Elevations in the Primary Sedimentation Basin (Unit K).

The pressure pipes connecting the effluent boxes of primary sedimentation basin with the collection-splitter box are 92 cm in diameter. Each pipe is designed to carry a peak design flow of 0.66 m³/s from each sedimentation basin. In case one sedimentation unit is out of service, half of the flow is diverted to aeration basin without primary treatment (see Sec. 12-6-2 for design criteria of primary sedimentation facility).

1. Compute head loss in the connecting pipe

    Velocity $$= \frac{Q}{A} = \frac{0.66 \text{ m}^3/\text{s}}{(\pi/4)\,(0.92 \text{ m})^2}$$
    $$= 0.99 \text{ m/s}$$

    Exit loss ($K = 1.0$) $$= \frac{1.0 \times (0.99 \text{ m/s})^2}{2 \times 9.81 \text{ /s}} = 0.05 \text{ m}$$

    Minor loss (4–90° elbows, two elbows in vertical plane and two elbows in horizontal plane, $K = 0.3$) $$= \frac{4 \times 0.3 \times (0.99 \text{ m/s})^2}{2 \times 9.81 \text{ m/s}^2}$$
    $$= 0.06 \text{ m}$$

    Friction loss (from the Hazen–Williams equation, $C = 100$ and $L = 30$ m) $$= 0.05 \text{ m}$$

    Entrance loss at the effluent box of primary sedimentation facility ($K = 0.5$) $$= \frac{0.5 \times (0.99 \text{ m/s})^2}{2 \times 9.81 \text{ m/s}^2}$$
    $$= 0.02 \text{ m}$$

    Assume miscellaneous losses $$= 0.03 \text{ m}$$

    Total head loss $$= \frac{0.05 \text{ m} + 0.06 \text{ m} + 0.05 \text{ m}}{+\ 0.02 \text{ m} + 0.03 \text{ m}}$$
    $$= 0.21 \text{ m}$$

2. Compute water surface and invert elevations in the effluent box of primary sedimentation facility

    Water surface elevation $= 124.85 \text{ m} + 0.21 \text{ m} = 125.06 \text{ m}$
    Invert elevation $\quad = 125.06 \text{ m} - 1.00 \text{ m} = 124.06 \text{ m}$

3. Compute water surface and other elevations in the primary sedimentation facility

    Head loss across primary sedimentation facility (Table 21-1) $= 0.99 \text{ m}$

Water surface elevation in the influent
channel of primary sedimentation tank = 125.06 m + 0.99 m
at peak design flow

$$= 126.05 \text{ m}$$

L. Water Surface and Invert Elevations in the Collection-Division Box (Unit L).* The
collection-division box receives flow from the grit removal facility and splits the flow
equally into two pipes leading to the primary sedimentation basins. The flow is controlled
by adjustable weirs. Sluice gates are provided to remove one sedimentation basin from
service and divert the flow into the aeration basin through a third weir and bypass pipe
(Figure 20-4).

1. Compute the head loss in the connecting pipe, and water surface elevation in the
    outer chamber

    Diameter of the connecting pipe = 0.76 m

    Peak design flow = 0.66 m³/s

    Velocity $= \dfrac{Q}{A} = \dfrac{0.66 \text{ m}^3/\text{s}}{\pi/4 \times (0.76 \text{ m})^2}$

    $= 1.46 \text{ m/s}$

    Exit loss at unit K ($K = 1.00$) $= \dfrac{1.00 \times (1.46 \text{ m/s})^2}{2 \times 9.81 \text{ m/s}^2} = 0.11 \text{ m}$

    Minor losses 4–90° elbows in $= \dfrac{4 \times 0.3 \times (1.46 \text{ m/s})^2}{2 \times 9.81 \text{ m/s}^2}$
    vertical plane ($K = 0.3$)

    $= 0.13 \text{ m}$

    Entrance loss at unit L ($K = 0.3$) $= \dfrac{0.3 \times (1.46 \text{ m/s})^2}{2 \times 9.81 \text{ m/s}^2} = 0.03 \text{ m}$

    Friction loss (Hazen–Williams
    equation, $C = 100$, $L = 22$ m) $= 0.08 \text{ m}$

    Total head loss $= \begin{aligned}&0.11 \text{ m} + 0.13 \text{ m} + 0.03 \text{ m}\\&+ 0.08 \text{ m} = 0.35 \text{ m}\end{aligned}$

    Water surface elevation in the outer
    chamber $= 126.05 \text{ m} + 0.35 \text{ m}$

    $= 126.40 \text{ m}$

2. Compute elevation of weir crest and water surface elevation in the collection-
    division box

Free-fall allowance	= 0.08 m
Elevation of the weir crest	= 126.40 m + 0.08 m = 126.48 m
Provide weir length	= 2 m

*Note that the connecting pipe is a syphon. These pipes should be avoided as they collect solids.

Head over weir from Eq. (11-3)	= 0.32 m
Elevation of water surface	= 126.48 m + 0.32 m = 126.8 m
Depth of water in the box (assumed)	= 1.00 m
Invert elevation of the box	= 126.80 m − 1.00 m = 125.80 m

M. Water Surface and Other Elevations in Grit Chamber (Unit M). A 1.22-m diameter pressure pipe connects the collection-division box (unit L) with the effluent box of the grit channel.

1. Compute the head loss in the connecting pipe

Diameter	= 1.22 m
Peak design flow	= 1.321 m³/s

$$\text{Velocity} = \frac{1.321 \ \text{m}^3/\text{s}}{(\pi/4) \times (1.22 \ \text{m})^2} = 1.13 \ \text{m/s}$$

$$\text{Exit loss at unit L } (K = 0.6) = \frac{0.6 \times (1.13 \ \text{m/s})^2}{2 \times 9.81 \ \text{m/s}^2}$$
$$= 0.04 \ \text{m}$$

Minor loss (4–90° elbows in vertical plane, $K = 0.3$)
$$= \frac{4 \times 0.3 \times (1.13 \ \text{m/s})^2}{2 \times 9.81 \ \text{m/s}^2}$$
$$= 0.08 \ \text{m}$$

Friction loss (Hazen–Williams equation, $C = 100$, $L = 10$ m) $= 0.01$ m

Entrance loss at effluent box of grit chamber ($K = 0.5$)
$$= \frac{0.5 \times (1.13 \ \text{m/s})^2}{2 \times 9.81 \ \text{m/s}^2} = 0.03 \ \text{m}$$

Total head loss
$$= 0.04 \ \text{m} + 0.08 \ \text{m} + 0.01 \ \text{m} + 0.03$$
$$= 0.16 \ \text{m}$$

2. Compute the water surface and other elevations in the grit chamber

Surface elevation in the effluent box of grit chamber $= 126.80$ m $+ 0.16$ m

$$= 126.96 \ \text{m}$$

Invert elevation of the effluent box $= 126.96$ m $- 1.5$ m $= 125.46$ m

Various elevations are adapted from Figure 11-5
Water surface elevation in the influent channel of grit removal facility at peak design flow when only one unit is in service = 126.96 m + 1.07 m = 128.03 m

N. Water Surface and Other Elevations in the Pumping Stations (Unit N). The pumping station includes wet well, dry well, valves fittings, and specials, Venturi meter, and transmission line. The head loss calculations were performed in Chapter 9.

The following elevations are taken from Table 9-7, Figure 9-12, and Chapter 10.

Floor elevation of wet and dry pit	= 114.46 m
Low-water surface elevation in wet well	= 116.90 m
High-water surface elevation in wet well	= 118.12 m
Floor elevation of screen chamber at the wet well	= 119.97
Elevation of energy line at the pumps at maximum station capacity (Table 9-7)	= 118.12 m + 12.67 m
	= 130.79 m*

O. Water Surface and Other Elevations in the Bar Screen (Unit O)

Hydraulic profile through the bar screen is given in Figure 8-6.

Floor elevation of the screen chamber = 119.97 m – 0.69 m

= 119.28 m

Water surface elevation below screen at peak design flow when one unit is out of service = 120.53 m

Water surface elevation above screen at peak design flow when one chamber is out of service = 120.56 m (clean screen)

Water surface elevation in the interceptor = 120.54 m

Invert elevation of the interceptor = 119.36 m

P. Water Surface and Other Elevations in the Junction Box (Unit P)

The water surface and invert elevations in the connecting sewer lines and junction box were calculated in Chapter 7 (Figure 7-9).

Water surface elevation in the junction box (unit P) = 120.55 m

Invert elevation in the junction box (unit P) = 119.36 m

Invert elevation of sewer line i at the junction box = 119.93 m

Water surface elevation in sewer line i at the junction box = 120.59 m

All the above elevations in various units are given in Figure 21-1.

21-5 PROBLEMS AND DISCUSSION TOPICS

21-1. A clarifier is added to an aerated lagoon. The crest elevation of the effluent weir in the lagoon is 61.00 m. The water flows over the weir and drops into an outlet box. The water surface elevation in the outlet box is 0.8 m below the weir crest. A 20-cm pipe connects the

---

*A final check of the pumping head can be made at this point. At maximum station capacity, total head loss in the force main should be equal to or slightly less than the difference in elevation of energy line at station common header at maximum station capacity, and the maximum water surface elevation in the grit chamber.

outlet box and the influent well of the circular clarifier. The pipe is an inverted syphon and has two 45° and two 90° bends, and straight length of 68 m. Assume K for 45° and 90° bends, entrance and exit conditions are 0.2, 0.3, 0.15, and 1.0 respectively. $C = 110$. Determine the water surface elevation and total number of 90° V notches in the final clarifier. Each V notch is 20 cm deep and has a freeboard of 5 cm at peak design flow of 0.015 m³/s.

21-2.   The hydraulic profile in Figure 20-1 is prepared at peak design flow when the largest unit is out of service. Prepare hydraulic profile at average design flow of 0.44 m³/s when all treatment units are operating. Consult Problems 7-3, 8-1, 11-6, 12-4, 13-12, 13-13, 14-5, and 15-2, to obtain head losses through different treatment units under average flow condition.

21-3.   As a design engineer you are required to prepare the piping layout and hydraulic profile of a wastewater treatment plant. Discuss various design considerations that are necessary in developing the piping layout and hydraulic profile through the treatment plant.

21-4.   A wastewater treatment plant has the following treatment units (a) bar screen, (b) velocity controlled grit channel, (c) primary sedimentation, (d) high rate, single stage trickling filter, (e) final clarifier, and (f) chlorination facility without the proportional weir. The connecting pipings between the units have head loss of approximately 20 percent of the average head loss across the upstream and downstream units. The outfall structure has the same head loss as the chlorination facility. Assume proper head losses across various treatment units, and justify your assumption with brief reasoning. Draw the hydraulic profile through the treatment plant having the pumping station after the primary treatment facility.

21-5.   A wastewater treatment plant was designed to provide secondary treatment. The head losses in the bar screen, aeration basin, final clarifier, and chlorination facility at peak design flow were 0.06, 0.5, 0.8 and 0.6 m respectively. Assume all units are directly connected by inverted syphon with 2-90° elbows in the connecting pipings between the units. If the diameter, straight length, and C of the connecting pipings are 20 cm, 50 m and 110 respectively, draw the hydraulic profile through the plant. Assume that the outfall pipe has submerged discharge. The high flood level in the receiving water is 100 m. The flow through all connecting piping is 0.02 m³/s.

21-6.   Obtain the hydraulic profile of the wastewater treatment plant is your community. Tabulate the following.
   (a)   Head losses in each treatment unit
   (b)   Head losses in the junction boxes
   (c)   Head losses in the connecting pipings

REFERENCES

1. Chow, V. T., *Open-Channel Hydraulics*, McGraw-Hill Book Co., New York, 1959.
2. Henderson, F. M., *Open Channel Flow*, Macmillan, New York, 1966.
3. Daugherty, R. L., and J. B. Franzini, *Fluid Mechanics with Engineering Applications*, McGraw-Hill Book Co., New York, 1977.
4. Benefield, L. D., J. F. Judkins, and A. D. Parr, *Treatment Plant Hydraulics for Environmental Engineers*, Prentice-Hall, Inc., Englewood Cliffs, N.J., 1984.

22

Instrumentation and Controls

22-1 INTRODUCTION

Manual measurements of process control variables are often infrequent, may lack the desired precision, and often cannot be made within the short span of time for proper operation of the facility. Therefore, the use of instrumentation and control in wastewater treatment facilities is increasing. This trend is not surprising because proper use of instrumentation and automatic controls can reduce labor, chemicals, and energy consumption, and improve treatment process efficiency and reliability.

   The subject of instrumentation and controls is very complex, requiring an extensive background for those involved in design, analysis, and selection of instrumentation for wastewater treatment facilities. The purpose of this chapter is to present information on instrumentation and control devices that can be used by the wastewater treatment plant design engineers, the plant operators, and maintenance personnel. Therefore, the discussion is divided into two parts: (1) to present the basic types of systems that have general applications in the wastewater area and (2) to list the instrumentation and control devices selected for various treatment processes presented in the Design Example.

22-2 INSTRUMENTATION AND CONTROL SYSTEMS

A process control system utilizes (1) process variables and (2) associated controls. The process variables are measured by the sensor of the control systems. The measured value is transmitted, displayed, or recorded to guide the operator in taking the proper corrective process adjustments. In other automatic control systems, the process variable signals are compared with the value of the variable against a preset reference and then the controller implements the selected actions for the desired results. The control variables and the associated controls are discussed below. The typical control system components are illustrated in Figure 22-1.[1]

22-2-1 Control Variables

In a wastewater treatment plant many variables are measured and controlled. These variables fall into three categories: physical, chemical, and biological. Common examples of these variables are listed below:

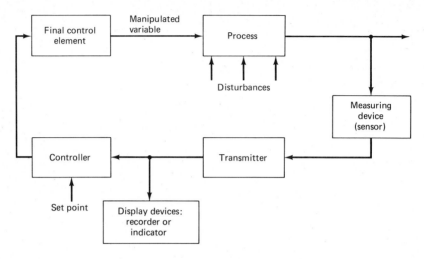

Figure 22-1   Typical Control-System Components. [From Ref. 1.]

Physical: flow, pressure, level, temperature, etc.
Chemical: pH, turbidity, specific conductance, dissolved oxygen, etc.
Biological: oxygen consumption rate, TOC reduction rate, sludge growth rate, etc.

22-2-2 Associated Controls

The automatic control system is made up of six components:[1,2]

1. The measurement section to detect the change in the variables
2. Signal-transmitting device
3. Data display or readout
4. Control loop
5. Controller
6. Computer and central control room

Each of these components is discussed below.

Measurement Section or Sensing Devices.   The sensing devices include instruments that sense, measure, or compute the process variables. The sensing device may be on-line or off-line, continuous or intermittent. Common on-line process measurements devices and their applications are summarized in Table 22-1.

Signal-Transmitting Devices.   The purpose of the signal-transmitting device is to transmit the process variable signals from the sensing instrument to the readout device or a controller. The transmission may be accomplished mechanically, pneumatically, or electrically. Each type is discussed below.

TABLE 22-1 Common On-Line Process Measurement Devices and Their Applications in Wastewater Treatment

Measured Variables	Primary Device	Measured Signal	Common Application
Flow[a]	Venturimeter	Differential pressure	Gas, liquids
	Flow nozzle meter	Differential pressure	Gas, liquids
	Orifice meter	Differential pressure	Gas, liquids
	Electromagnetic meter	Magnetic field and voltage change	Liquids, sludges
	Turbine meter	Propeller rotation	Clean liquid
	Acoustic meter	Sound waves	Liquids, sludges
	Parshall flume	Differential elevation of water surface	Liquids
	Palmer-Bowlus flume	Differential elevation of water surface	Liquids
	Weirs	Head over weir	Clean liquids
Pressure	Liquid-to-air diaphragm	Balance pressure across a metal diaphragm	Pressure 0–200 kN/m^2
	Strain gauge	Dimensional change in sensor	Pressure 0–350,000 kN/m^2
	Bellows	Displacement of mechanical linkage connected to the indicator	Pressure 0–20,000 kN/m^2
	Bourdon tube	Uncurling motion of noncircular cross sectional area of a curved tube	Pressure 0–35,000 kN/m^2
Liquid level	Float	Movement of a float riding on surface of liquid	Liquid head 0–11 m
	Bubbler tube	Measurement of back pressure in the tube bubbling regulated air at slightly higher pressure than the static head outside	Liquid head 0–56 m
	Diaphragm bulb	Pressure change in air on one side of the diaphragm caused by the liquid pressure on the other side of the diaphragm	Liquid head 0–15 m

TABLE 22-1 Common On-Line Process Measurement Devices and Their Applications in Wastewater Treatment

Measured Variables	Primary Device	Measured Signal	Common Application
Sludge level	Photocell	Detection of light in a probe by a photocell across sludge blanket	Primary sedimentation Final clarifier Gravity thickener
	Ultrasonics	Detection of the ultrasonic signal transmitted between two transducers	Primary sedimentation Final clarifier Gravity thickener
Temperature	Thermocouple	Current flow in a circuit made of two different metals	Anaerobic digester, hot-water boiler
	Thermal bulb	Absolute pressure of a confined gas is proportional to the absolute temperature	Sludge lines, water lines
	Resistance temperature detector	Change in electrical resistance of temperature-sensitive element	Bearing and winding temperatures of electrical machinery, anaerobic digesters, hot-water boilers
Speed	Tachometer, (generator or drag cup type)	Voltage, current	Variable-speed pump, blower, or mixer
Weight	Weight beam, hydraulic load cell or strain gauge	Lever mechanism or spring, pressure transmitted across diaphragm, dimensional change in sensor	Chemicals
Density	Gamma radiation	Absorption of gamma rays by the liquid between radiation source and the detector	Mixed liquor suspended solids concentration; returned, thickened and digested sludge solids

Parameter	Sensor	Principle	Location
	Ultrasonic sensor	Loss of ultrasonic signal by the liquid between the ultrasonic transmitter and receiver	Mixed liquor suspended solids concentration; returned, thickened and digested sludge solids
pH	Selective ion electrode	Voltage produced by hydrogen ion activity	Influent, chemical solution, anaerobic digester, dewatering, effluent
Oxidation-reduction potential	Electrode	Change in potential due to oxidation or reduction	Influent, maintenance of proper DO in aeration basin, anaerobic digester
Total dissolved salts	Conductivity	Flow of electrical current across the solution	Influent, effluent
Dissolved oxygen	Membrane electrode	Electric current due to reduction of molecular oxygen	Influent, aeration basin, plant effluent
Total organic carbon	Carbon analyzer	CO_2 produced from combustion of sample	Influent to the plant, influent to aeration basin, plant effluent
Chlorine residual	Sensor	Electrical output	Chlorine contact tank, plant effluent
Gases O_2, NH_3, CT_2, H_2S, CH_4	Sensors	Various types of sensor modules utilize electrical impulses	Detection of hazardous condition in the room or around covered treatment units
Oxygen uptake rate	Respirometers using sensor	Decrease in DO level with respect to time	Aeration basin
Anaerobic biological condition	Sensor, combustion	CO_2 and CH_4 production rate	Anaerobic sludge digester

Source: Adapted in part from Refs. 2 and 4.

[a]Additional details may be found in Chapter 10. psi × 6.895 = kN/m^2.

Mechanical Transmission.   The mechanical transmission is done by drive of a pen or indicator, or by a float or cable. This method is generally limited to on-site display or location of controller.

Pneumatic Transmission.   The pneumatic transmission system consists of a detector and an amplifier. The detector is a flapper-nozzle unit. Regulated input air is fed to the nozzle through a reducing tube. The pressure change is transmitted as the flapper is moved away or toward the nozzle. Small changes in back pressure at the nozzle become proportional to the movement of the flapper. The pressure change is amplified and transmitted to the receiver or controller.

The advantage of pneumatic transmissions over electric transmissions include no electrical hazard; less effect of temperature and humidity; and no freezing problem. Also, they are reliable, less complicated, and easy to maintain. The limitations of pneumatic transmissions, however, are the signal lag in long tubing, the relatively short distance necessary between the detector and the controller (up to 300 m), the requirement of clean dry air and air leakage problems. Detailed discussion on pneumatic transmission may be found in Refs. 5 and 6.

Electric Transmission.   The electric transmission of signals is achieved by voltage and current, pulse duration, or tone. In voltage and current transmission the signals are transmitted by milliamp direct current (maDC) or by voltage signals.

In pulse duration or time pulse transmission, the length of time a voltage is transmitted is in proportion to the measured data. In tone transmission, standard telephone lines are normally used to allow signals to be transmitted. The signals are transmitted by turning the transmitter on and off, or by slightly changing the frequency, or pitch, of the electronic tone.[2]

Recent advances in reliable radio and microwave equipment have encouraged the use of radio/microwave transmission where signals are transmitted over an assigned frequency per FCC regulations.* The radio/microwave transmission is particularly suited where the data gathering points are scattered over a large area and where telephone lines are not available or are too expensive. Although radio/microwave transmission is expensive at present, its use will continue to grow, particularly for large systems.

The electronic control systems are rapidly gaining popularity for a number of reasons. The basic advantages are[5] (1) the electrical signals operate over great distances without contributing time lags, (2) the electrical signals can easily be made compatible with a digital computer, (3) the electronic units can easily handle multiple-signal inputs, (4) intrinsic safety techniques have virtually eliminated electrical hazards, and (5) electronic devices can take less space, are less expensive to install, can handle almost all process instruments, and can be essentially maintenance-free.

Data Display Readout.   The transmitted information is displayed at a convenient location in a manner usable by the operating personnel. The most common type of readout devices are indicators, recorders, and totalizers. These devices may be driven mechanically from the sensing devices, or from the pneumatic or electric signals. The receiving

*Federal Communications Commission.

instruments may display the movement, pressure charge, or current change signals directly, or it may be servo-operated.[2] General types of pointers and scales commonly used in wastewater treatment works are shown in Figures 22-2 and 22-3.

Control Loop. *Loop* is defined by the Instrument Society of America as follows: "A combination of one or more interconnected instruments to measure or control a process or both."[7] The control system is a series of related or unrelated control loops for detection of variables that manipulates the process for desired results. The automatic control system is made up of three parts:[1]

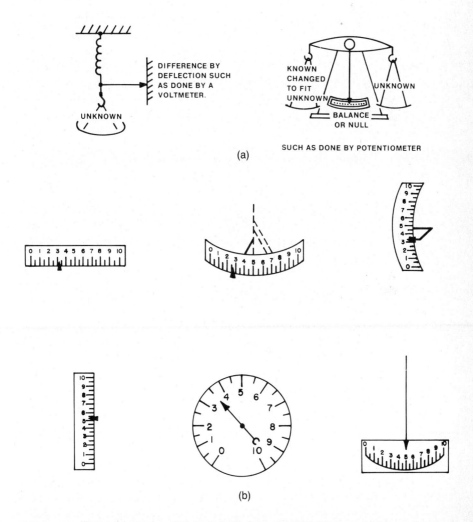

Figure 22-2   Different Types of Pointer Indicators. (a) Weighing pan balance. (b) Types of movable-pointer indicators, straight scale, horizontal arc scale, vertical arc scale, vertical straight scale, circular scale, and segmental scale. [From Ref. 2. Used with permission of Water Pollution Control Federation.]

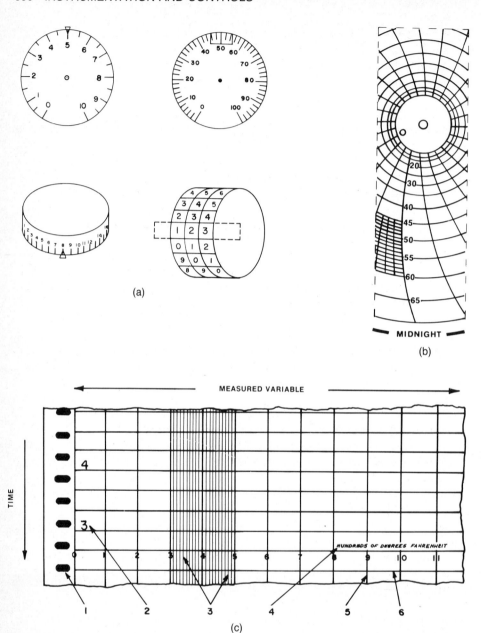

(a)

(b)

(c)

Figure 22-3   Different Types of Movable-Scale Indicators and Circular and Strip Charts. (a) Movable-scale indicators. (b) Section of a 30 cm (12-in) circular chart. (c) Section of a 28 cm (11 in) strip chart: (1) humidity compensating holes, (2) major time markings, (3) minor graduations, (4) identifying calibration, (5) major graduations, and (6) minor time markings. [From Ref. 2. Used with permission of Water Pollution Control Federation.]

1. A measurement section to detect the change in the variable
2. A reference source to compare the value of the variable with a preset reference, and
3. A mechanism to manipulate the variable until the measured signal reaches the reference value

The control circuit or loop is composed of control elements and comes in two general types: closed or open loop. A closed loop or "feed back loop" contains several automatic control units and the process is connected so as to provide a signal path that includes a forward path, a feed back path, and a summing point. The controlled variable is consistently measured, and if it deviates from the preset value, corrective action is applied to the final element to return the controlled variable to the desired value. An example of closed loop is an automatic chlorinator discussed in Chapter 14. An open loop or feed forward loop is without a feedback system. The addition of chemical solution in proportion to flow is a good example of an open loop.

The control system consists of a series of individual control loops, each of which is relatively simple. However, these loops may be combined into independent, interdependent, or dependent open and closed control loops to constitute an elaborate and complex system. Additional discussion of control systems may be found in Refs. 2, 5–7.

Controller. The device that automatically performs the control function in regulating the control variable is a controller. It may be of different type depending upon the functional needs. Following is a brief description of some common types of controller functions. Detailed discussion may be found in Refs. 2, 5, and 6.

Two-Position or On-Off. The controller may be fully on or off. An example of such a system is the starting and stopping of a pump by a level controller.

Proportional-Only Control. The controller produces a variable signal in proportion to the measurement. Example of such controller is a variable-speed motor on a pumping unit actuated in proportion to the change in liquid level in the wet well. Such a system has been discussed in Chapter 9.

Proportional Speed Floating Control. The controller output is proportional to the process deviation from a set point or the error. The device thus performs the integration. An example of such a control is when a valve opens or closes at a speed proportional to the difference between the control setting and the actual flow rate.

Sample Data Control. The control action is delayed to provide a time lag for the effect of a previously applied correction to be sensed. An excellent example is a chlorination facility, where chlorine dosage is controlled by the residual measured after the contact period. Such a device is discussed in Chapter 14.

Timer Control. The timer control device operates on a set time schedule. An example of such a control system is sludge pumping at preset time intervals (Chapters 12, 13 and 16–18).

Two- and Three-Mode Control. Two or three modes of control may be combined with the basic controls to achieve variable and complex control operations.

Computers and Control Room

Computers. Most of the medium-size wastewater treatment plants have data acquisition systems for data logging.[1] These systems accumulate, format, record, and display large quantities of data effectively. Modern data acquisition systems can provide accurate, impartial documentation of all process measurements and operator actions. Computers can also provide a maintenance schedule to the plant operator based on actual operating time of a particular piece of equipment, such as pump or motor.

Although computer process control systems are in common use in many industries, they have not found wide application in small to medium size wastewater treatment works. These process control computers accumulate the process data, display them, and also produce the necessary process corrections for optimum operation. These corrections may include control of chemical solutions, air supply, scheduling of pumps and blowers, and the like.

Central Control Room. Central control is used to organize the plant operation in such a manner that all treatment information, important events, and alarms are displayed, indicated, and recorded at a centralized location. This location is called the central control room. In addition, most central facilities practice automatic or manual actuation of final control elements. Central control rooms reduce the number of personnel required to operate a large treatment facility.

22-3 MANUFACTURERS AND EQUIPMENT SUPPLIERS OF INSTRUMENTATION AND CONTROL SYSTEMS

The names and addresses of many manufacturers and equipment suppliers of instrumentation and control systems for wastewater treatment facilities are provided in Appendix D. Important considerations for selection of equipment suppliers and the responsibility of the design engineer in providing data and making selection decisions are discussed in Sec. 2-10.

22-4 INFORMATION CHECKLIST FOR DESIGN AND SELECTION OF INSTRUMENTATION AND CONTROL SYSTEMS

Specific uses of instrumentation and the extent of automation provided at a wastewater treatment facility is a decision that must be made by the design engineer and management personnel. The extent of instrumentation may depend on the size of the plant, number of operating personnel, and available construction funds. The following information must be developed and decisions made for the selection of proper instrumentation and control systems.

1. Hours of manned operation daily.
2. Availability of trained instrumentation maintenance personnel.

3. Based on the size and complexity of the plant, should there be no control, manual control, supervisory control, automatic control, or computer control?
4. If automatic controls are selected, decisions should be made concerning the number of control elements and loops, accuracy needed, operating range, response time of process variables, and frequency of operator input.
5. The economics of instrumentation and controls must be compared with the savings achieved by plant automation.

22-5 DESIGN EXAMPLE

22-5-1 Design Criteria Used

The basic design criteria for the instrumentation and controls selected for many treatment processes have been discussed in respective chapters. The criteria are based on providing an economical and reliable operation of various treatment facilities. These include (1) operational range, (2) accuracy desired, (3) operational flexibility and reliability, (4) space requirements, and (5) future expansions. The selected process systems incorporate proven industrial control equipment and techniques of contemporary design.

22-5-2 Selected Process Control Systems

The selected process control system will consist of the following: (1) It shall contain automatic and continuous analyzers of sensing devices at strategic locations throughout the plant for measurement of process control variables. Where a particular process control variable cannot be monitored, a systematic grab-sampling device shall be instituted for piping samples directly into the laboratory. (2) The automatic control systems will be installed to offer a high degree of accuracy and permit remote control of process equipment. The instrumentation will sense flows, absolute pressure differential, and temperature. The majority of the control equipment will be solid state electronic. (3) The central control room shall contain analog indicating controllers, annunciator alarms panels; motor control and valve operator switches; trend recorders, indicators, and totalizer; converters; transducers; operator's control console; master communication equipment; telemetry transmitters and receivers; auxillary power supply for emergency use and data-logging systems; and so on. The heating and air conditioning for this room will be designed to maintain a uniform temperature and to provide a clean atmosphere with a stable relative humidity. (4) Various control and analyzer panels shall be located throughout the plant to provide the data needed to optimize the plant efficiency.

Each of the process unit and major equipment shall be monitored and, to some extent, controlled from the central control room. The general descriptions of various functions to be monitored for each treatment process are summarized in Table 22-2. A total of 10 simplified loop diagrams are shown in Figure 22-4. These loop diagrams show the operational principles of (1) differential head control, (2) differential pressure control for flow measurement, (3) bubbler tube control in wet well for pump operation, (4) air flow monitoring in grit chamber, (5) scum pump control in primary sedimentation basin, (6) dissolved oxygen monitoring in aeration basin, (7) aeration blower control, (8) return sludge control, (9) waste sludge control, and (10) postchlorination control. Controls for

TABLE 22-2 Brief Description of Various Functions Monitored for Different Treatment Processes in the Design Example

Treatment Unit	Sensing Device	Purpose	Display		Description	Ref.
			Local	Central		
Bar screen	Limit switch	Monitor gate position		x	Display green light	Secs. 8-3-3, and 8-8-6
	Time clock	Operation of rake		x	Display green light	
	High-level override by float in stilling well (excessive head loss)	Emergency operation of rake		x	Display red light	
Pumping station	Bubbler system in wet well	Control pumping rate	x		Display of liquid level in wet well	Sec. 9-11-2, and Table 9-9
	Conductivity switch or float	Liquid level in sump of dry well		x	Trip alarm	
	Limit switches	Monitor gate position		x	Display green light	
	Pressure switch	Monitor seal water pressure for pump protection		x	Display red light	
	Flow switch	Monitor seal and water flow to pump for pump protection		x	Display red light	
	Pressure indicator	Monitor the discharge pressure of the pumps	x		Pressure gauges	
	Motor monitors (contacts provided in motor control center)	Monitor on-off condition of each major motor, and motor malfunction	x	x	Display green light	
			x	x	Display red light	

Process	Instrument	Function			Readout/Action	Reference
Flow measurement	Sampling and analysis	Sensors for pH, H_2S, CH_4, DO; Sampler to deliver water sample in the lab	x		Continuous recording on strip chart; Routine chemical analyses. TSS, BOD, TOC, P, N	
	Main circuit breaker, auxiliary contact	Monitor power outage	x		Display red light (standby power required)	Sec. 10-9
	Solids bearing Venturi tube in force main	Monitor influent flow to the plant	x		Display, record, and totalize continuously	
Aerated grit chamber	Orifice plate with a square root extractor and display	Measure air flow to each chamber	x	x	Meter display record and totalize flow continuously	Secs. 11-11-3
	Pressure indicators	Monitoring of discharge pressure of air blowers		x	Pressure gauges	
	Motor monitors (contacts provided in motor control center)	Monitor on-off condition of motors in blowers	x	x	Display green light	
	Torque monitors	Monitor spiral conveyor and bucket elevator	x	x	Display red light	
Primary sedimentation	Level detector	Control sludge and scum pumps, and air agitator in scum box	x	x	Display green light	Secs. 12-3-9 and 12-3-10
	Pressure indicator	Monitor discharge from each sludge pump	x	x	Display red light	Sec. 12-6-2, step F
	Magnetic flow meter	Sludge-pumping rate	x	x	Display, record, and totalize continuously	Sec. 12-8-5
	Motor monitor (contacts provided in motor control center)	Monitor on-off condition of each motor	x	x	Display green light	Secs. 12-8-3, 12-8-4, and 12-8-7

TABLE 22-2 Brief Description of Various Functions Monitored for Different Treatment Processes in the Design Example

Treatment Unit	Sensing Device	Purpose	Display		Description	Ref.
			Local	Central		
	Torque monitor	Monitor motor torque for sludge-raking mechanism		x	Alarm signal	Secs. 12-8-3, and 12-8-4
Aeration basin and final clarifiers	Dissolved oxygen measurement	Control aeration rate		x	Display and continuous recording	Sec. 13-7-4, steps E and F
	Flow meter, differential pressure type	Monitor air supply to each aeration basin	x		Meter display record and totalize flow continuously	Sec. 13-7-4, step G2
	Ultrasonic sludge blanket in final clarifier	Monitor the sludge level and actuate sludge return pumps		x	Display pump operation	Sec. 13-7-4, step H
	Magnetic flow meter	Control return sludge flow	x		Display, record and totalize flow continuously	Sec. 13-7-5, step H
	Suspended solids analyzer	Measure MLSS concentration in collection box, and actuate waste sludge pumps	x		Display and record TSS, record and totalize flow	Sec. 13-7-4, step H
	Pressure indicator	Monitor the discharge pressure of the pumps	x		Pressure gauges	
	Motor monitors (contacts provided in motor control) torque monitor	Monitor on-off condition of each motor		x	Display green light	

			Alarm signal		
		Monitor motor torque for sludge-raking mechanism	x		
Chlorine contact	Parshall flume	Measure flow upstream of chlorination facility to control chlorine dosage	x	Display, record and totalize flow continuously	Sec. 14-6-2, step F
	Chlorine residual at mid-length of contact basin	Measure and control chlorine dosage	x	Display, record and totalize flow continuously	Secs. 14-8-3–14-8-5
	Chlorine leak detection	Detection of free chlorine in storage and dispensing facility	x	Alarm when dangerous levels are reached	Secs. 14-8-6, 14-8-8 and 14-8-10
	Pressure indicators in chlorine tanks	Monitor high and low pressures	x	Actuate alarm	
Gravity thickener	Magnetic flow meter	Flow measurement of feed sludge	x	Display, record and totalize flow	Sec. 16-7-2, step C6, and Figure 16-5
	Venturi meter	Flow measurement of dilution water	x	Display, record and totalize flow	Sec. 16-7-2, step A2
	Ultrasonic sludge blanket detector	Monitor the sludge blanket level in the gravity thickener	x	Display sludge blanket level	Sec. 16-7-2, step E
	Ultrasonic density meter	Monitor solids concentration in feed and thickened sludge	x	Display and record	Sec. 16-7-2, step E
	Pressure indicator, motor monitor, torque monitor	Similar to those for primary and final clarifiers	x	Similar to those for primary and final clarifiers	
Anaerobic digestion	Magnetic flow meter	Flow measurement of feed sludge	x	Display, record, and totalize flow	Sec. 17-7-2, step E, and Figure 17-7

TABLE 22-2 Brief Description of Various Functions Monitored for Different Treatment Processes in the Design Example

Treatment Unit	Sensing Device	Purpose	Display		Description	Ref.
			Local	Central		
	Ultrasonic density meter	Monitor solids concentration in feed and digested sludge, and mixing	x		Similar to those for gravity thickeners	Sec. 17-9-2
	Thermocouple, selective ion electrodes	Monitor temperature, pH, and ORP of digester content and the recirculating sludge		x	Display and record	Sec. 17-8-3
	Level indicator	Monitor liquid level in the digester and level of floating cover	x		Indicator	Sec. 17-9-1
	Pressure and temperature	Control hot water heating system	x		Display temperature, pump operation	Sec. 17-9-3
	Orifice meter	Measurement of digester gas flow		x	Display, record and totalize	Sec. 17-7-2, step I
	Chromatograph and calorimeter	Determination of gas composition and calorific value		x	Tabulation of result and data logging	
	Venturi meter	Measurement of supernatant flow		x	Record and totalize	Sec. 17-7-2, step E4
	Pressure switch, flow switch pressure indicator, motor monitor	Monitor pump, motor, compressor as indicated above	x	x	Display red light	

	Sensing device	Function			Display	Reference
Filter press	Bourdon with cylindrical seal	Measurement of pressure exerted by filter presses	x		Dial display	Sec. 18-6-2, step C
	Magnetic flow meter	Monitor the flow of feed sludge and filling rate		x	Record and totalize	Sec. 18-6-2, step C, and Sec. 18-8-2
	Ultrasonic density meter	Monitor solids concentration in feed sludge		x	Record and totalize	Sec. 18-6-1
	Static scale	Measure weight of sludge cake	x		Record and log data	Sec. 18-6-2, step D
	Ohm meter, lab test	Determination of moisture content of the sludge cake	x		Record and log data	Sec. 18-6-2, step D
	Tape and float	Measure level of lime and polymer solution in storage tank	x		Record and log data	Sec. 18-6-2, step C
	Magnetic flow meter	Measure flow of lime and polymer solution	x		Record and log data	Sec. 18-6-2, step C
	Pressure switch, flow switch, pressure indicator, motor monitor	Monitor pump, motor, compressor, as indicated above	x		Display as indicated above	

many other applications, such as thickened sludge withdrawal; sludge recirculation, mixing, and heating; chemical feed; etc., can be easily developed from the examples of 10 loops shown in Figure 22-4. It should, however, be mentioned that these loops are grossly simplified. Many complex loop diagrams are generally needed in the design of wastewater treatment facilities. The instrumentation engineer should develop such diagrams in consultation with the manufacturers of various instrumentation and control devices, and the manufacturers of pollution control equipment.

22-6 PROBLEMS AND DISCUSSION TOPICS

22-1. A Venturi meter is used for continuous flow recording of industrial wastewater. The differential pressure is transmitted and recorded on a chart. Using Eq. (10-6), calculate the hourly flows, and totalize the flow. Use the following data

Time:	12 midnight	1	2	3	4	5	6	7	8	9	10	11
h,m H_2O:	0.20	0.21	0.25	0.26	0.30	0.40	0.50	0.70	1.20	1.80	2.20	2.30

Time:	12 noon	1	2	3	4	5	6	7	8	9	10	11	12 midnight
h,m H_2O:	2.00	1.80	1.60	2.10	2.80	3.00	2.50	2.00	1.80	0.80	0.40	0.30	0.20

22-2. An industrial plant operates from 8 am to 5 pm. The hourly wastewater flow and TOC concentrations are given below. Totalize the TOC results and express in kg per operating day

Time:	7	8	9	10	11	12	1	2	3	4	5	6	7
Flow, m^3/s	0	0.02	0.05	0.10	0.15	0.20	0.21	0.22	0.25	0.15	0.10	0.02	0
TOC, mg/l	0	80	120	180	300	350	300	280	250	200	150	80	0

22-3. Using the typical control-system components given in Figure 22-1, list various options available at each control-system components for maintaining constant temperature in an anaerobic digester. The sludge is recirculated through external heat exchanger for heating purposes.

22-4. Match the primary flow measuring devices with their applications.

Orifice meter   (a) sludge
Flow nozzle   (b) clean liquid
Parshall flume   (c) raw wastewater
Acoustic meter
Electromagnetic meter

22-5. Write measured signals for the following variables.
   (a) conductivity
   (b) pH
   (c) density
   (d) oxidation reduction potential
   (e) dissolved oxygen
   (f) temperature
   (g) pressure

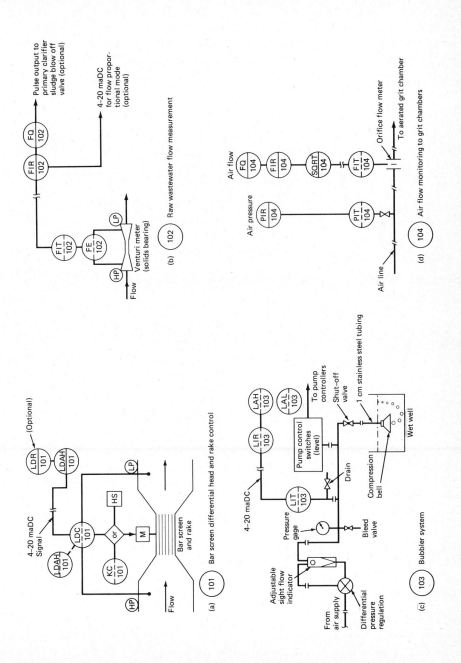

Figure 22-4   Simplified Control Loop Diagrams for Different Treatment Processes Designed in the Design Example.

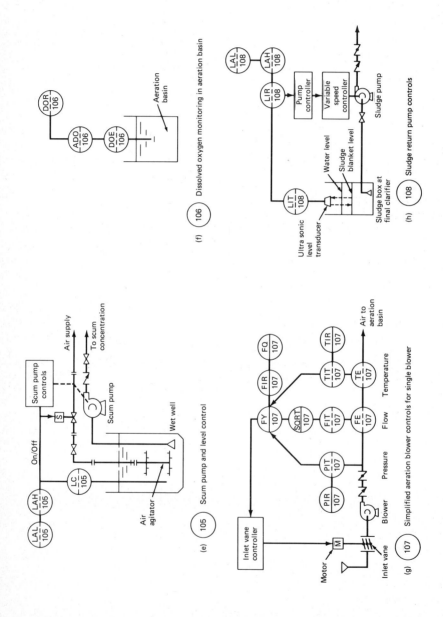

Figure 22-4   Simplified Control Loop Diagrams for Different Treatment Processes Designed in the Design Example.

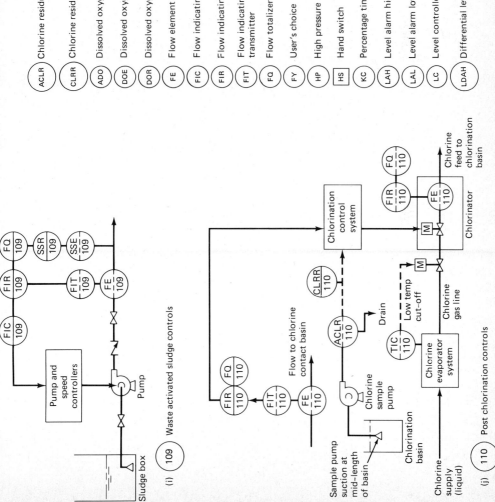

Symbol	Description	Symbol	Description
ACLR	Chlorine residual analyzer	LDC	Differential level control
CLRR	Chlorine residual recorder	LDR	Differential level recorder
ADO	Dissolved oxygen analyzer	LIR	Level indicating recorder
DOE	Dissolved oxygen element	L:T	Level indicating transmitter with P/I (pneumatic/current transducer)
DOR	Dissolved oxygen recorder	LP	Low pressure
FE	Flow element	M	Motor and controller
FIC	Flow indicating controller	PIR	Pressure indicating recorder
FIR	Flow indicating recorder	PIT	Pressure indicating transmitter
FIT	Flow indicating transmitter	S	Solenoid
FQ	Flow totalizer	SQRT	Squire root extractor
FY	User's choice	SSE	Suspended solids element
HP	High pressure	SSR	Suspended solids recorder
HS	Hand switch	TE	Temperature element
KC	Percentage timer	TIC	Temperature indicating controller
LAH	Level alarm high	TIR	Temperature indicating recorder
LAL	Level alarm low	TIT	Temperature indicating transmitter
LC	Level controller	⊖	Solid line denotes instrument mounted in central control panel
LDAH	Differential level high alarm	(- -)	Dashed line denotes instrument mounted in the field (local)

(i) 109   Waste activated sludge controls

(j) 110   Post chlorination controls

22-6. Flow scheme of a biological treatment process is given in Figure 13-11. Draw an integrated process mechanical and instrumentation diagram for aeration basin and final clarifier. Assume that there are single units of aeration basin, final clarifier, return sludge pump, waste sludge pump, blower, and air meter. Use the simplified control loop diagrams for different processes given in Figure 22-4.

22-7. Discuss various types of signal-transmitting devices. Give advantages and disadvantages, and applications of each type.

22-8. What is the basic difference between feed back and feed forward automatic control loops? Give three examples of each in wastewater treatment plant.

REFERENCES

1. Molvar, Allen E., et al., *Instrumentation and Automation Experiences in Wastewater Treatment Facilities,* U.S. Environmental Protection Agency, Municipal Environmental Research Laboratory, Cincinnati, Ohio, EPA-600/2-76-198, October 1976.

2. Committee on Instrumentation, *Instrumentation in Wastewater Treatment Plants,* Water Pollution Control Federation, Washington, D.C., MOP 21, 1978.

3. Joint Committee of the WPCF and ASCE, *Wastewater Treatment Plant Design,* MOP/8 Water Pollution Control Federation, 1977.

4. U.S. Environmental Protection Agency, *Process Design Manual for Sludge Disposal,* EPA 625/1-79-011, September 1979.

5. Anderson, N. A., *Instrumentation for Process Measurement and Control,* Chilton Company, Radnor, Pa., 1980.

6. Considine, D. M., ed., *Encyclopedia of Instrumentation and Control,* McGraw-Hill Book Co., New York, 1971.

7. Instrument Society of America, *Standard for Instrument Loop Diagram,* S5.4, 1976.

23

Design Summary

In this book the step-by-step design procedure for a medium-size wastewater treatment facility has been developed. One Design Example has been carried through 17 chapters (Chapters 6–23) to present the theory, design procedure, operation and maintenance, and equipment specifications for various components of the wastewater treatment facility. The purpose of this chapter is to consolidate the basic design data that have been developed in the Design Example. The design data are summarized in Table 23-1. Reference has been made to various sections in which the details may be found.

TABLE 23-1   Summary of Basic Design Data and Dimensions of the Wastewater Treatment Facility Designed in Chapters 6–21

Item	Design Value or Description	Ref.
A. *Existing Wastewater Treatment Facility*		Chap. 6
The existing facility includes three small plants: (1) trickling filter on east side, (2) stabilization pond on the west side, and (3) aerated lagoon on the northern end of town.		Table 6-4
Current sewered population served	46,750	Table 6-6
Current unsewered population	1,250	Table 6-6
Average flow to trickling filter, ℓ/s	67	Table 6-6
Average flow to stabilization pond, ℓ/s	95	Table 6-6
Average flow to aerated lagoon, ℓ/s	80	Table 6-6
Total flow treated, ℓ/s	242	
B. *Proposed Facility*		
The proposed facility includes construction of two gravity intercepting sewers to divert flows from trickling filter plant and stabilization pond area to the aerated lagoon site. Construct a new treatment plant at this location to treat the combined flows. The treatment flow scheme includes bar screen, pumping station, grit removal, primary sedimentation, activated		Sec. 6-10

sludge, disinfection, and river outfall. The sludge
treatment includes gravity thickening of combined
sludge, anaerobic digestion, and dewatering by filter
presses

Initial year	1985	Table 6-9
Design year	2000	Table 6-9
Design flows		Table 6-9
Average wastewater flow, ℓpcd	475	Table 6-9
Average flow, ℓ/s	440	Table 6-9
Peak flow, ℓ/s	1321	Table 6-9
Minimum flow, ℓ/s	220	Table 6-9
BOD_5, mg/ℓ	250	Table 6-9
TSS, mg/ℓ	260	Table 6-9
TS, mg/ℓ	910	Table 6-9
pH	7.2	Table 6-9
Total P, mg/ℓ	9.0	Table 6-9
Total N, mg/ℓ	48.0	Table 6-9

C. *Intercepting Sewer* — Chap. 7

Diversion sewer from trickling filter plant [line ii]		Table 7-5
Peak flow, m^3/s	0.385	Table 7-6
Diam, m	0.76	Table 7-6
Diversion sewer from stabilization pond area [line i]		Table 7-5
Peak flow, m^3/s	0.590	Table 7-6
Diam, m	0.91	Table 7-6
Existing intercepting sewer from central part of town [line iii]		Table 7-5
Peak flow, m^3/s	0.331	Table 7-6
Diam, m	0.76	Table 7-6
Final sewer to the plant [line iv]		Table 7-5
Peak flow, m^3/s	1.321	Table 7-6
Diam, m	1.53	Table 7-6
Bypass sewer to storage basin [line v]		Table 7-5
Peak flow, m^3/s	1.321	Table 7-6
Diam, m	1.22	Table 7-6

Junction box. Three incoming sewers join and one sewer carries the flow to the plant. Manually operated
stop gates are provided to close the line and divert
the flow into the relief sewer in case of power outages.   Sec. 7-5-2
                                                         and Figs.
                                                         7-8 and 7-9

D. *Bar Screen* — Chap. 8

Number of mechanically cleaned bar screens	2	Sec. 8-6-1
Design capacity of each screen, m^3/s	1.321	Sec. 8-6-1
Width of screen chamber, m	1.74	Sec. 8-6-2, step B1

Max. depth above floor, m	6.45	Fig.8-7
Length of screen chamber, m	11.5	Fig. 8-7
Slope of screen to horizontal, °	75	Fig. 8-7
Total number of clear openings	50	Fig. 8-5
Clear openings between bars, 5 mm	25	Fig. 8-5
Number of bars, each bar 10 mm × 50 mm (deep)	49	Fig. 8-5
Depth of flow in channel at peak design flow, m	1.28	Table 8-4
Velocity through rack at peak design flow, m/s	0.83	Table 8-4
Head loss clean bars, m	0.03	Table 8-4
Head loss 50 percent clogging, m	0.15	Table 8-4
Average quantity of screenings, m^3/d	0.76	Sec. 8-6-2, step E
Rake cleaning	Front-cleaned	Sec. 8-8-3
Control system	Time cycle and high-level over-ride	Sec. 8-8-6

E. *Pumping Station* — Chap. 9

Dry-well dimensions, m × m	15.24 × 7.62	Fig. 9-12
Wet-well dimensions, m × m	15.24 × 4.57	Fig. 9-12
Depth of pumping station above floor, m	11.38	Fig. 9-12
Total number of identical pumps, each vertical shafting, dry-pit, mixed flow, centrifugal pump with variable-speed drive.	5	Sec. 9-10-1 and Fig. 9-13
These pumps are arranged in parallel and discharge in a common header 92 cm diam. All pumps have 36 cm (14 in.) suction and discharge connections		
Number of standby units	1	Sec. 9-9-1
Max. static head (min. wet-well elev.), m	11.13	Fig. 9-9 and Fig. 21-1
Min. static head (max. wet-well elev.), m	9.91	Fig. 9-9 and Fig. 21-1
Min. station head and max. station capacity when four pumps in operation (max. wet-well level), m and m^3/s	12.67, 1.560	Table 9-7
Max. station head and min. station capacity when four pumps in operation (min. wet-well level), m and m^3/s	13.72, 1.425	Table 9-7
Motor power, each variable speed, kW	93	Sec. 9-9-2, step C
Motor efficiency range, percent	81.8–83.0	Table 9-8
Method of pump control	Liquid level bubbler	Table 9-9

F. *Flow Measurement*   Chap. 10

Number of solids bearing Venturi meter in the 92 cm   1   Sec. 10-7-1
    diam force main

Flow range, m^3/s 1.321–0.152   Sec. 10-7-1

Head loss at max. and min. flow ranges (1.321 and   0.45 &   Sec. 10-7-2
    0.152 m^3/s), m   0.006   Step C

Sensors, hydraulic type (pure null balance principle) uti-   Sec. 10-9-2
    lizing pure static liquid pressure on one side of a di-
    aphragm to oppose and sense the differential pres-
    sure of the solids bearing fluid at the inlet and throat
    sections. The output signal recorded on a strip chart

G. Aerated Grit Chambers   Chap. 11

Number of units   2   Sec. 11-9-1

Length, m   13.0   Sec. 11-9-2,
                                                         step A1

Width, m   3.5   Sec.11-9-2,
                                                        step A1

Average water depth, m   3.65   Sec. 11-9-2,
                                                        step A1

Freeboard, m   0.8   Sec. 11-9-2,
                                                       step A1

Detention time at peak design flow (1.321 m^3/s) when   4.2   Sec. 11-9-2,
    both units are in operation, min   step A2

Influent structure consists of a 1-m-wide submerged   Fig. 11-4
    channel with 1 m × 1 m orifice with sluice gates to
    divert the flow into one or both chambers

Effluent weir is a 2.5-m freefalling rectangular weir   Fig. 11-4

Air supply, ℓ/s per m length coarse bubble, swing type   7.8   Sec. 11-9-2,
    diffusers provided on one side of the chamber for   step B1
    spiral roll action

Number of blowers   2   Sec. 11-9-2,
                                                         step B1

Capacity of blowers, at 27.6 kN/m^2 (gauge), sm^3/min 20   Sec. 11-9-2,
                                                         step B1

Grit is pushed into a hopper by spiral conveyor and is   Sec. 11-11-4
    removed by bucket elevator   and Fig. 11-4

Average quantity of grit removed, m^3/d 1.14   Sec. 11-9-2,
                                                         step I

H. *Primary Sedimentation*   Chap. 12

Number of rectangular basins   2   Sec. 12-6-2,
                                                          step A1

Length, m   46.33   Sec. 12-6-2,
                                                       step A1

Width, m   11.58   Sec. 12-6-2,
                                                       step A1

Average water depth, m	3.10	Sec. 12-6-2, step A1
Freeboard, m	0.6	Sec. 12-6-2, step A1
Detention time		
At average flow, both basins in operation, h	2.1	Sec. 12-6-2, step A3
At peak design flow, both basins in operation, h	0.7	Sec. 12-6-2, step A3
Overflow rate		
At average flow, both basins in operation, $m^3/m^2 \cdot d$	35.4	Sec. 12-6-2, step A2
At peak design flow, both basins in operation, $m^3/m^2 \cdot d$	106.4	Sec. 12-6-2, step A2
Influent structure consists of 1-m-wide channel with eight submerged orifices		Sec. 12-6-2, step B1
Number of standard V-notch effluent weirs	505	Sec. 12-6-2, step C3
Total length of weir plate, m	101.72	Sec. 12-6-2, step C2
Average quantity of sludge produced, kg/d	6,227	Sec. 12-6-2, step F2
Longitudinal sludge collector consists of chain, sprockets, wheels, flights, scrapers, and drive unit		Sec. 12-8-3
Cross-collector consists of chain, sprocket, wheels, flights, and drive units		Sec. 12-8-3
Number of self-priming centrifugal nonclog pumps per basin for sludge pumping	1	Sec. 12-6-1, step F4, and Sec. 12-8-5
Skimmer consists of hand-operated scum trough, scum pit, and pump		Sec. 12-8-7
I. *Aeration Tanks*		Chap. 13
Number of completely mixed aeration basins	4	Sec. 13-7-4, step A5
Length, m	31.0	Sec. 13-7-4, step A5
Width, m	15.5	Sec. 13-7-4, step A5
Average water depth, m	4.5	Sec. 13-7-4, step A5
Freeboard, m	0.8	Sec. 13-7-4, step A5
Aeration period at design average flow, h	4.94	Sec. 13-7-4, step A8

Food to MO ratio, kg BOD_5/kg MLSS·d	0.312	Sec. 13-7-4, step A9
Organic loading, kg BOD_5/m³·d	0.97	Sec. 13-7-4, step A10
MLVSS, maintained, mg/ℓ	3000	Sec. 13-7-1
Ratio of MLVSS to MLSS	0.8	Sec. 13-7-1
Mean cell residence time, d	10	Sec. 13-7-1
Solids wasted from the aeration basin (MLSS), kg/d	2332	Sec. 13-7-4, step A6
Sludge volume wasted, m³/d	622	Sec. 13-7-4, step H
Number of identical waste sludge pumps	2	Sec. 13-7-4, step H
Solids lost in the effluent, kg/d	828	Sec. 13-7-4, step A6
Returned sludge rates, Q_r/Q	0.6	Sec. 13-7-4, step A7
Number of return sludge pumps	5	Sec. 13-7-5, step H
Number of standby units	1	Sec. 13-7-5, step H
Rated capacity of each pump, m³/s	0.182	Sec. 13-7-5, step H
Volume of air required m³/min per basin	148	Sec. 13-7-4, step E3
Diffusers are swing type in 10 rows along the width of the basin. The diffusers are Dacron sock, standard tube, discharging 0.21 m³ standard air per min per tube		Fig. 13-13
Total number of diffusers per basin	720	Sec. 13-7-4, step F2
Total number of blowers	5	Sec. 13-7-4, step G3
Total number of standby units	1	Sec. 13-7-4, step G3
Power requirement of each blower, kW	187	Sec. 13-7-4, step G4
Rated air supply capacity of each blower, m³/min	155	Sec. 13-7-4, step G3

J. *Final Clarifier*

Number of clarifiers	4	Sec. 13-7-1
Diam. of each clarifier, m	40.7	Sec. 13-7-5, step A4
Average water depth, m	3.5	Sec. 13-7-5, step B4

Freeboard, m	0.5	Sec. 13-7-5, step B4
Detention time		
At avg. design flow plus recirculation, h	6.6	Sec. 13-7-5, step C2
At peak design flow plus recirculation, h	3.1	Sec. 13-7-5, step C2
At peak design flow when one unit is out of service, h	2.4	Sec. 13-7-5, step C2
Overflow rate		
At average design flow plus recirculation, $m^3/m^2 \cdot d$	12.8	Sec. 13-7-5, step A5
At peak design flow plus recirculation, $m^3/m^2 \cdot d$	26.8	Sec. 13-7-5, step A7
At peak design flow plus recirculation when one unit is out of service, $m^3/m^2 \cdot d$	35.7	Sec. 13-7-5, step A7
Limiting solids flux		
At avg. design flow plus recirculation, $kg/m^2 \cdot d$	48.1	Sec. 13-7-5, step A8
At peak design flow plus recirculation, $kg/m^2 \cdot d$	100.4	Sec. 13-7-5, step A8
At peak design flow plus recirculation when one unit is out of service, $kg/m^2 \cdot d$	134	Sec. 13-7-5, step A8
K. *Chlorination*		Chap. 14
Number of chlorine contact chambers	2	Sec. 14-6-1
Length, m	33.75	Sec. 14-6-2, step A2
Width, m	2.5	Sec. 14-6-2, step A2
Depth variable, at peak design flow, m	3.2	Sec. 14-6-2, step A2
Freeboard, m	0.6	Sec. 14-6-2, step A2
Effluent structure consists of proportional weir for maintaining constant velocity through contact chamber, and two orifices at the bottom to flush the settled solids		
Contact time at peak design flow, min	20.2	Table 14-5
Contact time at average design flow, min	45.5	Table 14-5
Maximum chlorine usage at peak design flow and chlorine dosage of 8 mg/ℓ, kg/d	913.1	Sec. 14-6-2, step D2
Average chlorine usage at average design flow, and chlorine dosage of 5 mg/ℓ, kg/d	190	Sec. 14-6-2, step D1
Total number of 450 kg/d capacity, chlorinators	3	Sec. 14-6-2, step D3

Number of standby chlorinators	1	Sec. 14-6-2, step D3
Number of 907-kg chlorine containers attached to the header	5	Sec. 14-6-2, step E2
Instrumentation of chlorination facility include flow measurement (Parshall flume) and automatic chlorine residual control		Sec. 14-6-2, step F, and Fig. 14-8

L. *Outfall Structure* — Chap. 15

The outfall structures consist of a concrete lined trapezoidal channel, collection box, outfall pipe, and diffuser pipe

Outfall channel		
Bottom width, m	0.5	Fig. 15-1
Side slope	3H–1V	Fig. 15-1
Collection box,		
Length and width L $\times$ W, m	2 $\times$ 2	Fig. 15-1
Outfall pipe		
Diam, m	0.92	Sec. 15-4-2, step C
Diffuser pipe		
Diam, m	0.92	Sec. 15-4-2, step D3
Number of diffuser ports	6	Sec. 15-4-2, step D3
Diameter of diffuser ports, m	0.27	Sec. 15-4-2, step D3

M. *Sludge Thickeners* — Chap. 16

Number of gravity thickeners for combined primary and secondary sludge	2	Sec. 16-7-1
Diam, m	11.6	Sec. 16-7-2, step A3
Freeboard, m	0.6	Sec. 16-7-2, step B4
Sidewater depth including freeboard, m	4.5	Sec. 16-7-2, step B4
Depth at the center, m	5.5	Sec. 16-7-2, step B5
Solids loading, kg/m^2·d	42.8	Sec. 16-7-2, step A4
Hydraulic loading, m^3/m^2·d	9.0	Sec. 16-7-2, step A4
Dimensions of blending tank, depth $\times$ diam, m	3 $\times$ 8.2	Fig. 16-4
Thickened sludge withdrawal rate, m^3/d	62.3	Sec. 16-7-2, step E1

No. of sludge withdrawal pumps, plunger type with time clock control	2	Sec. 16-7-2, step E2
Thickened sludge solids, percent	6	Sec. 16-7-2, step E2

N. *Sludge Stabilization*   Chap. 17

Number of anaerobic sludge digesters	2	Sec. 17-7-1
Diameter, m	13.7	Sec. 17-7-2, step B1
Sidewater depth, m	8.5	Sec. 17-7-2, step B1
Digestion period		
At average flow, d	21.1	Sec. 17-7-2, step C1
At extreme high flow, d	10.9	Sec. 17-7-2, step C1
At extreme low flow, d	33.2	Sec. 17-7-2, step C1
Solids loading		
At average loading condition, kg VS/$m^3 \cdot$d	2.2	Sec. 17-7-2, step C2
At extreme minimum loading condition, kg VS/$m^3 \cdot$d	2.0	Sec. 17-7-2, step C2
At extreme high loading condition, kg VS/$m^3 \cdot$d	2.3	Sec. 17-7-2, step C2
Gas production, m^3/d	2560	Sec. 17-7-2, step D2
Diam of gas storage sphere, m	15	Sec. 17-7-2, step I1
Number of gas compressors	2	Sec. 17-7-2, step I2
Power of gas compressors, kW	15	Sec. 17-7-2, step I2
Digester mixing is achieved by gas recirculation		Sec. 17-7-1
Total number of compressors for gas mixing including one standby unit	3	Sec. 17-7-2, step J1
Compressor power each unit, kW	15	Sec. 17-7-2, step J1
Gas flow per digester, m^3/s	0.14	Sec. 17-7-2, step J2
Digested sludge		
Quantity of digested solids produced, kg/d	4096	Sec. 17-7-2, step E6
Volume of digested sludge, m^3/d	80	Sec. 17-7-2, step E6

O. *Sludge Dewatering*   Chap. 18

Filter press operation, h/d (5-day week)	8	Sec. 18-6-1
Total number of variable recessed plate filter presses (each filter unit has 20 chambers)	5	Sec. 18-6-2, step B2
Sludge solids processed, kg/h	718	Sec. 18-6-2, step B2
Lime (CaO), kg/h	36	Sec. 18-6-2, step B2
Polymer, kg/h	14	Sec. 18-6-2, step B2
Total	768	Sec. 18-6-2, step B2
Sludge cake moisture content, percent	25	Sec. 18-6-1
Total quantity of sludge cake, kg/h	719	Sec. 18-6-2, step D1
Volume of sludge cake, 5 d/week basis, m³/d	21.7	Sec. 18-6-2, step D2
Filtrate volume, m³/d	223.2	Sec. 18-6-2, step E1

P. *Sludge Disposal*   Chap. 19

Codisposal of sludge cake, screenings, grit and skimmings landfilled on the site		Sec. 19-5-1
Method of landfilling	Trenching	Sec. 19-5-1
Number of trenches located parallel to flood protection levees	10	Sec. 19-5-2, step A4
Dimensions of the trenches		
Length, m	120	Sec. 19-5-2, step A4
Bottom width, m	25	Sec. 19-5-2, step A4
Depth, m	2.5	Sec. 19-5-2, step A3
Side slope	1 : 1	Sec. 19-5-2, step A4
Top cover, m	1	Sec. 19-5-2, step A5
Equipment needed		
Number of tractor dozer and loader	1 each	Sec. 19-5-2, step D2
Life of fill, year	16.3	Sec. 19-5-2, step B3

24

Advanced Wastewater Treatment and Upgrading Secondary Treatment Facility

24-1 INTRODUCTION

Upgrading of a wastewater treatment facility may be necessary to meet the existing effluent quality and/or to meet the stricter future effluent quality requirements. An inability to meet the existing effluent quality may result from lack of proper plant operation and control, inadequate plant design, and increased hydraulic or organic loading due to a change in wastewater flow or characteristics. Such deficiencies can be overcome by staffing the plant with an adequate number of competent operators and laboratory personnel, and a proper maintenance program, and checking the key design components and conducting sufficient wastewater sampling and treatability studies. Existing facilities can be expanded to handle higher hydraulic and organic loads by process modifications or building additional treatment units.

Stricter effluent quality requirements are generally met by adding new treatment processes that provide the desired removal and efficiencies in conjunction with the existing secondary treatment facility. The stricter effluent quality may require organic and suspended solids removal beyond the capability of a secondary treatment plant. Under such requirements, additional treatment processes, such as filtration, carbon adsorption, chemical precipitation of phosphorus, and nitrification-denitrification, may be necessary.

The principles of plant operation and design procedure for secondary treatment plants have been extensively covered in the earlier chapters in this book. Additional information on plant upgrading may be found in Ref. 1.

In recent years advanced wastewater treatment has received great interest as many secondary treatment plants are being upgraded for removal of nitrogen, phosphorus, and other constituents that are not removed in conventional wastewater treatment plants. Therefore, this chapter is exclusively devoted to provide an overview of advanced wastewater treatment technology. Because land treatment of wastewater is becoming a widespread practice in the United States, a detailed discussion on land treatment of wastewater is also provided in this chapter.

24-2 OVERVIEW OF ADVANCED WASTEWATER TREATMENT TECHNOLOGY

Advanced wastewater treatment technology is designed to remove those constituents that are not adequately removed in the secondary treatment plants. These include nitrogen, phosphorus, and other soluble organic and inorganic compounds. Nitrogen and phosphorus constitute nutrients that accelerate the plants' growth in the receiving waters. Ammonia is also toxic to fish, exerts nitrogenous oxygen demand, and increases chlorine demand. Heavy metals, hydrogenated hydrocarbons, phenolic compounds, etc., are toxic to fish and other aquatic life, concentrate in the food chain, and may create taste and odor problems in water supplies. Many of these constituents must be removed to meet more stringent water quality standards, and also to reuse the effluent for municipal, industrial, irrigation, recreation, and other water needs. A brief discussion of effluent reuse is given in Chapter 15.

Most commonly used advanced wastewater treatment processes are chemical precipitation of phosphorus, nitrification, denitrification, ammonia stripping, breakpoint chlorination, filtration, carbon adsorption, ion exchange, reverse osmosis, and electrodialysis. Each of these unit operations and processes is discussed below. Additional information concerning principal applications and degree of removal achieved by these systems may be found in Tables 4-1 and 4-2.

24-2-1 Coagulation and Flocculation

Process Description.   Coagulation involves the reduction of surface charges and the formation of complex hydrous oxides. Flocculation involves combining the coagulated particles to form settleable floc. The coagulant (alum, ferric chloride, ferrous sulfate, ferric sulfate, etc.) is mixed rapidly then stirred to encourage formation of floc prior to settling. The objective is to improve the removal of BOD, TSS, and phosphorus. To accomplish this, chemicals have been added, prior to primary in biological treatment processes, and in separate facilities following the biological treatment processes. Polymers have also been used in conjunction with these chemicals. The chemical dosage is adjusted to give the desired amount of floc formation and BOD, TSS, or phosphorus removal. The exact application rate is determined by the *jar test*. The approximate average alum and ferric chloride dosages in municipal wastewater are 170 and 80 mg/ℓ, respectively. Lime may be needed in conjunction with iron salts.[2,3]

Proper mixing of chemicals at the point of addition and proper flocculation prior to clarification are essential for maximum effectiveness. Flocculation may be accomplished in a few minutes to half an hour in basins equipped with mixers, paddles, or baffles. The coagulants react with alkalinity to produce insoluble metal hydroxide for floc formation. If sufficient alkalinity is not present, lime or soda ash (sodium carbonnate) is added in the desired dosages.

In biological treatment units the addition of coagulants has marked influence on biota. Population of protozoans and higher animals is adversely affected. However, the BOD, TSS, and phosphorus removal is significantly improved. Overdosing of chemicals may cause toxicity to microorganisms.

Large quantities of sludge are produced from chemical precipitation. The chemical sludges may also cause serious handling and disposal problems.

Equipment.   Major equipment for a coagulation-flocculation system includes chemical storage, chemical feeders, pipings and control systems, flash mixer, flocculator, and sedimentation basin.

Design Parameters.   The important design parameters for coagulation and flocculation units are average and peak design flows, chemical dosages, flash mix time, and flocculation period. The design parameters for sedimentation devices have been presented in Chapters 12 and 13. The design of mixer paddles has been presented in Chapter 16.

24-2-2 Lime Precipitation

Process Description.   Lime reacts with bicarbonate alkalinity and orthophosphate causing flocculation. The objectives of lime addition are to increase removal of BOD, TSS, and phosphorus. Lime is added prior to primary sedimentation, in biological treatment, or in a separate facility after secondary treatment. Lime addition may be single stage or two stages.[4,5]

Single-Stage.   Lime in single stage is used in primary, biological treatment, or after secondary treatment. The procedure is the same as in coagulation. The pH of the wastewater is raised to about 10, the wastewater is flocculated and then settled. Normal lime dosage is about 180–250 mg/ℓ as CaO. The actual dosage depends primarily on phosphorus concentration, hardness, and alkalinity. The biological system is not adversely affected by lime addition in moderate amounts (80–120 mg/ℓ as CaO). The microbial production of carbon dioxide is sufficient to maintain a pH near neutral. High dosage may upset the biological process.

Two-Stage.   Lime addition prior to primary and after secondary treatment may be achieved in two stages. The pH of wastewater is raised to greater than 11, flocculated, and settled. Lime dosage up to 450 mg/ℓ as CaO may be necessary. The effluent is carbonated by adding CO_2 to lower the pH, then flocculated and settled. Higher BOD, TSS, and phosphorus removal is achieved in a two-stage lime process. However, use of excess lime causes scale formation in tanks, pipes, and other equipment. Also, handling and disposal of large quantities of lime sludge is a problem.

Equipment. Major equipment for a lime precipitation process includes lime storage; lime feeders, pipings, and control system; flash mixer; flocculator; sedimentation basin; and, for a two-stage system, a CO_2 source such as an incinerator or an internal combustion engine.

Design Parameters. Design parameters are the same as those listed for the coagulation-flocculation system.

24-2-3 Nitrification

Process Description. Nitrification converts ammonia to nitrate form, thus eliminating toxicity to fish and other aquatic life, and reducing the nitrogenous oxygen demand. Ammonia oxidation to nitrite and then to nitrate is performed by autotrophic bacteria. The reactions are shown by Eqs. (24-1) and (24-2).

$$NH_3^+ + \tfrac{3}{2} O_2 \xrightarrow{\text{\textit{Nitrosomonas}}} NO_2^- + H^+ + H_2O + \text{biomass} \tag{24-1}$$

$$NO_2 + \tfrac{1}{2} O_2 \xrightarrow{\text{\textit{Nitrobacter}}} NO_3^- + \text{biomass} \tag{24-2}$$

Temperature, pH, dissolved oxygen, and ratio of BOD and total Kjeldahl nitrogen (TKN) are important factors in nitrification. Nitrification can be achieved in suspended growth and attached growth systems. An excellent discussion of nitrification can be found in Ref. 2.

Suspended Growth System. In an activated sludge process, nitrification under normal operating conditions is seriously limited because of high BOD/TKN ratio. This is because the population of nitrifying bacteria is considerably lower than the population of heterotrophs. A second stage aeration basin followed by clarification produces excellent nitrification results. Second stage aeration is often more economical than in the same basin with organics removal. The MLSS concentration, sludge age, and air supply may be adjusted to provide maximum nitrification. Because alkalinity is destroyed during nitrification, lime may be added to maintain pH to a desirable level (8.2–8.6).

Attached Growth. Trickling filters, rotating biological contactors, or biological towers with plastic packing or redwood medium are used for nitrification. This can be achieved in single stage or separate stages. Separate stage is preferred for the best nitrification effects. In a single stage, the nitrifying organisms in sufficient number for nitrification develop only on the lower level of the media. By the time the wastewater reaches the lower level, sufficient BOD removal has occurred and the nitrification proceeds rapidly.

Equipment. The major equipment for suspended growth nitrification system includes aeration basin with aeration equipment, clarifier and return sludge system (see Chapter 13). For attached growth, trickling filter or biodisc assembly with clarifier and return flow systems are used.

Design Parameters. Important design parameters are average and peak design flow, hydraulic and organic loadings, nitrogen loading, and biological kinetic coefficients. The design procedure is similar to that for BOD removal and is presented in Chapter 13.

24-2-4 Denitrification

Process Description. Nitrite and nitrate are reduced to gaseous nitrogen by a variety of facultative heterotrophs in an anaerobic environment. An organic source, such as acetic acid, acetone, ethanol, methanol, or sugar, is needed to act as a hydrogen donor (oxygen acceptor) and to supply carbon for synthesis. Methanol is preferred because it is least expensive. Equations (24-3)–(24-5) express the basic reactions.

$$3O_2 + 2CH_3OH \rightarrow 2CO_2 + 4H_2O \qquad (24-3)$$
$$6NO_3^- + 5\,CH_3OH \rightarrow 3N_2 + 5CO_2 + 7H_2O + 6OH^- \qquad (24-4)$$
$$2NO_2^- + CH_3OH \rightarrow N_2 + CO_2 + H_2O + 2OH^- \qquad (24-5)$$

The amount of methanol required is calculated from Eq. (24-6).[2] Approximately 4–5 mg/ℓ methanol per mg/ℓ of NO_3-N is required.

$$\text{Methanol (mg/}\ell) = \frac{2.47 \times NO_3\text{-N (mg/}\ell) + 1.53 \times NO_2\text{-N (mg/}\ell) + 0.87}{\times \text{ dissolved oxygen (mg/}\ell)} \qquad (24-6)$$

Denitrification can be achieved in suspended growth or attached growth system.

Suspended Growth System. The suspended growth system consists of a completely mixed reactor with a series of compartments followed by a clarifier. The contents in the reactor are mixed by underwater stirrers similar to those in a coagulation tank. Covered tanks to minimize absorption of oxygen are preferred but airtight covers are not necessary. MLSS of 1000–2000 mg/ℓ is maintained by recycling the sludge from the clarifier. The organic carbon source (methanol) is added in the influent line.

Attached Growth System. The attached growth system uses submerged (anaerobic) upflow filters containing gravel, cinder, coal, or plastic media. The organic carbon source is added in the influent. Additional information on the denitrification system may be obtained in Refs. 2 and 3.

Equipment. The major equipment for suspended and attached growth denitrification system includes mixing basin with submerged paddle (covered basins are preferred) or attached growth upflow reactor with backwash system; methanol storage and feed system; and final clarifier and return sludge pump with piping.

Design Parameters. Important design parameters are average and peak design flow, nitrate concentration, detention period, and biological kinetic coefficients (see Chapter 13).

24-2-5 Biological Nutrient Removal

Biological phosphorus and nitrogen removal has received considerable attention in the recent years. Basic benefits reported for biological nutrient removal include monetary saving through reduced aeration capacity and the obviated expense for chemical treatment.

Biological nutrient removal involves anaerobic and anoxic treatment of return sludge prior to discharge into the aeration basin. Based on anaerobic, anoxic and aerobic treatment sequence, and internal recycling several processes have been developed. These include Bardenpho, A/O, and Phostrip processes, and sequencing batch reactors. Over 90 percent phosphorus, and high nitrogen removal (by nitrification and denitrification) has been reported by biological means. For detailed discussion on biological nutrient removal processes, equipment and design parameters the reader is referred to Refs. 7–10.

24-2-6 Ammonia Stripping

Process Description.   Ammonia gas can be removed from an alkaline solution by air stripping as expressed by Eq. (24-7).

$$NH_4^+ + OH^- \rightarrow NH_3 \uparrow + H_2O \qquad (24-7)$$

The process requires (1) raising the pH of the wastewater to about 11, (2) formation of droplets in stripping tower, and (3) providing air-water contact and droplet agitation by countercurrent circulation of large quantities of air through the tower. Ammonia-stripping towers are simple to operate and can be very effective in ammonia removal, but their efficiency is highly dependent on air temperature. As the air temperature decreases, the ammonia removal efficiency drops significantly. This process, therefore, is not recommended for a cold climate. A major operational disadvantage of stripping is the need for neutralization and calcium carbonate scaling on the tower. Also, there is some concern over discharge of ammonia into the atmosphere.[2-6]

Equipment.   The basic equipment for an ammonia-stripping system includes chemical feed, stripping tower, pump and liquid spray system, forced air draft, and recarbonation system.

Design Parameters.   Important design parameters for ammonia stripping system are average and peak design flow, surface loading rate, chemical dosage, tower height, and air-to-water ratio.

24-2-7 Breakpoint Chlorination

Process Description.   Oxidation of ammonia-nitrogen to nitrogen gas is achieved by breakpoint chlorination (Eq. 24-8).

$$2NH_3 + 3HOCl \rightarrow N_2 + 3H_2O + 3HCl \qquad (24-8)$$

The ratio of chlorine (Cl_2) to ammonia (NH_3) is 7 : 1–10 : 1. Optimum pH is in the range of 6–7. The advantages of breakpoint chlorination are low capital cost, high degree of efficiency and reliability to ammonia removal, disinfection, and insensitivity to cold weather. The disadvantages are the formation of high residual chloride concentration and chlorinated organic compounds. Often dechlorination is necessary.

Equipment.   Major equipment for breakpoint chlorination is the same as discussed in Chapter 14. These include flow measurement, chemicals and equipment for pH adjustment, chlorine storage, chlorine feed piping and mixers, and contact tank.

Design Parameters. The important design parameters are average and peak design flow, concentrations of ammonia and chlorine, and contact period.

24-2-8 Ion Exchange for Ammonia Removal

Process Description. Wastewater is passed through a bed of *Clinoptilolite* (a zeolite resin), which selectively removes the ammonium ion. When the resin becomes saturated, it is regenerated with a lime slurry containing sodium chloride. The lime solution after regeneration of the resin must be processed to remove the ammonium ions so that the solution can be reused. Air stripping of ammonia from the solution is a feasible method due to the small flow involved. The stripped ammonia gas is passed through an absorber material which has high selectivity for ammonia. Proper disposal of ammonia-bearing absorber material is necessary.

Equipment. Major equipment for an ion-exchange system includes an ion-exchange bed, bed regeneration system, ammonia-stripping tower, and ammonia absorber material.

Design Parameters. The design parameters are average and peak design flow, ion-exchange capacity of the resin bed, concentration of ammonia, percolation rate, etc.

24-2-9 Filtration

Process Description. Filtration has been used to polish secondary effluent or to produce effluent that is low in suspended solids. Filtration is not considered an advanced wastewater treatment process; instead, it is used as a pretreatment device prior to other advanced treatment processes.

Filtration consists of passing wastewater through a filtering medium that can strain out the colloidal particles. The filtering media may be fine sand, anthracite, mixed media, diatomaceous earth, or filter fabric. The filtration system may be gravity filter or pressure filter. The solids accumulate in the filter media and therefore backwashing using clean water in the opposite direction of flow must be accomplished periodically to control the head loss through the filtration system.

Microstrainers are commonly used for polishing the effluent from biological or physical-chemical treatment process. Microstrainers consist of a woven stainless steel or polyester plastic fabric having openings ranging from 15 to 60 μm, mounted over a rotating drum that is held horizontally in a tank. The influent enters the drum interior through one end and flows out through the filtering fabric. The solids are retained in the inside of the rotating screen and are washed by a jet of filtered water on the top of the entire length of the drum. The washwater is collected in a hopper and returned to the plant. The washwater is 3–5 percent of the total effluent volume.

Equipment. Major equipment for filtration devices includes filter housing, filter media, underdrain system, backwash system, etc. The pressure filtration and microstrainer assemblies are illustrated in Figure 24-1.

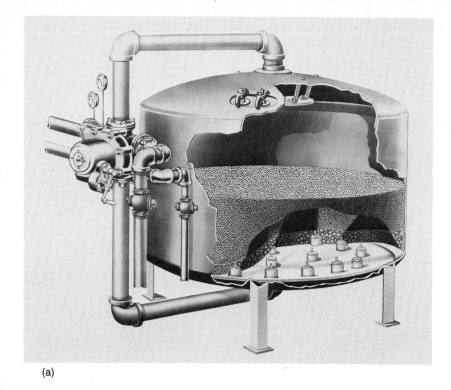

(a)

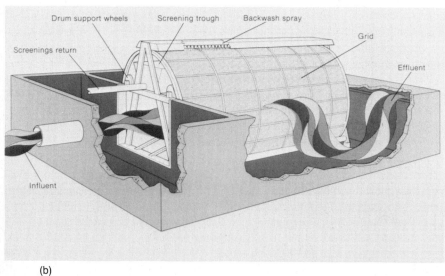

(b)

Figure 24-1   Design Details of Pressure Filter and Microstrainer. (a) Pressure filter assembly. (Courtesy The Permutit Company, Inc.) (b) Operational and equipment details of microstrainer. (Courtesy Envirex Inc., a Rexnord company.)

Design Parameters.   The basic design factors for the filtration device are average design flow, filtration media, filtration rate, applied head, allowable head losses, and backwash flow rate.

24-2-10 Carbon Adsorption

Process Description.   Carbon adsorption is used to remove soluble refractory organics. The process consists of entrapping organic material onto the carbon surface. The most common method currently used is granular activated carbon columns. The treated wastewater is percolated through the column until the column becomes saturated with organic material. It is then removed from service and the activated carbon is regenerated by burning off the organic material in a special furnace. Approximately 5 percent loss of carbon can be expected with each cycle; thus new material must be added.

   Activated carbon can effectively remove bacteria and viruses. It also removes organometallic compounds, pesticides, chlorinated compounds, chlorine, and many other compounds that are not removed in the conventional secondary treatment plants.

Equipment.   Following is the list of equipment commonly needed for carbon adsorption system: carbon column, granular activated carbon, feed and backwash pump and piping, and carbon regeneration system.

Design Parameters.   Common design parameters include average and design flow, influent characteristics, effluent quality, contact time, and adsorption capacity of carbon.

24-2-11 Ion Exchange

Process Description.   Ion exchange is a demineralization process in which the cations and anions in wastewater are selectively exchanged for ions in the insoluble resin material. When the resin capacity is used up, it is regenerated by using high concentrations of the original ion that is exchanged from the resin.

   Ion-exchange resins are cationic if they exchange positive ions and anionic if they exchange negative ions. In the hydrogen cation exchange process, hydrogen ions are exchanged for positive ions (Ca^{2+}, Mg^{2+}, Na^+, etc.). The hydrogen cation exchanger is regenerated with a mineral acid. Most widely used and the cheapest regenerant is sulfuric acid.[2,3,5]

   In anion exchangers the negative ions (Cl^-, NO_3^-, SO_4^{2-}, etc.) are exchanged by OH^- ions. Main types of anion exchangers are (1) weakly basic anion exchangers and (2) strongly basic anion exchangers. The weakly basic anion exchangers remove strongly ionized acids (HCl, H_2SO_4, etc.), but they will not remove weakly ionized acids (H_2CO_3, H_2SiO_3, etc.). These exchangers are regenerated with a solution of soda ash (Na_2CO_3). Strongly basic anion exchangers remove both strongly and weakly ionized acids. These resins are regenerated with caustic soda solution.[2,3]

   Ion exchangers are granular packed bed columns. The wastewater is trickled downward under pressure. When exchange capacity of the resin is exhausted, the column is backwashed to remove trapped solids and then the beds are regenerated. It is possible to produce high-quality demineralized water. Synthetic cation and anion resins perform better than the naturally occurring zeolites. The cation and anion resin beds for demineralization may be arranged in series in separate exchange columns or may be mixed in a single reactor. The typical bed depths are 1–2 m (3.3–6.5 ft), and the flow rate is 0.20–0.40 m^3/m^2·min (5–10 gal/ft^2·min). Total ion-exchange capacity of commercially available resins are 50,000–80,000 g/m^3 (as $CaCO_3$).

A high concentration of organic matter in effluent may plug the bed and blind the resin. Filtration and carbon absorption of secondary effluent may be necessary. A schematic flow diagram of an ion-exchange system is given in Figure 24-2.

Equipment.   Equipment commonly needed for the ion-exchange demineralization process includes gravity filters, carbon adsorption (may not be needed), ion-exchange beds, pressure pumps, regenerant solution and pumps, backwash, and rinse water system.

Design Parameters.   Common design parameters include average design flow, total dissolved solids, ion-exchange capacity of the resins, degree of demineralization needed, and flow rate through the beds.

24-2-12 Distillation

Distillation is the oldest demineralization process. It consists of evaporating a part or all of the water from a saline solution and subsequent condensation of mineral-free vapor. The energy requirements are very high for the system. One modification of the simple distillation process is multiple-effect evaporation, where water is evaporated in different stages at different pressures. The latent heat of condensation is used to preheat the incoming wastewater.[11,12] Another modification is distillation with vapor compression. This system utilizes the latent heat of the compressed steam to preheat and evaporate the incoming water. Both modifications conserve energy, but the energy requirements are quite large. These systems are complex and their application in wastewater treatment is

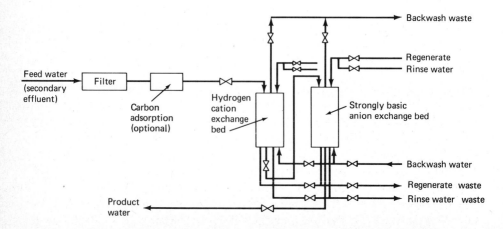

Figure 24-2   Typical Flow Diagram of Ion-Exchange Demineralization System.

not feasible. Solar distillation does offer the advantage of free energy, but land requirements are prohibitive. The process description, equipment details and design parameters
on distillation systems may be found in Refs. 11 and 12.

24-2-13 Reverse Osmosis and Ultrafiltration

Process Description. Reverse osmosis is a demineralization process applicable to the
production of high-quality water to meet the drinking water standards. The process
consists of permeating liquid through a semipermeable membrane at pressures up to
10,000 kN/m^2 (1500 lb/in.^{2}). The membranes reject most of the ions and molecules while
permitting acceptable rates of water passage.

Many types of membranes have been developed, but cellulose acetate and polyamide
(nylon) are currently the most widely used membrane materials. Reverse osmosis modules suitable for water treatment involve the arrangement of membranes and their supporting structures so that the feed water under high pressure can pass through the membrane
surface while product water is collected from the opposite face without brine contamination. Four different types of module designs have been developed: plate and frame, large
tubes, spiral wound, and hollow fine fiber. High-quality feed is required for efficient
operation of a reverse osmosis unit. Pretreatment of a secondary effluent with filtration
and carbon adsorption is usually necessary. pH adjustment (4.0–7.5), and hardness, iron
and manganese removal is necessary to decrease scale formation. The brine is returned to
the plant.[6]

Ultrafiltration is a process similar to reverse osmosis. It applies to dissolved particles
of larger size (0.002–10 mμ range). Ultrafiltration is actually a physical screening process
using a relatively coarse membrane. The applied pressure is normally below 1000
kN/m^2(150 lb/in.^{2}).[6]

Equipment. The major equipment for reverse osmosis system includes membrane
module with support system, pretreatment system, high-pressure pump and piping, and
brine-handling and disposal system.

Design Parameters. The design parameters for reverse osmosis system are average
design flow, salinity, applied pressure, product recovery rate, rejection rate, and influent
quality to the membrane and pretreatment requirements.

24-2-14 Electrodialysis

Process Description. Electrodialysis is also a demineralization process. In this process,
an electrical potential is used to transfer the ion through ion-selective membranes. This
process is most practical and widely used to treat brackish waters. The energy requirement
is directly proportional to the concentration of salts in the water being treated. An
electrical voltage is applied in cells (also called stacks). The cells contain alternatively
arranged cation- and anion-permeable membranes. The cations and anions in feed liquid
migrate to the opposite poles. They pass the cation- and anion-selective membranes, thus
producing clean water and brine streams. To avoid membrane fouling, the wastewater
must be pretreated as is done in reverse osmosis.[6]

Equipment. The major equipment includes membrane stack, pumping system, pretreatment, and brine-handling and disposal system.

Design Parameters. The basic design parameters are average design flow, type and amount of ions removed, electrical current and voltage, and rejection rate.

24-2-15 Land Treatment

Land treatment of municipal and industrial wastewaters includes the use of plants, land surface, and the soil matrix to remove various constituents in the wastewater by many physical, chemical, and biological means. The objectives of land treatment include irrigation, nutrient reuse, crop production, recharge of groundwater, and water reclamation for reuse. There are three basic methods of land application: (1) slow-rate irrigation, (2) rapid infiltration-percolation, and (3) overland flow. Each method can produce renovated water of different quality, can be adapted to different site conditions, and can satisfy different overall objectives. Table 24-1 provides a comparison of the design features, site characteristics, and effluent quality from the three primary land treatment systems and for two other methods of land treatment: wetlands and subsurface application.[2,13] These two methods are less developed and less adaptable to the general situations. Figure 24-3 illustrates the hydraulic pathway, methods of wastewater application, and recovery methods of renovated water in slow-rate irrigation, rapid infiltration, and overland flow systems. Brief descriptions of each of these methods of land treatment are given below.

Slow-Rate Irrigation. Irrigation is the most widely used form of land treatment system. The wastewater is applied to crops or vegetation either by sprinkling or by surface techniques. In this process surface runoff is not allowed. A large portion of water is lost by evapotranspiration whereas some water may reach groundwater table. Groundwater quality criteria may be a limiting factor for system selection. Some factors that are given consideration in design and selection of irrigation method are (1) availability of suitable site, (2) type of wastewater and pretreatment, (3) climatic conditions and storage needed, (4) soil type and organic or hydraulic loading rate, (5) crop production, (6) distribution methods, (7) application cycle, and (8) ground and surface water pollution.[2,13,14] Figure 24-4 illustrates the surface distribution and sprinkler application commonly used in irrigation systems.

Rapid Infiltration-Percolation. Rapid infiltration-percolation differs from the irrigation method in that wastewater percolates through the soils and treated effluent reaches the groundwater. Plants are not used for evapotranspiration as in irrigation systems. The objectives of rapid infiltration-percolation are (1) groundwater recharge, (2) natural treatment followed by withdrawal by pumping or underdrain systems for recovery of treated water, and (3) natural treatment with groundwater moving vertically and laterally in the soil, and recharging a surface water course.

The operational principles of rapid infiltration-percolation systems are shown in Figure 24-5. Sandy and loamy soils perform better. BOD_5 and TSS removals are comparable to those of irrigation system. However, nitrogen and phosphorus removals are

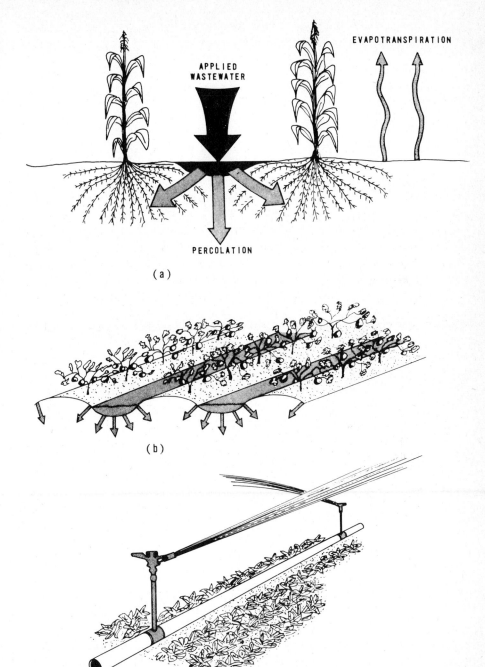

Figure 24-3   Hydraulic Pathway, and Methods of Wastewater Application in Slow-Rate Irrigation System. (a) Hydraulic pathway. (b) Surface distribution. (c) Sprinkler distribution. [From Ref. 13.]

TABLE 24-1 Comparison of Design Features, Site Characteristics, and Effluent Quality of Treated Wastewater from Land Treatment Processes

Conditions	Principal Processes			Other Processes	
	Slow-Rate Irrigation	Rapid Infiltration	Overland Flow	Wetlands	Subsurface
Design Features					
Application techniques	Sprinkler or surface[a]	Usually surface	Sprinkler or surface	Sprinkler or surface	Subsurface piping
Annual application rate, m	0.6–6.1	6.1–170	3.0–21	1.2–31	2.5–26.5
Field area required, hectare[b]	22.5–225	0.8–22.5	6.5–44.5	4.5–114	5.3–56.8
Typical weekly application rate, cm	1.3–10.2	10.2–305	6.4–15.3[c] 15.3–40.6[d]	2.5–63.5	5.0–50.1
Minimum preapplication treatment provided in U.S.	Primary sedimentation[e]	Primary sedimentation	Screening and grit removal	Primary sedimentation	Primary sedimentation
Disposition of applied wastewater	Evapotranspiration and percolation	Mainly percolation	Surface runoff and evapotranspiration with some percolation	Evapotranspiration, percolation, and runoff	Percolation with some evapotranspiration
Need for vegetation	Required	Optional	Required	Required	Optional
Site Characteristics					
Slope	Less than 20 percent on cultivated land; less than 40 percent on noncultivated land	Not critical; excessive slopes require much earthwork	Finish slopes 2–8 percent	Usually less than 5 percent	Not critical

Soil permeability	Moderately slow to moderately rapid	Rapid (sands, loamy sands)	Slow (clays, silts, and soils with imperme-able barriers)	Slow to moderate	Slow to rapid
Depth to groundwater	0.6 to 1.0 m	3 m (lesser depths are acceptable where underdrainage is provided)	Not critical	Not critical	Not critical
Climatic restrictions	Storage often needed for cold weather and precipitation	None (possibly modify operation in cold weather)	Storage often needed for cold weather	Storage may be needed for cold weather	None
Quality of Treated Wastewater[f]					
BOD$_5$ (mg/ℓ)	2 5	2 5	10 15	— —	— —
TSS (mg/ℓ)	1 5	2 5	10 20	— —	— —
Ammonia N (mg/ℓ)	0.5 2	0.5 2	0.8 2	— —	— —
Total N (mg/ℓ)	3 8	10 20	3 5	— —	— —
Total phosphorus (P) (mg/ℓ)	0.1 0.3	1 5	4 6	— —	— —

[a]Includes ridge-and-furrow and border strip.
[b]Field areas in hectares for 43.8 ℓ/s (1 mgd) flow. This area does not include land required for buffer zone, roads, or ditches.
[c]Range for application of screened wastewater.
[d]Range for application of a lagoon and secondary effluent.
[e]Depends on the use of the effluent and the type of crop.
[f]1. All values for slow rate are obtained for percolation of primary or secondary effluent through 1.4 m of soil.
2. All values for rapid infiltration are obtained for percolation of primary or secondary effluent through 4.5 m of soil.
3. All values for overland flow are obtained for runoff of comminuted wastewater over about 45 m of slope.
4. The first and second columns in slow-rate, rapid infiltration, and overland flow systems are the upper limits of average and maximum values, respectively.
1 in. = 2.54 cm
1 ft. = 0.305 m
1 acre = 0.405 hectare
Source: From Refs. 13–15

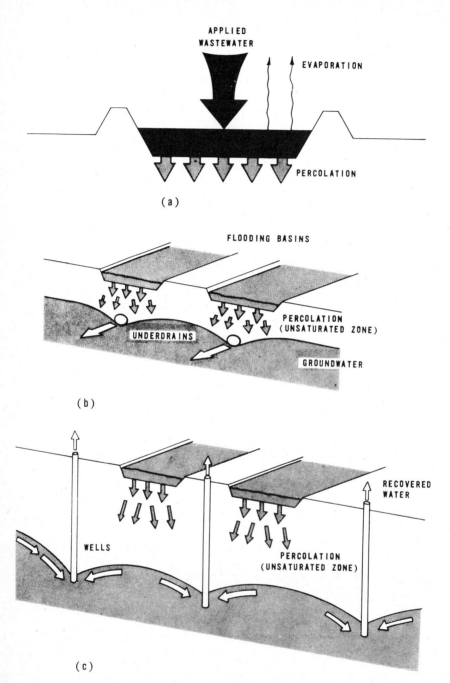

Figure 24-4   Hydraulic Pathway, Wastewater Application, and Recovery Methods of Renovated Water by Wells in Rapid Infiltration Percolation System. (a) Hydraulic pathway. (b) Recovery of renovated water by underdrains. (c) Recovery of renovated water by wells. [From Ref. 13.]

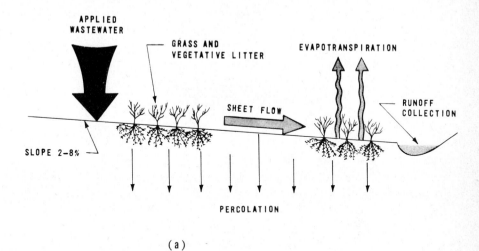

(a)

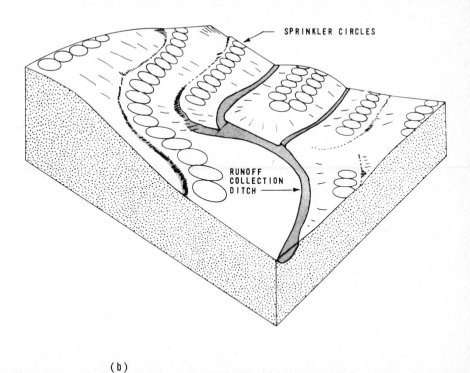

(b)

Figure 24-5   Hydraulic Pathway, and Method of Application in Overland Flow System. (a) Hydraulic pathway. (b) Pictorial view of sprinkler application. [From Ref. 13.]

not sufficient. Groundwater quality criteria may limit the use of this method. Table 24-1 gives a comparison of effluent quality and design considerations with other methods.[13]

Overland Flow.   In overland flow systems the wastewater is applied over the upper reaches of the sloped terraces and allowed to flow overland and is collected at the toe of the slopes. The collected effluent can be either reused or discharged to the receiving waters. Biochemical oxidation, sedimentation, filtration, and chemical adsorption are the primary mechanisms for removal of contaminants. Nitrogen removal is achieved through denitrification. Plant uptake of nitrogen and phosphorus are significant if crop harvesting is practiced. Table 24-1 summarizes the design features, site characteristics, and effluent quality. Figure 24-5 illustrates the operational principles and hydraulic pathway for overland system.

Other Systems.   Wastewater treatment and beneficial use of nutrients contained in wastewater may also be achieved through silviculture, aquaculture, and application in existing or artificial marshlands and wetlands. Various combinations of plants and animals from microscopic organisms to fish and water hyacinths have been attempted with some success in water quality improvements and other beneficial uses. In a subsurface application system a soil mound or a certain depth of porous soil is used. The wastewater is applied below the ground surface and the effluent after percolation may be recovered through an underdrain system. Design features of wetlands and subsurface application techniques are given in Table 24-1.

Design Considerations.   Design and selection of land treatment systems is complex. Many engineering, environmental, health, physical and social science, legal, and economic factors warrant detailed investigation. In recent years much emphasis has been given to land treatment as an alternative method of wastewater treatment and reuse to be evaluated in the overall wastewater treatment management program. Accumulation of toxic chemicals in soils and food chain, groundwater contamination, and odor problems have received wide concern. Table 24-2 provides a summary of many factors that must be considered in the evaluation and selection of land treatment systems. In all respects the land treatment of wastewater must comply with the regulations promulgated under the Resource Conservation and Recovery Act (as amended) to protect human health and the environment from the improper management of wastewater.[16] Many additional references are available on the technical and environmental factors and economics of land treatment of wastewater.[17–19]

24-3 PROBLEMS AND DISCUSSION TOPICS

24-1.   Draw a process train of a secondary wastewater treatment plant. Identify the possible locations where suitable coagulant may be added for precipitation of phosphorus. List the advantages and disadvantages of coagulation at each location.

24-2.   In a jar-test, municipal wastewater was coagulated using aluminum sulfate $Al_2(SO_4)_3$ $18H_2O$. The phosphorus concentration in the wastewater was 8 mg/ℓ as P. At alum dosages of 89, 136, and 216 mg/ℓ, soluble P remaining in the supernatants were 2, 1.5, and 0.4

TABLE 24-2   General Design Considerations That Must Be Studied in Design and Selection of Land Treatment Methods

Wastewater Characteristics	Climate	Geology	Soils	Plant Cover	Topography	Application	Environmental
Volume	Total precipitation and frequency	Groundwater	Type	Indigenous to region	Slope	Method	Odors
Constituent load BOD Nitrogen Phosphorus Heavy metals	Evapotranspiration Temperature Growing season Occurrence and depth of frozen ground Storage requirements Wind velocity and direction	Seasonal depth Quality Points of discharge Bedrock Type Depth Permeability	Gradation Infiltration/permeability Type and quantity of clay Cation-exchange capacity Phosphorus adsorption potential Heavy-metal adsorption potential pH Organic matter	Nutrient removal capability Toxicity levels Moisture and shade tolerance Marketability	Aspect of slope Erosion hazard Crop and farm management	Type of equipment Application rate Types of drainage	Legal Socioeconomic Health

Source: Adapted in part from Ref. 15.

mg/ℓ, respectively. Calculate the molar ratio Al : P for these removals. Also prepare a plot of molar ratio of Al : P versus log of percent P removed.

24-3. What is the purpose of recarbonation after two-stage line treatment? Give chemical equations to justify your reasoning.

24-4. Why does nitrification not occur concurrently with BOD reduction in an aeration basin?

24-5. Calculate methanol consumption (kg/d) for denitrification of 0.01 m^3/s well nitrified effluent containing 18 mg/ℓ nitrate nitrogen, and 5 mg/ℓ dissolved oxygen. Nitrite nitrogen is 0.3 mg/ℓ.

24-6. Discuss the major limiting factors of ammonia stripping.

24-7. Breakpoint chlorination is used for oxidation of 18 mg/ℓ ammonia-nitrogen in an effluent. Calculate chlorine dosage, and how many mg/ℓ chloride ions are added in the effluent.

24-8. List various types of filtration systems commonly used in polishing of secondary effluent. Discuss the advantages and disadvantages of microscreens.

24-9. Prepare a list of different chemicals that are removed from secondary effluent by carbon adsorption process. (Consult Sec. 3-4-2, and Ref. 27 in Chapter 3.)

24-10. Discuss the advantages and disadvantages of various demineralization processes (ion-exchange, distillation, and membrane processes) if used for recovery of potable water from secondary effluent.

24-11. List the environmental concerns of land treatment of municipal wastewater.

24-12. Describe the major design features, site characteristics, and effluent quality from an overland flow system.

24-13. Calculate the total land area required for a slow rate irrigation system to treat 0.08 m^3/s primary treated municipal wastewater. Assume buffer land is 300 percent of theoretical land requirement, and area needed for roads and ditches is 25 percent of theoretical land area. Sewage application rate is 1.5 m per year. Also calculate (a) theoretical field area required (ha for each 43.8 ℓ/s) and (b) BOD$_5$ loading kg/ha·d if influent has a BOD$_5$ of 140 mg/ℓ.

REFERENCES

1. U.S. Environmental Protection Agency, *Process Design Manual for Upgrading Existing Wastewater Treatment Plants*, Technology Transfer, EPA 625/1-71-004a, October 1974.

2. Metcalf and Eddy, Inc., *Wastewater Engineering: Treatment, Disposal, and Reuse*, Second Edition, McGraw-Hill Book Co., New York, 1979.

3. Clark, J. W., Warren Viessman and M. J. Hammer, *Water Supply and Pollution Control*, IEP-A Dun-Donnelley, New York, 1977.

4. U.S. Environmental Protection Agency, *A Guide to the Selection of Cost-Effective Wastewater Treatment System*, Technical Report EPA-430/9-75-002, Washington, D.C., July 1975.

5. Culp, R. L., and G. L. Culp, *Advanced Wastewater Treatment*, Van Nostrand Reinhold, New York, 1971.

6. Weber, W. J., *Physicochemical Processes for Water Quality Control*, Wiley-Interscience, New York, 1972.

7. Wanielista, M. P., and W. W. Eckenfelder (eds.) *Advances in Water and Wastewater Treatment, Biological Nutrient Removal*, Ann Arbor Science Publishers, Ann Arbor, Mich., 1978.

8. Irvine, R. L., *Sequencing Batch Reactor-Case Study*, Emerging I/A Technology Seminar, sponsored by U.S. Environmental Protection Agency, Dallas, Texas, May 19–20, 1982.

9. Barth, Edwin F., H. David Stensel, "International Nutrient Control Technology for Municipal Effluents," *Journal Water Pollution Control Federation*, Vol. 53, No. 12, December 1981, pp. 1691–1701.

10. Witherow, Jack L., "Phosphate Removal by Activated Sludge," *Proceeding of the 24th Industrial Waste Conference,* Purdue University, May 1969, pp. 1169–1184.

11. McCabe, W. L., and J. C. Smith, *Unit Operations of Chemical Engineering,* McGraw-Hill Book Co., New York 1967.

12. Geankoplis, G. J., *Transport Processes and Unit Operations,* Allyn and Bacon, Inc., Boston, 1978.

13. U.S. Environmental Protection Agency, *Process Design Manual for Land Treatment of Municipal Wastewater,* EPA/COE/USDA, EPA 625/1-77-008, October 1977.

14. U.S. Environmental Protection Agency, *Application of Sludges and Wastewaters on Agricultural Land: A Planning and Educational Guide,* EPA, MCD-35, March 1978.

15. U.S. Environmental Protection Agency, *Land Treatment of Municipal Wastewater Effluents-Design Factors-1,* EPA, 625/4-76-010, January 1976.

16. 40 CFR Part 260, *Hazardous Waste and Consolidated Permit Regulations, Federal Register,* Vol. 45, No. 98, May 19, 1980.

17. U.S. Environmental Protection Agency, *Land Application of Wastewater and State Water Law: An Overview,* EPA 600/2-77-232, November 1977.

18. U.S. Environmental Protection Agency, *Evaluation of Land Application Systems,* EPA 430/9-75-001, March 1975.

19. Sanks, R. L., and T. Asano (eds.), *Land Treatment and Disposal of Municipal and Industrial Wastewater,* Ann Arbor Science Publisher, Ann Arbor, Mich., 1976.

25

Avoiding Design Errors

25-1 INTRODUCTION

We often hear about the horror stories or the spectacular blunders that designers have made. No matter how humorous or silly these blunders may sound, it should be recognized that somehow they passed through undetected by the project team and the review process. In fact, in the midst of complex details in plans and specifications, little things can be easily overlooked and any designer can become the victim. As the saying goes, mistakes once committed are often difficult to detect in spite of a regress and careful review of plans and specifications.

The purpose of this chapter is to present many design errors and design deficiencies that have been discovered during construction, operation, and maintenance phases of various projects. Also in this chapter procedures have been presented that can be implemented by the design and review teams to reduce or eliminate many types of design errors.

25-2 EXAMPLES OF DESIGN ERRORS AND DEFICIENCIES

Some blunders are senseless, and their occurrence is perhaps due to an oversight or omission while details of the drawings are being worked out. Others may happen because a designer simply forgot to include some items in the design calculations, made a poor judgment or assumption, or perhaps overlooked the implications of the design criteria. Many other poor design provisions may not be the errors in the real sense, but may be least desirable from the point of view of plant construction or expansion, or may not provide the desired flexibility and convenience in plant operation and routine maintenance. Following is a list of many types of the design errors or blunders that have happened in many designs.

1. The flow scheme of a wastewater treatment facility showed the sidestreams from the sludge-processing area being returned to the primary sedimentation basins. Somehow, the BOD_5 and total suspended solids contributed by these flows were overlooked in the design of the primary and secondary facilities. As a result these units and the sludge handling facilities were underdesigned.

2. In one design the final clarifier was designed without the consideration of the

returned sludge. As a result, the actual detention time and overflow rate for the clarifier did not meet the design criteria under the average design flow conditions with recirculation.

3. In one design of a clarifier, Wright reported that the detailer neglected to show the steel bars from a cantilevered effluent launders hooked and embedded into the basin wall. No one detected this, and when the facility was first put into service, the launder dropped off.[1]

4. In another settling basin incidence, Wright cited that the designer correctly provided the corbels to support the weight of the effluent launder full of water. This condition could occur when the side gates on the effluent lines were closed and the basin drained. However, the designer inadvertently did not anchor the launder to the corbel against uplift. When the basin was first filled, the launders floated off the support due to buoyant force.[1]

5. In one case, Wright reported that a package-pumping station was installed in a flood plane. The station was provided with a submarine-type hatch and all electrical connections and leads were sealed against flooding. The pumps were controlled through a pneumatic system employing a bubbler tube for level sensing in the wet well. The bubbler tube was not extended above the flood stage. As a result, the station was flooded through the bubbler tube.[1]

6. Flooding of wastewater treatment facilities is a common occurrence in spite of the flood protection levees. Reported reasons for flooding have included the following: check valves or gates valves were not provided in the storm water or effluent pipes so the water backed into the plant area; storm water or effluent pumps were not provided; the storm drainage within the plant site was inadequately designed due to incorrect assumptions of rainfall intensity or runoff coefficient causing flash flooding; or flood protection levees were not high enough at some sections.

7. While preparing the unit layout, a design engineer lowered the foundation elevation of one treatment unit to avoid construction above the ground. However, he forgot to make proper changes in the hydraulic profile and relative elevations of the other units. As a result, the free water surface dropped below that of the other units that followed it.

Many similar incidences of design blunders can be cited where a change in the design assumption criteria or critical elevation was made, but the resulting effects were not carried throughout the entire design.

8. Wright reported that once while preparing an inked drawing, a tracer went down the line tracing one diagonal of X symbol of gate valves on the suction line. He never completed the other diagonal, and the pump station was constructed with check valves in the suction line.[1] Wright also gave an example of a detailer's error where 24 joists, 16-in. (41-cm) centers were incorrectly shown as 16 joists, 24-in. (61-cm) centers.[1]

9. In one design a chemical feed pump was connected to the chemical solution tank. The tank was not vented to the atmosphere. When the chemical feed pump withdrew the liquid, the tank collapsed due to atmospheric pressure.[1]

10. Many times the designed units flooded because the designer did not check the hydraulics at peak design flow under emergency conditions when the largest unit was out of service.

11. One rectangular primary sedimentation basin was constructed with endless

conveyor chains and the scraper mechanism running over sprockets. The scraper mechanism was not extended below the effluent launders to the far end wall. As a result, solids accumulated under the effluent structure and created serious odors and effluent quality problem. The equipment needed redesign.

12. Two screen chambers were designed for operation in parallel. Each chamber had gates in the front to stop the flow and remove one chamber from service for routine maintenance. Both channels discharged into a common box. Stop gates were not provided in the downstream box. As a result, the flow backed up from the effluent side and neither chamber could be completely drained. Similarly, in many designs, by pass pipes or gates were overlooked, and the unit could not be drained for maintenance purposes.

13. In one instance the designer misinterpreted the design criteria or inadvertently oversized the variable-speed pumps at a rate of 150 percent of the peak design flow. Naturally, the plant got flooded whenever the pumps operated at that flow.

14. Wright reported that once a flume was constructed to carry a raw water supply across a coulee. The piers were tall and unstable, but the structure would have been quite stable once the flume was in place and tied to the piers. During the construction phase, however, several piers fell over.[1]

15. In many designs the common wall between two units failed when one unit was drained. On many other occasions the bottom slab heaved upward because of hydrostatic pressure or the side wall collapsed due to the earth pressure when the units were drained. In all cases the designer forgot to check the structural safety under the most critical loading conditions.

16. In one instance, Wright reported that the discharge piping of a pump included a harnessed dresser coupling arrangement. The harnessed connection was to take the hydraulic thrust, and therefore the bolts were designed and specified to be high-tensile bolts. Instead, mild steel harness bolts were installed. When the pumps were tested, the harness bolts failed, resulting in pump base bending and the anchor bolts failing with consequent movement of the entire pumping unit away from the header.[1]

17. Many designs have been prepared with incorrect design criteria or design assumptions. As a result, extensive revisions had to be made later to meet the specific criteria supplied by the concerned regulatory agency. Examples of such criteria are overflow rates, detention time, solids loading, standby units, maximum capacity, chemical dosages, etc.

18. In one plant, Sanks reported that several fabricated vertical filters were spaced so closely that the operator had to squeeze sideways to get between them. In another plant he reported that the control panel required removal of face plate to replace the desiccant.[2]

19. Many designs have been prepared with little consideration of plant construction and future expansions. Other designs have overlooked the basic operation and maintenance needs of the facility. Common examples of such design deficiencies reported by Wright, Sanks, and others include:[1-3]

1. Not enough space was left for future expansion or another unit blocked the expansion.
2. A tee of a blind flange was not provided in a pipe at appropriate location for future connection.
3. Provisions were not made for future expansions of a building.

4. Enough work space was not left around a piece of machinery for servicing or putting small hoist for lifting heavy piece.
5. Bolted pipe fittings were casted into walls without enough clearance to install the bolt or put a wrench.
6. Valves were provided in lines without sufficient clearance for the handwheels or for the operators to reach them.
7. Ceiling hooks were not casted into concrete at appropriate places for chain hoist.
8. Dimensions of the equipment did not fit the unit size.
9. Floor drains were not provided or the floor was not properly sloped.
10. Door in the building was too small to bring in the equipment or take it out if the equipment was installed before the structure was completed.
11. Enough space for chemical storage was not provided or chemical feedings equipment was not properly designed.
12. Isolation of chemicals (chlorine feeders, hydrogen peroxide, oxygen, etc.) or isolation of noisy or heat producing equipment was not considered.
13. Freezing of the pipes, valves, and equipment in cold climate was disregarded, with disastrous consequences.

There is a long list of such small design deficiencies. The designer should be conscious of them, and proper considerations should be given to them during early stages of the design.

25-3 PROCEDURE TO AVOID OR REDUCE COMMON DESIGN ERRORS AND DEFICIENCIES

Once an error is made in plans and specifications, there is a good chance that it will escape detection. To avoid such incidents from happening, many design engineering firms have developed a comprehensive design review process. Although such review processes may vary from job to job based on availability of personnel within the organization, there are some basic items that must be considered under all conditions. Some of these items are listed below:

1. Several reviews of the design and checks of the plans and specifications are generally necessary at various stages of the design to eliminate errors and inconsistencies.
2. The preliminary design data of the wastewater treatment facility is developed as part of the facility-planning process (see Chapter 6). This includes design and initial flows, evaluation of alternatives and process selection, detailed process train, and preliminary unit sizing. Many states require justification and approval if major changes in the facility plan are expected. In the author's opinion, the facility plan must be used as a guide during the entire design calculation phase and during preparation of plans and specifications. However, a predesign review of the facility plan should be made by the designers, checkers, project engineer, and project manager. At this review the work plan, responsibilities, time schedules, budgets, and so on should be discussed. Any changes in design data developed in the facility plan (including process train) must be approved by every one concerned. Afterward, the design data should not be changed.
3. The second review should be made after the design of the individual treatment

units are complete. At this review the basic design criteria of the concerned regulatory agency, material mass balance analyses, flows and loadings under normal and emergency conditions (when largest unit is out of service), unit details, and the mechanical equipment for each unit should be checked. Considerations should be given to equipment compatibility, minimum and maximum chemical feed rates, operating pressures, ratings of the pumps, chemical feeders, etc. Any changes in meeting the necessary space, flexibility requirements, and equipment compatibility and performance should be made at this time.

4. The third review should be made after the layout plan, yard pipings, and hydraulic profile are complete. At this stage the review team should include the design engineer and checker, project manager or principles, and perhaps one or two experienced designers who have not been involved in the preliminary reviews. Small consulting engineering firms may not have enough in-house expertise. In such circumstances a competent consultant may be used to check the design.

This review should be made with a limited number of design items at a time. These items include (1) plant schematics and unit layout, (2) future expansion, (3) hydraulic profile and elevations of the treatment units, (4) space around equipment, (5) flexibility, (6) freezing, and (7) plant operation and maintenance. Each of these items and review considerations is discussed below.

1. The plant schematics and unit layout should be thoroughly checked. Special consideration should be given to the following items:

> Forward and return flow lines
> Material mass balance
> Basins and their volumes
> Pumps, piping, valves, gates, bypasses, collection and splitter boxes, etc.

2. Space for future expansion of the plants should be identified on the layout. This includes future basins, buildings, appurtenances, etc. Most designers prefer to mark these on the layout plan in dotted lines. This way, any unit or building obstructing the future expansion can be easily detected.

Pipe connections for future expansion should also be checked. Many designers prefer to provide tees and crosses at appropriate places; the blocked end is used for future connections.

3. Hydraulic calculations and profile should be thoroughly checked. Unit elevations, ground elevations, and water surface elevations at critical flow conditions should be marked on the hydraulic profile sheet. Omissions and errors in the hydraulic profiles can have disastrous consequences.

4. Adequate space around the equipment should be provided to go around, and place small hoist and other tools for equipment maintenance. Headrooms, doors and accesses, ceiling hooks for chain hoists, and the like should be carefully checked.

5. Plant flexibility is important to operate the plant properly under high and low flow conditions. Pipe channels, valves, and structures should be sized for peak design flow when the largest unit is out of operation. The chemical feeders should be sized for

extreme conditions. Space for chemical storage and future feeders should be properly planned.

6. Freeze protection of equipment and pipes should be checked if freezing is expected. Channels, pipes, sludge lines, and drains that flow intermittently should have enough earth cover or be designed to drain completely. Moving parts, such as sprockets, chain, etc., should be kept below freeze depth in the basin.

7. Operation and maintenance considerations should not be overlooked by the reviewing team. Some designers prefer writing an O&M manual during the design phase to include the general operational flexibility of the plant.[3] Following is a list of basic operation and maintenance items that should be checked by the review team:

Arrangements to bypass a unit should be made for routine maintenance.

Floor drains and proper slope should be provided for draining and flushing the unit.

Chlorine solution lines should be provided to wash the basin walls, weirs, channels, for routine operational needs.

Sampling points for each process should be clearly identified on the O&M manual. In large plants samples from different units are piped directly to the laboratory. This way the representative fresh samples can be obtained by the laboratory staff directly.

The review team should evaluate the equipment and controls for its complexity and maintenance needs. A simple system should always be preferred over a complex system.

Hand wheels and valves should have operating clearance.

The operating panels should be accessible, and strip charts, recording pens, and desiccant containers should be easily replaceable.

Hazardous chemicals (chlorine, ozone, etc.) should be isolated. Noisy and heat-producing equipment should also be isolated. Arrangement should be made for dry feeder equipment for easy filling.

All stairs and walkways should have proper hand rails.

Laboratory and workshops should have adequate space, equipment and tools, lavatory and showers.

As mentioned earlier each review process may vary from project to project and the availability of resources of the design firm. Each firm should develop the review procedure that may best fit its resources and the needs of the project. The review process given above should be used only as a guide, as it is neither complete nor definitive.

25-4 DISCUSSION TOPICS

25-1. Information checklists have been presented in Chapters 7 through 19 for the design of various treatment processes. These checklists were highly specialized for individual processes. Using this information, develop a generalized design checklist for an entire wastewater treatment plant that a designer must use as a guide during the entire design phase of a project.

25-2. Develop a comprehensive checklist for design review that must be used by a review team. Arrange the checklist items under the following headings:

    (a)   facility plan information that must be carried out in the plans and specifications
    (b)   process train, instrumentation
    (c)   plant layout and piping
    (d)   buildings and equipment
    (e)   plant hydraulics
    (f)   roads, drainage and landscaping
    (g)   operation and maintenance flexibility
    (h)   expansion flexibility

25-3. You have been hired as an environmental engineer by a small consulting engineering company to complete the final plans and specifications of a medium-size wastewater treatment plant. There is one additional civil-environmental engineer in your organization. The facility plan for the project was prepared by another engineer who is no more with this organization. Discuss how you would organize the project components including revisions to the facility plan, design tasks and review process, review team, and project time table. The project must be completed in 18 months. The project involves upgrading of the existing primary treatment facility and pumping station, and adding a new activated sludge treatment facility.

25-4. Complete Problem 25-3 assuming that a large consulting engineering company has hired you as a junior design engineer to assist a project manager who prepared the facility plan.

REFERENCES

1. Wright, James R., "Design Blunders," Chapter 31 in *Water Treatment Design for Practicing Engineers,* (Robert L. Sanks, ed.), Ann Arbor Science Publisher, Ann Arbor, Mich., 1978.
2. Sanks, Robert L., "How to Avoid Design Blunders," Chapter 33, *Water Treatment Plant Design for Practicing Engineers,* (Robert L. Sanks, ed.), Ann Arbor Science Publisher, Ann Arbor, Mich., 1978.
3. Regan, Terry M., "O & M Manuals and Operator Training," Chapter 32, *Water Treatment Plant Design for Practicing Engineers,* (Robert L. Sanks, ed.), Ann Arbor Science Publishers, Ann Arbor, Mich., 1978.

Physical and Chemical Properties of Water

The principal physical and chemical properties of water commonly used in wastewater treatment plant design are summarized in this appendix. Most of the information is developed from Refs. 1–5.

REFERENCES

1. Lange, N. A. (ed.), *Handbook of Chemistry*, 10th ed., McGraw-Hill Book Co., New York, 1961.
2. Morris, H. M., and J. H. Wiggert, *Applied Hydraulics in Engineering*, Ronald Press, New York, 1972.
3. Janna, W. S., *Introduction to Fluid Mechanics*, Brooks/Cole, Monterey, California, 1983.
4. Qasim, S. R., *Laboratory Manual for CE 3252, Water Quality Measurement and Pollution Control*, Department of Civil Engineering, The University of Texas at Arlington, 1982.
5. APHA, AWWA, and WPCF, *Standard Methods for Examination of Water and Wastewater*, American Public Health Association, 15th ed., Washington, D.C., 1980.

Table A-1   Properties of Water

Molecular formula	=	H_2O
Molecular weight	=	18
Ionization constant (K_w) at 25°C	=	10^{-14}
Specific weight, γ at 4°C	=	9.81 kN/m³ (62.43 lb_f/ft^3)
1 U.S. gal weighs (average)	=	8.34 lb
Density at 4°C	=	1 g/cm³ (1.94 slug/ft³)
Specific heat	=	1 cal/g°C (1 Btu/lb°F)
Dynamic viscosity, μ	=	1×10^{-3} N·s/m² (2.089×10^{-5} $lb_f·s/ft^2$
Kinematic viscosity, ν	=	1.002×10^{-6} m²/s (1.078×10^{-5} ft²/s)
One atm (760 mm Hg or 14.7 psi)	=	10.33 m (33.9 ft) water
Boiling point at 1 atm	=	100°C (212°F)
Melting point at 1 atm	=	0°C (32°F)
Heat of fusion at 0°C	=	80 cal/g (144 Btu/lb)
Heat of vaporization at 100°C	=	540 cal/g (973 Btu/lb)
Modulus of elasticity at 15°C	=	2.15 kN/m² (312×10^3 lb_f/in^2)

Temperature T		Density ρ, g/cm³	Dynamic Viscosity $\mu \times 10^3$, N·s/m²	Kinematic Viscosity $\nu \times 10^6$, m²/s	Surface Tension Against Air $\sigma \times 10^3$, N/m	Vapor Pressure ρ, mm Hg
°C	°F					
0	32	0.9998	1.787	1.787	75.60	4.579
3.98	39.2	1.0000	1.568	1.568	75.00	6.092
5	41	0.9999	1.519	1.519	74.90	6.543
10	50	0.9997	1.307	1.307	74.22	9.209
15	59	0.9991	1.139	1.140	73.49	12.788
20	68	0.9982	1.002	1.004	72.75	17.535
25	77	0.9970	0.890	0.893	71.97	23.756
30	86	0.9957	0.798	0.801	71.18	31.824
35	95	0.9941	0.719	0.724	70.37	42.175
40	104	0.9922	0.653	0.658	69.56	55.324
45	113	0.9903	0.596	0.602	68.74	71.88
50	122	0.9881	0.547	0.553	67.91	92.51

Table A-2   Solubility of Oxygen and Other Gases in Water

Temperature		Dissolved Oxygen Saturation[a] C_s, mg/ℓ					Decrease in Oxygen Concentration per 100 mg Chloride	Solubility of Other Gases, mg/ℓ		
°C	°F	Chloride Concentration, mg/ℓ						Nitrogen	Carbon Dioxide	Air
		0	5000	10,000	15,000	20,000				
0	32	14.62	13.79	12.97	12.14	11.32	0.017	22.81	1.00	37.50
5	41	12.80	12.09	11.39	10.70	10.01	0.014	20.16	0.83	32.94
10	50	11.33	10.73	10.13	9.55	8.98	0.012	17.93	0.70	29.27
15	59	10.15	9.65	9.14	8.63	8.14	0.010	16.15	0.59	26.25
20	68	9.17	8.73	8.30	7.86	7.42	0.009	14.72	0.51	23.74
25	77	8.38	7.96	7.56	7.15	6.74	0.008	13.55	0.43	21.58
30	86	7.63	7.25	6.86	6.49	6.13	0.008	12.53	0.38	19.60

[a]The solubility of oxygen (C_s) exposed to air containing 20.9 percent oxygen by volume at one atmosphere (760 mm Hg). Under any other barometric pressure the solubility can be calculated from the equation

$$C'_s = \frac{C_s (P-p)}{(760 - p)}$$

where p = vapor pressure at the temperature of water, P = atmospheric pressure, mm Hg.

Table A-3   Barometric Pressure with Altitude

Elevation above Sea Level		Absolute Pressure in Head of Water (H_{abso})	
m	ft	m	ft
0	0	10.3	33.9
305	1000	10.0	32.8
457	1500	9.8	32.1
610	2000	9.6	31.5
1219	4000	8.9	29.2
1829	6000	8.3	27.2
2438	8000	7.7	25.2
3048	10,000	7.1	23.4
4572	15,000	5.9	19.2

Head Loss Constants in Open Channels and Pressure Pipes

In this appendix the constants commonly used to calculate minor head losses into open channels and pressure conduits are given. Many of these constants have been used in the Design Example to calculate the losses due to contraction and expansion, and exit and entrance in the appurtenances. The values of most of these constants may be obtained in many hydraulics texts or handbooks. The readers are referred to the references cited at the end of this appendix.

A computer program for computing the depth of flow in a flume, effluent trough or launder receiving flow from a free falling weir [Eq. (11-4)] is also given in this appendix. Discussion on this subject may be found in Chapters 11–13.

MINOR LOSSES IN OPEN CHANNELS

Sudden Contraction or Inlet Losses

Sharp-cornered entrance $\qquad$ $0.5 \left(\dfrac{v_1^2}{2g} - \dfrac{v_2^2}{2g} \right)$

Round-cornered entrance $\qquad$ $0.25 \left(\dfrac{v_1^2}{2g} - \dfrac{v_2^2}{2g} \right)$

Bell-mouthed entrance $\qquad 0.05 \left(\dfrac{v_1^2}{2g} - \dfrac{v_2^2}{2g} \right)$

where v_1 and v_2 are velocities downstream and upstream of the contraction.

Sudden Enlargement or Outlet Losses

Sharp-cornered outlet $\qquad 0.2\text{–}1.0 \qquad \left(\dfrac{v_2^2}{2g} - \dfrac{v_1^2}{2g} \right)$

Bell-mouthed outlet $\qquad 0.1 \qquad \left(\dfrac{v_2^2}{2g} - \dfrac{v_1^2}{2g} \right)$

where v_1 and v_2 are velocities downstream and upstream of the enlargement.

Syphons

Head loss $\qquad 2.78 \ \dfrac{v^2}{2g}$

Manhole, Junction, and Division Boxes

Head loss in manhole, junction, and division boxes depend on the size of the appurtenances, change in direction, and the contour of the bottom.

In a straight-through manhole where there is no change in pipe size

Manhole losses $\qquad 0.05 \qquad \dfrac{v^2}{2g}$

Terminal manhole $\qquad 1 \qquad \dfrac{v^2}{2g}$

In large junction-division boxes in which the velocity is small

Exit loss $\qquad 1.0 \ \dfrac{v^2}{2g}$

Entrance loss $\qquad 0.5 \ \dfrac{v^2}{2g}$

If sewer size changes, or change in direction occurs in a manhole or junction box

$45°$ no shaping $\qquad 0.4 \qquad \dfrac{v^2}{2g}$

$45°$ with shaping $\qquad 0.3 \qquad \dfrac{v^2}{2g}$

$90°$ no shaping $\qquad 1.3 \qquad \dfrac{v^2}{2g}$

90° with shaping   1.0 $\dfrac{v^2}{2g}$

MINOR LOSSES IN PRESSURE CONDUITS

Gate Valve

Fully open   0.19 $\dfrac{V^2}{2g}$

One fourth closed   1.15 $\dfrac{V^2}{2g}$

One half closed   5.6 $\dfrac{V^2}{2g}$

Three fourths closed   24.0 $\dfrac{V^2}{2g}$

Butterfly Valve

Fully open   0.3 $\dfrac{V^2}{2g}$

Angle closed, 20°   1.4 $\dfrac{V^2}{2g}$

Angle closed, 40°   10 $\dfrac{V^2}{2g}$

Angle closed, 60°   94 $\dfrac{V^2}{2g}$

Check Valve

Swing check   (0.6–2.3) $\dfrac{V^2}{2g}$

Swing check fully open   2.5 $\dfrac{V^2}{2g}$

Sluice Gate

Fully open   (0.2–0.8) $\dfrac{V^2}{2g}$

Plug Globe Valve

Fully open   4.0 $\dfrac{V^2}{2g}$

Diaphragm Valve

Fully open $2.3 \; \dfrac{V^2}{2g}$

Sudden Contraction

d/D = one fourth $0.42 \; \dfrac{V^2}{2g}$

d/D = one half $0.33 \; \dfrac{V^2}{2g}$

d/D = three fourths $0.19 \; \dfrac{V^2}{2g}$

Sudden Enlargement

d/D = one fourth $0.92 \; \dfrac{V^2}{2g}$

d/D = one half $0.56 \; \dfrac{V^2}{2g}$

d/D = three fourths $0.19 \; \dfrac{V^2}{2g}$

Elbows

90° elbow, regular $(0.2\text{–}0.3) \; \dfrac{V^2}{2g}$

90° elbow, long $(0.14\text{–}0.23) \; \dfrac{V^2}{2g}$

45° elbow, use three fourths of losses for 90° elbow
$22\frac{1}{2}°$ elbow, use one half of losses for 90° elbow

Entrance

Pipe projecting into tank $0.83 \, \dfrac{V^2}{2g}$

End of pipe flushed with tank $0.50 \, \dfrac{V^2}{2g}$

Slightly rounded $0.23 \, \dfrac{V^2}{2g}$

Bell-mouthed $0.04 \, \dfrac{V^2}{2g}$

Exit

From pipe into still water $1.0 \, \dfrac{V^2}{2g}$

Venturi Meter

Angle of divergence 5° $(0.10\text{--}0.15) \, \dfrac{V^2}{2g}$ (throat velocity)

Angle of divergence 15° $(0.06\text{--}0.33) \, \dfrac{V^2}{2g}$ (throat velocity)

COMPUTER PROGRAM

```
C    THIS PROGRAM IS TO CALCULATE THE DEPTH OF FLOW IN
C              A CHANNEL FOR A GIVEN SUBMERGENCE.

C    THE COMPUTATIONAL PROCEDURE FOR OBTAINING THE DEPTH OF FLOW IN
C    THE FLUME OR EFFLUENT TROUGH IS GIVEN IN SEC.11-9-2 STEP E6.
C    THE INPUT DATA FOR DESIGNING THE EFFLUENT TROUGH IN SEC.13-7-4
C    STEP C4 IS GIVEN BELOW. THE RESULTS OF COMPUTER ANALYSIS ARE
C    SUMMARIZED IN TABLE 13-8.

C    N          = MANNING COEFFICIENT
C    W          = WIDTH OF THE CHANNEL
```

```
C   Q               = TOTAL FLOW IN THE CHANNEL
C   DX              = HORIZONTAL DISTANCE BETWEEN SECTIONS  1 AND 2
C   RL              = LENGTH OF THE CHANNEL
C   S               = SLOPE OF THE BOTTOM OF THE CHANNEL
C   G               = GRAVITY CONSTANT
C   X1 AND X2 = DISTANCES FROM UPPER END AT SECTIONS 1 AND 2
C   Q1 AND Q2 = DISCHARGES AT SECTIONS 1 AND 2
C   V1 AND V2 = VELOCITIES AT SECTIONS 1 AND 2
C   DRO             = ASSUMED DROP IN WATER SURFACE ELEVATION
C                       BETWEEN SECTIONS 1 AND 2
C   INC             = INCREMENT OF DRO
C   DX              = HORIZONTAL DISTANCE BETWEEN SECTIONS 1 AND 2
C   SAVG            = AVERAGE SLOPE OF THE ENERGY LINE
C   STA             = LENGTH OF THE EFFLUENT TROUGH FROM UPPER END
C   Y1 AND Y2 = DEPTH OF FLOW AT SECTIONS 1 AND 2
C   DC              = DROP IN WATER SURFACE ELEVATION BETWEEN
C                       SECTIONS 1 AND 2
C   YC              = WATER DEPTH AT THE CONTROL POINT(LOWER END
C                       OF THE CHANNEL)
C   ACC             = TOLERANCE LEVEL OF  DRO-DC
C   LEV             = LIMITED VALUE OF DRO
```

```
 1          REAL N,INC
 2          WRITE(6,999)
 3      999 FORMAT('1',T11,'STA',T18,'Y2',T25,'DP',T32,'Y1',T39,'Q2',
            $T46,'Q1',T53,'V2',T60,'V1',T67,'DC',/)
 4          XS=0.
 5          READ(5,*)N,W,RL,S,G,INC,Q2,YC
 6          X2=RL
 7          Y2=YC
 8          READ(5,*) DRO,ACC,LEV
 9       10 DP=DRO
10          READ(5,*)Q1,X1
11          DX=X2-X1
12       20 DY=(S*DX)-DP
13          Y1=Y2-DY
14          V2=Q2/(W*Y2)
15          V1=Q1/(W*Y1)
16          R1=(W*Y1)/((2*Y1)+W)
17          R2=(W*Y2)/((2*Y2)+W)
18          RAVG=(R1+R2)/2
19          SAVG=((N**2)*((V1+V2)**2))/(4.00*(RAVG**1.333))
20          TERM1=(Q1*(V1+V2))/(G*(Q1+Q2))
21          DQ=Q2-Q1
22          DV=V2-V1
23          TERM2=DV+((V2/Q1)*DQ)
24          DC=(TERM1*TERM2)+(SAVG*DX)
25          IF(ABS(DP-DC).LT.ACC) GO TO 30
26          IF(DP.GT.LEV) GO TO 50
27          DP=DP+INC
28          GO TO 20
29       30 X1=X2-DX
30          WRITE(6,99)X2,Y2,DP,Y1,Q2,Q1,V2,V1,DC
31          IF(X1.LE.DX)GO TO 60
32          XS=XS+DX
33          X2=X1
```

```
34              Y2=Y1
35              Q2=Q1
36              GO TO 10
37         60 DP=(2*(V1**2))/(2*G)
38              DC=DP
39              X2=X1
40              Y2=Y1
41              V2=V1
42              Q2=Q1
43              V1=0.
44              DF=X2-XS
45              DY=(S*DF)-DP
46              Y1=Y2-DY
47              WRITE(6,99)X2,Y2,DP,Y1,Q2,Q1,V2,V1,DC
48              GO TO 199
49         50 WRITE(6,55) LEV
50         55 FORMAT(5X,'DP IS GREATER THAN',I2)
51         99 FORMAT(7X,9(1X,F6.3),/)
52        199 WRITE(6,88)
53         88 FORMAT('1')
54              WRITE(6,333)N,W,RL,S,G
55        333 FORMAT('1','        *******PROBLEM  DATA********',//,
              $        '              MANNINGS COEFFICIENT=',F6.3,/,
              $        '              CHANNEL   WIDTH      =',F6.3,/,
              $        '              CHANNEL   LENGTH     =',F6.3,/,
              $        '              BOTTOM CHANNEL SLOPE=',F6.3,/,
              $        '              GRAVITY CONSTANT    =',F6.3,//,
              $        ' Q1 AND X1 ARE THE VALUES AT EACH SECTION',/,
              $        '    AS SHOWN IN THE RESULTS.')
56              WRITE(6,77)
57         77 FORMAT('1')
58              STOP
59              END
       $ENTRY
```

```
INPUT DATA FOR DESIGN EXAMPLE IN SEC.13-7-4 STEP C4
     .013,1.0,14.5,0.,9.81,0.00002,.538,.44
     .00009,.0001,10
     .47075,14
     .47075,12.5
     .4035,12.0
     .4035,10.5
     .3362,10.0
     .3362,8.5
     .269,8.0
     .269,6.5
     .20175,6.0
     .20175,4.5
     .1345,4.0
     .1345,2.5
     .06725,2.0
     .06725,0.5
```

REFERENCES

1. FMC Corporation, *Hydraulics and Useful Information,* Chicago Pumps, Chicago, Ill., 1973.
2. Brater, E. F., and H. W. King, *Handbook of Hydraulics,* McGraw-Hill Book Co., New York, 1976.
3. Shaw, B. V., and A. W. Loomis, *Cameron Hydraulic Data,* Compressed Air Magazine Co., Phillipsburg, N.J., 1965.
4. Morris, H. M., and J. M. Wiggert, *Applied Hydraulics in Engineering,* Ronald Press, New York, 1972.
5. American Iron and Steel Institute, *Modern Sewer Design,* Washington, D.C., 1980.
6. Joint Committee of the ASCE and WPCF, *Design and Construction of Sanitary and Storm Sewers,* MOP/9, Water Pollution Control Federation, Washington, D.C., 1970.
7. Benefield, L. D., J. F. Judkins, and A. D. Parr, *Treatment Plant Hydraulics for Environmental Engineers,* Prentice-Hall, Englewood Cliffs, N.J., 1984.

Treatment Plant Cost Curves

C

The generalized construction as well as operation and maintenance cost curves for primary and secondary treatment unit operations and processes are given in this appendix. These cost curves are taken from EPA *Areawide Assessment Procedures Manual: Performance and Cost.*[1] The cost estimates based on these curves are valid only for comparison and evaluation of alternatives in the facility plans. There may be a considerable difference between cost estimates for comparative purposes and the actual construction costs of the facility that are developed after completion of plans and specifications.

All construction costs have been indexed to September 1976. To adjust for other time periods, appropriate costs indexes, such as those in the following list, should be used when appropriate:

- ENR Building Cost Index—appears weekly in *Engineering News Record,* McGraw-Hill.
- ENR Construction Cost Index—appears weekly in *Engineering News Record,* McGraw-Hill.
- EPA Treatment Plant and Sewer Construction Cost Index—appears monthly in the *WPCF Journal.*
- BLS Labor Cost Index—appears monthly in *Employment and Earnings,* Bureau of Labor Statistics.
- BLS Wholesale Price Index—appears monthly in *Wholesale Prices and Price Indexes,* Bureau of Labor Statistics.

Each cost curve given in this appendix includes a summary of the design basis, assumptions, and the cost basis. Generalized adjustment factors are also provided to modify cost data for any desired solids and hydraulic loadings, detention times, solids concentrations, periods of operation, and the like.

Table C-1 provides the readers with the generalized basis of all cost curves contained in this appendix. The construction costs do not include piping, electrical, instrumentation, site preparation, engineering and construction, supervision, and contingencies. These

655

Table C-1   General Cost and Design Basis for Cost Curves

Basis of Costs

1. ENR = 2475, September 1976
2. Labor rate, including fringe benefits = $7.50/h   *Note:* Labor costs are based on a man-year of 1500 h. This represents a 5-day work week; an average of 29 days for holidays, vacations, and sick leave; and $6\frac{1}{2}$ hours of productive work time per day.
3. Energy costs

a. Electric power	=	$0.02/kW · h
b. Fuel oil	=	$0.37/gal
c. Gasoline	=	$0.60/gal

4. Land   = $1000/acre
5. Chemical costs

a. Chlorine	150-lb cylinder	=	$360/ton
	1-ton cylinder	=	$260/ton
	Tank car	=	$160/ton
b. Quicklime		=	$25/ton
c. Hydrated lime		=	$30/ton (as CaO)
d. Polymer (dry)		=	$1.50/lb
e. Ferric chloride		=	$100/ton
f. Alum		=	$72/ton
g. Sulfuric acid		=	$50/ton

Design Basis

1. Construction costs and operation and maintenance costs are based on design average flow unless otherwise noted.
2. Operation and maintenance costs include:
    a. Labor costs for operation, preventive maintenance, and minor repairs
    b. Materials costs include replacement parts and major repair work (normally performed by outside contractors)
    c. Chemical costs
    d. Fuel costs
    e. Electrical power costs
3. Construction costs do not include land costs (except for landfilling), external piping, electrical, instrumentation, site work, contingency, engineering and construction-supervision, or miscellaneous structures.

---

costs are added as a percentage of installed construction costs of the components specific to the unit or system being evaluated for comparative cost estimates. Table C-2 represents a general format for identifying the costs, with representative percentages for the items indicated. The cost curves presented in this appendix include:

- Gravity sewers, Figure C-1
- Low-lift pump station, Figure C-2
- Preliminary treatment, Figure C-3

Table C-2   Development of Capital Costs

	Avg.	Range[b]	
Total cost of unit processes specific to each cost estimate			$ _____
Misc. structures (Figure C-17)[a]			$ _____
Subtotal 1			$ _____
Piping	10%	8–15%	$ _____
Electrical	8%	5–12%	$ _____
Instrumentation	5%	3–10%	$ _____
Site preparation	5%	1–10%	$ _____
Subtotal 2			$ _____
Engineering and construction-supervision @ 15%[c]			$ _____
Contingencies @ 15%[c]			$ _____
Total capital cost[d]			$ _____

[a]Miscellaneous structures include administrative offices, laboratories, shops, and garage facilities.
[b]Range due to level of complexity, degree of instrumentation, subsoil conditions, configuration of site, etc.
[c]Percentage of subtotal 2.
[d]See Tables 6-11 and 6-12 for calculation procedure.

- Primary clarifier with sludge pumps, Figure C-4
- Conventional activated sludge aeration with diffused air, Figure C-5
- Final clarifier (flocculator-type), used with aeration basin, Figure C-6
- High-rate trickling filter, Figure C-7
- Clarifier for high-rate trickling filter (includes recycle pumps), Figure C-8
- Chlorination (disinfection), Figure C-9
- Sludge pumping, Figure C-10
- Gravity thickener, Figure C-11
- Aerobic digesters, Figure C-12
- Two-stage anaerobic digesters, Figure C-13
- Sludge-drying beds, Figure C-14

- Filter press (biological sludge), Figure C-15
- Landfilling (biological sludge, excluding transportation), Figure C-16
- Miscellaneous structures, Figure C-17
- Support personnel, Figure C-18

For the cost estimates of many other unit operations and processes, such as chemical coagulation, nitrification–denitrification, carbon adsorption, ammonia stripping, ion exchange, membrane processes, etc., the reader should consult the EPA *Areawide Assessment Procedures Manual: Performance and Cost.*[1] Many other sources listed in the references can also be used for similar cost curves.[2–4] Several computer programs for obtaining costs and performance data for various treatment units are also available in the literature.[5–7]

GRAVITY SEWERS

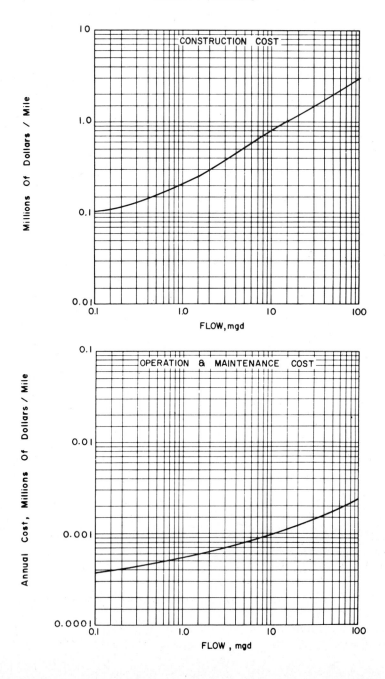

Figure C-1   Gravity Sewers ($\ell/s \times 0.0228 = mgd$).

Service Life: 50 years

Design Basis:
1. *Peaking factor allowance ranges from 3 for average flows of 43.8 ℓ/s (1 mgd) and less, to 2 for average flows greater than 438 ℓ/s (10 mgd)*
2. *Average slope of 0.002 to 0.005*
3. *Velocity of not less than 0.61 mps (2 fps) when flowing full at peak flow; minimum sewer size = 20 cm (8 inches)*
4. *Repaving of road surface required for 10 percent of distance*
5. *Does not include right-of-way or aerial crossing, etc.*

LOW-LIFT PUMP STATION

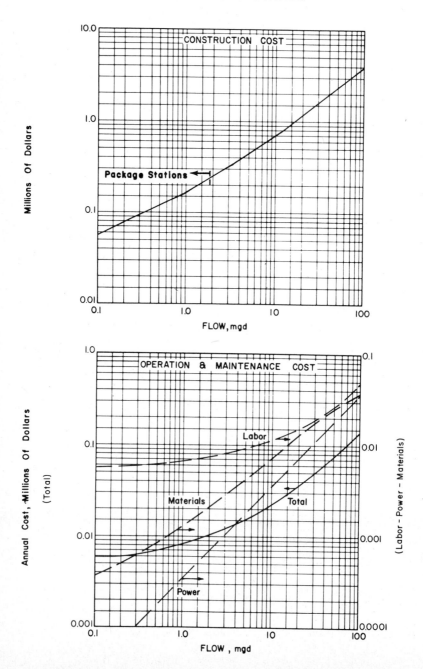

Figure C-2   Low-Lift Pump Station (ℓ/s $\times$ 0.0228 = mgd).

Service Life: 15 years

Design Basis:
1. *Construction costs include:*
   a. *Fully-enclosed wet well/dry well structure*
   b. *Pumping equipment capable of meeting peak pumping requirements of 2Q with largest unit out of service*
   c. *Standby pumping facilities*
   d. *Piping and valves within structure*
   e. *Bar screens—mechanically cleaned*
2. *Package pumping stations are used in a flow range of 3.2 to 75.8 ℓ/s (0.072 to 1.73 mgd)*
3. *Curves developed for TDH = 3.05 m (10 ft)*

Adjustment Factor:
   To adjust operating power costs for TDH other than 3.05 m, enter curves at effective flow, Q_E

$$Q_E = Q_{DESIGN} \times \frac{TDH}{3.05 \text{ m (10 ft)}}$$

PRELIMINARY TREATMENT

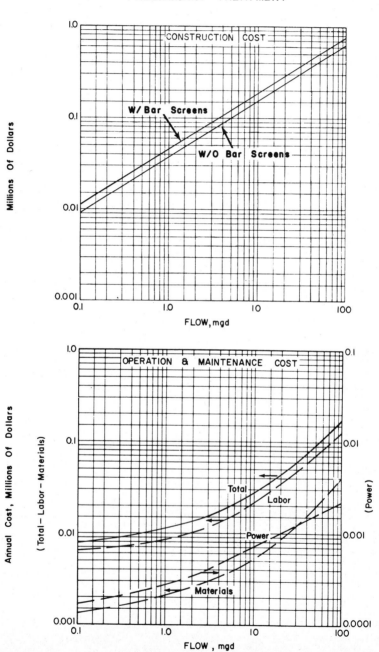

Figure C-3   Preliminary Treatment ($\ell/s \times 0.0228 = mgd$).

Service Life: 30 years

Design Basis:
1. *Construction costs include:*
   a. *Flow channels and superstructures*
   b. *Bar screens (mechanical)*
   c. *Grinders for screenings*
   d. *Gravity grit chamber with mechanical grit-handling equipment*
   e. *Parshall flume and flow-recording equipment*
2. *Operation and maintenance costs do not include cost for grit disposal*
3. *Screenings $7.5 - 22.5$ $m^3/10^6 \cdot m^3$ $(1 - 3$ $ft^3/10^6 \cdot gal.)$*
4. *Grit $15 - 37.5$ $m^3/10^6 \cdot m^3$ $(2 - 5$ $ft^3/10^6 \cdot gal.)$*
5. *If low-lift pumping is used prior to preliminary treatment, the cost for bar screens can be subtracted from the unit cost of preliminary treatment, because they are included in the low-lift pumping station*

Process Performance:
Suspended material greater than 1.5 cm (5/8 inch) and grit coarser than 0.208 mm (65 mesh) will be removed. However, removal of BOD and TSS by pretreatment is assumed negligible.

PRIMARY CLARIFIER WITH SLUDGE PUMPS

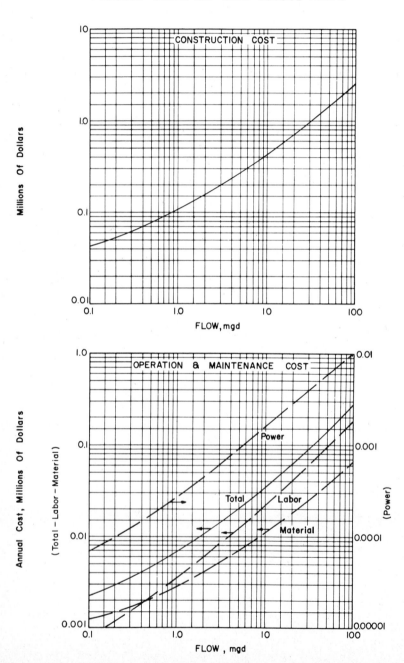

Figure C-4   Primary Clarifier with Sludge Pumps ($\ell/s \times 0.0228 =$ mgd).

Service Life: 50 years

Design Basis:
1. *Clarifier designed for surface overflow rate of 32.6 m³/m²·d (800 gpd/ft²) at average Q*
2. *Costs include primary sludge pumps. Sludge concentration of 4 percent solids. Pump head assumed as 3.05 m (10 ft)*

Adjustment Factor:
To adjust costs for alternative surface overflow rate, enter flow at effective flow (Q_E)

$$Q_E = Q_{DESIGN} \times \frac{32.6 \text{ m}^3/\text{m}^2 \cdot \text{d (800 gpd/ft}^2)}{\text{New design surface overflow rate}}$$

CONVENTIONAL ACTIVATED SLUDGE
AERATION WITH DIFFUSED AIR

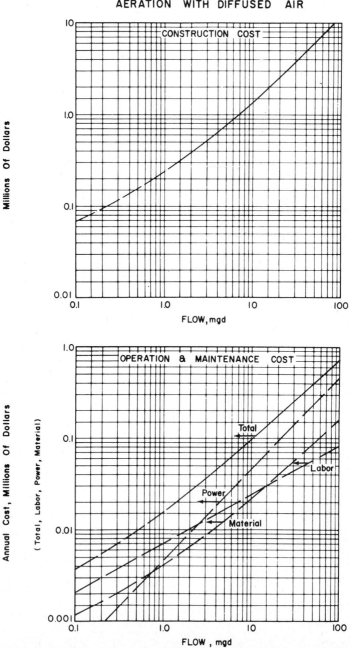

Figure C-5   Conventional Activated Sludge Aeration with Diffused Air (ℓ/s × 0.0228 = mgd).

Service Life: 40 years

Design Basis:
1. *Construction costs include costs for basins, air supply equipment and piping, and blower building. Clarifier and recycle pumps are not included. Clarifier costs are found in Figure C-6.*
2. *Diffused aeration*
3. *1.2 g O_2 supplied per g of BOD_5 removed*
4. *MLVSS = 2,000 mg/ℓ*
5. *F/M = 0.5*
6. *Detention time = 6 hours*

FINAL CLARIFIER (FLOCCULATOR TYPE)

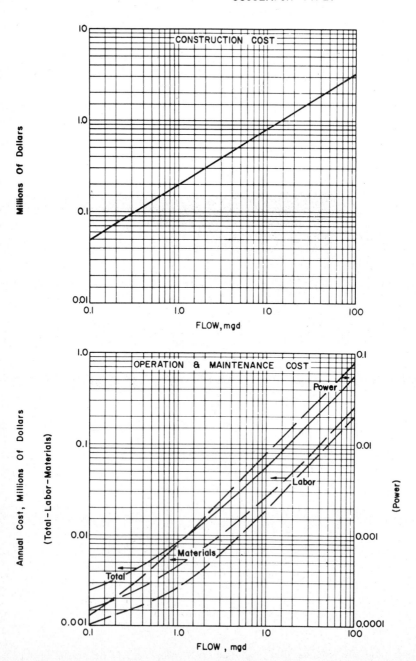

Figure C-6   Final Clarifier (Flocculator Type), Used with Aeration Basin (ℓ/s $\times$ 0.0228 = mgd).

Service Life: 40 years

Design Basis:
1. *Flocculator-type clarifier*
2. *Overflow rate of 24.5 m^3/m^2·d (600 gpd/ft^2) used for development of costs*
3. *Costs include sludge return and waste pumps at sludge concentration of 1 percent solids, pump TDH of 3.05m (10 ft), pump capacity 50 percent of plant capacity, nonclog centrifugal pumps, and spare pumps included as necessary.*
4. *Rectangular units when surface area is less than 46.45 m^2 (500 ft.^{2})*
5. *Maximum clarifier diameter = 61m (200 ft)*

Adjustment Factor:
To adjust capital cost for alternative overflow rates, enter the curve at effective flow (Q_E)

$$Q_E = Q_{DESIGN} \times \frac{24.5 \ m^3/m^2 \cdot d \ (600 \ gpd/ft^2)}{New \ design \ overflow \ rate}$$

HIGH-RATE TRICKLING FILTER

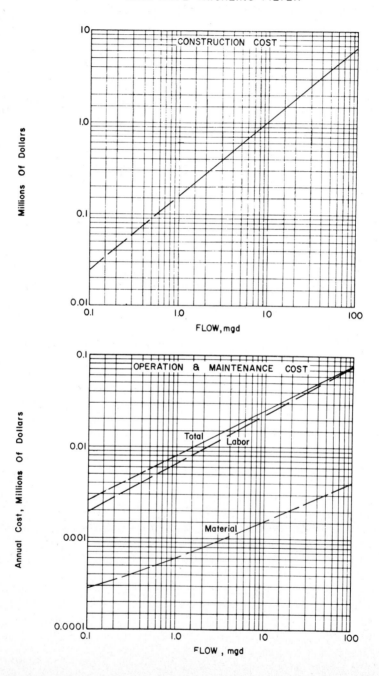

Figure C-7   High Rate Trickling Filter ($\ell/s \times 0.0228 = mgd$).

Service Life: 50 years

Design Basis:
1. Construction costs include circular filter units with rotating distributor arms, synthetic media 1.8 m deep (6 ft), and underdrains. Clarifiers and recycle equipment not included (see Figure C-8)
2. Organic loading rate: High rate = 0.72 kg/m^3 d (45 lb BOD$_5$/10^3 ft^3·d)
3. Hydraulic loading rate: High Rate = 28.3 m^3/m^2·d (693 gpd/ft^2) at 3 : 1 recycle rate
4. Electrical power not required (included in clarifier cost)

CLARIFIER FOR HIGH RATE TRICKLING FILTER
(INCLUDES RECYCLE PUMPS)

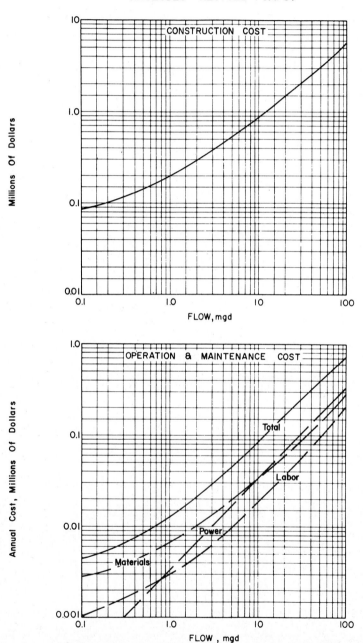

Figure C-8 Clarifier for High-Rate Trickling Filter (Includes Recycle Pumps) (ℓ/s × 0.0228 = mgd).

Service Life: 40 years

Design Basis:
1. *Construction costs include: sludge pumps, effluent recycle pumps, clarifier mechanisms and internal piping*
2. *Over flow rate = 32.6 m^3/m^2·d (800 gpd/ft^2) at average design flow*
3. *Recycle pumping capacity = 3 time average wastewater flow*
4. *Curve is to be used in conjunction with high-rate trickling filter only*
5. *Rectangular units when surface area is less than 46.45 m^2 (500 ft^2). Circular units when surface area is greater than 46.45 m^2 (500 ft^2)*
6. *Maximum clarifier diameter = 61 m (200 ft)*

Adjustment Factor:
To adjust construction cost for alternative loading rates, enter curve at effective flow (Q_E)

$$Q_E = Q_{DESIGN} \times \frac{32.6 \text{ m}^3/\text{m}^2 \cdot \text{d (800 gpd/ft}^2)}{\text{New design overflow rate}}$$

CHLORINATION (DISINFECTION)

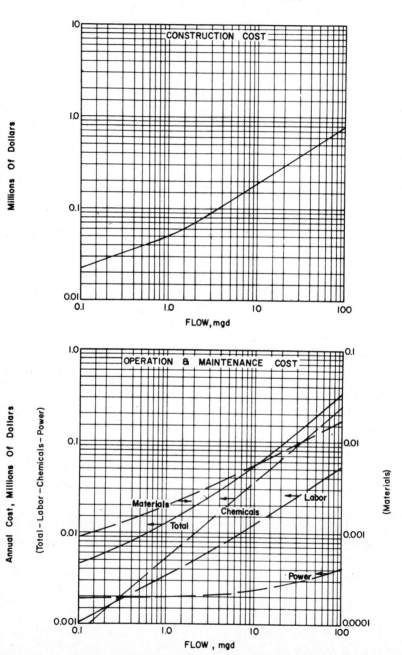

Figure C-9   Chlorination (Disinfection) (ℓ/s × 0.0228 = mgd).

Service Life: 15 years

Design Basis:
1. *Construction costs include:*
    a. *Chlorination building*
    b. *Chlorine storage and handling facilities including hoists, etc.*
    c. *Chlorinators*
    d. *Plug flow contact chamber*
2. *Average chlorine dosage = 10 mg/ℓ*
3. *Chlorination contact time = 30 min for average flow*
4. *Chlorine residual = 1 mg/ℓ*

Adjustment Factor:
To adjust costs, enter curves at effective flow (Q_E)

$$\text{Chemical Cost (O\&M): } Q_E = Q_{DESIGN} \times \frac{\text{New chlorine dosage, mg/}\ell}{10 \text{ mg/}\ell}$$

SLUDGE PUMPING

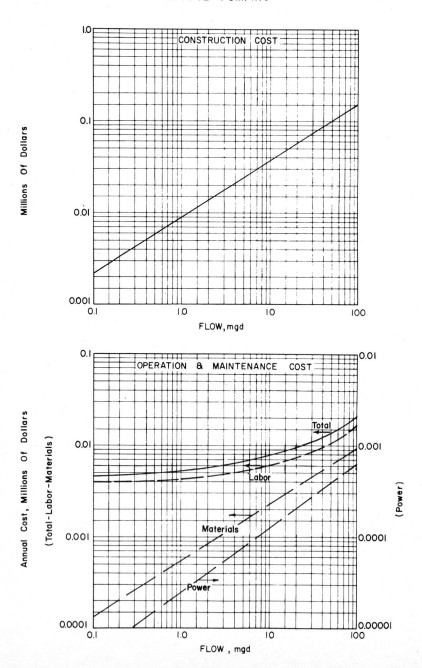

Figure C-10   Sludge Pumping (ℓ/s × 0.0228 = mgd).

Service Life: 10 years

Design Basis:
1. *Costs are based on a sludge loading of 0.227 kg/m³ (1,900 lb/10⁶· gal.) at 4 percent solids, i.e., 5.7 ℓ/m³ (5,700 gallons of sludge/10⁶·gal.)for combined primary and secondary sludge after thickening*
2. *Non-clog centrifugal pumps*
3. *TDH = 9.15 m (30 ft)*

Adjustment Factor:
To adjust costs for alternative sludge quantities and characteristics, enter curves at effective flow (Q_E):

$$Q_E = Q_{DESIGN} \times \frac{\text{New design sludge mass}}{0.227 \text{ kg/m}^3 \ (1,900 \text{ lb/10}^6 \cdot \text{ gal.})}$$

$$\times \frac{4 \text{ percent}}{\text{New design concentration (percent)}}$$

GRAVITY THICKENER

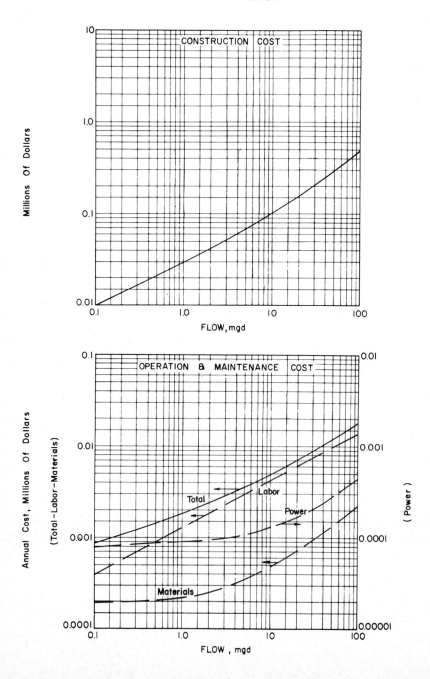

Figure C-11   *Gravity Thickener (ℓ/s × 0.0228 = mgd).*

Service Life: 50 years

Design Basis:
1. *Construction costs include thickener and all related mechanical equipment. Pumps are not included. (See sludge pumping, Figure C-10)*
2. *Costs are based on thickening of secondary sludge 0.098 kg/ m^3 (820 lb/10^6·gal.); loading = 29.3 kg/m^2·d (6 lb/ft^2·d). See adjustment factors for other sludge loadings.*
3. *O & M costs do not include polymer or metal addition.*

Adjustment Factor:
To adjust costs for alternative sludge quantities, concentrations, and thickening properties, enter curves at effective flow (Q_E)

$$Q_E = Q_{DESIGN} \times \frac{29.3 \text{ kg/}m^2 \cdot d \text{ (6 lb/}ft^2 \cdot d)}{\text{New design mass loading}} \times \frac{\text{New design sludge mass}}{0.098 \text{ kg/}m^3 \text{ (820 lbs/}10^6 \cdot \text{gal.)}}$$

AEROBIC DIGESTERS

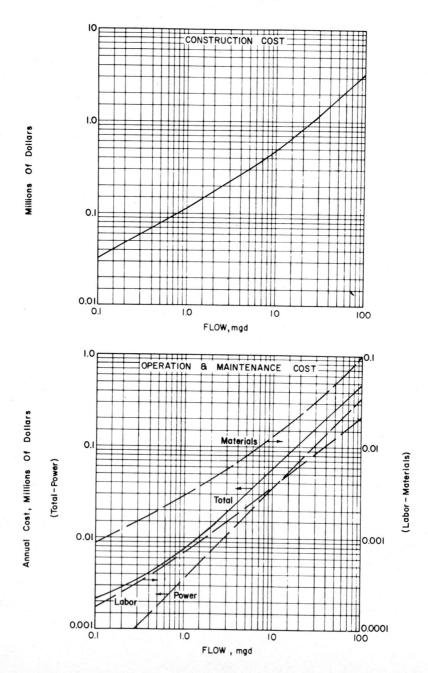

Figure C-12   Aerobic Digesters (ℓ/s $\times$ 0.0228 = mgd).

Service Life: 40 years

Design Basis:
1. *Construction costs include:*
   a. *Basins (20 day detention time). Sludge flow = 5.7 ℓ/m^3 (5,700 gal./$10^6 \cdot$gal.), or 0.227 kg/m^3 (1900 lb/$10^6 \cdot$gal.) at 4 percent*
   b. *Floating mechanical aerators*
2. *Power for mixing requirements = 26.3 W/m^3 (134 HP/$10^6 \cdot$gal)*
3. *Oxygen requirements = 1.6 g O_2/g VSS destroyed*

Adjustment Factor:
To adjust costs for design factors different from those above, enter curves at effective flow (Q_E)

$$Q_E = Q_{DESIGN} \times \frac{20 \text{ days}}{\text{New design retention time}} \times \frac{\text{New design sludge mass}}{0.227 \text{ kg/m}^3 \text{ (1,900 lb/}10^6 \cdot \text{ gal.)}}$$

$$\times \frac{4 \text{ percent}}{\text{New design sludge concentration}}$$

TWO STAGE ANAEROBIC DIGESTERS

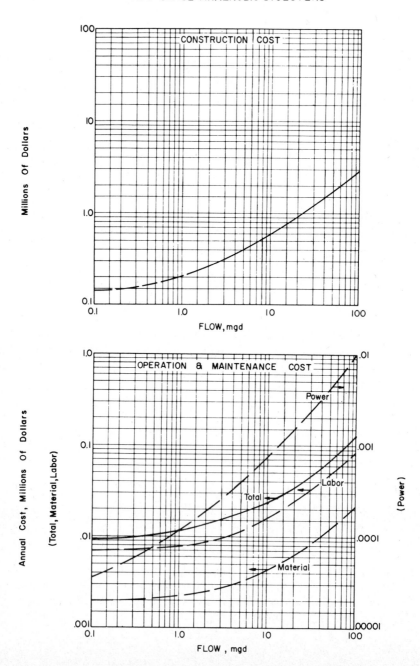

Figure C-13   Two-Stage Anaerobic Digester ($\ell/s \times 0.0228 = mgd$).

Service Life: 50 years

Design Basis:
1. *Capital costs include: digester, heat-exchanger, gas-collection equipment, control building*
2. *Feed to digesters is combined thickened sludge.*
3. *Feed = 0.227 kg/m³ (1,900 lb/10⁶·gal at 4 percent solids (75 percent volatile matter)*
4. *Effluent from digesters = 0.108 kg/m³ (900 lb/10⁶·gal) at 2.5 percent solids*
5. *Loading rate = 2.56 kg/m³·d (0.16 lb/ft³·d)*
6. *Operating temperature = 29 to 43°C*
7. *Digester gas is utilized for heating. Excess gas is not utilized.*

Adjustment Factor:

To adjust costs for loading rates different than those presented here, enter curve at effective flow (Q_E)

$$Q_E = Q_{\text{DESIGN}} \times \frac{\text{New design sludge mass}}{0.227 \text{ kg/m}^3 \ (1{,}900 \text{ lb}/10^6 \cdot \text{gal.})}$$

SLUDGE DRYING BEDS

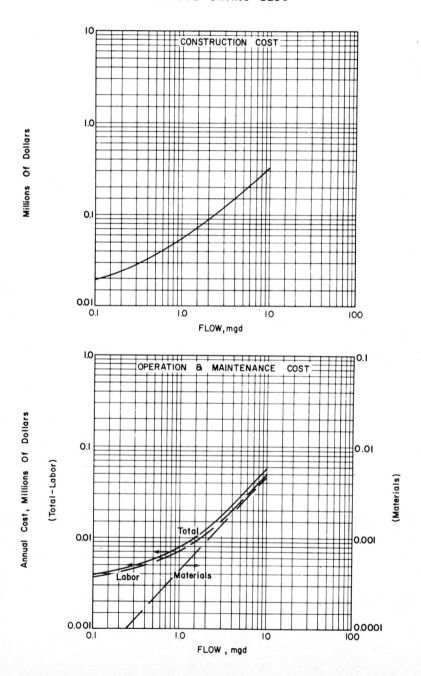

Figure C-14   Sludge-Drying Beds (ℓ/s $\times$ 0.0228 = mgd).

Service Life: 20 years

Design Basis:
1. *Construction costs include: sand beds, sludge inlets, underdrains, cell dividers, sludge piping, underdrain return, and other structural elements of the beds.*
2. *Bed loading: 0.108 kg/m^3 (900 lb of sludge/10^6·gal); 97.6 kg/m^2 per year (20 lb/ft^2 · yr)*

Adjustment Factor:

To adjust costs for bed loading rates, sludge quantities, or characteristics, enter curve at effective flow (Q_E)

$$Q_E = Q_{DESIGN} \times \frac{\text{New design sludge mass}}{0.108 \text{ kg/m}^3 \text{ (900 lb/10}^6 \cdot \text{ gal)}}$$

$$\times \frac{97.6 \text{ kg/m}^2 \cdot \text{yr (20 lb/ft}^2 \cdot \text{ yr)}}{\text{New design bed loading}}$$

FILTER PRESS (BIOLOGICAL SLUDGE)

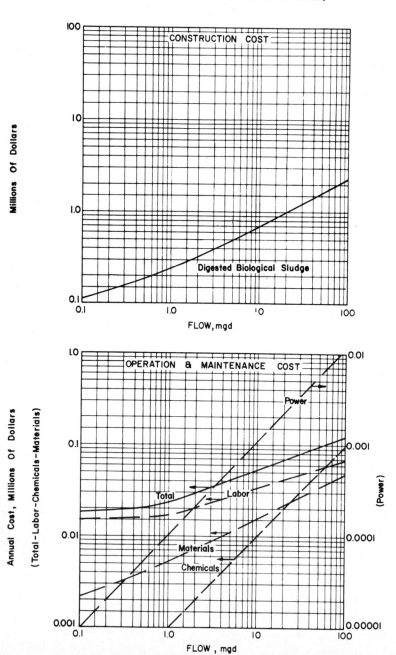

Figure C-15   Filter Press (Biological Sludge) (ℓ/s × 0.0228 = mgd).

Service Life: 15 years

Design Basis:
1. Construction costs include: filter presses, pressure pumps, conveyor equipment, chemical feed and storage facilities, conditioning tanks, sludge storage tanks, and building.
2. Sludge loading: digested primary + secondary = 0.108 kg/m^3 (900 lb/10^6·gal) at 2.5 percent
3. Cake characteristics: density = 1040 kg/m^3 (65 lb/ft^3); solids content = 40 percent
4. Operations: For 4.4 ℓ/s to 44 ℓ/s (0.1 to 1 mgd) plant = 20 cycles/week; For 44 ℓ/s to 440 ℓ/s (1 to 10 mgd) plant = 48 cycles/week; For 440 ℓ/s to 4400 ℓ/s (10 to 100 mgd) plant = 84 cycles/week
5. Conditioning chemicals: FeCl$_3$ = 4.2 g/m^3 (35 lb/10^6·gal), CaO = 10.8 g/m^3 (90 lb/10^6·gal)

Adjustment Factor:

To develop cost for sludge quantities, concentrations, characteristics or cycles per week different than those used to develop these curves, enter curve at effective flow (Q_E)

$$Q_E = Q_{DESIGN} \times \frac{\text{New design sludge mass}}{0.108 \text{ kg/m}^3 \text{ (900 lb/10}^6 \cdot \text{gal.)}}$$

$$\times \frac{\text{Original design cycles per week}}{\text{New design cycles per week}} \times \frac{\text{New design cycle time}}{2 \text{ h}}$$

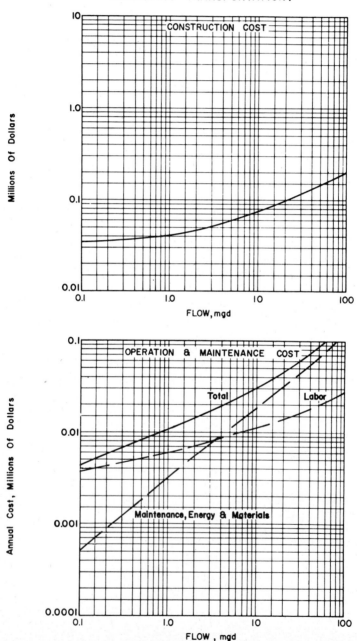

Figure C-16   Landfilling (Biological Sludge, Excluding Transportation) (ℓ/s $\times$ 0.0228 = mgd).

Service Life: 20 years

Design Basis:
1. *Construction costs do not include cost for land. Average land requirement is 1.17 $m^2/m^3 \cdot d$ (1.1 acres/mgd including allowance for 25 percent of volume for cover and 10 percent loss of surface area for roads and buffer zone).*
2. *Construction costs include: site preparation, front end loaders, monitoring wells, fencing, leachate collection and treatment.*
3. *Operation and maintenance costs include:*
   a. Labor costs for operation, preventive maintenance, and minor repairs
   b. Material costs include replacement parts, soil for mixing with sludge (1 part soil to 3 parts sludge), fuel for equipment operation, and repair work performed by outside contractors.
4. *Costs are for landfilling of dewatered and digested biological sludge. Costs for landfilling of other predominantly biological sludge. Costs for landfilling of other predominantly biological sludges can be obtained by using the adjustment factor.*
5. *Sludge quantity = 0.108 kg/m^3 (900 $lb/10^6 \cdot gal$) at 20 percent solids; 545.3 $m^3/10^6 \, m^3$ (2.7 $yd^3/10^6 \cdot gal$). Note: 20 percent solids concentration may require blending with soil.*

Adjustment Factor:

For sludge quantities and concentrations other than those listed above, enter curve at effective flow (Q_E)

$$Q_E = Q_{DESIGN} \times \frac{\text{New design sludge mass}}{0.108 \text{ kg/m}^3 \text{ (900 lb/10}^6 \cdot \text{gal)}}$$

$$\times \frac{20 \text{ percent}}{\text{New design sludge concentration}}$$

MISCELLANEOUS STRUCTURES

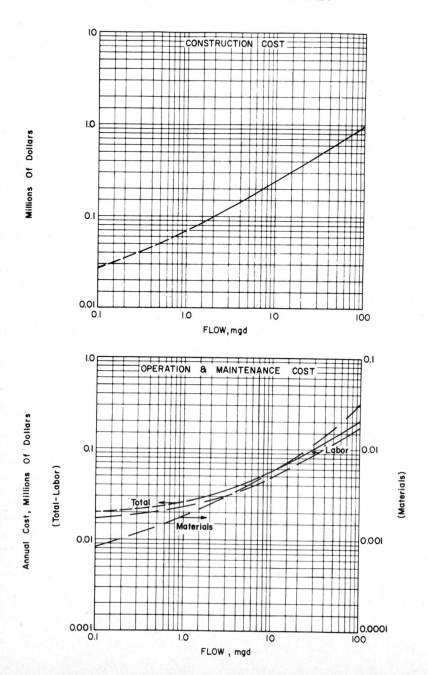

CONSTRUCTION COST

OPERATION & MAINTENANCE COST

Figure C-17 Miscellaneous Structures (ℓ/s × 0.0228 = mgd).

Service Life: 50 years

Design Basis:
Construction cost includes:
*   a. Administrative offices*
*   b. Laboratories*
*   c. Shop and garage facilities*
Note: Full cost of miscellaneous structures might not be applicable for small-flow systems.

SUPPORT PERSONNEL

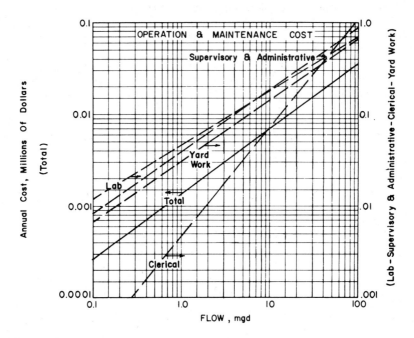

Figure C-18   Support Personnel (ℓ/s × 0.0228 = mgd).

Design Basis:
1. Costs include the manpower required to support the operation of a wastewater treatment facility. These support functions include supervision and administration, clerical work, laboratory work, and administrative costs.
2. The labor costs are based on 1500 hour man-year. This includes a five-day work week; an average of 29 days for holidays, vacations and sick leave; and $6\frac{1}{2}$ hours of productive work per 8 hour day.

REFERENCES

1. U.S. Environmental Protection Agency, *Areawide Assessment Procedures Manual: Performance and Cost,* Appendix H, Municipal Environmental Research Laboratory, Cincinnati, Ohio, EPA 600/9-76-014, July 1976.
2. U.S. Environmental Protection Agency, *Cost-Effectiveness Wastewater Treatment Systems,* EPA-430-/9-75-002, Office of Water Program Operations, Washington, D.C., 20460.
3. U.S. Environmental Protection Agency, *Construction Costs for Municipal Wastewater Treatment Plants: 1973–1977,* EPA MCD-37, Office of Water Program Operations, Washington, D.C., January 1979.
4. U.S. Environmental Protection Agency, *Analysis of Operations and Maintenance Costs for Municipal Wastewater Treatment Systems,* MCD-39, Office of Water Program Operations, Washington, D.C., May 1978.

5. Eilers, R. G., and Robert Smith, *Wastewater Treatment Plant Cost Estimating Program Documentation,* U.S. Environmental Protection Agency, Distributed by NTIS PB-222 762, U.S. Department of Commerce, Springfield, Va., March 1973.

6. Male, J. W., and S. P. Graef, *Applications of Computer Programs in the Preliminary Design of Wastewater Treatment Facilities,* Short Course Proceedings Section I and II, EPA-600/2-78-185, EPA, Cincinnati, Ohio, September 1978.

7. U.S. Army Corps of Engineers and Environmental Protection Agency, *Computer Assisted Procedure for the Design and Evaluation of Wastewater Treatment Facilities (CAPDET),* Waterway Experiment Station, Vicksburg, Mississippi, 1979.

Manufacturers and Suppliers of Wastewater Treatment Plant Equipment

An alphabetical listing of the manufacturers of water pollution control equipment for design of wastewater treatment plants is provided in this appendix. This list is developed from the *1984 Catalog and Buyers Guide of Pollution Equipment News,* with permission from *Pollution Equipment News.* The keywords for the listing are based on major functions, pollutants, and class of equipment. Each listing includes the complete address and phone number. Toll-free phone numbers have also been listed when available.

The *Catalog and Buyers Guide of Pollution Equipment News* is a complete product reference and telephone directory in the pollution abatement field. Other pollution control equipment buyers guides are published by *Water Pollution Control Federation, Water and Waste Digest, Industrial Wastes, Water Engineering and Management, Industrial Water Engineering, Plant Engineering, Specifying Engineering, Instrument Society of America, Thomas Register,* and many others.

LISTING BY TYPE

Aeration Systems
(Aerators and Diffusers)

Aeration Inds Inc
Air O Lator Corp
Aqua-Aerobic Sys, Inc
Ashbrook Simon Hartley

Clow Corp Pump Div
Dorr Oliver
E P I Inc

EIMCO Process Equipment Co
Endurex Corp
Envirex Inc, a Rexnord Co
Environmental Dynamics Inc
FMC Corp Matl Hadlg Systems Div
Ferro Corp Composites Div
Gray Engineering Group Inc
H & H Pump Co, Inc
Hinde Engineering Co
Keene Corp Water Pollution Ctl
Lakeside Equipment Corp
Lightning Aerators
Neptune Microfloc Inc
Parker Hannifin
Pollution Control Systems Inc
Sanitaire Water Poll Ctl Corp
Schramm Inc
Semblex Custom Assemblers
Smith & Loveless Inc
United Industries Inc
Walker Process Corp
Welles Products Corp
Wyss Inc
Zimpro Inc

Blowers and Compressors

Dresser Inds Roots Operations
Elliot Co
FMC Corp Matl Hdlg Systems Div
Fuller Co
Gardner Denver Compressors
Gast Mfg Corp
Ingersoll-Rand Co
Spencer Turbine Co
Sullair Corp

Centrifuges

Bird Machine Co Inc
Dorr Oliver
Hoffman Air & Filtration Systems
J W I Inc
Sharples Stokes Div Pennwalt
Treatment Technologies Inc

Chlorination Equipment

Air O Lator Corp
Capital Controls Co Inc
Chlorinators Inc
Eltech Systems Corp
Fischer & Porter Co
Force Flow Equipment
Ionics Inc
Orion Research Inc
Pollution Control Systems Inc
Wallace & Tiernan Div Pennwalt

Clarifiers and Sedimentation Basins

Bird Machine Co Inc
Clow Corp Pump Div
Dorr Oliver
EIMCO Process Equipment Co
Envirex Inc, a Rexnord Co
Environmental Elements Corp
FMC Corp Matl Hdlg Systems Inc
General Clarifier Corp
Infilco Degremont Inc
Keene Corp Water Pollution Ctl
Lakeside Equipment Corp
Munters Corp
Neptune Microfloc Inc
Pacesetter Separator Co
Parkson Corp
Pollution Control Systems Inc
Richards of Rockford Inc
Sanitaire Water Poll Ctl Corp
Sharples Stokes Div Pennwalt
Smith & Loveless Inc
Trac-Vac Inc
Treatment Technologies Inc
United Industries Inc
Walker Process Corp

Digesters

Carter Ralph B Co
Clow Corporation Pump Div

Dorr Oliver
EIMCO Process Equipment Co
Envirex Inc, a Rexnord Co
FMC Corp Matl Hdlg Systems Div
Lakeside Equipment Corp
Walker Process Corp
Welles Products Corp

Filter Presses

Ashbrook Simon Hartley
Carter Ralph B Co
Clow Corp Waste Trmnt Div
Envirex Inc a Rexnord Co
Komline-Sanderson Engineering Corp
Netzsch Inc
Parkson Corp
Permutit Co Inc
Roediger Pittsburgh Inc
Treatment Technologies Inc
Zimpro Inc

Flow Instrumentation, Closed Pipe and Open Channel

American Sigma
B I F a Unit of General Signal
Badger Meter Inc Indust Prod
Bernhard Inc
Bristol Babcock Inc
Capital Controls Co Inc
Dynasonics Inc
Fischer & Porter Co
Foxboro Co
ISCO Environmental Div
Leeds & Northrup Insts
Leupold & Stevens Inc
Manning Technologies Inc
Mapco Inc Process Controls Div
Marsh-McBirney Inc
Montedoro Whitney
Pollution Control Systems Inc
Polysonics Inc
Rockwell Intl M & U Div
Signet Scientific Co

Sparling Instruments Co Inc
Taylor Instrument Co
Teledyne Gurley
Wallace & Tiernan Div Pennwalt

Grit Handling Equipment

Beaumont Birch Co
Dorr Oliver
EIMCO Process Equipment Co
Envirex Inc, a Rexnord Co
FMC Corp Matl Hdlg Systems Div
Infilco Degremont Inc
Keene Corp Water Pollution Ctl
Krebs Engineers
Lakeside Equipment Corp
Process Equipment Division Wilco
    Machinery Corp
Smith & Loveless Inc
Super Products Corp
Walker Process Corp
Wemco

Process Control Instrumentation

Acromag Inc
Autocon Industries Inc
Automation Products Co
Badger Meter Inc Indust Prod
Bristol Babcock Inc
Calgon Carbon Corp
Capital Controls Co Inc
Davis Water & Wastewater Industries
    Inc
Envirex Inc, a Rexnord Co
FMC Corp Matl Hdlg Systems Div
Fischer & Porter Co
Foxboro Co
Hach Chemicals Co
IN-SITU Inc
ISCO Environmental Div
Keene Corp Water Pollution Ctl
Leupold & Stevens Inc
Manning Technologies Inc
Martek Instruments Inc
Murphy Frank W Mfg Co

Neptune Microfloc Inc
Taylor Instrument Co
Wallace & Tiernan Div Pennwalt

Pumps and Pumping Stations

Abel Pumps Corp
Allweiler Pump Inc
Apex Chemical Equipment Inc
Aurora Pump Unit of Gen Signal
Barrett, Haetjens & Co
Byron Jackson Pumps
CPC Engineering Corp
Carter Ralph B Co
Clow Corp Pump Div
Crane Co Chempump Div
Crisafulli Pump Co
Dorr Oliver
Enpo Pump Co
Flygt Corp
Gorman-Rupp Co
Goulds Pumps Inc
H & H Pump Co
Hayward Gordon Inc
Homa Pump Inc
Hydromatic/The Marley Pump Co
ITT Marlow Pumps
Keene Corp Water Pollution Ctl
Komline-Sanderson Engineering Corp
Lakeside Equipment Corp
Morris Pumps Inc
Myers FE Co
Netzsch Inc
Peabody Barnes
Robbins & Myers Moyno Products
S P P Pumps
Smith & Loveless Inc
Smith EC & Associates Inc
Vanton Pump & Equipment Corp
Vaughan Co Inc
Warren Rupp Co
Waukesha Div Abex Corp
Weil Pump Co
Wemco
Worthington Grp McGraw Edison
Zimpro Inc

Screens/Strainers

Crane Co Chempump Div
Dorr Oliver
Envirex Inc, a Rexnord Co
FMC Corp Matl Hdlg Systems Div
Hycor Corp
Keene Corp Water Pollution Ctl
LYCO
Lakeside Equipment Corp
Parkson Corp
Walker Process Corp
Wemco

Sludge Concentrating and Thickening Equipment

Arus-Andritz
Ashbrook Simon Hartley
Bird Machine Co
Carter Ralph B Co
Dorr Oliver
EIMCO Process Equipment Co
Envirex Inc, a Rexnord Co
H & H Pump Co Inc
Hendrick Fluid Systems Div
Infilco Degremont Inc
Komline-Sanderson Engineering Corp
Krofta Engineering Corp
Parkson Corp
Permutit Company Inc
Roediger Pittsburgh Inc
Sharples Stokes Div Pennwalt
Treatment Technologies Inc
Walker Process Corp

Sludge Handling

Ashbrook Simon Hartley
Bird Machine Co
Calgon Carbon Corp
Carter Ralph B Co
Dorr Oliver
Envirex Inc, a Rexnord Co
FMC Corp Matl Hdlg Systems Div
H & H Pump Co Inc

Hinde Engineering Co
Infilco Degremont Inc
Keene Corp Water Pollution Ctl
Komline-Sanderson Pump Div
Krofta Engineering Corp
Leupold & Stevens Inc
Neptune Microfloc Inc
Netzsch Inc
Parkson Corp
Process Equipment Division Wilco
   Machinery Corp
Sharples Stokes Div Pennwalt
Walker Process Corp
Wemco
Zimpro Inc

Suspended Solids Instrumentation

Automation Products Co
Eur Control USA Inc Env Prod
H F Inst Div of shaban Mfg Inc
Hach Chemicals Co
ISCO Environmental Div
Leeds & Northrup Insts
Martek Instruments Inc
National Sonics Xertex
Rexnord Electronic Products

ADDRESSES AND PHONE NUMBERS

Abel Pumps Corp
79 N Industrial Bldg 205 Bay 7
Sewickley PA 15143
(412) 741–3222

Acromag Inc
30765 Wixom Rd
Wixom MI 48096
(313) 624–1541

Aeration Inds Inc
Hazeltine Gates
Chaska MN 55318
(612) 448–6789

Air O Lator Corp
8100–04 Paseo St
Kansas City MO 64131
(816) 363–4242   (800) 821–3177

Allweiler Pump Inc
5410 Newport Dr Ste 40
Arlington Heights IL 60008
(312) 392–9194

American Sigma
14 Elizabeth St
Middleport NY 14105
(716) 735–3616

Apex Chemical Equipment Inc
1647 Elmhearst Rd
Arlington Heights IL 60007
(312) 364–1590

Aqua-Aerobic Systems Inc
6306 N Alpine Rd
P.O. Box 2026
Rockford IL 61130
(815) 654–2501

Arus-Andritz
1010 Commercial Blvd S
Arlington TX 76017
(817) 465–5611

Ashbrook Simon Hartley
Box 16327
Houston TX 77222
(713) 449–0322

Aurora Pump Unit of Gen Signal
800 Airport Rd
North Aurora IL 60542
(312) 859–7000

Autocon Industries Inc
2300 Berkshire Ln N
Minneapolis MN 55441
(612) 553–4150   (800) 328–3351

Automation Products Co
11705 Research Box 9429
Austin TX 78766
(512) 258–1651   (800) 531–5308

B I F A Unit of General Signal
1600 Division Rd
West Warwick RI 02893
(401) 885–1000

Badger Meter Inc Indust Prod
PO Drawer 1
Tulsa OK 74112
(918) 836–8411

Barrett, Haetjens & Co
225 N Cedar Box 488
Hazleton PA 18201
(717) 455–7711

Beaumont Birch Co
3900 River Rd
Camden NJ 08110
(609) 663–6440

Bernhard Inc
114 E Doe Run Rd
Kennett Square PA 19348
(215) 444–1949

Bird Machine Co Inc
100 Neponset St
S Walpole MA 02071
(617) 668–0400

Bristol Babcock Inc
40 Bristol St
Waterbury CT 06708
(203) 575–3209

Byron Jackson Pumps
Box 2017 Terminal Annex
Los Angeles CA 90051
(213) 587–6171

C P C Engineering Corp
Rte 20 Box 36
Sturbridge MA 01566
(617) 347–7344

Calgon Carbon Corp
Box 1346
Pittsburgh PA 15230
(412) 777–8000

Capital Controls Co Inc
Box 211
Colmar PA 18915
(215) 822–2901   (800) 523–2553

Carter Ralph B Co
192 Atlantic St
Hackensack NJ 07602
(201) 342–3030

Chlorinators Inc
733 NE Dixie Hwy
Jensen Beach FL 33457
(305) 344–8070   (800) 327–9761

Clow Corp Pump Div
1999 N Ruby St
Melrose Park IL 60160
(312) 344–9600

Crane Co Chempump Div
175 Titus Ave
Warrington PA 18976
(215) 343–6000

Crisafulli Pump Co
Crissafulli Dr Box 1051
Glendive MT 59330
(406) 365–3393

Davis Water and Waste Industries Inc
1828 Metcalf Ave
Thomasville GA 31792
(912) 226–5733

Dorr Oliver
77 Havemeyer Ln
Stamford CT 06904
(203) 358–3200

Dresser Inds Roots Operations
900 W Mount St
Connersville IN 47331
(317) 825–2181

Dynasonics Inc
522 W 5th Ave
Naperville IL 60540
(312) 355–3055

E P I Inc
Erle Rd Ellerson Ind Pk
Mechanicsville VA 23111
(804) 746–5238

EIMCO Process Equipment Co
Box 300
Salt Lake City UT 84110
(801) 526–2000

Elliott Co
N 4th St
Jeannette PA 15644
(412) 527–8680

Eltech Systems Corp
470 Center St
Chardon OH 44024
(216) 286–9511

Endurex Corp
415 Wards Corner Rd
Loveland OH 45140
(513) 831–9040   (800) 543–0489

Enpo Pump Co
420 E 3rd St
Piqua OH 45356
(513) 773–2442

Envirex Inc, a Rexnord Co
1901 S Prairie Ave
Waukesha WI 53186
(414) 547–0141

Environmental Dynamics Inc
4501 I-70 Dr SE
Columbia MO 65201
(314) 474–9456

Environmental Elements Corp
Box 1318
Baltimore MD 21203
(301) 368–7346

Eur Control USA Inc
2579 Park Central Blvd
Decatur GA 30035
(404) 981–3998

FMC Corp Matl Hadlg Systems Div
2050 North Board St, Box 482
Lansdale PA 19446
(215) 368–6600

Ferro Corp Composites Div
34 Smith St
Norwalk CT 06852
(203) 853–2123   (800) 321–1414

Fischer & Porter Co
111 Warminster Rd
Warminster PA 18974
(215) 674–6000

Flygt Corp
129 Glover Ave
Norwalk CT 06856
(203) 846–2051

Force Flow Equipment
3467 Golden Gate Way
Lafayette CA 94549
(415) 284–2200

Foxboro Co
38 Neponset Ave
Foxboro MA 02035
(617) 543–8750

Fuller Co
2966 E Victoria St
Compton CA 90224
(213) 639–7600

Gardner Denver Compressors
1800 Gardner Expwy
Quincy, IL 62301
(217) 222–5400

Gast Mfg Corp
Box 97
Benton Harbor MI 49022
(616) 926–6171

General Clarifier Corp
2000 Eastern Pkwy
Brooklyn NY 11233
(718) 385–1800

Gorman-Rupp Co
305 Bowman St
Mansfield OH 44903
(419) 755–1011

Goulds Pumps Inc
East Bayard St
Seneca Falls NY 13148
(315) 568–2811

Gray Engineering Group Inc
633 Denison St
Markham Ont L3R 1B8
(416) 475–9160

H & H Pump Co Inc
Box 486
Clarksdale MS 38614
(601) 627–9631

H F Inst Div of Shaban Mfg Inc
3052 Metro Pkwy SE
Fort Myers FL 33901
(813) 337–2116

Hach Chemicals Co
Box 389
Loveland CO 80537
(303) 669–3050   (800) 525–5940

Hayward Gordon Inc
1051 Clinton St
Buffalo NY 14206
(716) 856–4636

Hendrick Fluid Systems Div
7th Ave & Clidco Dr
Carbondale PA 18407
(717) 282–1010

Hinde Engineering Co
654 Deerfield Box 188
Highland Park IL 60035
(312) 432–6031

Hoffman Air & Filtration Sys
6035 Corporate Box 548
E Syracuse NY 13057
(315) 437–0311

Homa Pump Inc
5 Landmark Square
Stamford CT 06901
(203) 327–6365

Hycor Corp
29850 N Hwy 41
Lake Bluff IL 60044
(312) 473–3700   (800) 323–9033

Hydromatic/The Marley Pump Co
1900 Johnson Dr
Shawnee Mission KS 66205
(913) 722–1485

I T T Marlow Pumps
Box 200
Midland Park NJ 07432
(201) 444–6900

In-Situ Inc
209 Grand Avenue
Laramie WY 82070
(307) 742–8213

Infilco Degremont Inc
Box 29599
Richmond VA 23229
(804) 281–7600

Ingersoll-Rand Co
Woodcliff Lake NJ 07675
(201) 573–0123

Ionics Inc
65 Grove St
Boston MA 02172
(617) 926–2500

ISCO Environmental Div
Box 82531
Lincoln NE 68501
(402) 474–2233   (800) 228–4373

J W I Inc
Box 9A
Holland MI 49423
(616) 399–9130

Keene Corp Water Pollution Ctl
1740 Molitor Rd
Aurora IL 60507
(312) 898–6900

Komline-Sanderson Engineering Corp
Box 257
Peapack NJ 07977
(201) 234–1000

Krebs Engineers
1205 Chrysler Dr
Menlo Park CA 94025
(415) 325–0751

Krofta Engineering Corp
101 Yokun Ave
Lenox MA 01240
(413) 637–0740

Lakeside Equipment Corp
1022 E Devon Box T
Bartlett IL 60103
(312) 837–5640

Leeds & Northrup Insts
Sumneytown Pike
North Wales PA 19454
(215) 643–2000

Leupold & Stevens Inc
Box 688
Beaverton OR 97075
(503) 646–9171

Lightning Aerators
Div of Mixing Equip Co
210 Rount Read Blvd
Rochester NY 14603
(716) 482–9640

LYCO
Box 5058
Williamsport PA 17701
(717) 323–9861

Manning Technologies Inc
100 Technology Cir
Santa Cruz CA 95066
(408) 438–3900

Mapco Inc Process Controls Div
11391 E Tecumseh Box 21418
Tulsa OK 74121
(918) 438–1010

Marsh-McBirney Inc
8595 Grovemont Cir
Gaithersburg MD 20877
(301) 869–4700   (800) 368–2723

Martek Instruments Inc
17302 Daimler Box 16487
Santa Ana CA 92713
(714) 540–4435

Montedoro Whitney
2741 E McMillan Rd
San Luis Obispo CA 93401
(805) 543–1233   (800) 235–4104

Morris Pumps Inc
31 E Genesee St
Baldwinsville NY 13027
(315) 635–3931

Munters Corp
Box 6428
Fort Myers FL 33911
(813) 936–1555

Murphy Frank W Mfg Co
Box 45248
Tulsa OK 74145
(918) 627–3550

Myers F E Co
400 Orange St
Ashland OH 44805
(419) 289–1144

National Sonics Xertex
250 Marcus Blvd
Hauppauge NY 11787
(516) 273–6600

Neptune Microfloc Inc
Box 612
Corvallis OR 97339
(503) 754–7654

Netzsch Inc
119 Pickering Way
Exton PA 19341
(215) 363–8010

Orion Research Inc
840 Memorial Dr
Boston MA 02139
(617) 864–5400   (800) 225–1480

Pacesetter Separator Co
Box 9637
Corpus Christi TX 78408
(512) 289–1541

Parker Hannifin
30240 Lakeland Blvd
Wickliffe OH 44092
(216) 943–5700

Parkson Corp
2727 NW 62nd St Box 9388
Fort Lauderdale FL 33310
(305) 974–6610

Peabody Barnes
651 N Main St
Mansfield OH 44902
(419) 522–1511

Permutit Co Inc
E 49 Midland Ave
Paramus NJ 07652
(201) 967–6000

Pollution Control Systems Inc
County Rd 550 S Box 17
Laotto IN 46763
(219) 637–3137

Polysonics Inc
3221 Marquart St
Houston TX 77027
(713) 623–2134

Process Equipment Division
Wilco Machinery Corp
65 Midcounty Dr
Orchard Park NY 14127
(716) 662–2100

Rexnord Electronic Products
45 Great Valley Pkwy
Malvern PA 19355
(215) 647–7200

Richards of Rockford Inc
Box 5247
Rockford IL 61125
(815) 398–0771

Robbins & Myers Moyno Products
Box 960
Springfield OH 45501
(513) 327–3553

Rockwell Intl M & U Div
400 N Lexington Ave
Pittsburgh PA 15208
(412) 247–3387

Roediger Pittsburgh Inc
RJ Casey Ind Pk
Columbus & Preble Aves
Pittsburgh PA 15233
(412) 231–7979

S P P Pumps
22 Corporate Dr
North Haven CT 06473
(203) 234–1000

Sanitaire Water Poll Ctl Corp
Box 744
Milwaukee WI 53201
(414) 228–1515

Schramm Inc
800 E Virginia Ave
W Chester PA 19380
(215) 696–2500

Semblex Custom Assemblers
1355 B North Nias
Springfield MO 65802
(417) 866–1035

Sharples Stokes Div Pennwalt
955 Mearns Rd
Warminster PA 18974
(215) 443–4000

Signet Scientific Co
3401 Aerojet Box 5770
El Monte CA 91734
(213) 571–2770

Smith & Loveless Inc
14040 Santa Fe Trail
Shawnee Mission KS 66215
(913) 888–5201

Smith E C & Associates Inc
60 E 42nd St
New York NY 10165
(212) 682–6890   (800) 847–4300

Sparling Instruments Co Inc
4097 N Temple City Blvd
El Monte CA 91731
(213) 444–0571

Spencer Turbine Co
600 I Day Hill Rd
Windsor CT 06095
(203) 668–8361

Sullair Corp
3700 E Michigan Blvd
Michigan City IN 46360
(219) 879–5451

Super Products Corp
Box 27225
Milwaukee WI 53227
(414) 784–7100   (800) 558–6190

Taylor Instrument Co
95 Ames St Box 110
Rochester NY 14692
(716) 235–5000

Teledyne Gurley
514 Fulton St
Troy NY 12181
(518) 272–6300

Trac-Vac Inc
Box 1597
Englewood CO 80150
(303) 761–7334

Treatment Technologies Inc
RD 4 Poplar Rd
Honey Brook PA 19344
(215) 273–2977

United Industries Inc
Box 3838
Baton Rouge LA 70821
(504) 292–5527

Vanton Pump & Equipment Corp
201 Sweetland Ave
Hillside NJ 07205
(201) 688–4216

Vaughan Co Inc
364 Monte-Elma Rd
Montesano WA 98563
(206) 249–4042

Walker Process Corp
840 N Russell Ave
Aurora IL 60507
(312) 892–7921

Wallace & Tiernan Div Pennwalt
25 Main St
Newark NJ 07109
(201) 759–8000

Warren Rupp Co
Box 1568
Mansfield OH 44901
(419) 524–8388

Waukesha Div Abex Corp
1300 Lincoln Ave
Waukesha WI 53186
(414) 542–0741

Weil Pump Co
5921 W Dickens
Chicago IL 60639
(312) 642–4960

Welles Products Corp
11765 Main St
Roscoe IL 61073
(815) 623–2111   (800) 435–8551

Wemco
1796 Tribute-Ste 100-Box 15619
Sacramento CA 95852
(916) 929–9363   (800) 221–3333

Worthington Grp McGraw Edison
270 Sheffield St
Westfield NJ 07092
(201) 654–3300

Wyss Inc
2493 CR 65
Ada OH 45810
(419) 634–1971   (800) 592–0143

Zimpro Inc
Military Rd
Rothschild WI 54474
(715) 359–7211

E

Selected Chemical Elements

Table E-1 Chemical Symbols, Atomic Numbers, Atomic Weights, and Valences of Common Elements[a]

Element	Symbol	Atomic Number	Atomic Weight	Valence
Aluminum	Al	13	26.9815	3
Antimony	Sb	51	121.75	3,5
Argon	Ar	18	39.948	—
Arsenic	As	33	74.9216	3,5
Barium	Ba	56	137.34	2
Beryllium	Be	4	9.0122	—
Bismuth	Bi	83	208.980	3,5
Boron	B	5	10.811	3
Bromine	Br	35	79.909	1,3,5,7
Cadmium	Cd	48	112.40	2
Calcium	Ca	20	40.08	2
Carbon	C	6	12.01115	4
Chlorine	Cl	17	35.453	1,3,5,7
Chromium	Cr	24	51.996	2,3,6
Cobalt	Co	27	58.9332	2,3
Copper	Cu	29	63.54	1,2
Fluorine	F	9	18.9984	1
Gold	Au	79	196.967	1,3
Helium	He	2	4.0026	—
Hydrogen	H	1	1.00797	1
Iodine	I	53	126.9044	1,3,5,7
Iron	Fe	26	55.847	2,3
Lead	Pb	82	207.19	2,4

Table E-1   Chemical Symbols, Atomic Numbers, Atomic Weights, and Valences of Common Elements[a]

Element	Symbol	Atomic Number	Atomic Weight	Valence
Lithium	Li	3	6.939	1
Magnesium	Mg	12	24.312	2
Manganese	Mn	25	54.9380	2,3,4,6,7
Mercury	Hg	80	200.59	1,2
Molybdenum	Mo	42	94.94	3,4,6
Neon	Ne	10	20.183	—
Nickel	Ni	28	58.71	2,3
Nitrogen	N	7	14.0067	3,5
Oxygen	O	8	15.9994	2
Phosphorus	P	15	30.9738	3,5
Platinum	Pt	78	195.09	2,4
Potassium	K	19	39.102	1
Radium	Ra	88	226.05	—
Silicon	Si	14	28.086	4
Silver	Ag	47	107.870	1
Sodium	Na	11	22.9898	1
Strontium	Sr	38	87.62	2
Sulfur	S	16	32.064	2,4,6
Tin	Sn	50	118.69	2,4
Titanium	Ti	22	47.90	—
Tungsten	W	74	183.85	3,4
Uranium	U	92	238.03	4,6
Zinc	Zn	30	65.37	2

[a] For complete list, consult a handbook of chemistry and physics

F

Abbreviations and Symbols, Useful Constants, Unit Conversions, Design Parameters, and Units of Expression for Wastewater Treatment

Table F-1 Abbreviations and Symbols

Abbreviations and Symbols		Abbreviations and Symbols	
Acceleration due to gravity (LT^{-2})	g	Gallons per minute (L^3T^{-1})	gal/min or gpm
Area (L^2)	A	Gallons per second (L^3T^{-1})	gal/s or gps
Atmosphere	atm	Horsepower	HP
British thermal unit	Btu	Horsepower-hour	HP·h
Calorie	cal	Hour	h
Centimeter	cm	Inch	in.
Centipoise	cp	Joule	J
Cubic centimeter	cm^3	Kilocalorie	kcal
Cubic feet per minute	cfm or ft^3/min	Kilonewton per cubic meter	kN/m^3
Cubic feet per second	cfs or ft^3/s	Liter (L^3)	ℓ
Cubic meter	m^3	Million gallons per day (L^3T^{-1})	mgd
Degree Celsius	°C		
Degree Fahrenheit	°F	Pound force	lb_f
Degree Kelvin	°K	Pound mass	lb_m
Degree Rankine	°R	Standard temperature and pressure	STP (273°K, 760 mmHg)
Density (ML^{-3})	ρ		
Discharge (L^3T^{-1})	Q	Watt	W
Energy	E		
Feet per minute (LT^{-1})	fpm or ft/min		
Feet per second (LT^{-1})	fps or ft/s		
Gallon (L^3)	gal		

Table F-2 Values of Useful Constants

Acceleration due to gravity, $g = 9.807$ m/s^2 (32.174 ft/s^2)

Standard atmosphere $= 101.325$ kN/m^2 (14.696 lb$_f$/in^2)
$= 101.325$ kPa (1.013 bars)
$= 10.333$ m (33.899 ft) of water

Molecular weight of air $= 29.1$

Density of air (15°C and 1 atm) $= 1.23$ kg/m^3 (0.0766 lb$_m$/ft^3)

Dynamic viscosity of air, μ (15°C and 1 atm) $= 17.8 \times 10^{-6}$ N·s/m^2 (0.372 $\times$ 10^{-6} lb$_f$·s/ft^2)

Kinematic viscosity of air, ν (15° and 1 atm) $= 14.4 \times 10^{-6}$ m^2/s (155 $\times$ 10^{-6} ft^2/s)

Specific heat of air $= 1005$ J/kg°·K (0.24 Btu/lb$_m$°·R)

Table F-3   Conversion Factors, Mixed Units

LENGTH (L)

mile	yard	ft	in	m	cm
1	1760	5280	6.336×10^4	1.609×10^3	1.609×10^5
5.68×10^{-4}	1	3	36	0.9144	91.44
1.894×10^{-4}	0.333	1	12	0.3048	30.48
1.578×10^{-5}	0.028	0.083	1	0.0254	2.54
6.214×10^{-4}	1.094	3.281	39.37	1	100

AREA (A)

mile^2	acre	yard^2	ft^2	in^2	m^2
1	640	3.098×10^6	2.788×10^7	4.014×10^9	2.59×10^6
1.563×10^{-3}	1	4840	43,560	6.27×10^6	4047
3.228×10^{-7}	2.066×10^{-4}	1	9	1296	0.836
3.587×10^{-8}	2.3×10^{-5}	0.111	1	144	0.093
2.491×10^{-10}	1.59×10^{-7}	7.716×10^{-4}	6.944×10^{-3}	1	6.452×10^{-4}
3.861×10^{-7}	2.5×10^{-4}	1.196	10.764	1550	1

VOLUME (V)

acre·ft	U.S. gal	ft^3	in^3	ℓ	m^3	cm^3
1	325,851	43,560	75.3×10^6	1.23×10^6	1230	1.23×10^9
3.07×10^{-6}	1	0.134	231.552	3.785	3.875×10^{-3}	3875.412
2.3×10^{-5}	7.481	1	1728	28.317	0.028	28,316.846
1.33×10^{-8}	4.329×10^{-3}	5.787×10^{-4}	1	0.016	1.639×10^{-5}	16.387
8.1×10^{-7}	0.264	0.035	61.024	1	1×10^{-3}	1000
8.13×10^{-4}	264.2	35.31	6.10×10^4	1000	1	10^6

TIME (T)

yr	mon	d	h	min	s
1	12	365	8760	525,600	3.1536×10^7

VELOCITY (L/T)

ft/s	ft/min	m/s	m/min	cm/s
1	60	0.3048	18.29	30.48
0.017	1	5.08×10^{-3}	0.3048	0.5080
3.281	196.8	1	60	100
0.055	3.28	0.017	1	1.70
0.032	1.969	0.01	0.588	1

DISCHARGE (L^3/T)

mgd	gpm	ft^3/s	ft^3/min	ℓ/s	m^3/d
1	694.4	1.547	92.82	43.747	3.78×10^3
1.44×10^{-3}	1	2.228×10^{-3}	0.134	0.063	5.45
0.646	448.9	1	60	28.317	2446.59
0.011	7.481	0.017	1	0.472	40.78
0.023	15.851	0.035	2.119	1	86.41
2.64×10^{-4}	0.183	4.09×10^{-4}	0.025	0.012	1

MASS (M)

ton	lb$_m$	grain	ounce (oz)	kg	g
1	2000	1.4×10^7	32,000	907.185	907,184.70
0.0005	1	7000	16	0.454	454
7.14×10^{-8}	1.429×10^{-4}	1	2.29×10^{-3}	6.48×10^{-5}	0.065
3.125×10^{-5}	0.0625	437.56	1	0.028	28.35
1.10×10^{-3}	2.205	1.54×10^4	35.274	1	1000
1.10×10^{-6}	2.20×10^{-3}	15.43	0.035	10^{-3}	1

TEMPERATURE (T)

°F	°C	°K	°R
°F	$\frac{5}{9}(°F - 32)$	$\frac{5}{9}°F + 255.38$	°F + 459.69
$\frac{9}{5}°C + 32$	°C	°C + 273.16	$\frac{9}{5}°C + 491.69$
$\frac{9}{5}°K - 459.69$	°K − 273.16	°K	$\frac{9}{5}°K$
°R − 459.69	$\frac{5}{9}°R - 273.16$	$\frac{5}{9}°R$	°R

DENSITY (M/L^3)

lb/ft^3	lb/gal (U.S.)	kg/m^3	kg/ℓ	g/cm^3
1	0.1337	16.019	0.01602	0.01602
7.48	1	119.815	0.1198	0.1198
0.0624	8.345×10^{-3}	1	0.001	0.001
62.43	8.345	1000	1	1

PRESSURE (F/L^2)

lb/in^2	ft water	in Hg	atm	mm Hg	kg/cm^2	N/m^2
1	2.307	2.036	0.068	51.71	0.0703	6894.8
0.4335	1	0.8825	0.0295	22.414	0.0305	2989
0.4912	1.133	1	0.033	25.40	0.035	3386.44
14.70	33.93	29.92	1	760	1.033	1.013×10^5
0.019	0.045	0.039	1.30×10^{-3}	1	1.36×10^{-3}	133.34
14.225	32.783	28.96	0.968	744.657	1	98,070
1.45×10^{-4}	3.35×10^{-4}	2.96×10^{-4}	9.87×10^{-6}	7.50×10^{-3}	1.02×10^{-5}	1

FORCE (F)

lb_f	N	dyne
1	4.448	4.448×10^5
0.225	1	10^5
2.25×10^{-6}	10^{-5}	1

ENERGY (E)

kW h	HP·h	BTU	J	kJ	calories
1	1.341	3412	3.6×10^6	3600	8.6×10^5
0.7457	1	2545	2.684×10^6	2684.5	6.4×10^5
2.930×10^{-4}	3.929×10^{-4}	1	1054.8	1.055	252
2.778×10^{-7}	3.72×10^{-7}	9.48×10^{-4}	1	0.001	0.239
2.778×10^{-4}	3.72×10^{-4}	0.948	1000	1	239
1.16×10^{-6}	1.56×10^{-6}	3.97×10^{-3}	4.186	4.18×10^{-3}	1

POWER *(P)*

kW	BTU/min	HP	ft·lb/s	kg·m/s	cal/min
1	56.89	1.341	737.6	102	14,330
0.018	1	0.024	12.97	1.793	252
0.746	42.44	1	550	76.09	10,690
1.35×10^{-3}	0.077	1.82×10^{-3}	1	0.138	19.43
9.76×10^{-3}	0.558	0.013	7.233	1	137.55
6.977×10^{-5}	3.97×10^{-3}	9.355×10^{-5}	0.0514	7.12×10^{-3}	1

VISCOSITY

DYNAMIC ABSOLUTE VISCOSITY (μ)

cp	$lb_f \cdot s/ft^2$	$lb_m/ft \cdot s$	g/cm·s	$N \cdot s/m^2$	kg/m·s	dp
1	2.09×10^{-5}	6.72×10^{-4}	0.01	1×10^{-3}	1×10^{-3}	1×10^{-3}
4.78×10^4	1	32.15	478.47	47.85	47.85	47.85
1488.09	0.031	1	14.881	1.488	1.488	1.488
100	2.09×10^{-3}	0.0672	1	0.10	0.10	0.10
1000	0.021	0.672	10	1	1	1

KINEMATIC VISCOSITY (ν)

Centistoke	ft^2/s	cm^2/s	m^2/s	myriastoke
1	1.076×10^{-5}	0.01	1.0×10^{-6}	1.0×10^{-6}
9.29×10^4	1	929.368	0.093	0.093
100	1.076×10^{-3}	1	1.0×10^{-4}	1.0×10^{-4}
10^6	10.76	10^4	1	1

Table F-4 Design Parameter and Units of Expressions for Wastewater Treatment Plant Design

Design Parameter	Typical Design Values SI Units	SI Units	Factor K Divide by K ← to Obtain / Multiply by K → to Obtain	U.S. Customary Units
SCREENING				
m³ screening/10³ m³ flow	$3.4\times10^{-3} - 8.0\times10^{-2}$	m³/10³ m³	133.68	ft³/10⁶·gal
GRIT REMOVAL				
Volume, m³ grit/10³ m³ flow	$7.4\times10^{-3} - 9.0\times10^{-2}$	m³/10³ m³	133.68	ft³/10⁶·gal
Surface overflow rate, m³ flow/m² surface·d	898 – 2040	m³/m²·d	24.55	gpd/ft²
Detention time, s	10 – 100	—	—	—
FLOW EQUALIZATION				
Power for mixing, kW/m³ tank volume	0.004 – 0.008	kW/m³	5.20	HP/10³·gal
Air supply, m³ air/m³ tank volume·min	0.01 – 0.015	m³/m³·min	133.68	ft³/10³·gal min
SEDIMENTATION				
Surface overflow rate, m³ flow/m² surface area·d	20 – 82	m³/m²·d	24.55	gpd/ft²
Solids loading rate, kg solids/m² surface area·d (final clarifier)	40 – 100	kg/m²·d	0.205	lb/ft²·d

Detention time, h	0.5 – 5		—	—
Weir loading rate,				
m^3 flow/m weir length·d	103 – 500	$m^3/m \cdot d$	80.52	gpd/ft
Volume of sludge,				
m^3 sludge/10^3 m^3 flow	2.2 – 11.22	$m^3/10^3 m^3$	133.68	$ft^3/10^6 \cdot gal$
Weight of dry sludge solids,				
g dry solids/m^3 flow	60 – 240	g/m^3	8.35	$lb./10^6 \cdot gal$
Sludge concentration,				
percent total solids in sludge	1 – 10	—	—	1–10
ACTIVATED SLUDGE				
BOD_5 loading				
kg BOD_5/m^3 aeration tank·d	0.4 – 1.6	$kg/m^3 \cdot d$	62.43	$lb/10^3 ft^3 \cdot d$
g BOD_5/g volatile solids·d	0.1 – 1.0	$g/g \cdot d$	1	$lb/lb \cdot d$
kg BOD_5/m^2 surface area·d	0.015 – 0.15	$kg/m^2 \cdot d$	0.205	$lb/ft^2 \cdot d$
Air requirement,				
m^3 air/kg BOD_5 removed	19 – 95	m^3/kg	16.02	ft^3/lb
m^3 air/m^3 wastewater treated	3.7 – 15	m^3/m^3	0.134	ft^3/gal
Oxygen requirement,				
kg O_2/kg BOD_5 applied·d	0.7 – 2.4	$kg/kg \cdot d$	1	$lb/lb \cdot d$
Oxygen transfer rate,				
kg O_2 transferred/kW h	1.2 – 2.4	$kg/kW \cdot h$	1.644	$lb/HP \cdot h$
kg O_2 transferred/m^3 wastewater·h	—	$kg/m^3 \cdot h$	0.063	$lb/ft^3 \cdot h$
Return sludge				
Return flow/Incoming flow	0.1 – 0.5	—	—	—
BIOLOGICAL FILTERS				
(Trickling filter and biological contactor)				
Hydraulic loading				
Low rate, m^3 flow/m^2·d surface area	1 – 4	$m^3/m^2 \cdot d$	24.54	gpd/ft^2
High rate, m^3 flow/m^2·d surface area	8 – 40	$m^3/m^2 \cdot d$	24.54	gpd/ft^2

Table F-4 Design Parameter and Units of Expressions for Wastewater Treatment Plant Design

Design Parameter	Typical Design Values SI Units	SI Units	Factor K Divide by K ← to Obtain / Multiply by K → to Obtain	U.S. Customary Units
BOD$_5$ loading				
Low rate, kg BOD$_5$/m^3·d filter volume	0.08 – 0.4	kg/m^3·d	62.43	lb/10^{3}ft^3·d
High rate, kg BOD$_5$/m^3·d filter volume	0.3 – 1.0	kg/m^3·d	62.43	lb/10^{3}ft^3·d
Recirculation				
Low rate (percent of incoming flow)	0 – 50	—	—	—
High rate (percent of incoming flow)	25 – 300	—	—	—
SAND FILTER				
Hydraulic loading				
m^{3}flow/m^2·min surface area	0.08 – 0.48	m^3/m^2·min	24.54	gpm/ft^2
OXIDATION POND				
BOD$_5$ loading				
kg BOD$_5$/ha·d surface area	15 – 160	kg/ha·d	0.892	lb/acre·d
CHLORINATION				
Dosage				
mg/ℓ of chlorine applied (feed rate)	4 – 40	mg/ℓ	8.35	lb/10^6·gal
SLUDGE THICKENING				
Sludge loading (gravity),				
kg. dry solids/m^2·d surface area	20 – 100	kg/m^2·d	0.205	lb/ft^2·d

Surface overflow rate, m³ wastewater/m²·d surface area	14–36	m³/m²·d	24.54	gal/ft²·d
DIGESTION				
Organic loading, kg volatile solids/m³·d digester volume	0.8–8	kg/m³·d	62.43	lb/10³ft³·d
BOD$_5$ loading rate, kg BOD$_5$/m³·d digester volume	0.16–1.6	kg/m³·d	62.43	lb/10³ft³·d
Sludge heating, W/m²·surface area°C	0.60–5.0	W/m²°C	0.1763	Btu/h·ft²·°F
SLUDGE DEWATERING				
Sludge drying bed kg dry solids/m² area·yr	100–400	kg/m²·yr	0.205	lb/ft²·yr
Feed depth in m/yr wet sludge	0.9–3.6	m/yr	3.281	ft/yr
m²·area/persons	0.12–0.30	m²/persons	10.76	ft²/person
Pressure or vacuum filtration, kg dry solid/m²·h filter surface	6–60	kg/m²·h	0.205	lb/ft²·h
m³ wet sludge/m²·h surface area	0.2–1.2	m³/m²·h	3.281	ft³/ft²·h
kN/m² (kPa) pressure applied	300–2000	kPa, kN/m²	0.145	lb/in²
kN/m² (kPa) vacuum applied	40–70	kPa, kN/m²	0.296	in Hg
Heat drying, kJ heat energy required/kg water evaporated	600–1000	kJ/kg	0.430	Btu/lb
kg water evaporated/h.	—	kg/h	2.205	lb/h
kg wet sludge/m²heating·h	—	kg/m²h	0.205	lb/ft²·h
Incineration, kg sludge/m² effective hearth area·h	36–60	kg/m² h	0.205	lb/ft²·h
LAND DISPOSAL OF WASTEWATER AND SLUDGE				
kg solids/ha field area	—	kg/ha	0.892	lb/acre
m³ liquid/ha field area·d	—	m³/ha·d	106.91	gal/acre·d

Index

Acoustic meters, 226, 585
Activated sludge
  definition, 305
  design parameters, 310–313
  process modifications, 308, 310–314
    complete mix, 308, 311, 314
    contact stabilization, 308, 312, 314
    conventional, 308, 310, 314
    extended aeration, 308, 311, 314
    pure oxygen, 308, 313, 314
    step aeration, 308, 310, 314
    tapered aeration, 308, 310
  removal efficiency, 52
  return sludge, 322, 330
  solids removal, 320, 322, 331, 359–369
    design calculations, 359–369
    design criteria, 331
    design values, 322
  starting a new plant, 369
  waste activated sludge, 322
ADP (adenosine diphosphate), 304
Advanced wastewater treatment, 56, 95, 615
Aerated grit chamber, 240–243, 251
Aerated lagoon, 48
Aeration basin, 320, 338–359
  design calculations, 338–359
  design factors, 320
  oxygen requirement, 351
Aeration equipment
  activated sludge, 315–321
  grit chamber, 259
Aeration system
  air velocities in pipes, 353
  blowers, 320, 355
  coarse bubble, 315
  design calculations, 352–359
  diffused air, 309, 315
  fine bubble, 315
  head loss calculations, 352–357
  installation, 319
  jet, 309, 316, 318
  mechanical, 309, 316, 320, 321
  medium bubble, 315
  radial flow, 317
  submerged turbine, 317

transfer efficiency, 315
transfer rate, 315
tubular, 316
Aerobic digestion of sludge
  air requirement, 461
  application, 64
  design parameters, 460, 461
  solids retention time, 460
  supernatant quality, 462
Air diffuser, 309, 315, 316, 318, 319, 352–355
  bubble, 309, 318
  head loss, 352–355
  jet, 309, 316, 318
  tubular, 309, 318
Ammonia nitrogen removal, 51, 53, 620
Anaerobic digestion of sludge
  application, 64
  description, 451, 453
  design and operation, 452
  design capacity, 452
  design criteria, 454
  design details, 468
  digester cover, 459, 484
  digester heating, 455, 456, 473, 485
  digester mixing, 455, 457, 485
  digester startup, 481
  digestion period, 452
  gas collection system, 458
  gas production, 457, 458, 469
  gas utilization, 457
  mixing arrangement, 457
  operation and maintenance checklist, 483, 484
  solids retention time, 452
  supernatant quality, 459
  volumetric loading, 453
  yield coefficient, 457
Aquaculture, 2, 99, 412, 413
Arithmetic growth, 9, 11
ATP (adenosine triphosphate), 304
Attached growth biological treatment
  definition, 326
  rotating biological contactor (*see* Rotating biological contactor)
  trickling filter (*see* Trickling filter)
Axial-flow pumps, 188, 191

Bar racks, 155–158, 159–175
  cleaning rake, 174
  design calculations, 163–172
  design criteria, 163
  design details, 172, 173
  design example, 163–172
  design factors, 156, 158
  efficiency coefficient, 164
  equipment manufacturers, 162
  head losses, 160, 563, 565
  hydraulic profiles, 171
  manually cleaned, 158, 161
  mechanically cleaned, 158, 161
Bar screens, 48
Batch reactor, 18
Best efficiency point (BEP), 188
Biodegradable organics, 304
Biological kinetic equations, 305–307
Biological nutrient removal, 619
Biological sludge, 54
Biological treatment process, 99, 303, 305, 326
  activated sludge, 305
  attached growth, 326
Birth cohart, 9, 13
Breakpoint chlorination, 51, 53, 384, 620
Buchner funnel test, 490, 496, 497

Carbon adsorption
  process description, 623
  purpose, 51
  removal efficiency, 53
Carcinogenicity, 17
Categorical standards, 16
Cavitation, 187, 188
Centrifugal pumps, 177, 178, 180, 181, 184–189, 204
Centrifugation of sludge, 64, 65, 434
Characteristic curves, pump, 185, 186, 189, 199
  classification, 189
Chemical–biological sludge, 55
Chemical conditioning of sludge, 64, 490
Chemical oxidation of sludge, 64
Chemical precipitation, 54
Chlorination, 379, 381, 391
  chlorine chemistry, 379, 381
  design criteria, 391
  equipment, 391
  facility, 379
Chlorination system
  chlorine feed, 386, 387
  chlorine supply, 386
  control system, 386, 390
  design calculations, 391–405
  mixing arrangements, 389
Chlorinator, 407

Chlorine
  compounds, 380
  gas, 380
  liquid, 380
Chlorine contact chamber, 387
  contact time, 387, 391
  details of, 393
  hydraulic profile, 405, 407
  velocity, 391
Chlorine demand, 384
Chlorine diffusers, 388
Chlorine dioxide, 381
Chlorine dosages, 385
Chlorine residual
  combined, 381
  free, 381, 384
  total, 385, 391, 407
Chlorine storage and handling, 386
Clean Water Act, 1, 15, 17, 19
Clinoptiloite bed, 51
Coagulation, 50, 52
  and flocculation, 616
Coarse screens, 155, 157
Combined chlorine residual, 381
Comminutors, 155, 157, 159
Complete-mix activated sludge process, 308, 311, 314
Complete-mix biological reactor, 306
Composting, 65, 67, 521, 522
Conditioning of sludge
  chemical, 67, 491
  description, 489–491
  elutriation, 64, 67, 491
  freezing and thawing, 491
  heat treatment, 67, 491
  irradiation, 491
  physical, 491
  thermal, 491, 492
  ultrasonic vibration, 491
Constant-speed pumps, 193, 195
Construction costs, 104–109, 119
Contact stabilization process, 308, 312, 314
Continuous flow reactor, 18
Conventional activated sludge process, 308, 310, 314
Cost comparison of alternatives, 102, 103
Cost curves, 655–692
Cost-effective analysis, 20, 47, 101, 103

Darcy–Weisbach equation, 182, 309
Decreasing rate of increase, 9, 11
Degree of treatment, 52, 53
Demineralization process, 51
Denitrification
  process description, 619

purpose, 51
removal efficiency, 53
separate-stage, 55
Design errors, 636
Design examples
bar rack, 163
biological reactor, 330
chlorination facility, 391
flow measurement, 230
grit removal facility, 247
hydraulic profile, 564
instrumentation and control systems, 592
outfall structure, 418
plant layout, 554, 555
primary sedimentation tank, 284
pumping station, 197
sewer, 141
sludge dewatering facility, 506
sludge landfill (see Sludge landfills)
sludge stabilization facility, 464
solids separation facility, 330
thickeners, 435
Design factors, 5
Design flow, 127
Design parameters, 18, 715–717
Design period, 6, 84, 95, 96, 127
Design plans and specifications, 73, 77
Design population, 9, 127
Design summary, 605–614
Design year, 6, 18
Dewatering of sludge
capillary, 499, 503
centrifugation, 67, 490, 494, 495
design calculations, 507–515
design criteria, 506
drying beds, 67, 489, 491
drying lagoons, 489, 491
filter press, 67, 490, 498, 499
gravity bag filter, 499, 505
horizontal belt filter, 67, 490
mechanical systems, 499
moving screen concentrator, 499, 503
rotating gravity concentrator, 499, 503
screw press, 499, 505
twin-roll press, 499, 503
vacuum filter, 67, 490, 494, 496, 497
Digested sludge, 428
Disinfection
chemical, 379, 380
physical, 379
purpose, 50
radiation, 379
chlorine, 380
Dissolved air flotation of sludge, 64, 432, 433
Dissolved organics, 49, 50

Distillation, 624
Diversion sewer, 123, 141
Dry weather diurnal flow, 90
Dry weather flow, 95, 96
Drying beds, 8, 65, 67, 489, 491

Effectiveness evaluation, 103, 110, 111
Effluent disposal and reuse
aquaculture, 99, 412, 413
discharge into natural waters, 99, 414
groundwater recharge, 99, 412
industrial uses, 99, 413
irrigational uses, 99, 413
methods of, 412
municipal uses, 99, 414
natural evaporation, 99, 412
recreational lakes, 99, 413
Effluent limitations, 84
Effluent limited, 16, 75
Effluent quality, 58, 60
activated sludge process, 58
advanced treatment, 58, 60
preliminary treatment, 58
primary treatment, 58
trickling filter process, 58
Effluent standards, 15, 16
Electrodialysis, 51, 53, 616, 625
Electromagnetic meter, 226, 585
Elutriation, 64, 67, 491
Endogenous decay coefficient, 307, 458
Endogenous phase, 304
Enforcement, 16
Engineering News Record (ENR), 103
Enzymes, 304
Equipment manufacturers, 694–705
Equipment selection, 19
Extended aeration process, 308, 311, 314

Facility planning, 3, 17, 73, 74, 83
Federal Water Pollution Control Act, 1
Filter leaf test, 490, 496, 497
Filter media, 517
Filter press, 65, 67, 489, 490, 500, 502, 503, 509
Filtration, 51, 53, 621
Fine screens, 155
Flood hazard analysis, 113
Flood protection, 8
Flow equalization, 48
Flow measurement devices, 48, 90, 219–229
acoustic meter, 221, 226, 229
current meter, 221
depth measurement, 221, 227, 229
elbow meter, 221
electromagnetic meter, 221, 226, 229
float actuated liquid level recorder, 225

flow nozzle meter, 221–223, 229
flow recorder, 228
flow sensor, 228
Kennison nozzle, 225
magnetic flow meter, 225
open flow nozzle, 229
operational characteristics of, 220
orifice meter, 221, 223, 226, 229, 585
Palmer–Bowlus flume, 221, 223, 226, 228, 229
Parshall flume, 221, 225, 226, 228, 229
pitot tube, 221
rotameter, 221
turbine meter, 221, 226, 229
ultrasonic, 225
Venturi meter, 221, 223, 228–231, 233
weirs, 221, 227, 229
Flow reduction measures, 94
Flow scheme, 18, 47, 61, 69
Flow sensor probes, 228
Flow sheets, 18, 47
Food-to-MO ratio, 305, 307
Force main, 183, 198
Francis–Vane area, 191, 192
Friction head, 182
Froth control system, 320

Gamma radiation, 380
Geometric growth, 9, 11
Graphical comparison, 9, 10, 12
Gravity thickening of sludge, 64, 429–431, 440, 448
Grit, 54, 238, 245, 257
Grit disposal, 245
Grit removal systems, 48, 50, 238–260
aerated grit chamber, 240–243, 251
centrifuge, 244
conveyor and elevator, 259
cyclone, 244
design calculations, 248–257
design criteria, 247
design factors, 240, 243
detritus tanks, 244
facility location, 238, 239
hydraulic profile, 252, 257
velocity controlled grit channel, 239–241
Groundwater recharge, 2, 412

Hazen–Williams equation, 182, 183, 199, 309
Head capacity curves, pump, 185, 188
Head losses in treatment units, 563–565
Heat drying, 65, 521, 522
Heat treatment, 64, 463
Horizontal belt filter, 65, 67
Hydraulic profile, 19, 561–582
Hydrogen peroxide, 380

Incineration, 65, 521
Industrial cost recovery (ICR), 79, 80
Industrial pretreatment standards, 16
Infiltration/inflow, 17, 18, 29–33, 91–94, 123, 127
Information checklist for design
anaerobic digester, 483
bar rack, 162
biological treatment and solids clarification facilities, 329
chlorination facility, 390
flow-measuring device, 228
grit removal facilities, 247
instrumentation and control systems, 592
outfall structure, 418
pumping station, 196
sanitary sewer, 139
sedimentation basin, 283
sludge conditioning and dewatering facilities, 505
sludge landfills, 537
sludge stabilization facility, 463
thickener, 434
Initial period, 95
Initial year, 6, 18
Inorganic solids, specific gravity, 427
Instrumentation and controls, 583–603
measurement devices, 585
sensing devices, 584, 594
Intercepting sewer, 123, 124, 127, 141, 149
Ion exchange, 51, 53, 621, 623
Irrigation, 18, 413, 626, 627

Kinetic coefficients, 3, 305, 308, 329

Land application of sludge, 65, 67, 520, 523
Land treatment, 48, 98, 626, 628, 633
Lime-biological sludge, 55
Lime precipitation, 51, 52, 617
Lime recalcining, 65
Lime stabilization of sludge, 64
Log-growth phase, 304
Logistic curve fitting, 9, 11

Manholes, 129, 130, 132, 136, 152
drop, 130, 132
typical line, 130
Manning equation, 132, 133, 563
Manufacturers and suppliers, 694–705
Mass-loading variations, 48
Material mass balance, 334–341
Mathematical curve fitting, 9, 11
Mean cell residence time, 307
Mesophilic, 452
Microorganisms, 304, 305
Microstrainers, 158, 621, 622
Minor losses, 182, 563, 647–651

Mixed-flow pumps, 188, 191, 192
Mixed liquor, 305
Model facility plan, 73, 83
Modified pump curve, 188, 190, 201–204
Motors, pump, 215
Moving screen, 156, 158
Municipal sludge, 427
Municipal water demand, 23–27 (*see also* Water demand)
Mutagenicity, 17

National Clay Pipe Institute, 128, 135, 140
National Register of Historic Places, 8
Net positive suction head (NPSH), 187, 188, 197, 208
Nitrification, 51, 53, 55, 618
Nonclog pump, 191
NPDES, 1, 16, 75, 84

Occupational Safety and Health Act (OSHA), 78
Operating problems
    activated sludge system, 370, 371
    anaerobic sludge digestion facility, 481
    bar rack, 172
    filter press facility, 515
    primary sedimentation facility, 294
    pumping station, 212
    sludge-thickening facility, 447
Operation and maintenance
    activated sludge, 369
    aerated grit removal, 257
    aeration basin, 371
    anaerobic sludge digestion, 480, 482
    bar rack, 172
    chlorination, 406
    filter press, 515, 516
    outfall structure, 424
    primary sedimentation, 294, 297
    pumping station, 210
    sanitary sewer, 148
    secondary clarifier, 371
    sludge landfills, 536, 546
    sludge-thickening, 446, 447
    treatment facility, 112
    Venturi tube flow meter, 233
Outfall
    design calculations, 418–424
    design criteria, 418
    design factors, 417
    diffusers, 417–424
    estuarine, 415
    lake, 415
    ocean, 417
    river, 415
Overland flow system, 631, 632

Oxidation ditch process, 314
Oxygen transfer efficiency, aerator, 315–317
Oxygen transfer rate, aerator, 315–317
Oxygen utilization rate, 307
Ozone, 380

Painting, 215, 260, 300
Pasteurization, 379
Physical–chemical treatment system, 98
Pilot plant studies, 3, 18
Planning phase report, 73
Plant hydraulics, 562
Plant layout, 19, 549–559
Plant utilities, 553
Population density, 10
Population equivalent, 43, 44
Population forcasting, 9–14
Predesign phase, 73
Preliminary treatment, 52
Primary flow element, 234
Primary sedimentation tank
    circular, 264, 266
    depths, 269
    design calculations, 284
    design criteria, 284
    design factors, 268
    detention period, 269
    effluent structure, 273–276, 287, 289, 299
    hydraulic profile, 291, 292
    influent structure, 271–273, 285
    overflow rate, 268, 269
    rectangular, 264
    removal efficiency, 52, 263, 268, 270
    scum removal, 281
    sludge collection, 274, 277, 281 (*see also* Sludge collection equipment)
    surface-loading rate, 268
    types of, 263
    weir-loading rate, 270
Primary sludge, 54, 291, 427, 428
Process train, 18
Prohibited discharges, 16
Proportional weir, 171, 239
Public law 92-500, 1, 14
Public law 95-217, 1, 14
Publicly owned treatment works, 16–17
Public participation program, 115–117
Public Works Journal Corporation Magazine, 134
Pump analysis
    characteristic curves, 180, 183, 185, 188
    efficiency, 180, 183, 184, 207
    power input, 207
    specific speed (*see* Specific speed)
    total dynamic head (TDH), 182–184, 186
    work power, 180, 183

Pump class, 191
Pump combinations, 188, 190, 201
Pump drive, 193, 214
Pumping stations, 31, 48, 177, 181, 193–195, 197, 208, 209
  design calculations, 197
  design criteria, 197
  details of, 208
  dry well, 177, 194, 197
  force main (*see* Force main)
  layout, 208, 209
  pipe arrangement, 208, 209
  selection of, 194
  site selection, 194
  wet well, 177, 187, 194, 195, 197
Pumps
  air lift, 179–181
  centrifugal (*see* Centrifugal pumps)
  axial flow, 188, 191
  mixed flow, 188, 191
  radial flow, 188, 191
  diaphragm, 179, 180
  impeller, 214
  installation, 211
  kinetic, 177, 178, 180
  peripheral, 178, 180
  plunger, 179, 180
  pneumatic ejector, 179–181
  positive displacement, 177, 180
  propeller, 191
  rotary, 178, 180
  screw, 178, 180, 181
  self priming, 181, 196
  submersible, 177, 181
  suspended, 177, 181
  torque-flow, 178, 180
  vortex, 178, 180
Pyrolysis, 65

Rack chamber, 165
Racks (*see* Bar racks)
Radial-flow pumps, 188, 191
Rapid infiltration percolation, 626, 630
Reactor
  arbitrary flow, 305
  complete mix, 305
  plug flow, 305
  types of, 305
Recalcination, 67, 521, 522
Recreation lake, 2, 413
Refractory organics, 51
Resource Conservation and Recovery Act, 100
Reverse osmosis, 51, 53, 616, 625
Rotating biological contactor, 326, 327

Sanitary landfilling, 54, 65, 67, 102
Sanitary sewers, 1, 88, 123, 125, 132, 139, 148
  common problems, 148
  maintenance program, 149
  pressure, 125
  slopes, 129
  specifications, 139
  troubleshooting, 148
  vacuum, 125
Satellite treatment, 2
Screen chamber, 157, 163
Screening, 50, 54, 155
Screenings, 161, 162, 171, 172
Screens
  coarse, 155, 157
  control system, 175
  fine, 155
  fixed, 155
  movable, 155
  racks (*see* Bar racks)
Scum, 54, 281, 282, 294
Secondary clarifier
  design calculations, 359–369
  design criteria, 331
  details of, 365
  solids-settling rate, 332
Secondary treatment, 49, 303, 333
Sedimentation tank, primary (*see* Primary sedimentation tank; Secondary clarifier)
Sewer
  appurtenances, 129, 143
  branch, 123, 124, 126
  building, 123, 124, 126, 132
  cleaning, 150
  combined, 31, 123
  construction, 136, 139, 153
  construction materials, 133, 137
  corrosion, 148
  design calculations, 141–148
  design criteria, 141
  hammer taps, 148
  hydrogen sulfide gas, 129
  hydraulic profile, 142, 143, 145–148
  inspection, 149
  joints, 136
  lateral, 123, 124, 126, 127
  layout plan, 126, 142
  leakage test, 136
  main, 123, 124, 126, 127
  methane, 129
  profiles, 126
  protruding tap, 148, 149
  repairs, 151
  sanitary, 1, 123, 125, 132, 139, 148
  size distribution, 125

specifications, 151
stoppage clearing, 151
storm, 31, 123
submain, 123, 127
trunk, 123, 126, 127
Sewerage system, 123, 125, 126
Single-stage lime sludge, 55
Site selection of treatment plants, 7
Skimmer, 299
Sludge
    conditioning (*see* Conditioning of sludge)
    conversion processes, 520–522
    dewatering (*see* Dewatering of sludge)
    disposal, 64, 65, 67
    drying beds (*see* Sludge drying beds)
    drying lagoons (*see* Sludge drying lagoons)
    filterability, 496
    flow schemes, 69
    selection of processes, 70, 71
    specific resistance, 496
    stabilization (*see* Stabilization of sludge)
    thickening (*see* Thickening of sludge)
    vacuum filters (*see* Vacuum filtration)
Sludge blending tank, 439, 441, 445
Sludge collection equipment, 277–280,
    298
Sludge drying beds, 489, 492, 493
Sludge drying lagoons, 489, 492, 493
Sludge landfills, 523–547
Sludge processing and disposal alternatives, 99,
    428
Sludge pumping, 193, 299, 330
Sludge removal, 281
Sludge volume ratio (SVR), 429
Sludge wasting, 305, 306, 330
Soil conditioner, 65
Solids flux curve, 360
Solubility of gases in water, 645
Specific speed, 185, 191
Specific substrate utilization rate, 307
Specifications
    aeration system, 372–374
    anaerobic digesters, 484
    bar rack, 174
    chlorination facility, 407
    filter press, 516
    gravity thickeners, 448
    grit removal facility, 258
    outfall structure, 424
    primary sedimentation basin, 297
    pumping station, 213
    sanitary sewers, 139, 151
    Venturi tube flow meter, 233
Stabilization of sludge
    aerobic digestion, 67, 451 (*see also* Aerobic
        digestion of sludge)

anaerobic digestion, 67, 451 (*see also* Anaero-
    bic digestion of sludge)
biological, 451
chemical, 451
chlorine oxidation, 67, 462
design calculations, 465–480
design criteria, 464
fly ash, 490
heat treatment, 67, 463
lime stabilization, 67, 451, 462
physical, 451
thermal conditioning, 451, 463
Stabilization ponds, 48, 305, 323–325
Standard oxygen requirement (SOR), 350
Step-aeration, 308, 310, 314
Sterilization, 379
Suspended growth biological reactor (*see* Acti-
    vated sludge)

Tapered aeration, 308, 310
Ten States Standards, 20, 127
Tertiary treatment, 49, 615
Theoretical oxygen requirement, aeration basin,
    350
Thermal conditioning, 64
Thermophilic, 452
Thickening of sludge
    centrifugation, 67 (*see also* Centrifugation of
        sludge)
    flotation, 67
    gravity, 67 (*see also* Gravity thickening of
        sludge)
Total dynamic head (TDH), 182–184, 186, 198,
    201
Total static head, 182, 187
Treatability studies, 18
Trickling filter, 52, 326, 328, 563
Troubleshooting
    activated sludge treatment facility, 369
    aerated grit removal facility, 257
    anaerobic sludge digestion facility, 480
    bar rack, 172
    filter press facility, 515
    outfall structure, 424
    primary sedimentation facilities, 294
    pumping station, 210
    sanitary sewers, 148
    sludge-thickening facility, 446, 447
Two-stage lime sludge, 55

Ultrafiltration, 51, 625
Ultraviolet radiation, 379
Unit operations, 18, 47, 50, 51, 56
Unit processes, 18, 47, 50, 51, 56

Vacuum filtration, 65, 489, 494, 497

Variable speed pumps, 193, 195
Venturi meter, 222–234

Waste activated sludge, 428
Wastewater
    chemical quality, 37–43
    fixture unit-load method, 31
    flow reduction, 32, 33
    flow variation, 29, 30
    microbiological quality, 42
    minimum dry weather flow, 31, 32
    peak dry weather flow, 31, 32
    population equivalent, 43, 44
    sustained flow, 32
    sustained peak mass loadings, 43
Wastewater sources and flowrates
    commercial, 127
    dry weather, 18
    industrial, 127
    infiltration/inflow, 17, 18, 29, 30, 123, 127
    residential, 127
    sustained, 18
    wet weather, 18
Water, properties of, 643, 644
Water conservation, 32, 33
Water demand, 23–29
Water quality limited, 16
Water quality management plan, 15
Weirs, 227, 229, 585
Wet oxidation, 65, 521
Wet weather peak flows, 96
Worthington Pump Corporation, 192, 204, 210

Yard piping, 561
Yield coefficient, 307, 308, 457

Zeolite resin, 51
Zimmerman process, 521